Periodic Table of the Elements with the Gmelin System Numbers

Main table (atomic number, symbol, Gmelin System Number):

I	II																	He
1 H 2																		2 He 1
1 H 2																		
3 Li 20	4 Be 26												5 B 13	6 C 14	7 N 4	8 O 3	9 F 5	10 Ne 1
11 Na 21	12 Mg 27												13 Al 35	14 Si 15	15 P 16	16 S 9	17 Cl 6	18 Ar 1
19 * K 22	20 Ca 28	21 Sc 39	22 Ti 41	23 V 48	24 Cr 52	25 Mn 56	26 Fe 59	27 Co 58	28 Ni 57	29 Cu 60	30 Zn 32	31 Ga 36	32 Ge 45	33 As 17	34 Se 10	35 Br 7	36 Kr 1
37 Rb 24	38 Sr 29	39 Y 39	40 Zr 42	41 Nb 49	42 Mo 53	43 Tc 69	44 Ru 63	45 Rh 64	46 Pd 65	47 Ag 61	48 Cd 33	49 In 37	50 Sn 46	51 Sb 18	52 Te 11	53 I 8	54 Xe 1
55 Cs 25	56 Ba 30	57** La 39	72 Hf 43	73 Ta 50	74 W 54	75 Re 70	76 Os 66	77 Ir 67	78 Pt 68	79 Au 62	80 Hg 34	81 Tl 38	82 Pb 47	83 Bi 19	84 Po 12	85 At 8a	86 Rn 1
87 Fr 25a	88 Ra 31	89*** Ac 40	104 71	105 71													

* NH₄ 23 → NH_4 23

Lanthanides 39:

58 Ce	59 Pr	60 Nd	61 Pm	62 Sm	63 Eu	64 Gd	65 Tb	66 Dy	67 Ho	68 Er	69 Tm	70 Yb	71 Lu
58 Ce	59 Pr	60 Nd	61 Pm	62 Sm	63 Eu	64 Gd	65 Tb	66 Dy	67 Ho	68 Er	69 Tm	70 Yb	71 Lu

Actinides:

90 Th 44	91 Pa 51	92 U 55	93 Np 71	94 Pu 71	95 Am 71	96 Cm 71	97 Bk 71	98 Cf 71	99 Es 71	100 Fm 71	101 Md 71	102 No 71	103 Lr 71
Th	Pa	U	Np	Pu	Am	Cm	Bk	Cf	Es	Fm	Md	No	Lr

A Key to the Gmelin System is given on the Inside Back Cover

Gmelin Handbuch der Anorganischen Chemie

Achte Auflage und Ergänzungswerk

Main Series, 8th Edition, and New Supplement Series

The following Index volumes for individual System Numbers of Gmelin Handbook are published up today:

System Number 3 „Sauerstoff"	Formula Index and Alphabetical Index (comprised in Section 8)
System Number 13 „Boron Compounds"	Formula Index
System Number 16 „Phosphor"	Formula Index and Alphabetical Index
System Number 21 „Natrium"	Formula Index and Alphabetical Index
System Number 34 „Quecksilber"	Formula Index and Alphabetical Index (comprised in Section B 4)
System Number 48 „Vanadium" System Number 49 „Niob" System Number 50 „Tantal"	Formula Index and Alphabetical Index (commonly for System Numbers 48, 49, 50, as well as for New Supplement Series volume 2)
System Number 71 „Transuranium Elements"	Alphabetical Index of Subjects and Substances

Other Indexes are comprised in various volumes of Main Series and New Supplement Series

Zu einzelnen System-Nummern des Gmelin Handbuchs sind bisher folgende Register erschienen:

System-Nr. 3 „Sauerstoff"	Formel- und Schlagwortregister (in Lieferung 8 enthalten)
System-Nr. 13 „Borverbindungen"	Formula Index
System-Nr. 16 „Phosphor"	Formel- und Schlagwortregister
System-Nr. 21 „Natrium"	Formel- und Schlagwortregister
System-Nr. 34 „Quecksilber"	Formel- und Schlagwortregister (in Lieferung B 4 enthalten)
System-Nr. 48 „Vanadium" System-Nr. 49 „Niob" System-Nr. 50 „Tantal"	Formel- und Schlagwortregister, gemeinsam für die System-Nr. 48, 49, 50 sowie für Band 2 des Ergänzungswerks zur 8. Auflage
System-Nr. 71 „Transurane"	Alphabetical Index of Subjects and Substances

Weitere Register sind in verschiedenen Bänden des Haupt- und Ergänzungswerks enthalten.

Gmelin Handbuch der Anorganischen Chemie

Achte völlig neu bearbeitete Auflage

BEGRÜNDET VON

Leopold Gmelin

ACHTE AUFLAGE BEGONNEN

im Auftrag der Deutschen Chemischen Gesellschaft
von R. J. Meyer

FORTGEFÜHRT VON

E. H. E. Pietsch und A. Kotowski
Margot Becke-Goehring

HERAUSGEGEBEN VOM

Gmelin-Institut für Anorganische Chemie
der Max-Planck-Gesellschaft zur Förderung der Wissenschaften
Direktor: Ekkehard Fluck

Springer-Verlag Berlin Heidelberg GmbH

Gmelin Handbuch der Anorganischen Chemie

Achte Auflage und Ergänzungswerk

Main Series, 8th Edition, and New Supplement Series

Index

Formula Index

Volume 12

O–Zr

Elements 104 to 132

AUTHORS
(BEARBEITER)

Ramona Becker, Helga Hartwig, Herbert Köppe, Hans Vanecek, Paul Velić, Rudolf Warncke, Anna Zelle

EDITOR
(REDAKTEUR)

Rudolf Warncke

Springer-Verlag Berlin Heidelberg GmbH

The volumes of the Main Series are evaluated up to the end of 1974, the volumes of the New Supplement Series up to the end of 1973.

Es sind die Bände des Hauptwerkes bis zum Erscheinungstermin Ende 1974, die des Ergänzungswerkes bis Ende 1973 berücksichtigt.

Die vierte bis siebente Auflage dieses Werkes erschien im Verlag von Carl Winter's Universitätsbuchhandlung in Heidelberg

Library of Congress Catalog Card Number: Agr 25-1383

ISBN 978-3-662-09032-9   ISBN 978-3-662-09030-5 (eBook)
DOI 10.1007/978-3-662-09030-5

© by Springer-Verlag Berlin Heidelberg 1980
Originally published by Springer-Verlag, Berlin · Heidelberg · New York in 1980
Softcover reprint of the hardcover 8th edition 1980

Wiesbadener Graphische Betriebe GmbH, Wiesbaden

Introduction

Volume 12 completes the alphabetical Formula Index. The index covers all volumes of the Eighth Edition through 1974 for the Main Series and through 1973 for the New Supplement Series. In all, there are more than 142,000 entries.

The index includes all definite elements, compounds, ions, and systems discussed in the volumes covered. The substances are arranged by their empirical formulas. In the second column the usual formula is given. The elements are entered by their symbols. Transuranium elements that do not yet have an official IUPAC name are listed separately at the end of this volume.

Supplements will keep this index up-to-date. The first supplement, which will cover the years 1974 through 1979, is already in preparation.

Frankfurt/Main, January 1980

Rudolf Warncke

Vorwort

Mit dem vorliegenden Band 12 ist das alphabetische Formelregister zum Gmelin Handbuch abgeschlossen, das mit nahezu 142000 Einträgen alle Bände der 8. Auflage des Handbuchs umfaßt, soweit sie bis Ende 1974 (Hauptwerk) bzw. Ende 1973 (Ergänzungswerk) erschienen sind.

Das Formelregister enthält alle in den vorgenannten Bänden besprochenen definierten Elemente, Verbindungen, Ionen und Systeme. Als Haupt-Ordnungsbegriff dient die Summenformel; neben jeder Summenformel ist die entsprechende Formel in üblicher Schreibweise aufgeführt. Die Elemente selbst sind unter ihrem Symbol eingeordnet; diejenigen Transurane, die noch keinen von der IUPAC anerkannten Namen besitzen, sind im Anhang des vorliegenden Bandes zusammengestellt.

In einer Ergänzung zum Formelregister werden die in den letzten Jahren erschienenen Handbuch-Bände, die im vorliegenden Register noch nicht berücksichtigt werden konnten, erfaßt. Die Ergänzung für die Jahre 1974 bis 1979 ist in Vorbereitung.

Frankfurt/Main, Januar 1980

Rudolf Warncke

Introduction to Volume 1

The extraordinary quantity of material included in the Gmelin Handbook of Inorganic Chemistry can make it extremely difficult to find a specific compound using only a general understanding of the basic rules for the arrangement of the subject matter (the so-called "System of the Last Position"). Consequently, subject indexes have been provided for the past few years to individual Handbook volumes, covering only a single section, or in some cases, a complete system number. During this period a comprehensive subject index—the Gmelin Index—was being prepared, whose first volume is now available.

The broad distribution and extensive utilization enjoyed by the Gmelin Handbook in English-speaking countries has led to a decision to publish the Gmelin Index in the English language. This decision was partly based on the fact that since the English language today occupies a preeminent position in the field of chemistry, a German-speaking user could utilize an English text much more readily than would be the case for an English-speaking user and a German text.

The core of the Gmelin Index is formed by a Formula Index which includes all of the definite elements, compounds, ions, and systems which are discussed or mentioned anywhere in the entire Eight Edition of the Gmelin Handbook. All volumes of the Main Series which were published before the end of 1974 are included, as well as volumes 1 through 12 of the New Supplement Series. The Formula Index will consist of about 12 volumes which will appear at intervals of four to six months.

Subsequent index volumes will contain accessible entries for items, which could not be considered in the Formula Index, for example, minerals, alloys, trivial names, and names of classes of compounds.

The Gmelin Index will be printed—for the first time in the history of the Gmelin Handbook—not by using lead type but rather by optical printing. It has thus become possible to store the Index text on magnetic tape, thereby having it available for future expansion when newly-published Handbook volumes can be readily included. It also has become possible to collect subject areas from various specific points of view.

It is not possible to include in a single published index *all* of the material which is scattered throughout the Handbook text. Such an undertaking would vastly exceed the capacity of a printed index, and could only be done utilizing the capabilities of an electronic computer system. At the present time, the Gmelin Institute is endeavoring to work out a project of this type.

The Gmelin Index should give an user better access to the information contained in the Gmelin Handbook, and should put him in position better to utilize the advantages of the Handbook than heretofore. These advantages, for example, include a review of the entire literature of chemistry going back to the very origins of this field and a collection not merely of literature citations but also of the data themselves—critically evaluated, sifted, and arranged in exhaustive monographs.

We wish to thank the Ministry for Research and Technology for their financial support during development and preparation of the Gmelin Index.

Frankfurt/Main, August 1975

Rudolf Warncke

Vorwort zu Band 1

Die überaus große Fülle des im Gmelin Handbuch dargebotenen Materials macht es selbst bei Kenntnis des zugrundeliegenden Ordnungsprinzips (»System der letzten Stelle«) mitunter recht schwierig, eine bestimmte Verbindung aufzufinden. Es wurden daher schon seit mehreren Jahren einigen Handbuch-Bänden Register beigegeben, die jeweils eine Lieferung oder auch einzelne System-Nummern umfaßten, und gleichzeitig wurde ein Gesamtregister, das Gmelin Register, in Angriff genommen, dessen erste Lieferung heute vorliegt.

Die große Verbreitung, die das Gmelin Handbuch im angelsächsischen Sprachraum gefunden hat, führte zu dem Entschluß, das Gmelin Register in englischer Sprache herauszugeben. Damit wird zugleich der Tatsache Rechnung getragen, daß im Bereich der Chemie heute die englische Sprache an erster Stelle steht und daß infolgedessen ein deutschsprachiger Benutzer leichter einen englischen Text zu lesen vermag als dies umgekehrt der Fall ist.

Das Kernstück des Gmelin Registers bildet ein Formelregister, das alle in der 8. Auflage des Gmelin Handbuchs besprochenen bzw. erwähnten definierten Elemente, Verbindungen, Ionen und Systeme umfaßt. Berücksichtigt sind dabei alle Bände des Hauptwerks, soweit sie bis zum Ende des Jahres 1974 erschienen sind, sowie die Bände 1 bis 12 des Ergänzungswerks. Das Formelregister wird etwa 12 Lieferungen umfassen, die in Abständen von 4 bis 6 Monaten erscheinen sollen.

In zusätzlichen Register-Bänden sollen Angaben zugänglich gemacht werden, die im Formelregister nicht berücksichtigt werden können, z. B. Mineralien, Legierungen, Trivialnamen und Verbindungsgruppen-Namen.

Die Herstellung des Gmelin Registers erfolgt – erstmalig für das Gmelin Handbuch – nicht im Bleisatz, sondern im Lichtsatz. Damit ist es möglich geworden, den Register-Text auf Magnetband zu speichern und für eine spätere Erweiterung unter Einbeziehung der in der Zwischenzeit neu erschienenen Handbuch-Bände zur Verfügung zu halten. Zugleich wird damit die Möglichkeit geschaffen, Stoffzusammenstellungen nach bestimmten Gesichtspunkten vorzunehmen.

Es ist nicht möglich, alle im Handbuch-Text verstreuten Angaben durch ein Register zu erfassen. Ein derartiges Vorhaben würde den Rahmen eines gedruckten Registers bei weitem überschreiten und wäre nur in der Form eines elektronischen Informationsspeichers durchführbar. Das Gmelin-Institut ist zur Zeit bemüht, die Voraussetzungen für ein derartiges Projekt herauszuarbeiten.

Das Gmelin Register soll dem Benutzer einen besseren Zugang zu den im Gmelin Handbuch enthaltenen Informationen geben und ihn in die Lage versetzen, die Vorteile des Handbuchs – zum Beispiel Berücksichtigung der gesamten chemischen Literatur, Zurückgehen bis in die Anfänge der Chemie, Zusammenstellung nicht nur von Literaturstellen, sondern von Daten, die kritisch gesichtet und in größere Zusammenhänge eingefügt werden – besser als bisher zu nutzen.

Dem Bundesministerium für Forschung und Technologie danken wir für die finanzielle Unterstützung bei der Entwicklung und Vorbereitung des Gmelin Registers.

Frankfurt am Main, August 1975

Rudolf Warncke

Instructions for the user

The first column of the Formula Index contains the empirical formulas which provide the principal basis for the sequence in which the entries are arranged. The formulas are written in their more normal form in the second column, together with extensions and subdivisions. The citations themselves are given in the third column.

In the *empirical formula* listing (first column), the element symbols for each compound are arranged in alphabetical order; C and H are not treated separately here.

H_2O is included in these empirical formulas only when it is bound in the compound as a complex, as shown in the formulations in the second column. In all other cases, H_2O is not considered.

Isotopes are not shown in the empirical formulas, but are replaced by the normal symbol for the element. However, ionic charges are included in the empirical formulas.

For the elements themselves, the symbol of the element is used as the empirical formula—specifically, in the monatomic form (e.g., Br); entries which definitely refer to polyatomic molecules can be found under the corresponding empirical formula (e.g., Br_2, Br_3).

In the empirical formulas the letters x, y, and z are used for indefinite subscripts. High polymers of the type $(AB)_n$ are listed under the empirical formula of the monomeric compound.

Multicomponent systems (solid solutions, melts, etc.) can be found under the empirical formula of their inorganic components, while solutions are only listed under the formula of the dissolved substance.

The arrangement of the empirical formulas follows in alphabetic sequence and in terms of increasing subscripts; indefinite subscripts follow at the end of each subscript sequence (e.g., $Al_2Br_{x+6}Cu_x$ comes after $Al_2Br_{31}Sb_5$). Ions follow the corresponding neutral compound, and are arranged sequentially in terms of increasing positive, and then negative, charges. The transuranium elements for which no element symbol has been assigned can be found under their atomic number at the end of the Formula Index.

The formulas used in the *second column* of the Index are generally those which are employed in the body of the Handbook text; if more than one formulation is given for a single compound, these are all shown in the Index. In some cases—especially for isomeric compounds—due to limited space formulas are used which deviate from the Handbook text style so as to better illustrate the structure of the compound.

Compounds with the same empirical formula are arranged in the second column in the following sequence: the compound–isotopic compounds–polymeric compounds–hydrates. Isomeric compounds, if any, then follow in the same sequence, and, finally, multicomponent systems corresponding to this empirical formula. For example, the following arrangement can be found in the second column under the empirical formula ClNa:

$$NaCl$$
$$Na^{37}Cl$$
$$(NaCl)_2$$
$$NaCl \cdot 2\,H_2O$$
...
$$NaCl \text{ glasses:}$$
$$NaCl\text{–}B_2O_3$$
...

NaCl solid solutions:
Na(Br,Cl)
...
NaCl solutions:
$NaCl-H_2O$
...
NaCl systems:
$NaCl-As_2O_3-H_2O$
...

If it is not possible to distinguish the isomers of a given composition by different formulas, they are labelled by successive numbers.

The term "system" is used in a very narrow sense in the Index; it includes those equilibria which are characterized by phase diagrams. Ionic systems are treated as systems of the corresponding compounds. Data on solutions are given under the formula of the compound itself for those compounds which have only rather short text passages.

For the elements, the name of the element is utilized in the second column (e.g., "Silver"). The previously given sequence also applies here, but it has been modified so that the ions (in general, i.e., "Silver ions") and the isotopes (i.e., "Silver isotopes") are inserted before the multicomponent systems.

The hydrates are sequenced in terms of increasing water content.

In the case of multicomponent systems, the components are listed in the sequence: inorganic components–organic components–H_2O; the entries then follow in alphabetic order, but the organic components are considered in order of increasing C and H content as well. For elements appearing as components in multicomponent systems, the monatomic element symbol is always used. Isotopic compounds as components are sequenced directly behind the corresponding usual compounds.

Elements and compounds for which the associated Handbook text is voluminous, are subdivided in the Index by subject headings; the degree of subdivision depends on the amount of text material involved.

The concepts "solubility", "solutions", and "systems" partly overlap; in this case, all three positions must be searched. The same is true for the concepts "diffusion" and "systems" as well as for "sorption" and "systems".

With respect to the *citations* (third column of the Index), the following abbreviations are utilized:

Hb   =   Main Series Volume
Eb   =   Supplement Volume to Main Series
Ab   =   Appended Volume to Main Series
Erg. W. =   New Supplement Series Volume.

For the first three of the above cases, the citation includes the System Number, the system element, the nature of the volume (Main Series Volume, Supplement Volume, or Appended Volume to Main Series), the Part or Section (if any), and then, finally, following a hyphen, the page or pages involved. All volumes of the Main Series which are not Supplement Volumes or Appended Volumes to Main Series, are designated as Main Series volumes in the citation. The citation "61 (Ag): Hb/A3–70/106" has the following meaning: pages 70 to 106 in Section 3 of Part A of System Number 61, Silver. The System Number 1, Rare Gases, and the System Number 39, Rare Earth Elements, are designated by "1 (EG)" and "39 (SE)" respectively.

Introduction

Volume 12 completes the alphabetical Formula Index. The index covers all volumes of the Eighth Edition through 1974 for the Main Series and through 1973 for the New Supplement Series. In all, there are more than 142,000 entries.

The index includes all definite elements, compounds, ions, and systems discussed in the volumes covered. The substances are arranged by their empirical formulas. In the second column the usual formula is given. The elements are entered by their symbols. Transuranium elements that do not yet have an official IUPAC name are listed separately at the end of this volume.

Supplements will keep this index up-to-date. The first supplement, which will cover the years 1974 through 1979, is already in preparation.

Frankfurt/Main, January 1980

Rudolf Warncke

Vorwort

Mit dem vorliegenden Band 12 ist das alphabetische Formelregister zum Gmelin Handbuch abgeschlossen, das mit nahezu 142000 Einträgen alle Bände der 8. Auflage des Handbuchs umfaßt, soweit sie bis Ende 1974 (Hauptwerk) bzw. Ende 1973 (Ergänzungswerk) erschienen sind.

Das Formelregister enthält alle in den vorgenannten Bänden besprochenen definierten Elemente, Verbindungen, Ionen und Systeme. Als Haupt-Ordnungsbegriff dient die Summenformel; neben jeder Summenformel ist die entsprechende Formel in üblicher Schreibweise aufgeführt. Die Elemente selbst sind unter ihrem Symbol eingeordnet; diejenigen Transurane, die noch keinen von der IUPAC anerkannten Namen besitzen, sind im Anhang des vorliegenden Bandes zusammengestellt.

In einer Ergänzung zum Formelregister werden die in den letzten Jahren erschienenen Handbuch-Bände, die im vorliegenden Register noch nicht berücksichtigt werden konnten, erfaßt. Die Ergänzung für die Jahre 1974 bis 1979 ist in Vorbereitung.

Frankfurt/Main, Januar 1980

Rudolf Warncke

Introduction to Volume 1

The extraordinary quantity of material included in the Gmelin Handbook of Inorganic Chemistry can make it extremely difficult to find a specific compound using only a general understanding of the basic rules for the arrangement of the subject matter (the so-called "System of the Last Position"). Consequently, subject indexes have been provided for the past few years to individual Handbook volumes, covering only a single section, or in some cases, a complete system number. During this period a comprehensive subject index—the Gmelin Index—was being prepared, whose first volume is now available.

The broad distribution and extensive utilization enjoyed by the Gmelin Handbook in English-speaking countries has led to a decision to publish the Gmelin Index in the English language. This decision was partly based on the fact that since the English language today occupies a preeminent position in the field of chemistry, a German-speaking user could utilize an English text much more readily than would be the case for an English-speaking user and a German text.

The core of the Gmelin Index is formed by a Formula Index which includes all of the definite elements, compounds, ions, and systems which are discussed or mentioned anywhere in the entire Eight Edition of the Gmelin Handbook. All volumes of the Main Series which were published before the end of 1974 are included, as well as volumes 1 through 12 of the New Supplement Series. The Formula Index will consist of about 12 volumes which will appear at intervals of four to six months.

Subsequent index volumes will contain accessible entries for items, which could not be considered in the Formula Index, for example, minerals, alloys, trivial names, and names of classes of compounds.

The Gmelin Index will be printed—for the first time in the history of the Gmelin Handbook—not by using lead type but rather by optical printing. It has thus become possible to store the Index text on magnetic tape, thereby having it available for future expansion when newly-published Handbook volumes can be readily included. It also has become possible to collect subject areas from various specific points of view.

It is not possible to include in a single published index *all* of the material which is scattered throughout the Handbook text. Such an undertaking would vastly exceed the capacity of a printed index, and could only be done utilizing the capabilities of an electronic computer system. At the present time, the Gmelin Institute is endeavoring to work out a project of this type.

The Gmelin Index should give an user better access to the information contained in the Gmelin Handbook, and should put him in position better to utilize the advantages of the Handbook than heretofore. These advantages, for example, include a review of the entire literature of chemistry going back to the very origins of this field and a collection not merely of literature citations but also of the data themselves—critically evaluated, sifted, and arranged in exhaustive monographs.

We wish to thank the Ministry for Research and Technology for their financial support during development and preparation of the Gmelin Index.

Frankfurt/Main, August 1975

Rudolf Warncke

NaCl solid solutions:
Na(Br,Cl)

...

NaCl solutions:
NaCl–H$_2$O

...

NaCl systems:
NaCl–As$_2$O$_3$–H$_2$O

...

Können zu einer Summenformel gehörende Isomere nicht durch ihre Schreibweise unterschieden werden, so werden sie fortlaufend numeriert.

Der Begriff »Systeme« wird im Register im engen Sinne verwendet; er umfaßt Gleichgewichte, wie sie durch das Zustandsdiagramm charakterisiert sind. Ionensysteme sind als Systeme der entsprechenden Verbindungen erfaßt. Angaben über Lösungen sind bei Verbindungen mit geringem Textumfang unter der Formel der Verbindung selbst zu finden.

Bei Elementen wird in der zweiten Spalte der Name des Elements eingesetzt (z.B. »Silber«). Auch hier gilt die oben angeführte Reihenfolge, jedoch mit der Änderung, daß vor der Gruppe der Mehrstoffsysteme die Ionen (allgemein, z.B. »Silber-Ionen«) und die Isotope (z.B. »Silber-Isotope«) eingereiht werden.

Hydrate werden in sich nach steigendem Wassergehalt gereiht.

Bei Mehrstoffsystemen werden die Komponenten in der Reihenfolge »Anorganische Komponenten – Organische Komponenten – H$_2$O« aufgeführt; die Reihung erfolgt dann in alphabetischer Folge, wobei jedoch bei organischen Komponenten der C-Gehalt und danach der H-Gehalt maßgeblich ist. Für Elemente als Komponenten in Mehrstoffsystemen wird stets das einatomige Element-Symbol verwendet. Isotope Verbindungen als Komponenten sind unmittelbar hinter den entsprechenden normalen Verbindungen eingeordnet.

Elemente und Verbindungen, bei denen der zugehörige Handbuch-Text umfangreich ist, werden im Register durch Sachverhaltsangaben unterteilt; der Grad der Unterteilung richtet sich dabei nach dem Textumfang.

Die Begriffe »Löslichkeit«, »Lösungen« und »Systeme« überlappen sich teilweise; hier muß ggf. bei allen drei Stellen nachgelesen werden. Dasselbe gilt für die Begriffe »Diffusion« und »Systeme« sowie für »Sorption« und »Systeme«.

Bei den *Band- und Seitenangaben* (dritte Spalte des Registers) werden folgende Abkürzungen verwendet:

Hb   Hauptband des Hauptwerks der 8. Auflage
Eb   Ergänzungsband des Hauptwerks der 8. Auflage
Ab   Anhangband des Hauptwerks der 8. Auflage
Erg. W.   Ergänzungswerk zur 8. Auflage.

Für die Bände des Hauptwerks wird zunächst die System-Nummer und das System-Element angegeben, danach folgt die Angabe der Art des Bandes (Hauptband, Ergänzungsband oder Anhangband) und gegebenenfalls des Teils und der Lieferung und schließlich hinter einem Bindestrich die Seitenzahl bzw. Seitenzahlen. Alle Bände des Hauptwerks, die nicht Ergänzungsband oder Anhangband sind, werden dabei als Hauptband bezeichnet. Die Angabe »61 (Ag): Hb/A3–70/106« bedeutet beispielsweise: Seiten 70 bis 106 in der Lieferung 3 des Teils A der System-Nummer 61 »Silber«. Die System-Nummer 1 »Edelgase« wird mit »1 (EG)«, die System-Nummer 39 »Seltenerdelemente« mit »39 (SE)« bezeichnet.

Die Bände des Ergänzungswerks werden nur mit der Bandnummer zitiert, es bedeutet beispielsweise »Erg. W. 12–53«: Seite 53 im Band 12 des Ergänzungswerks.

Die Ergänzungsbände zu den Lieferungen 3 bis 5 des Teils A der System-Nummer 59»Eisen« (Gmelin-Durrer) werden mit »59 (Fe): Eb/GD1a«, »59 (Fe): Eb/GD1b« usw. zitiert, die Bände »Magnetische Werkstoffe« der gleichen System-Nummer mit »59 (Fe): D/1. Eb« bzw. »59 (Fe): D/2. Eb«.

Bei *Verweisungen* wird auf die Angaben in der ersten und der zweiten Spalte des Registers verwiesen, also z. B. »*see* $Al_2Na_2O_4 \ldots Na_2O \cdot Al_2O_3$«. Bei Verweisungen innerhalb von Sachverhaltsangaben wird lediglich der Sachverhalt angeführt, z. B. »*see Deposits*«.

Im Formelregister beziehen sich die Angaben der Handbuch-Textstellen auf die alte Anordnung der Handbuch-Bände nach System-Nummern (Hauptwerk) bzw. laufenden Nummern (Ergänzungswerk). Nach der im Januar 1978 eingeführten neuen Anordnung werden die Handbuch-Bände, Hauptwerk und Ergänzungswerk gemeinsam, alphabetisch nach den Element-Symbolen abgestellt. Für die im Formelregister angeführten Zitate bedeutet dies:

1. Hauptwerk:
   Jeder Band steht bei dem Element-Symbol, das im Zitat hinter der System-Nummer angeführt ist, also z. B. der Band 61 (Ag) beim Symbol Ag.
   Ausnahmen:
   Es gehören die Bände mit dem Zitat

1 (EG)	zum Symbol He
8 (J)	zum Symbol I
39 (SE)	zum Symbol Sc
69 (Ma)	zum Symbol Tc

2. Ergänzungswerk:
   Die mit „Erg. W." zitierten Bände gehören zu folgenden Element-Symbolen:

Erg. W. 1:	He
Erg. W. 2:	V
Erg. W. 3:	Cr
Erg. W. 4, 7a, 7b, 8:	Np
Erg. W. 5, 6:	Co
Erg. W. 9, 12:	F
Erg. W. 10:	Zr
Erg. W. 11:	Hf

Die Bände Erg. W. 2 und 3 sowie 10 und 11 sind zusammengebunden.

Formula Index
O–Zr
Elements 104–132

Formula Index O–Zr

$O_{0.03}Ta$ . . . . $TaO_{0.03}$ . . . . 50 (Ta):   Hb/B1–30
$O_{0.4}Sm$ . . . . $SmO_{0.4}$ . . . . 39 (SE):   Hb/C1–314
$O_{0.53}V$ . . . . $VO_{0.53}$. . . . . 48 (V):   Hb/B1–20
$O_{0.6}Sm$ . . . . $SmO_{0.6}$ . . . . 39 (SE):   Hb/C1–314
O . . . . Oxygen
                              see subdivisions:
                                  O ... Air
                                  O ... Oxygen
                                  O ... Ozone

–. . . . . Air:
                              Acoustic properties. . . . . 3 (O):   Hb/4–949/56
                              Boiling point . . . . 3 (O):   Hb/4–861/3
                              Chemical reactions . . . . 3 (O):   Hb/4–819
                              Composition . . . . 3 (O):   Hb/4–819/20
                              Compressibility . . . . 3 (O):   Hb/4–827/41
                              Critical constants . . . . 3 (O):   Hb/4–858/9
                              Density . . . . 3 (O):   Hb/4–820/7
                              Detection and determination. . . . . . 3 (O):   Hb/4–819
                              Diffusion. . . . . 3 (O):   Hb/4–912/21
                              Electrical properties . . . . 3 (O):   Hb/4–967/76
                              Entropy . . . . 3 (O):   Hb/4–891/6
                              Equation of state . . . . 3 (O):   Hb/4–837/41
                              Expansion coefficient. . . . . 3 (O):   Hb/4–853/6
                              Expansion effects . . . . 3 (O):   Hb/4–841/53
                              Heat content. . . . . 3 (O):   Hb/4–887/91
                              Heat of vaporization . . . . 3 (O):   Hb/4–863/5
                              Internal energy . . . . 3 (O):   Hb/4–876, 884/7
                              Isotherms . . . . 3 (O):   Hb/4–827/37
                              Joule-Thomson effect . . . . 3 (O):   Hb/4–842/52
                              Liquefaction. . . . . 3 (O):   Hb/2–182/98
                              Magnetic properties . . . . 3 (O):   Hb/4–966
                              Mechanical properties . . . . 3 (O):   Hb/4–820/948
                              Melting point . . . . 3 (O):   Hb/4–863
                              Optical properties. . . . . 3 (O):   Hb/4–956/66
                              Pressure coefficient. . . . . 3 (O):   Hb/4–856/8
                              Rectification. . . . . 3 (O):   Hb/2–191/3
                              Solubility . . . . 3 (O):   Hb/4–921/35
                              Sorption . . . . 3 (O):   Hb/4–935/48

See the "Instructions for the user"
at the beginning of the volume

O . . . . Air:
            Specific cohesion . . . . 3 (O):   Hb/4−903/4
            Specific heat . . . . 3 (O):   Hb/4−865/84
            Spectra . . . . 3 (O):   Hb/4−963/6
            States of aggregation. . . . . 3 (O):   Hb/4−819/20
            Surface tension . . . . 3 (O):   Hb/4−903/4
            Thermal conductivity. . . . . 3 (O):   Hb/4−896/903
            Thermal properties . . . . 3 (O):   Hb/4−820/903
            Vapor pressure . . . . 3 (O):   Hb/4−859/61
            Viscosity. . . . . 3 (O):   Hb/4−904/12
−. . . . . Oxygen:
            Absorption of light . . . . 3 (O):   Hb/3−527/30
            − of X- and gamma rays . . . . 3 (O):   Hb/3−559/61
            Acoustic properties. . . . . 3 (O):   Hb/3−522/5
            Active oxygen. . . . . 3 (O):   Hb/4−976/84
            Association . . . . 3 (O):   Hb/3−362
            Atom. . . . . 3 (O):   Hb/3−319/47
            Atomic nucleus . . . . 3 (O):   Hb/3−319
            Atomic radius . . . . 3 (O):   Hb/3−324/33
            Atomic scattering factors. . . . . . . 3 (O):   Hb/3−374/80
            Atomic volume . . . . 3 (O):   Hb/3−333/42
             Atomic weight . . . . 3 (O):   Hb/3−342/6
            Behavior at the anode . . . . 3 (O):   Hb/3−614/21
            − at the cathode. . . . . 3 (O):   Hb/3−621/3
            − at the dropping mercury
              electrode . . . . 3 (O):   Hb/3−621/3
            Birefringence, electrical . . . . 3 (O):   Hb/3−531/2
            −, magnetic . . . . 3 (O):   Hb/3−530/1
            Boiling point . . . . 3 (O):   Hb/3−409/11
            Boiling point elevation . . . . 3 (O):   Hb/3−411
            Bond energy . . . . 3 (O):   Hb/3−364/5
            Breakdown voltage . . . . 3 (O):   Hb/3−586
            Cells . . . . 3 (O):   Hb/3−604/14
            Chemical constant . . . . 3 (O):   Hb/3−430/1
            Chemical reactions . . . . 3 (O):   Hb/4−819
            − with hydrogen
              see O ... Oxygen mixtures: O−H
            Color. . . . . 3 (O):   Hb/3−525
            Combustion process, history . . . . . 3 (O):   Hb/1−1/33
            Compressibility . . . . 3 (O):   Hb/3−387/98
            Compton scattering. . . . . 3 (O):   Hb/3−560/1
            Cosmochemistry . . . . 3 (O):   Hb/2−91/6, 129
            Critical constants . . . . 3 (O):   Hb/3−401/3
            Crystal form. . . . . 3 (O):   Hb/3−372
            Crystallographic properties . . . . . . 3 (O):   Hb/3−371/80
            Debye temperature . . . . 3 (O):   Hb/3−430
            Density . . . . 3 (O):   Hb/3−381/7
            Depolarization of light . . . . 3 (O):   Hb/3−533

O . . . . Oxygen:

Detection	3 (O):	Hb/4−819
− in iron	59 (Fe):	Hb/FI−42, 57/8
Determination.	3 (O):	Hb/4−819
− in iron	59 (Fe):	Hb/FI−41/103
		Hb/FII2−344
Dielectric constant	3 (O):	Hb/3−584/5
Diffusion.	3 (O):	Hb/3−446/55
Discovery	3 (O):	Hb/1−49/77
Dissociation	3 (O):	Hb/3−365/71
Ebullioscopic constant	3 (O):	Hb/3−411/2
Effusion	3 (O):	Hb/3−446
Electrical conductivity	3 (O):	Hb/3−586/7
Electrical moment	3 (O):	Hb/3−363/4
Electrical properties	3 (O):	Hb/3−584/93
Electrochemical behavior	3 (O):	Hb/3−593/623
Electron affinity	3 (O):	Hb/3−588/90
Electron configuration	3 (O):	Hb/3−319/20
Electronegativity	3 (O):	Hb/3−347
Electrostriction	3 (O):	Hb/3−585/6
Entropy	3 (O):	Hb/3−427/30
Equation of state	3 (O):	Hb/3−391/5
Expansion coefficient.	3 (O):	Hb/3−399/400
Expansion effects	3 (O):	Hb/3−398/9
Faraday effect.	3 (O):	Hb/3−530/1
Free energy	3 (O):	Hb/3−426/7
Geochemistry	3 (O):	Hb/2−96/101, 111/81
Handling.	3 (O):	Hb/2−225
Heat content.	3 (O):	Hb/3−426/7
Heat of fusion	3 (O):	Hb/3−415/6
− of sublimation.	3 (O):	Hb/3−415
− of transformation	3 (O):	Hb/3−416
− of vaporization	3 (O):	Hb/3−413/5
History.	3 (O):	Hb/1−1/82
		Hb/2−183/4
		Hb/3−343/5
Industrial hazards.	3 (O):	Hb/2−226
Internal energy	3 (O):	Hb/3−426
Internal pressure	3 (O):	Hb/3−439
Ionization potentials	3 (O):	Hb/3−590/2
Joshi effect	3 (O):	Hb/3−587/8
Kerr effect	3 (O):	Hb/3−531/2
Lattice structure.	3 (O):	Hb/3−372/4
Liquefaction	3 (O):	Hb/3−314
Magnetic moment	3 (O):	Hb/3−324, 363
Magnetic properties	3 (O):	Hb/3−578/84
Magnetostriction	3 (O):	Hb/3−583/4
Mass absorption coefficient	3 (O):	Hb/3−559

*See the "Instructions for the user"
at the beginning of the volume*

1*

O . . . . Oxygen:

Mass fluorescence coefficient	3 (O):	Hb/3−559/60
Mass scattering coefficient.	3 (O):	Hb/3−560
Mechanical properties	3 (O):	Hb/3−381/525
Melting point	3 (O):	Hb/3−412/3
Molar volume	3 (O):	Hb/3−360/1
Molecular radius	3 (O):	Hb/3−354/9
Molecular structure.	3 (O):	Hb/3−348/51
Molecular weight	3 (O):	Hb/3−361/2
Molecule.	3 (O):	Hb/3−348/71
Moment of inertia.	3 (O):	Hb/3−362/3
Occurrence, extraterrestrial	3 (O):	Hb/2−91/6, 129
–, terrestrial	3 (O):	Hb/2−96/181
– in the atmosphere	3 (O):	Hb/2−96/101
– in the biosphere.	3 (O):	Hb/2−179/81
– in the hydrosphere	3 (O):	Hb/2−111/9, 161/79
– in the lithosphere	3 (O):	Hb/2−119/28, 129/38, 145/61
Optical properties.	3 (O):	Hb/3−525/78
Oxygen electrode	3 (O):	Hb/3−595/604
Oxyhydrogen gas electrode	3 (O):	Hb/3−604
Parachor.	3 (O):	Hb/3−439/41
Phlogiston theory	3 (O):	Hb/1−33/49
Photoelectron emission	3 (O):	Hb/3−593
Polarizablitity	3 (O):	Hb/3−320/3, 364
Polymorphism	3 (O):	Hb/3−371/2
Position in the electrochemical series	3 (O):	Hb/3−593/4
– in the periodic system	3 (O):	Hb/3−346
Potentials	3 (O):	Hb/3−594/5
Preparation	3 (O):	Hb/3−301/14
–, industrial	3 (O):	Hb/2−182/226
–, –, from air.	3 (O):	Hb/2−182/202
–, –, from substances rich in oxygen .	3 (O):	Hb/2−221/4
–, –, from water	3 (O):	Hb/2−202/21
Pressure coefficient.	3 (O):	Hb/3−400/1
Purification	3 (O):	Hb/2−224 Hb/3−311/3
Purity testing	3 (O):	Hb/2−224/5 Hb/3−313
Refraction	3 (O):	Hb/3−525/7
Rheochor	3 (O):	Hb/3−446
Scattering of light.	3 (O):	Hb/3−533
– of X- and gamma rays	3 (O):	Hb/3−559/61
Solubility	3 (O):	Hb/3−455/94
– in gels and hydrosols.	3 (O):	Hb/3−481
– in ice	3 (O):	Hb/3−473
– in inorganic solvents	3 (O):	Hb/3−475/82

Hb = Main Series Volume; GD = Gmelin-Durrer;   Ab = Appended Volume to Main Series;
Eb = Supplement Volume to Main Series;   Erg. W. = New Supplement Series Volume

O . . . . Oxygen:

Solubility in iron . . . . 59 (Fe): Hb/A5–856
Eb/GD1a–57/8
Eb/GD1b–11
  – in natural waters . . . . 3 (O): Hb/3–473/5
  – in organic solvents . . . . 3 (O): Hb/3–481/94
  – in steam . . . . 3 (O): Hb/3–473
  – in tin . . . . 46 (Sn): Hb/B–351
  – in water . . . . 3 (O): Hb/3–456/75
Sorption on carbon . . . . 3 (O): Hb/3–495/508
  – on metal oxides . . . . 3 (O): Hb/3–511/6
  – on organic substances . . . . 3 (O): Hb/3–521
  – on palladium . . . . 65 (Pd): Hb/2–257/9
  – on platinum . . . . 68 (Pt): Hb/C1–26/41
  – on silica gel . . . . 3 (O): Hb/3–508/10
  – on other inorganic substances . . . . 3 (O): Hb/3–516/21
Sound absorption . . . . 3 (O): Hb/3–524/5
Sound velocity . . . . 3 (O): Hb/3–522/4
Specific heat . . . . 3 (O): Hb/3–416/24
Spectra:
  Arc spectrum . . . . 3 (O): Hb/3–534/46
  Band spectrum . . . . 3 (O): Hb/3–563/78
  Continuous spectrum . . . . 3 (O): Hb/3–578
  Microwave spectrum . . . . 3 (O): Hb/3–562/3
  Raman spectrum . . . . 3 (O): Hb/3–533/4
  Spark spectra . . . . 3 (O): Hb/3–546/57
  X-ray spectrum . . . . 3 (O): Hb/3–557/9
Standard potential . . . . 3 (O): Hb/3–593
Structure . . . . 3 (O): Hb/3–372/4
Surface tension . . . . 3 (O): Hb/3–436/9
Susceptibility . . . . 3 (O): Hb/3–578/83
Thermal conductivity . . . . 3 (O): Hb/3–431/6
Thermal properties . . . . 3 (O): Hb/3–387/436
Thermochor . . . . 3 (O): Hb/3–441
Toxicity . . . . 3 (O): Hb/2–298
Transformation points . . . . 3 (O): Hb/3–413
Triple point . . . . 3 (O): Hb/3–408/9, 412/3
Valence . . . . 3 (O): Hb/3–346/7
Vapor pressure . . . . 3 (O): Hb/3–403/9
Vibration frequencies . . . . 3 (O): Hb/3–354, 430
Viscosity . . . . 3 (O): Hb/3–441/6
Working dangers . . . . 3 (O): Hb/2–226

–. . . . . $O_2 \cdot xH_2O$ . . . . 3 (O): Hb/5–1191

–. . . . . Ozone:

Chemical reactions . . . . 3 (O): Hb/4–1147/66
  see also Decomposition
  – with air . . . . 3 (O): Hb/4–1147

See the "Instructions for the user"
at the beginning of the volume

O . . . . Ozone:

Chemical reactions with elements. . .	3 (O):	Hb/4–1148/57
– with inorganic compounds.	3 (O):	Hb/4–1147/8, 1157/66
– with organic compounds.	3 (O):	Hb/4–1166
– with water.	3 (O):	Hb/4–1147/8
Decomposition	3 (O):	Hb/4–988/1082
– in corpuscular radiation	3 (O):	Hb/4–1077/82
– in electrical discharges	3 (O):	Hb/4–1038/77
– in photochemical processes	3 (O):	Hb/4–1026/38
– in thermal processes	3 (O):	Hb/4–999/1019
Decomposition equilibria	3 (O):	Hb/4–988/94
Detection	3 (O):	Hb/4–819
Determination.	3 (O):	Hb/4–819
Electrical properties	3 (O):	Hb/4–1139
Electrochemical behavior	3 (O):	Hb/4–1139/47
Formation	3 (O):	Hb/4–988/1100
–, formation equilibria	3 (O):	Hb/4–988/94
–, thermodynamic data.	3 (O):	Hb/4–1109/10
– in chemical reactions	3 (O):	Hb/3–640/1, 778, 782
		Hb/4–1092/9
– in corpuscular radiation	3 (O):	Hb/4–1077/82
– in electrical discharges	3 (O):	Hb/4–1038/77
– in electrolytic processes	3 (O):	Hb/4–1082/91
– in photochemical processes	3 (O):	Hb/4–1020/5
– in thermal processes	3 (O):	Hb/4–995/9
– in other processes	3 (O):	Hb/4–1099/100
History.	3 (O):	Hb/4–984/7
Magnetic properties	3 (O):	Hb/4–1139
Mechanical properties	3 (O):	Hb/4–1117/24
Molecule.	3 (O):	Hb/4–1111/7
Occurrence	3 (O):	Hb/2–101/11
Optical properties.	3 (O):	Hb/4–1124/39
Preparation	3 (O):	Hb/4–1100/9
–, industrial	3 (O):	Hb/2–226/32
Solubility	3 (O):	Hb/4–1121/3
Sorption	3 (O):	Hb/4–1123/4
Spectra	3 (O):	Hb/4–1127/39
Thermal properties	3 (O):	Hb/4–1117/24
Toxicity	3 (O):	Hb/2–298/300
Uses	3 (O):	Hb/2–232

–.	$^{16}O^{16}O^{18}O$	3 (O):	Hb/4–1111
–.	$^{16}O^{18}O^{16}O$	3 (O):	Hb/4–1111
–.	$^{16}O^{18}O^{18}O$	3 (O):	Hb/4–1111
–.	$^{18}O_3$	3 (O):	Hb/4–1111
–.	$O_3 \cdot H_2O = H_2O_4$	3 (O):	Hb/8–2626/7

–. . . . . Oxygen ions
                       see $O^{n+} \ldots O^{n+}$, etc.

O . . . . Oxygen isotopes:
    Atomic nucleus . . . . 3 (O):   Hb/3−319
    Determination in water . . . . 3 (O):   Hb/6−2019/23
    Enrichment and separation . . . . . . 3 (O):   Hb/3−314/8
    Occurrence . . . . 3 (O):   Hb/2−83/91

−. . . . . Oxygen mixtures:
    Air−Ar−CO . . . . 14 (C):   Hb/C2−105
    Air−CH₄ . . . . 3 (O):   Hb/4−902/3
    Air−CH₄−CO . . . . 14 (C):   Hb/C2−99, 106
    Air−CH₄−CO−H . . . . 14 (C):   Hb/C2−99, 107
    Air−CO . . . . 3 (O):   Hb/4−902/3
                                 14 (C):   Hb/C1−276
                                           Hb/C2−94/7, 103/4
    Air−CO−CO₂ . . . . 14 (C):   Hb/C2−105
    Air−CO−C₂H₂ . . . . 14 (C):   Hb/C2−106
    Air−CO−C₂H₄ . . . . 14 (C):   Hb/C2−99, 106
    Air−CO−C₂H₄−H . . . . 14 (C):   Hb/C2−107
    Air−CO−C₂H₆ . . . . 14 (C):   Hb/C2−99, 106
    Air−CO−C₃H₈ . . . . 14 (C):   Hb/C2−99
    Air−CO−H . . . . 14 (C):   Hb/C2−98, 106
    Air−CO−He . . . . 14 (C):   Hb/C2−105
    Air−CO−N . . . . 14 (C):   Hb/C2−105
    Air−CO₂ . . . . 3 (O):   Hb/4−886/7, 890/1
                                           14 (C):   Hb/C1−528/31
    Air−C₂H₂ . . . . 3 (O):   Hb/4−902/3
    Air−C₂H₄ . . . . 3 (O):   Hb/4−912
    Air−C₂H₅OH . . . . 3 (O):   Hb/4−896
    Air−CH₃COCH₃ . . . . 3 (O):   Hb/4−825, 884, 903
    Air−(C₂H₅)₂O . . . . 3 (O):   Hb/4−825, 884
    Air−C₆H₆ . . . . 3 (O):   Hb/4−825
    Air−C₈H₁₈ . . . . 3 (O):   Hb/4−896
    Air−C₈H₁₈−H₂O . . . . 3 (O):   Hb/4−896
    Air−H . . . . 3 (O):   Hb/4−825, 884/7,
                                           890/1, 903, 912
    Air−H₂O . . . . 3 (O):   Hb/4−886/7, 890/1,
                                           896, 902/3, 912
    Air−He . . . . 3 (O):   Hb/4−845
    Air−NH₃ . . . . 3 (O):   Hb/4−902/3
    O−Ar−CO . . . . 14 (C):   Hb/C2−109
    O−Ar−N . . . . 4 (N):   Hb/1−151
    O−CO . . . . 3 (O):   Hb/3−510, 512, 515
                                           14 (C):   Hb/C1−268/9
                                           Hb/C2−75/111
    O−CO−CO₂ . . . . 14 (C):   Hb/C2−95/6, 108/9
    O−CO−C₂H₄ . . . . 14 (C):   Hb/C1−284
    O−CO−Cl . . . . 14 (C):   Hb/C2−109
    O−CO−H . . . . 14 (C):   Hb/C2−97/8
    O−CO−H₂O . . . . 14 (C):   Hb/C2−98/9

See the "Instructions for the user"
at the beginning of the volume

O . . . . Oxygen mixtures:

O–CO–He. . . . . 14 (C):   Hb/C2–109
O–CO–N . . . . 4 (N):   Hb/1–143
                                   14 (C):   Hb/C2–95/6, 108
O–CO–N–H₂O . . . . 14 (C):   Hb/C2–99
O–CO₂. . . . . 3 (O):   Hb/3–510, 515
                                    14 (C):   Hb/C1–520/1
O–CO₂–H–N . . . . 14 (C):   Hb/C1–528
O–CO₂–N . . . . 14 (C):   Hb/C1–527/8
                                                     Hb/C3–79
O–C₂H₄ . . . . 3 (O):   Hb/3–395, 446
O–H:
    Compression coefficient. . . . . . . 3 (O):   Hb/5–1190
    Heat conductivity. . . . . 3 (O):   Hb/3–432, 436
                                                     Hb/5–1190
    Hydrogen flames. . . . . 3 (O):   Hb/3–768/803
    –, flame temperature. . . . . 3 (O):   Hb/3–783/9
    –, flame velocity
        and flame propagation . . . . . 3 (O):   Hb/3–781/803
    –, general properties. . . . . 3 (O):   Hb/3–768/70
    –, spectrum and chemism . . . . . . 3 (O):   Hb/3–770/82
    Ignition limits. . . . . 3 (O):   Hb/3–804/16
    Ignition temperature . . . . 3 (O):   Hb/3–816/8
    Oxyhydrogen gas coulometer. . . . 3 (O):   Hb/3–620/1
    Oxyhydrogen gas electrode . . . . . 3 (O):   Hb/3–604
    Photochemical reaction . . . . 3 (O):   Hb/3–639/51
    Reaction in presence
        of electrical discharges . . . . . . 3 (O): Hb/3–633/9
    – – of foreign O or H atoms . . . . . 3 (O):   Hb/3–625/30
    – – of ionizing radiation . . . . 3 (O):   Hb/3–630/3
    Sound velocity . . . . 3 (O):   Hb/3–523
    Surface reaction . . . . 3 (O):   Hb/3–651/92
    – on metals. . . . . 3 (O):   Hb/3–655/85
    – on oxides . . . . 3 (O):   Hb/3–685/91
    – on other materials . . . . 3 (O):   Hb/3–691/2
    Thermal reaction. . . . . 3 (O):   Hb/3–692/768
    –, detonation limits . . . . 3 (O):   Hb/3–695/737
    –, detonation peninsula . . . . 3 (O):   Hb/3–702/6
    –, reaction mechanism. . . . . 3 (O):   Hb/3–737/59
    –, sensitizers and inhibitors . . . . . 3 (O): Hb/3–759/68
    –, steady reaction . . . . 3 (O):   Hb/3–714/31
    Viscosity . . . . 3 (O):   Hb/3–445/6

O–H–N . . . . 4 (N):   Hb/1–142
O–H₂O. . . . . 3 (O):   Hb/3–523
                                                     Hb/6–1750
O–He . . . . 3 (O):   Hb/3–523
                                                     Hb/5–1185
O–N . . . . 3 (O):   Hb/3–510, 512
                                    4 (N):   Hb/1–142, 148, 168

Hb = Main Series Volume; GD = Gmelin-Durrer;   *Ab = Appended Volume to Main Series;*
Eb = Supplement Volume to Main Series;   *Erg. W. = New Supplement Series Volume*

O . . . . Oxygen mixtures:
          O$-NH_3$. . . . . 4 (N):   Hb/3$-$639/84

−. . . . . Oxygen solutions:
          Ozone$-CO_2$ . . . . 3 (O):   Hb/4$-$1167
          Ozone$-Cl$ . . . . 3 (O):   Hb/4$-$1167
          Ozone$-H_2O$ . . . . 3 (O):   Hb/4$-$1167/83
          Ozone$-$inorganic solvents . . . . 3 (O):   Hb/4$-$1183/4
          Ozone$-N$ . . . . 3 (O):   Hb/4$-$1167
          Ozone$-NH_3$ . . . . 3 (O):   Hb/4$-$1167
          Ozone$-O$ . . . . 3 (O):   Hb/4$-$1166/7

−. . . . . Oxygen systems:
          O$-$Ag . . . . 61 (Ag):   Hb/B1$-$33/53
          O$-$Ag$-Ag_2O-B_2O_3$ . . . . 61 (Ag):   Hb/B3$-$249
          O$-$Ag$-Cr_2O_3$ . . . . 61 (Ag):   Hb/B4$-$288
          O$-$Ag$-Cu$ . . . . 61 (Ag):   Hb/B4$-$438/9
          O$-$Ag$-Mn$ . . . . 61 (Ag):   Hb/B4$-$359
          O$-$Ag$-S$. . . . . 9 (S):   Hb/B1$-$185
                                     61 (Ag):   Hb/B3$-$61/2
          O$-$Al$-Cu$ . . . . 60 (Cu):   Hb/B3$-$1169/70
          O$-$Al$-Ta$ . . . . 50 (Ta):   Hb/B2$-$197
          O$-$Am . . . . Erg. W. 4$-$32/4
          O$-$Am$-Cm$ . . . . Erg. W. 4$-$79
          O$-$Ar. . . . . 3 (O):   Hb/5$-$1185/8
          O$-$Ar$-CO-N$ . . . . 14 (C):   Hb/C1$-$276
          O$-$Ba$-Cu$ . . . . 60 (Cu):   Hb/B3$-$1139
          O$-$Ba$-Ni$ . . . . 57 (Ni):   Hb/B3$-$1125/6
          O$-$Ba$-S$. . . . . 9 (S):   Hb/B1$-$183
          O$-BaO-BaO_2$. . . . . 30 (Ba):   Hb$-$93/4
          O$-$Be$-Ta$ . . . . 50 (Ta):   Hb/B2$-$172
          O$-$Bi. . . . . 19 (Bi):   Eb$-$464
          O$-$Bi$-Cu-Fe$ . . . . 60 (Cu):   Hb/B3$-$1315
          O$-$Bi$-Nb$ . . . . 49 (Nb):   Hb/B1$-$422/3
          O$-$Bi$-Pb$ . . . . 47 (Pb):   Hb/C3$-$966
          O$-$Bi$-S$ . . . . 9 (S):   Hb/B1$-$182
                                      19 (Bi):   Eb$-$745
          O$-$Bi$-Ta$ . . . . 50 (Ta):   Hb/B1$-$275
          O$-$Bi$-Te$ . . . . 19 (Bi):   Eb$-$817
          O$-$C$-Fe$. . . . . 59 (Fe):   Hb/A5$-$861/74
                                               Hb/A8$-$1703/16
                                               Hb/B1$-$21/6
                                               Eb/GD1a$-$58/64
                                               Eb/GD1b$-$11/2
          O$-$C$-Mo$ . . . . 53 (Mo):   Hb$-$196
          O$-$C$-N-Nb$. . . . . 49 (Nb):   Hb/B1$-$387
          O$-$C$-N-Ta$ . . . . 50 (Ta):   Hb/B1$-$250
          O$-$C$-Nb$ . . . . 49 (Nb):   Hb/B1$-$381/2
          O$-$C$-Pu$. . . . . Erg. W. 4$-$224/5
          O$-$C$-Ta$ . . . . 50 (Ta):   Hb/B1$-$248

See the "Instructions for the user"
at the beginning of the volume

O . . . . Oxygen systems:
        O–C–Ti . . . . 41 (Ti):   Hb–368
        O–C–V . . . . 48 (V):   Hb/B1–335/6
        O–C–Zn . . . . 32 (Zn):   Eb–977
        O–CO$_2$ . . . . 14 (C):   Hb/C1–516/20
        O–Ca–Ni . . . . 57 (Ni):   Hb/B3–1118
        O–Ca–S . . . . 9 (S):   Hb/B1–182/3
                                  28 (Ca):   Hb/B3–657/60
        O–CaS . . . . 28 (Ca):   Hb/B3–658
        O–Cd . . . . 33 (Cd):   Eb–418
        O–Ce–N . . . . 39 (SE):   Hb/C2–179/80
        O–Cl–S–Ta . . . . 50 (Ta):   Hb/B1–178
        O–Cm . . . . Erg. W. 4–35
        O–Co . . . . 58 (Co):   Eb/A–475
        O–Co–CoO . . . . 58 (Co):   Eb/A–475/6
        O–Co–Co$_3$O$_4$ . . . . 58 (Co):   Eb/A–476
        O–Co–Cu . . . . 60 (Cu):   Hb/B3–1263
        O Co H . . . . 58 (Co):   Hb/A2–224
        O–Co–S . . . . 9 (S):   Hb/B1–183
                                  58 (Co):   Eb/A–626
        O–Co–V . . . . 58 (Co):   Eb/A–873
        O–Co–W . . . . 58 (Co):   Eb/A–877
        O–CoO–Co$_3$O$_4$ . . . . 58 (Co):   Eb/A–476
        O–Cr . . . . 52 (Cr):   Hb/B–7/9
        O–Cr–Cu . . . . 60 (Cu):   Hb/B3–1215/7
        O–Cu . . . . 60 (Cu):   Hb/B1–23/34
        O–Cu–Fe . . . . 60 (Cu):   Hb/B3–1270
        O–Cu–Fe–Mn . . . . 60 (Cu):   Hb/B3–1330
        O–Cu–Fe–S . . . . 60 (Cu):   Hb/B3–1295
        O–Cu–Fe$_2$O$_3$ . . . . 60 (Cu):   Hb/B3–1270
        O–Cu–Mn . . . . 60 (Cu):   Hb/B3–1242/3
        O–Cu–Mn–S . . . . 60 (Cu):   Hb/B3–1246
        O–Cu–Mo . . . . 60 (Cu):   Hb/B3–1230
        O–Cu–Ni . . . . 60 (Cu):   Hb/B3–1250/2
        O–Cu–P . . . . 60 (Cu):   Hb/B2–918
        O–Cu–Pb . . . . 60 (Cu):   Hb/B3–1197/8
        O–Cu–S . . . . 9 (S):   Hb/B1–184/5
        O–Cu–U . . . . 60 (Cu):   Hb/B3–1240
        O–Cu–W . . . . 60 (Cu):   Hb/B3–1235
        O–Cu–Zn . . . . 60 (Cu):   Hb/B3–1145/6
        O–Er . . . . 39 (SE):   Hb/C1–414/5
        O–Eu . . . . 39 (SE):   Hb/C1–323/5
        O–Eu–N . . . . 39 (SE):   Hb/C2–180
        O–Fe . . . . 59 (Fe):   Hb/A5–856/60
                                        Hb/A8–1692/8
                                        Hb/B1–12/5
                                        Eb/GD1a–53/7
                                        Eb/GD1b–9/11

Hb = Main Series Volume; GD = Gmelin-Durrer;   *Ab = Appended Volume to Main Series;*
Eb = Supplement Volume to Main Series;   *Erg. W. = New Supplement Series Volume*

O . . . . Oxygen systems:

O–Fe–H . . . . 59 (Fe):   Hb/A8–1700/2
                                                      Hb/B1–17/21
                                                      Eb/GD1a–58/64
                                                      Eb/GD1b–11/2

O–Fe–Mn . . . . 59 (Fe):   Hb/A5–881/923
O–Fe–Ni . . . . 59 (Fe):   Hb/A5–942/3
O–Fe–Ni–Si . . . . 59 (Fe):   Hb/A5–943
O–Fe–P . . . . 59 (Fe):   Hb/A5–923/35
O–Fe–S . . . . 9 (S):   Hb/B1–183/4
O–Fe–Si . . . . 59 (Fe):   Hb/A5–876/81
O–Ga–Si–V . . . . 48 (V):   Hb/B2–592
O–Ga–V . . . . 48 (V):   Hb/B2–591/2
O–Gd–N . . . . 39 (SE):   Hb/C2–181
O–H . . . . 3 (O):   Hb/5–1190/1
O–H–Nb–H_2O . . . . 49 (Nb):   Hb/B1–33, 52
O–H_2O . . . . 3 (O):   Hb/6–1750
O–He . . . . 3 (O):   Hb/5–1185
O–Hf–Nb . . . . 49 (Nb):   Hb/B3–218
O–Hf–Ni . . . . 57 (Ni):   Hb/B1–118
O–K–Nb . . . . 49 (Nb):   Hb/B2–1/2
O–Kr . . . . 3 (O):   Hb/5–1188/9
O–La–N . . . . 39 (SE):   Hb/C2–179
O–Li–Ta . . . . 50 (Ta):   Hb/B2–1, 126
O–Mg–Ta . . . . 50 (Ta):   Hb/B2–172
O–Mn . . . . 56 (Mn):   Hb/C1–7/14
O–Mn–H_2O . . . . 56 (Mn):   Hb/C1–367/8, 414
O–Mn–Si . . . . 59 (Fe):   Hb/A5–941/2
O–Mo–Ni . . . . 57 (Ni):   Hb/B3–1217
O–N–Nb . . . . 49 (Nb):   Hb/B1–127
O–N–Pm . . . . 39 (SE):   Hb/C2–180
O–N–Sm . . . . 39 (SE):   Hb/C2–180
O–Na–Ta . . . . 50 (Ta):   Hb/B2–141
O–Nb . . . . 49 (Nb):   Hb/B1–14/23
O–Nb–Ta . . . . 50 (Ta):   Hb/B2–245/6
O–Nb–Ti . . . . 49 (Nb):   Hb/B3–178/9
O–Nb–Zn . . . . 49 (Nb):   Hb/B3–94
O–Nb–Zr . . . . 49 (Nb):   Hb/B2–128/9, 171/2
                                                      Hb/B3–209
O–Nd . . . . 39 (SE):   Hb/C1–296/7
O–Ne . . . . 3 (O):   Hb/5–1185
O–Ni . . . . 57 (Ni):   Hb/B2–377/8
O–Ni–S . . . . 9 (S):   Hb/B1–183
                                            57 (Ni):   Hb/B2–676/7
O–Ni–Ti . . . . 57 (Ni):   Hb/B1–97
O–Ni–V . . . . 57 (Ni):   Hb/B1–159
                                                      Hb/B3–1198
O–Ni–W . . . . 57 (Ni):   Hb/B3–1225

See the "Instructions for the user"
at the beginning of the volume

O . . . . Oxygen systems:
               O–Ni–Zr . . . . 57 (Ni):   Hb/B1–114
               O–Np–Sc . . . . Erg. W. 4–45/6
               O–ozone . . . . 3 (O):   Hb/4–1166/7
               O–Pb . . . . 47 (Pb):   Hb/C1–51/2
               O–Pb–S . . . . 9 (S):   Hb/B1–183
                                              47 (Pb):   Hb/C2–533/4
               O–Pb–Sb . . . . 47 (Pb):   Hb/C3–926/9
               O–Pb–Ta . . . . 50 (Ta):   Hb/B2–232
               O–Pb–Te . . . . 47 (Pb):   Hb/C2–678/9
               O–Pd–S . . . . 9 (S):   Hb/B1–185
               O–PrO_x . . . . 39 (SE):   Hb/C1–275/8
               O–Pt . . . . 68 (Pt):   Hb/C1–26/41
               O–Pu . . . . Erg. W. 4–10/2
               O–Pu–H_2O . . . . Erg. W. 4–10/2
               O–Pu–U . . . . Erg. W. 4–59/60
               O–Ru . . . . 63 (Ru):   Eb–232/3
               O–Ru–H_2O . . . . 63 (Ru):   Eb–232/3
                O–S–Sb . . . . 9 (S):   Hb/B1–182
                O–S–Sr . . . . 9 (S):   Hb/B1–183
                O–S–Ti . . . . 41 (Ti):   Hb–343
               O–Sc . . . . 39 (SE):   Hb/C1–180
                O–Si–Ta . . . . 50 (Ta):   Hb/B1–263
                O–Sm . . . . 39 (SE):   Hb/C1–313/4
                O–Sn . . . . 46 (Sn):   Hb/C1–24/5
                O–Ta . . . . 50 (Ta):   Hb/B1–17/23
                O–Ta–Th . . . . 50 (Ta):   Hb/B2–229
                O–Ta–Ti . . . . 50 (Ta):   Hb/B2–221
                O–Ta–Ti–Zr . . . . 50 (Ta):   Hb/B2–229
                O–Ta–V . . . . 50 (Ta):   Hb/B2–244
                O–Ta–Y . . . . 50 (Ta):   Hb/B2–205
                O–Ta–Zr . . . . 50 (Ta):   Hb/B2–228
                O–TbO_x . . . . 39 (SE):   Hb/C1–391/4
                O–Ti . . . . 41 (Ti):   Hb–211/5
                O–Ti–V . . . . 48 (V):   Hb/B2–658
                O–U . . . . 55 (U):   Hb–82
                O–V . . . . 48 (V):   Hb/B1–11/2, 18/9
                O–Xe . . . . 3 (O):   Hb/5–1190
                O–Y . . . . 39 (SE):   Hb/C1–188/9
                O–Yb . . . . 39 (SE):   Hb/C1–426
                O–Zn . . . . 32 (Zn):   Eb–809
                O–Zr . . . . 42 (Zr):   Hb–208/12
                Ozone–KOH . . . . 22 (K):   Hb/1–245/6
                Ozone–O . . . . 3 (O):   Hb/4–1166/7
O^{n+} . . . . O^{n+} (n = 1 to 7):
               Electron configuration . . . . 3 (O):   Hb/3–320
               Formation . . . . 3 (O):   Hb/3–370/1

Hb = Main Series Volume; GD = Gmelin-Durrer;   Ab = Appended Volume to Main Series;
Eb = Supplement Volume to Main Series;   Erg. W. = New Supplement Series Volume

O^{n+} . . . . O^{n+} (n = 1 to 7):

        Ionic radius . . . . 3 (O):   Hb/3−326/7

        Ionic volume . . . . 3 (O):   Hb/3−335

        Ionization potentials . . . . 3 (O):   Hb/3−590/2

        Magnetic moment . . . . 3 (O):   Hb/3−324

        Spectra . . . . 3 (O):   Hb/3−546/57

O^- . . . . O^- . . . . 3 (O):   Hb/3−320, 324,
                                   335, 557

O^{2-} . . . . O^{2-} . . . . 3 (O):   Hb/3−320, 321/3,
                                   324/33, 335/42

OOs . . . . OsO . . . . 66 (Os):   Hb−25

−. . . . . $OsO \cdot H_2O$ . . . . 66 (Os):   Hb−25

OP. . . . . PO . . . . 16 (P):   Hb/C−64/6

−. . . . . $(PO)_x$ . . . . 16 (P):   Hb/C−67

OP^+ . . . . $[PO]^+$ . . . . 16 (P):   Hb/C−66

OP_2 . . . . P_2O . . . . 16 (P):   Hb/C−59

OP_2Zn. . . . . . . ZnP_2O . . . . 32 (Zn):   Hb−276

OP_3 . . . . P_3O . . . . 16 (P):   Hb/C−59

OP_4 . . . . P_4O . . . . 16 (P):   Hb/C−58/64

−. . . . . $P_4O \cdot 2\,H_2O$ . . . . 16 (P):   Hb/C−62

OPb . . . . PbO:

        Chemical reactions . . . . 47 (Pb):   Hb/C1−105/19

        Colloidal forms . . . . 47 (Pb):   Hb/C1−62/3

        Crystallographic properties . . . . . 47 (Pb):   Hb/C1−68/74

        Electrical properties . . . . 47 (Pb):   Hb/C1−83/95

        Electrochemical behavior . . . . . . 47 (Pb):   Hb/B2−414
                                   Hb/C1−98/105

        Formation . . . . 47 (Pb):   Hb/C1−57/61, 65/7

        History. . . . . 47 (Pb):   Hb/A1−100/11

        Magnetic properties . . . . 47 (Pb):   Hb/C1−84/5

        Mechanical properties . . . . 47 (Pb):   Hb/C1−74/6

        Molecule. . . . . 47 (Pb):   Hb/C1−67/8

        Optical properties. . . . . 47 (Pb):   Hb/C1−95/8

        Preparation . . . . 47 (Pb):   Hb/C1−57/65

        −, industrial . . . . 47 (Pb):   Hb/C1−63/5

        − of special forms. . . . . 47 (Pb):   Hb/C1−61/3

        Purity testing . . . . 47 (Pb):   Hb/C1−64/5

        Spectra . . . . 47 (Pb):   Hb/C1−68

        Thermal properties . . . . 47 (Pb):   Hb/C1−76/83

−. . . . . $PbO \cdot 0.33\,H_2O = 3\,PbO \cdot H_2O$
         = $Pb_3O_2(OH)_2$. . . . . 47 (Pb):   Hb/C1−121/2

−. . . . . $8\,PbO \cdot 3\,H_2O$ . . . . 47 (Pb):   Hb/C1−119, 121

−. . . . . $PbO \cdot 0.4\,H_2O = 5\,PbO \cdot 2\,H_2O$ . . . . . . 47 (Pb):   Hb/C1−121

−. . . . . $PbO \cdot 0.5\,H_2O = 2\,PbO \cdot H_2O$
         = $Pb_2O(OH)_2$ . . . . 47 (Pb):   Hb/C1−120

−. . . . . $PbO \cdot 0.6\,H_2O = 5\,PbO \cdot 3\,H_2O$ . . . . . . 47 (Pb):   Hb/C1−120

−. . . . . $PbO \cdot nH_2O$. . . . . 47 (Pb):   Hb/C1−119/22

See the "Instructions for the user"
at the beginning of the volume

OPb . . . . PbO glasses:

PbO$-$AlF$_3-$B$_2$O$_3$	47 (Pb):	Hb/C4$-$1295/6
PbO$-$Al$_2$O$_3-$B$_2$O$_3$.	47 (Pb):	Hb/C4$-$1295
PbO$-$Al$_2$O$_3-$B$_2$O$_3-$CeO$_2$	47 (Pb):	Hb/C4$-$1375
PbO$-$Al$_2$O$_3-$B$_2$O$_3-$Na$_2$O	47 (Pb):	Hb/C4$-$1302
PbO$-$Al$_2$O$_3-$SiO$_2$	47 (Pb):	Hb/C4$-$1297/300
PbO$-$B$_2$O$_3-$Bi$_2$O$_3$	47 (Pb):	Hb/C3$-$978
PbO$-$B$_2$O$_3-$Cs$_2$O	47 (Pb):	Hb/C3$-$1156
PbO$-$B$_2$O$_3-$K$_2$O$-$SiO$_2$	47 (Pb):	Hb/C3$-$1109/10
PbO$-$B$_2$O$_3-$La$_2$O$_3$.	47 (Pb):	Hb/C4$-$1375
PbO$-$B$_2$O$_3-$Li$_2$O	47 (Pb):	Hb/C3$-$992/3
PbO$-$B$_2$O$_3-$Na$_2$O$-$SiO$_2$	47 (Pb):	Hb/C3$-$1047
PbO$-$B$_2$O$_3-$SiO$_2$	47 (Pb):	Hb/C3$-$858/60
PbO$-$BaO$-$SiO$_2$	47 (Pb):	Hb/C3$-$1208/10
PbO$-$Bi$_2$O$_3$	47 (Pb):	Hb/C3$-$978
PbO$-$Bi$_2$O$_3-$P$_2$O$_5$	47 (Pb):	Hb/C3$-$978
PbO$-$Bi$_2$O$_3-$SiO$_2$	47 (Pb):	Hb/C3$-$978
PbO$-$CaO$-$SiO$_2$.	47 (Pb):	Hb/C3$-$1191/2
PbO$-$CdO$-$SiO$_2$	47 (Pb):	Hb/C4$-$1257
PbO$-$GeO$_2$	45 (Ge):	Eb$-$501
	47 (Pb):	Hb/C4$-$1475/6
PbO$-$K$_2$O$-$Li$_2$O$-$SiO$_2$	47 (Pb):	Hb/C3$-$1116
PbO$-$K$_2$O$-$Na$_2$O$-$SiO$_2$.	47 (Pb):	Hb/C3$-$1120/2
PbO$-$K$_2$O$-$SiO$_2$.	47 (Pb):	Hb/C3$-$1102/6
PbO$-$La$_2$O$_3-$SiO$_2$.	47 (Pb):	Hb/C4$-$1376
PbO$-$Li$_2$O$-$Na$_2$O$-$SiO$_2$.	47 (Pb):	Hb/C3$-$1052
PbO$-$Li$_2$O$-$SiO$_2$	47 (Pb):	Hb/C3$-$993
PbO$-$Li$_2$O$-$TeO$_2$	47 (Pb):	Hb/C3$-$992/3
PbO$-$Na$_2$O$-$SiO$_2$.	47 (Pb):	Hb/C3$-$1039/46
PbO$-$Na$_2$O$-$TeO$_2$.	47 (Pb):	Hb/C3$-$1034
PbO$-$P$_2$O$_5$.	47 (Pb):	Hb/C3$-$864/5
PbO$-$P$_2$O$_5-$ThO$_2$	47 (Pb):	Hb/C4$-$1471
PbO$-$Pr$_2$O$_3-$SiO$_2$.	47 (Pb):	Hb/C4$-$1376
PbO$-$Rb$_2$O$-$SiO$_2$	47 (Pb):	Hb/C3$-$1144/5
PbO$-$Sb$_2$O$_3-$SiO$_2$.	47 (Pb):	Hb/C3$-$941
PbO$-$SiO$_2$:		
Chemical reactions.	47 (Pb):	Hb/C3$-$835/6
Enamel	47 (Pb):	Hb/C3$-$836
Formation and preparation	47 (Pb):	Hb/C3$-$824/5
Physical properties.	47 (Pb):	Hb/C3$-$827/34
Solder glass	47 (Pb):	Hb/C3$-$836
Structure	47 (Pb):	Hb/C3$-$825/6
PbO$-$SiO$_2-$SrO	47 (Pb):	Hb/C3$-$1199
PbO$-$SiO$_2-$ThO$_2$	47 (Pb):	Hb/C4$-$1471
PbO$-$SiO$_2-$TiO$_2$	47 (Pb):	Hb/C4$-$1392/4
PbO$-$SiO$_2-$ZnO.	47 (Pb):	Hb/C4$-$1231
PbO$-$TeO$_2$.	47 (Pb):	Hb/C2$-$678/9

OPb . . . . PbO melts:
　　　　　　　　　　PbO$-B_2O_3-Na_2O-SiO_2$ . . . . 47 (Pb):　　Hb/C3$-$1047
　　　　　　　　　　PbO$-B_2O_3-Tl_2O$ . . . . 47 (Pb):　　Hb/C4$-$1354
　　　　　　　　　　PbO$-$MgO. . . . . 47 (Pb):　　Hb/C3$-$1175
　　　　　　　　　　PbO$-Na_2B_4O_7$. . . . . 47 (Pb):　　Hb/C3$-$1034
　　　　　　　　　　PbO$-Na_2O$ . . . . 47 (Pb):　　Hb/C3$-$1004/5
　　　　　　　　　　PbO$-SiO_2$ . . . . 47 (Pb):　　Hb/C3$-$810/23
　　　　　　　　　　PbO$-SiO_2-Tl_2O$ . . . . 47 (Pb):　　Hb/C4$-$1354

$-$. . . . . PbO phosphors:
　　　　　　　　　　PbO$-SiO_2-$ZnO. . . . . 47 (Pb):　　Hb/C4$-$1231

$-$. . . . . PbO solid solutions:
　　　　　　　　　　PbO$-Al_2O_3-Ga_2O_3$ . . . . 47 (Pb):　　Hb/C4$-$1306
　　　　　　　　　　PbO$-$BaO$-HfO_2-Nb_2O_5$ . . . . 49 (Nb):　　Hb/B3$-$319
　　　　　　　　　　PbO$-Fe_2O_3$ . . . . 59 (Fe):　　D/2. Eb$-$441
　　　　　　　　　　PbO$-K_2O-Nb_2O_5-Ta_2O_5$ . . . . . . 50 (Ta):　　Hb/B2$-$265
　　　　　　　　　　PbO$-Na_2O-Nb_2O_5-Ta_2O_5$. . . . . . 50 (Ta):　　Hb/B2$-$265
　　　　　　　　　　PbO$-Sb_2O_5-SnO_2$ . . . . 47 (Pb):　　Hb/C4$-$1595

$-$. . . . . PbO solutions:
　　　　　　　　　　PbO$-$NaOH$-$ZnO$-H_2O$ . . . . 47 (Pb):　　Hb/C4$-$1230

$-$. . . . . PbO systems:
　　　　　　　　　　PbO$-Ag_2O-$CuO . . . . 61 (Ag):　　Hb/B4$-$441
　　　　　　　　　　PbO$-Ag_2O-Cu_2O$ . . . . 61 (Ag):　　Hb/B4$-$441
　　　　　　　　　　PbO$-Al_2O_3$ . . . . 47 (Pb):　　Hb/C4$-$1291
　　　　　　　　　　PbO$-Al_2O_3-B_2O_3-Y_2O_3$ . . . . 47 (Pb):　　Hb/C4$-$1375/6
　　　　　　　　　　PbO$-Al_2O_3-SiO_2$ . . . . 47 (Pb):　　Hb/C4$-$1296/7
　　　　　　　　　　PbO$-As_2O_3$ . . . . 47 (Pb):　　Hb/C3$-$889
　　　　　　　　　　PbO$-As_2O_3-H_2O$ . . . . 47 (Pb):　　Hb/C3$-$889
　　　　　　　　　　PbO$-As_2O_5$ . . . . 47 (Pb):　　Hb/C3$-$890/1
　　　　　　　　　　PbO$-As_2O_5-H_2O$ . . . . 47 (Pb):　　Hb/C3$-$890/1
　　　　　　　　　　PbO$-B_2O_3$. . . . . 47 (Pb):　　Hb/C2$-$685/91
　　　　　　　　　　PbO$-B_2O_3-$BaO$-TiO_2$ . . . . 47 (Pb):　　Hb/C4$-$1412/3
　　　　　　　　　　PbO$-B_2O_3-Bi_2O_3-$CdO . . . . 47 (Pb):　　Hb/C4$-$1258
　　　　　　　　　　PbO$-B_2O_3-$CdO . . . . 47 (Pb):　　Hb/C4$-$1256/7
　　　　　　　　　　PbO$-B_2O_3-Na_2O$ . . . . 47 (Pb):　　Hb/C3$-$1034
　　　　　　　　　　PbO$-B_2O_3-Nb_2O_5$ . . . . 49 (Nb):　　Hb/B4$-$313/4
　　　　　　　　　　PbO$-B_2O_3-PbTiO_3-PbZrO_3$ . . . . . . 47 (Pb):　　Hb/C4$-$1454
　　　　　　　　　　PbO$-B_2O_3-SO_3$. . . . . 47 (Pb):　　Hb/C2$-$697
　　　　　　　　　　PbO$-B_2O_3-SiO_2$ . . . . 47 (Pb):　　Hb/C3$-$857/60
　　　　　　　　　　PbO$-B_2O_3-TeO_2$ . . . . 47 (Pb):　　Hb/C2$-$697
　　　　　　　　　　PbO$-B_2O_3-TiO_2$ . . . . 47 (Pb):　　Hb/C4$-$1391
　　　　　　　　　　PbO$-B_2O_3-$ZnO . . . . 47 (Pb):　　Hb/C4$-$1227/8
　　　　　　　　　　PbO$-$BaO$-$MgO$-SiO_2$ . . . . 47 (Pb):　　Hb/C3$-$1211
　　　　　　　　　　PbO$-$BaO$-P_2O_5-TiO_2$ . . . . 47 (Pb):　　Hb/C4$-$1413
　　　　　　　　　　PbO$-$BaO$-SiO_2$. . . . . 47 (Pb):　　Hb/C3$-$1207/8
　　　　　　　　　　PbO$-Bi_2O_3$ . . . . 47 (Pb):　　Hb/C3$-$966/7
　　　　　　　　　　PbO$-Bi_2O_3-TiO_2$ . . . . 47 (Pb):　　Hb/C4$-$1394

See the "Instructions for the user"
at the beginning of the volume

OPb . . . . PbO systems:

$PbO-Bi_2O_3-V_2O_5$	48 (V):	Hb/B2–573
$PbO-CO_2$	47 (Pb):	Hb/C2–697/702
$PbO-CaO-SiO_2$.	47 (Pb):	Hb/C3–1191/2
$PbO-CaO-SiO_2-ZnO$	47 (Pb):	Hb/C4–1231
$PbO-CrO_3-H_2O$	52 (Cr):	Hb/B–931
$PbO-Cr_2O_3$	52 (Cr):	Hb/B–910
$PbO-CuO$	60 (Cu):	Hb/B3–1198
$PbO-Cu_2O$	60 (Cu):	Hb/B3–1198
$PbO-Fe_2O_3$	59 (Fe):	Hb/B5–1130/1
$PbO-GeO_2$	47 (Pb):	Hb/C4–1474/5
$PbO-GeO_2-Ta_2O_5$	50 (Ta):	Hb/B2–243
$PbO-KBr$	47 (Pb):	Hb/C3–1086
$PbO-KCl$	47 (Pb):	Hb/C3–1081
$PbO-KF$	47 (Pb):	Hb/C3–1068
$PbO-KI$	47 (Pb):	Hb/C3–1092
$PbO-KNaC_4H_4O_6-H_2O$	47 (Pb):	Hb/C3–1120
$PbO-KNaC_4H_4O_6-NaOH-H_2O$	47 (Pb):	Hb/C3–1120
$PbO-K_2CO_3$	47 (Pb):	Hb/C3–1097
$PbO-K_2C_4H_4O_6-H_2O$	47 (Pb):	Hb/C3–1098
$PbO-K_2O-Nb_2O_5-H_2O$	49 (Nb):	Hb/B3–248/9
$PbO-K_2O-SiO_2$	47 (Pb):	Hb/C3–1100/1
$PbO-K_2SO_4$	47 (Pb):	Hb/C3–1096
$PbO-La_2O_3$	47 (Pb):	Hb/C4–1373
$PbO-La_2O_3-Nb_2O_5-TiO_2$	49 (Nb):	Hb/B3–302
$PbO-Li_2CO_3$.	47 (Pb):	Hb/C3–993
$PbO-Li_2SO_4$	47 (Pb):	Hb/C3–992
$PbO-MgO-SiO_2$	47 (Pb):	Hb/C3–1177
$PbO-MoO_3$	53 (Mo):	Hb–307
$PbO-N_2O_5-H_2O$	47 (Pb):	Hb/C1–243
$PbO-NaBr$.	47 (Pb):	Hb/C3–1026
$PbO-NaCl$.	47 (Pb):	Hb/C3–1025
$PbO-NaCl-NaOH$	47 (Pb):	Hb/C3–1025
$PbO-NaCl-PbCl_2-H_2O$	47 (Pb):	Hb/C3–1025
$PbO-NaF$	47 (Pb):	Hb/C3–1017
$PbO-NaOH$	47 (Pb):	Hb/C3–1005/6
$PbO-NaOH-Nb_2O_5-H_2O$	49 (Nb):	Hb/B3–248
$PbO-NaPO_3$.	47 (Pb):	Hb/C3–1047/9
$PbO-Na_2CO_3$	47 (Pb):	Hb/C3–1034
$PbO-Na_2C_4H_4O_6-H_2O$	47 (Pb):	Hb/C3–1037
$PbO-Na_2O-H_2O$	47 (Pb):	Hb/C3–1006/7
$PbO-Na_2O-SiO_2$	47 (Pb):	Hb/C3–1038/46
$PbO-Na_2O-SiO_2-TiO_2$	47 (Pb):	Hb/C4–1392/3
$PbO-Na_2O-ZnO-H_2O$	47 (Pb):	Hb/C4–1229/30
$PbO-Na_2SO_4$	47 (Pb):	Hb/C3–1030
$PbO-Nb_2O_5$	49 (Nb):	Hb/B3–223/5
$PbO-Nb_2O_5-P_2O_5$	49 (Nb):	Hb/B4–361
$PbO-Nb_2O_5-TeO_2$	49 (Nb):	Hb/B4–307

OPb . . . . PbO systems:
                         $PbO-Nb_2O_5-TiO_2$ . . . . 49 (Nb):   Hb/B3-284
                         $PbO-P_2O_5$ . . . . 47 (Pb):   Hb/C3-863/4
                         $PbO-P_2O_5-H_2O$ . . . . 47 (Pb):   Hb/C3-870
                         $PbO-P_2O_5-SiO_2$ . . . . 47 (Pb):   Hb/C3-885/6
                         $PbO-P_2O_5-V_2O_5$ . . . . 48 (V):   Hb/B2-572
                         $PbO-P_2O_5-ZrO_2$ . . . . 47 (Pb):   Hb/C4-1431
                         $PbO-Pb$ . . . . 47 (Pb):   Hb/C1-52/4
                         $PbO-PbBr_2$ . . . . 47 (Pb):   Hb/C2-369/70
                         $PbO-PbBr_2-H_2O$ . . . . 47 (Pb):   Hb/C2-371
                         $PbO-Pb(CH_3COO)_2-H_2O$ . . . . 47 (Pb):   Hb/C2-771/2
                         $PbO-PbCl_2$ . . . . 47 (Pb):   Hb/C1-333
                         $PbO-PbCl_2-H_2O$ . . . . 47 (Pb):   Hb/C1-336
                         $PbO-PbCrO_4$ . . . . 52 (Cr):   Hb/B-928
                         $PbO-PbF_2$ . . . . 47 (Pb):   Hb/C1-280
                         $PbO-PbI_2$ . . . . 47 (Pb):   Hb/C2-404
                         $PbO-PbS$ . . . . 47 (Pb):   Hb/C2-534
                         $PbO-PbSO_4$ . . . . 47 (Pb):   Hb/C2-582/3
                         $PbO-PbSO_4-SiO_2$ . . . . 47 (Pb):   Hb/C3-856
                         $PbO-PbSe$ . . . . 47 (Pb):   Hb/C2-632
                         $PbO-PbSeO_3$ . . . . 47 (Pb):   Hb/C2-634/5
                         $PbO-Pb_3(AsO_4)_2$ . . . . 47 (Pb):   Hb/C3-898/9
                         $PbO-Pb_3O_4$ . . . . 47 (Pb):   Hb/C1-54
                         $PbO-RbCl$ . . . . 47 (Pb):   Hb/C3-1142
                         $PbO-SO_2-H_2O$ . . . . 47 (Pb):   Hb/C2-535
                         $PbO-SO_3-TeO_2$ . . . . 47 (Pb):   Hb/C2-681
                         $PbO-Sb_2O_3$ . . . . 47 (Pb):   Hb/C3-927/8
                         $PbO-Sb_2O_3-Sb_2S_3$ . . . . 47 (Pb):   Hb/C3-939
                         $PbO-Sb_2O_4$ . . . . 47 (Pb):   Hb/C3-928/9
                         $PbO-Sb_2O_5$ . . . . 47 (Pb):   Hb/C3-929
                         $PbO-SeO_2-TeO_2$ . . . . 47 (Pb):   Hb/C2-684
                         $PbO-SiO_2$ . . . . 47 (Pb):   Hb/C3-808/10
                         $PbO-SiO_2-ZrO_2$ . . . . 47 (Pb):   Hb/C4-1431
                         $PbO-SnO_2$ . . . . 47 (Pb):   Hb/C4-1581
                         $PbO-Ta_2O_5$ . . . . 50 (Ta):   Hb/B2-232/3
                         $PbO-TiO_2$ . . . . 47 (Pb):   Hb/C4-1378/9
                         $PbO-TiO_2-V_2O_5$ . . . . 48 (V):   Hb/B2-574
                         $PbO-TiO_2-ZrO_2$ . . . . 47 (Pb):   Hb/C4-1439/40
                         $PbO-V_2O_5$ . . . . 48 (V):   Hb/B2-567/8
                         $PbO-WO_3$ . . . . 54 (W):   Hb-317
                         $PbO-ZrO_2$ . . . . 47 (Pb):   Hb/C4-1423

OPb₂ . . . . Pb_2O . . . . 47 (Pb):   Hb/C1-106
OPb₂²⁺ . . . . $[Pb_4O_2]^{4+}$ . . . . 47 (Pb):   Hb/C1-199
OPd . . . . PdO . . . . 65 (Pd):   Hb/2-260/3
−. . . . . $PdO \cdot H_2O$ . . . . 65 (Pd):   Hb/2-263/4
−. . . . . $PdO \cdot xH_2O$ . . . . 65 (Pd):   Hb/2-263/5
OPd₂ . . . . Pd_2O . . . . 65 (Pd):   Hb/2-259
OPm . . . . PmO . . . . 39 (SE):   Hb/C1-312

See the "Instructions for the user"
at the beginning of the volume

OPo	PoO	12 (Po):	Hb−133
OPo^{2+}	[PoO]$^{2+}$	12 (Po):	Hb−102
OPr	PrO	39 (SE):	Hb/C1−87/92, 96,
			98/102
OPr$^+$	PrO$^+$	39 (SE):	Hb/C1−102
OPt	PtO	68 (Pt):	Hb/C1−26/8, 42
−.	PtO · H$_2$O	68 (Pt):	Hb/C1−26, 42/5
−.	PtO · 2 H$_2$O	68 (Pt):	Hb/C1−26, 42/5
OPtS	PtOS	68 (Pt):	Hb/C1−109
−.	PtOS · 0.5 H$_2$O	68 (Pt):	Hb/C1−109
−.	PtOS · H$_2$O	68 (Pt):	Hb/C1−109
OPu$_{0.5}$SU$_{0.5}$.	U$_{0.5}$Pu$_{0.5}$OS	Erg. W. 4−173	
OPu	PuO	Erg. W. 4−12, 19	
−.	PuO solid solutions:		
	Pu(C,O)	Erg. W. 4−224/5	
	PuO−PuC−PuN	Erg. W. 4−226/8	
	PuO−(U,Pu)C−(U,Pu)N	Erg. W. 4−226/8	
OPuS	PuOS	Erg. W. 4−172/3	
OPuSe	PuOSe	Erg. W. 4−189, 191	
ORb$_2$	Rb$_2$O	24 (Rb):	Hb−105/6
−.	Rb$_2$O glasses:		
	Rb$_2$O−PbO−SiO$_2$	47 (Pb):	Hb/C3−1144/5
−.	Rb$_2$O melts:		
	Rb$_2$O−V$_2$O$_5$	48 (V):	Hb/B2−486
−.	Rb$_2$O systems:		
	Rb$_2$O−CO$_2$−CaO	28 (Ca):	Hb/B3−1337
	Rb$_2$O−Nb$_2$O$_5$	49 (Nb):	Hb/B4−187/8
ORe	ReO · H$_2$O	70 (Re):	Hb−96
ORe$_2$	Re$_2$O · 2 H$_2$O	70 (Re):	Hb−96
ORh	RhO	64 (Rh):	Hb−46
ORh$_2$	Rh$_2$O	64 (Rh):	Hb−45/6
ORu	RuO	63 (Ru):	Hb−25
			Eb−234
ORu$^+$	[RuO]$^+$	63 (Ru):	Eb−233
ORu^{2+}	[RuO]$^{2+}$	63 (Ru):	Eb−248/9
ORu$^-$	[RuO]$^-$	63 (Ru):	Eb−233
ORu$_2$$^{6+}$	[RuORu]$^{6+}$	63 (Ru):	Eb−248/9
OS	SO	9 (S):	Hb/B1−58/9, 167/78
−.	(SO)$_x$	9 (S):	Hb/B1−167
OS$^+$.	SO$^+$	9 (S):	Hb/B1−262
OS^{2+}	SO^{2+}	9 (S):	Hb/B1−310
OSSi	SiOS	15 (Si):	Hb/B−752
OSTh	ThOS	44 (Th):	Hb−282/3
OSZn	ZnSO	32 (Zn):	Hb−219
OSZn$_2$.	ZnS · ZnO	32 (Zn):	Hb−219
OSZr	ZrOS	42 (Zr):	Hb−332/3
OS$_2$	S$_2$O	9 (S):	Hb/B1−167

Hb = Main Series Volume; GD = Gmelin-Durrer;   *Ab = Appended Volume to Main Series;*
Eb = Supplement Volume to Main Series;   *Erg. W. = New Supplement Series Volume*

OS_2Sb_2 . . . . $Sb_2OS_2 = 2\,Sb_2S_3 \cdot Sb_2O_3$ . . . . 18 (Sb):   Hb/A3−305
                                                   Hb/B3−540

OS_2V^- . . . . $[VOS_2]^-$ . . . . 48 (V):   Hb/B1−281
OS_3 . . . . S_3O . . . . 9 (S):   Hb/B1−167
OS_3V^{3-} . . . . $[VOS_3]^{3-}$ . . . . 48 (V):   Hb/B1−281
OS_4Sr_2 . . . . $SrO \cdot SrS_4 \cdot 12\,H_2O$ . . . . 29 (Sr):   Hb−171
OS_4Zn_5 . . . . $4\,ZnS \cdot ZnO$ . . . . 32 (Zn):   Hb−219
OS_5Sb_4 . . . . $Sb_4OS_5 = 5\,Sb_2S_3 \cdot Sb_2O_3$ . . . . . . 18 (Sb):   Hb/B3−540
$OS_6V_2^{4-}$ . . . . $[V_2OS_6]^{4-}$ . . . . 48 (V):   Hb/B1−281
OSb . . . . SbO . . . . 18 (Sb):   Hb/B2−348/9
OSb_2 . . . . Sb_2O . . . . 18 (Sb):   Hb/B2−348
OSc . . . . ScO . . . . 39 (SE):   Hb/C1−85/92,.98/102,
                                                   180/1
OSc^+ . . . . ScO^+ . . . . 39 (SE):   Hb/C1−102
OSc_2 . . . . Sc_2O . . . . 39 (SE):   Hb/C1−102/3
OSe . . . . SeO . . . . 10 (Se):   Hb/B−23/5
OSi . . . . SiO . . . . 15 (Si):   Hb/B−258/69
OSi^+ . . . . SiO^+ . . . . 15 (Si):   Hb/B−267
OSm . . . . SmO . . . . 39 (SE):   Hb/C1−90/2, 98/102,
                                                   106, 314/5

−. . . . . SmO solid solutions:
                SmO_xN_{1-x}. . . . . 39 (SE):   Hb/C2−180

OSn . . . . SnO:
                Chemical reactions . . . . 46 (Sn):   Hb/C1−42/8
                Electrochemical behavior . . . . . . 46 (Sn):   Hb/C1−41/2
                Formation . . . . 46 (Sn):   Hb/C1−25/8
                History. . . . . 46 (Sn):   Hb/A−164/5
                Molecule. . . . . 46 (Sn):   Hb/C1−28/35
                Physical properties . . . . 46 (Sn):   Hb/C1−31/41
                Preparation . . . . 46 (Sn):   Hb/C1−25/7

−. . . . . Sn_2O_2. . . . . 46 (Sn):   Hb/C1−34/5, 104
−. . . . . Sn_3O_3. . . . . 46 (Sn):   Hb/C1−34/5, 104
−. . . . . Sn_4O_4. . . . . 46 (Sn):   Hb/C1−34/5, 104
−. . . . . $5\,SnO \cdot 2\,H_2O$ . . . . 46 (Sn):   Hb/C1−49/54
−. . . . . $12\,SnO \cdot 5\,H_2O$. . . . . 46 (Sn):   Hb/C1−49
−. . . . . $SnO \cdot 0.5\,H_2O$ . . . . 46 (Sn):   Hb/C1−49
−. . . . . $3\,SnO \cdot 2\,H_2O$ . . . . 46 (Sn):   Hb/C1−49
−. . . . . SnO solutions:
                $SnO-NaOH-H_2O$. . . . . 46 (Sn):   Hb/C1−55/7

−. . . . . SnO systems:
                $SnO-CoO$ . . . . 58 (Co):   Eb/A−871
                $SnO-H_2O$ . . . . 46 (Sn):   Hb/C1−48
                $SnO-NbO_2$ . . . . 49 (Nb):   Hb/B3−221
                $SnO-Nb_2O_5$. . . . . 49 (Nb):   Hb/B3−221/2
                $SnO-Ta_2O_5$ . . . . 50 (Ta):   Hb/B2−231
OSn_2^{2+} . . . . $[Sn_2O]^{2+}$ . . . . 46 (Sn):   Hb/B−384

See the "Instructions for the user"
at the beginning of the volume

OSr . . . . SrO:

          Chemical reactions . . . . 29 (Sr):   Hb$-$74/6
                                                  Eb$-$159/61
          Electrochemical behavior . . . . 29 (Sr):   Eb$-$159
          Formation . . . . 29 (Sr):   Hb$-$71/2
                                                  Eb$-$152/3
          Physical properties . . . . 29 (Sr):   Hb$-$72/4
                                                  Eb$-$153/9
          Preparation . . . . 29 (Sr):   Hb$-$71/2
                                                  Eb$-$152
          $-$, industrial . . . . 29 (Sr):   Hb$-$17/33
                                                  Eb$-$84
          Purification . . . . 29 (Sr):   Hb$-$33
          Solubility . . . . 29 (Sr):   Hb$-$76/8
          Toxicity . . . . 29 (Sr):   Eb$-$146
          Uses . . . . 29 (Sr):   Eb$-$84

$-$. . . . . SrO glasses:

          SrO$-$PbO$-$SiO$_2$ . . . . 47 (Pb):   Hb/C3$-$1199

$-$. . . . . SrO solid solutions:

          SrO$-$BaO . . . . 30 (Ba):   Eb$-$531/2
          SrO$-$Fe$_2$O$_3$ . . . . 59 (Fe):   D/2. Eb$-$440/1

$-$. . . . . SrO systems:

          SrO$-$Al$_2$O$_3$$-SiO_2$ . . . . 35 (Al):   Hb/B2$-$603
          SrO$-$As$_2$O$_3$$-H_2$O . . . . 29 (Sr):   Eb$-$287
          SrO$-$As$_2$O$_5$$-H_2$O . . . . 29 (Sr):   Eb$-$289/90
          SrO$-$B$_2$O$_3$ . . . . 29 (Sr):   Hb$-$188
                                                  Eb$-$263
          SrO$-$BaCl$_2$$-BaO-$SrCl$_2$ . . . . 30 (Ba):   Hb$-$388
          SrO$-$BaO . . . . 30 (Ba):   Eb$-$531/2
          SrO$-$BaO$-$TiO$_2$ . . . . 41 (Ti):   Hb$-$457
          SrO$-$Bi$_2$O$_3$ . . . . 29 (Sr):   Eb$-$295
          SrO$-$C$_{12}$H$_{22}$O$_{11}$$-H_2$O . . . . 29 (Sr):   Hb$-$208/9
          SrO$-$CaCrO$_4$ . . . . 52 (Cr):   Hb/B$-$833
          SrO$-$CaO . . . . 29 (Sr):   Eb$-$303
          SrO$-$CaO$-$SiO$_2$ . . . . 29 (Sr):   Eb$-$306
          SrO$-$HCl$-$H$_2$O . . . . 29 (Sr):   Hb$-$141
          SrO$-$H$_2$O . . . . 29 (Sr):   Hb$-$76/9
                                                  Eb$-$161/2
          SrO$-$MgO . . . . 29 (Sr):   Eb$-$303
          SrO$-$Nb$_2$O$_5$ . . . . 49 (Nb):   Hb/B3$-$26
          SrO$-$P$_2$O$_5$$-H_2$O . . . . 29 (Sr):   Hb$-$213
          SrO$-$S$-$SO$_2$$-SrS-$SrSO$_4$ . . . . 29 (Sr):   Eb$-$257
          SrO$-$SiO$_2$ . . . . 29 (Sr):   Hb$-$210
                                                  Eb$-$280/1
          SrO$-$SiO$_2$$-H_2$O . . . . 29 (Sr):   Eb$-$282
          SrO$-$SrBr$_2$$-H_2$O . . . . 29 (Sr):   Hb$-$155
          SrO$-$SrCl$_2$ . . . . 29 (Sr):   Hb$-$140
                                                  Eb$-$217/8

Hb = Main Series Volume; GD = Gmelin-Durrer;   *Ab = Appended Volume to Main Series;*
Eb = Supplement Volume to Main Series;   *Erg. W. = New Supplement Series Volume*

OSr . . . . SrO systems:
                $SrO-SrCl_2-H_2O$ . . . . 29 (Sr):   Hb–140/1
                $SrO-SrCl_2-SrS$ . . . . 29 (Sr):   Eb–249
                $SrO-SrCl_2-SrSe$ . . . . 29 (Sr):   Eb–261
                $SrO-SrI_2-H_2O$ . . . . 29 (Sr):   Hb–164
                $SrO-Ta_2O_5$ . . . . 50 (Ta):   Hb/B2–181
                $SrO-ZrO_2$ . . . . 42 (Zr):   Hb–439

OSr$_2$ . . . . Sr_2O . . . . 29 (Sr):   Hb–70/1
                                                                          Eb–152

OTa . . . . TaO . . . . 50 (Ta):   Hb/A1–8
                                                                          Hb/A2–327, 498
                                                                          Hb/B1–21, 26/8

– . . . . TaO solid solutions:
                TaO–TaN . . . . 50 (Ta):   Hb/B1–81

OTa$^+$ . . . . $[TaO]^+$ . . . . 50 (Ta):   Hb/A2–407
                                                                         Hb/B1–59

OTa^{3+} . . . . $[TaO]^{3+}$ . . . . 50 (Ta):   Hb/B1–59
OTa$_2$ . . . . Ta_2O . . . . 50 (Ta):   Hb/B1–25/6, 30
– . . . . Ta$_2$O solid solutions:
                $Ta_2(O_xN_{1-x})$ . . . . 50 (Ta):   Hb/B1–81

OTa$_2$$^+$ . . . . $[Ta_2O]^+$ . . . . 50 (Ta):   Hb/A2–407
OTa$_{3.56}$ . . . . $Ta_{3.56}O = Ta_{32}O_9$ . . . . 50 (Ta):   Hb/B1–24/5
OTa$_4$ . . . . Ta_4O . . . . 50 (Ta):   Hb/B1–24, 51
OTb . . . . TbO . . . . ! . . 39 (SE):   Hb/C1–87, 91/2,
                                                                                       97/102

OTe . . . . TeO . . . . 11 (Te):   Hb–273/4
OTeTh . . . . ThOTe . . . . 44 (Th):   Hb–295
OTh . . . . ThO . . . . 44 (Th):   Hb–201/2
OTi . . . . TiO . . . . 41 (Ti):   Hb–216/21
– . . . . TiO solid solutions:
                (Ti,V)O . . . . 48 (V):   Hb/B2–558

– . . . . TiO systems:
                TiO–TiC . . . . 41 (Ti):   Hb–368/9
                TiO–VN . . . . 48 (V):   Hb/B2–560

OTi^{2+} . . . . TiO^{2+} . . . . 41 (Ti):   Hb–270, 471, 479
OTl . . . . TlO . . . . 38 (Tl):   Hb/2–219
OTl$_2$ . . . . Tl_2O . . . . 38 (Tl):   Hb/2–219/21
– . . . . Tl$_2$O melts:
                $Tl_2O-B_2O_3-PbO$ . . . . 47 (Pb):   Hb/C4–1354
                $Tl_2O-PbO-SiO_2$ . . . . 47 (Pb):   Hb/C4–1354

– . . . . Tl$_2$O systems:
                $Tl_2O-B_2O_3$ . . . . 38 (Tl):   Hb/3–377/8

OTm . . . . TmO . . . . 39 (SE):   Hb/C1–87, 91/2,
                                                                                       97/100

OU^{2+} . . . . $[UO]^{2+}$ . . . . 55 (U):   Hb–136

See the "Instructions for the user"
at the beginning of the volume

OV . . . . VO $= V_2O_2$:

    Chemical reactions . . . . 48 (V): Hb/B1–31

    Determination. . . . . 48 (V): Hb/A2–714

    Electrochemical behavior . . . . . . 48 (V): Hb/A2–465/6, 498

    Formation and preparation . . . . . 48 (V): Hb/A2–20/3

    History. . . . . 48 (V): Hb/A1–7, 10

    Occurrence, extraterrestrial . . . . . 48 (V): Hb/A1–14

    Physical properties . . . . 48 (V): Hb/A2–379/80, 391

                     Hb/B1–23/31

–. . . . . VO solid solutions:

    (V,Ti)O . . . . 48 (V): Hb/B2–558

–. . . . . VO systems:

    VO–NbO . . . . 49 (Nb): Hb/B3–321/2

    VO–TiO$_2$–ZrO$_2$ . . . . 48 (V): Hb/B2–564

    VO–V . . . . 48 (V): Hb/B1–12/3

    VO–VN . . . . 48 (V): Hb/B1–177

    VO–V$_2$O$_5$ . . . . 48 (V): Hb/B1–13

OV$^+$ . . . . VO$^+$:

    Detection . . . . 48 (V): Hb/A2–646

    Determination. . . . . 48 (V): Hb/A2–676/7

    Electrochemical behavior . . . . . . 48 (V): Hb/A2–466/8, 486/7

    Transition reactions . . . . 48 (V): Hb/A2–550/640

OV^{2+} . . . . VO^{2+}:

    Chemical reactions . . . . 48 (V): Hb/A2–533/48

                     Hb/B1–161, 298

    Detection . . . . 48 (V): Hb/A2–548, 646/7

    Determination. . . . . 48 (V): Hb/A2–676/7

    Electrochemical behavior . . . . . . 48 (V): Hb/A2–465/9, 474/6,

                       498/505, 509/11

    Exchange reactions. . . . . 48 (V): Hb/A1–259

                     Hb/A2–549/50

    Properties of the ion . . . . 48 (V): Hb/A2–340, 391

                     Hb/B1–70/2

    Sorption . . . . 48 (V): Hb/A1–259

                     Hb/A2–549/50

    Transition reactions . . . . 48 (V): Hb/A2–550/640

–. . . . . (VO)$_2^{4+}$ . . . . 48 (V): Hb/A2–601

–. . . . . [VO(H$_2$O)]$^{2+}$ = V(OH)$_2^{2+}$ . . . . 48 (V): Hb/A2–534/5

–. . . . . [VO(H$_2$O)$_5$]$^{2+}$ . . . . 48 (V): Hb/A2–391, 534, 536

                     Hb/B1–71, 296/7

OV^{3+} . . . . [VO]$^{3+}$ . . . . 48 (V): Hb/A2–511, 638

                     Hb/B1–157/60, 163, 236

OV$_2$ . . . . V$_2$O . . . . 48 (V): Hb/B1–13

OV$_2^{2+}$ . . . . [V$_2$O]$^{2+}$ . . . . 48 (V): Hb/A2–475

OV$_2^{3+}$ . . . . [V$_2$O]$^{3+}$ . . . . 48 (V): Hb/A2–475

OV$_2^{4+}$ . . . . [VOV]$^{4+}$ . . . . 48 (V): Hb/A2–600

OV$_3$Zr$_3$ . . . . V$_3$Zr$_3$O . . . . 48 (V): Hb/B2–673

Hb = Main Series Volume; GD = Gmelin-Durrer;   *Ab = Appended Volume to Main Series;*
Eb = Supplement Volume to Main Series;   *Erg. W. = New Supplement Series Volume*

OV₄ . . . . V₄O . . . . 48 (V):   Hb/B1−21
                                                                             Hb/B2−524
OW . . . . WO . . . . 54 (W):   Hb−107, 111
OXe . . . . XeO . . . . Erg. W. 1−93
OXe⁺ . . . . XeO⁺ . . . . Erg. W. 1−101
OY . . . . YO . . . . 39 (SE):   Hb/C1−85/93, 98/102,
                                                                                    188/9
−. . . . . Y₂O₂ . . . . 39 (SE):   Hb/C1−91, 100
OY⁺ . . . . YO⁺ . . . . 39 (SE):   Hb/C1−102
OY₂ . . . . Y₂O . . . . 39 (SE):   Hb/C1−102/3
OYb . . . . YbO . . . . 39 (SE):   Hb/C1−91/2, 98/100, 426
−. . . . . YbO solid solutions:
                     YbO−EuO . . . . 39 (SE):   Hb/C1−430
OZn . . . . ZnO:
                     Chemical reactions . . . . 32 (Zn):   Hb−131/6
                                                                           Eb−809/17
                     Commercial products. . . . . 32 (Zn):   Eb−787/9
                     Crystallographic properties . . . . . 32 (Zn):   Hb−127
                                                                           Eb−795/8
                     Electrical properties . . . . 32 (Zn):   Hb−130
                                                                           Eb−806/8
                     Electrochemical behavior . . . . 32 (Zn):   Eb−808/9
                     Formation . . . . 32 (Zn):   Hb−125, 127
                                                                           Eb−772/3, 794/5
                     Magnetic properties . . . . 32 (Zn):   Hb−130
                                                                           Eb−805/6
                     Mechanical properties . . . . 32 (Zn):   Hb−128
                                                                           Eb−798/802
                     Optical properties. . . . . 32 (Zn):   Hb−128/30
                                                                           Eb−804/5
                     Pigments . . . . 32 (Zn):   Eb−788/9
                     Preparation . . . . 32 (Zn):   Hb−125
                                                                           Eb−773/6
                     −, industrial . . . . 32 (Zn):   Hb−126
                                                                           Eb−776/87
                     Purification . . . . 32 (Zn):   Hb−92/3
                                                                           Eb−789/91
                     Purity testing . . . . 32 (Zn):   Eb−791/2
                     Solubility . . . . 32 (Zn):   Hb−130/1
                                                                           Eb−816/7
                     Special forms . . . . 32 (Zn):   Eb−787/8, 792/4
                     Thermal properties . . . . 32 (Zn):   Hb−128
                                                                           Eb−802/3
−. . . . . ZnO · xH₂O . . . . 32 (Zn):   Eb−833
−. . . . . ZnO glasses:
                     ZnO−B₂O₃ . . . . 32 (Zn):   Eb−971/2
                     ZnO−Na₂O−SiO₂ . . . . 32 (Zn):   Eb−1008
                     ZnO−PbO−SiO₂ . . . . 47 (Pb):   Hb/C4−1231

See the "Instructions for the user"
at the beginning of the volume

OZn . . . . ZnO melts:
                $ZnO-B_2O_3-V_2O_5$ . . . . 48 (V):   Hb/B2-536

−. . . . . ZnO phosphors:
                $ZnO-PbO-SiO_2$. . . . . 47 (Pb):   Hb/C4-1231

−. . . . . ZnO solid solutions:
                $ZnO-CoO$ . . . . 58 (Co):   Hb/A2-461
                                                                    Eb/A-849

−. . . . . ZnO solutions:
                $ZnO-NaOH-H_2O$ . . . . 32 (Zn):   Eb-1004
                $ZnO-NaOH-PbO-H_2O$ . . . . 47 (Pb):   Hb/C4-1230

−. . . . . ZnO systems:
                $ZnO-Al_2O_3$ . . . . 35 (Al):   Hb/B2-608
                $ZnO-Al_2O_3-CoO$ . . . . 58 (Co):   Eb/A-866
                $ZnO-Al_2O_3-Na_2O-SiO_2$ . . . . 35 (Al):   Hb/B2-611
                $ZnO-Al_2O_3-NiO$ . . . . 57 (Ni):   Hb/B3-1162
                $ZnO-Al_2O_3-SiO_2$ . . . . 35 (Al):   Hb/B2-611
                $ZnO-B_2O_3$. . . . . 32 (Zn):   Eb-970/1
                $ZnO-B_2O_3-PbO$ . . . . 47 (Pb):   Hb/C4-1227/8
                $ZnO-B_2O_3-SiO_2$ . . . . 32 (Zn):   Eb-994
                $ZnO-CaO-PbO-SiO_2$ . . . . 47 (Pb):   Hb/C4-1231
                $ZnO-CdO$ . . . . 33 (Cd):   Eb-711
                $ZnO-CrO_3-H_2O$. . . . . 52 (Cr):   Hb/B-856
                $ZnO-CuO-Fe_2O_3$ . . . . 60 (Cu):   Hb/B3-1325
                $ZnO-GeO_2-Nb_2O_5$ . . . . 49 (Nb):   Hb/B3-221
                $ZnO-H_2O$ . . . . 32 (Zn):   Eb-818
                $ZnO-K_2O-SiO_2$. . . . . 32 (Zn):   Eb-1015
                $ZnO-MgO$. . . . . 32 (Zn):   Eb-1023
                $ZnO-NH_4Cl-H_2O$ . . . . 32 (Zn):   Eb-874
                $ZnO-NH_4Cl-ZnCl_2-H_2O$ . . . . 32 (Zn):   Eb-1017
                $ZnO-N_2O_5-H_2O$ . . . . 32 (Zn):   Eb-843
                $ZnO-NaOH-H_2O$. . . . . 32 (Zn):   Eb-1001/3
                $ZnO-Na_2O-H_2O$ . . . . 32 (Zn):   Hb-294/5
                $ZnO-Na_2O-PbO-H_2O$ . . . . 47 (Pb):   Hb/C4-1229/30
                $ZnO-Nb_2O_5$ . . . . 49 (Nb):   Hb/B3-95
                $ZnO-Nb_2O_5-P_2O_5$ . . . . 49 (Nb):   Hb/B4-361
                $ZnO-Nb_2O_5-SiO_2$ . . . . 49 (Nb):   Hb/B4-359
                $ZnO-Nb_2O_5-V_2O_5$ . . . . 49 (Nb):   Hb/B3-329
                $ZnO-NiO$ . . . . 57 (Ni):   Hb/B3-1134/5
                $ZnO-NiO-TiO_2$ . . . . 57 (Ni):   Hb/B3-1184/5
                $ZnO-P_2O_5-H_2O$. . . . . 32 (Zn):   Hb-277
                $ZnO-P_2O_5-Ta_2O_5$. . . . . 50 (Ta):   Hb/B2-349
                $ZnO-SiO_2$ . . . . 32 (Zn):   Eb-989/90
                $ZnO-SiO_2-Ta_2O_5$. . . . . 50 (Ta):   Hb/B2-349
                $ZnO-Ta_2O_5$ . . . . 50 (Ta):   Hb/B2-193
                $ZnO-TiO_2$ . . . . 41 (Ti):   Hb-460
                $ZnO-VO_2$ . . . . 48 (V):   Hb/B2-533
                $ZnO-VO_2-V_2O_5$. . . . . 48 (V):   Hb/B2-534

Hb = Main Series Volume; GD = Gmelin-Durrer;   Ab = Appended Volume to Main Series;
Eb = Supplement Volume to Main Series;   Erg. W. = New Supplement Series Volume

OZn . . . . ZnO systems:
                     $ZnO-V_2O_5$ . . . . 48 (V):   Hb/B2−534
                     $ZnO-ZnCl_2-H_2O$ . . . . 32 (Zn):   Eb−879
                     $ZnO-ZnSO_4-H_2O$ . . . . 32 (Zn):   Eb−963/4
                     $ZnO-ZrO_2$ . . . . 42 (Zr):   Hb−442
OZn$_2$ . . . . Zn_2O . . . . 32 (Zn):   Hb−124
OZr . . . . ZrO . . . . 42 (Zr):   Hb−212/5
OZr^{2+} . . . . ZrO^{2+}:
                     Chemical reactions . . . . 42 (Zr):   Hb−172/91
                     − with elements . . . . 42 (Zr):   Hb−173
                     − with inorganic compounds . . . . . 42 (Zr):   Hb−173/80
                     − with organic compounds . . . . . . 42 (Zr):   Hb−180/91
                     Electrochemical behavior . . . . 42 (Zr):   Hb−157
O$_{1.27}$V . . . . $VO_{1.27}$ . . . . 48 (V):   Hb/B1−12/3, 31/2
O$_{1.41}$Pb . . . . $PbO_{1.41} = Pb_{12}O_{17}$ . . . . 47 (Pb):   Hb/C1−139/44
O$_{1.5}$V . . . . $VO_{1.5}$
                     see O_3V_2 ... V_2O_3
O$_{1.54}$Pb . . . . $PbO_{1.54}$ . . . . 47 (Pb):   Hb/C1−139/44
O$_{1.55}$Pb . . . . $PbO_{1.55}$ . . . . 47 (Pb):   Hb/C1−144/7
O$_{1.67}$Pb . . . . $PbO_{1.67} = Pb_6O_{10}$ . . . . 47 (Pb):   Hb/C1−144/7
O$_{1.67}$V . . . . $VO_{1.67}$
                     see O_5V_3 ... V_3O_5
O$_{1.71}$Pr . . . . $PrO_{1.714} = Pr_7O_{12}$ . . . . 39 (SE):   Hb/C1−284/5
O$_{1.71}$Tb . . . . $TbO_{1.714} = Tb_7O_{12}$ . . . . 39 (SE):   Hb/C1−387, 394/5
O$_{1.75}$Tb . . . . $TbO_{1.75}$
                     see O_7Tb_4 ... Tb_4O_7
O$_{1.75}$V . . . . $VO_{1.75} = V_4O_7$ . . . . 48 (V):   Hb/B1−13, 30, 50/1
O$_{1.76}$Ta . . . . $TaO_{1.76}$ . . . . 50 (Ta):   Hb/B1−30
O$_{1.78}$Pr . . . . $PrO_{1.778} = Pr_9O_{16}$ . . . . 39 (SE):   Hb/C1−285
O$_{1.8}$Pr . . . . $PrO_{1.8} = Pr_{10}O_{18}$ . . . . 39 (SE):   Hb/C1−285
O$_{1.8}$V . . . . $VO_{1.80} = V_5O_9$ . . . . 48 (V):   Hb/B1−13, 17, 50/1
O$_{1.82}$Pr . . . . $PrO_{1.818} = Pr_{11}O_{20}$ . . . . 39 (SE):   Hb/C1−285
O$_{1.82}$Tb . . . . $TbO_{1.818} = Tb_{11}O_{20}$ . . . . 39 (SE):   Hb/C1−387, 395/6
O$_{1.83}$Pr . . . . $PrO_{1.833}$
                     see $O_{11}Pr_6$ ... Pr_6O_{11}
O$_{1.83}$Tb . . . . $TbO_{1.833} = Tb_6O_{11}$ . . . . 39 (SE):   Hb/C1−387
O$_{1.83}$V . . . . $VO_{1.833} = V_6O_{11}$ . . . . 48 (V):   Hb/B1−13, 30, 50/1
O$_{1.86}$V . . . . $VO_{1.86} = V_7O_{13}$ . . . . 48 (V):   Hb/B1−13, 30, 50/1
O$_{1.87}$V . . . . $VO_{1.87} = V_8O_{15}$ . . . . 48 (V):   Hb/B1−13, 17, 50/1
O$_2$ . . . . O_2 . . . . 3 (O):   Hb/3−348/51
              see also O ... Oxygen
−. . . . . $O_2 \cdot xH_2O$ . . . . 3 (O):   Hb/5−1191
O$_2$$^+$ . . . . O_2^+ . . . . 3 (O):   Hb/3−351, 359/60, 367,
                                           572/5, 638
O$_2$$^{2+}$ . . . . O_2^{2+} . . . . 3 (O):   Hb/3−592

See the "Instructions for the user"
at the beginning of the volume

O_2^-	O_2^-	3 (O):	Hb/3−351/2, 359/61, 371, 590
			Hb/8−2626
		21 (Na):	Eb/6−338
O_2^{2-}	O_2^{2-}	3 (O):	Hb/3−360/1, 371, 590
O_2Os	OsO_2	66 (Os):	Hb−26/7
−.	$OsO_2 \cdot H_2O$	66 (Os):	Hb−29
−.	$OsO_2 \cdot 2\,H_2O$	66 (Os):	Hb−27/9
−.	$OsO_2 \cdot 5\,H_2O$	66 (Os):	Hb−29
$O_2PS_2Tl_3$	$Tl_3PO_2S_2$	38 (Tl):	Hb/3−432
$O_2P_2Pb_3S_6$	$Pb_3(POS_3)_2$	47 (Pb):	Hb/C3−884
$O_2P_2S_3$	$P_2S_3O_2$	16 (P):	Hb/C−583
O_2Pb	PbO_2:		
	Chemical reactions	47 (Pb):	Hb/C1−189/95
	Crystallographic properties	47 (Pb):	Hb/C1−154/8
	Electrical properties	47 (Pb):	Hb/C1−160/2
	Electrochemical behavior	47 (Pb):	Hb/C1−163/89
	Estimation.	47 (Pb):	Hb/C1−151
	Formation	47 (Pb):	Hb/C1−147/50
	History.	47 (Pb):	Hb/A1−125/7
	Magnetic properties	47 (Pb):	Hb/C1−160
	Mechanical properties	47 (Pb):	Hb/C1−158/9
	Optical properties.	47 (Pb):	Hb/C1−162/3
	Preparation	47 (Pb):	Hb/C1−147/54
	− of special forms.	47 (Pb):	Hb/C1−151/4
	Purity testing	47 (Pb):	Hb/C1−151
	Thermal properties	47 (Pb):	Hb/C1−158/9
−.	$PbO_2 \cdot xH_2O$	47 (Pb):	Hb/C1−195/6
−.	PbO_2 systems:		
	$PbO_2-Cr_2O_3$	52 (Cr):	Hb/B−910
	$PbO_2-Pb_3O_4$.	47 (Pb):	Hb/C1−55/6
	$PbO_2-Pb_{12}O_{19}$.	47 (Pb):	Hb/C1−55
O_2Pb^{2-}	PbO_2^{2-}	47 (Pb):	Hb/B2−11
			Hb/C1−196/9
O_2PbSr	$SrPbO_2$	47 (Pb):	Hb/C3−1195
$O_2Pb_4^{4+}$	$[Pb_4O_2]^{4+}$	47 (Pb):	Hb/C1−199
O_2Pd	PdO_2	65 (Pd):	Hb/2−266
−.	$PdO_2 \cdot 2\,H_2O$	65 (Pd):	Hb/2−266/7
−.	$PdO_2 \cdot xH_2O$	65 (Pd):	Hb/2−266/8
O_2Po	PoO_2	12 (Po):	Hb−133
O_2Pr.	PrO_2	39 (SE):	Hb/C1−103/4, 148, 290/4
−.	$Pr_2O_4 \cdot nH_2O$	39 (SE):	Hb/C2−106/8, 112
−.	$Pr_6O_{12} \cdot nH_2O$.	39 (SE):	Hb/C2−106
−.	PrO_2 solid solutions:		
	PrO_2-CeO_2	39 (SE):	Hb/C1−296
	PrO_2-TbO_2	39 (SE):	Hb/C1−404

Hb = Main Series Volume; GD = Gmelin-Durrer;   *Ab = Appended Volume to Main Series;*
Eb = Supplement Volume to Main Series;   *Erg. W. = New Supplement Series Volume*

O$_2$Pr. . . . . PrO$_2$ systems:
      PrO$_2$–Dy$_2$O$_3$ . . . . 39 (SE):   Hb/C1–295
      PrO$_2$–Er$_2$O$_3$ . . . . 39 (SE):   Hb/C1–295
      PrO$_2$–Gd$_2$O$_3$. . . . . 39 (SE):   Hb/C1–295
      PrO$_2$–La$_2$O$_3$ . . . . 39 (SE):   Hb/C1–295
      PrO$_2$–Nd$_2$O$_3$. . . . . 39 (SE):   Hb/C1–295
      PrO$_2$–Sm$_2$O$_3$ . . . . 39 (SE):   Hb/C1–295

O$_2$Pr$^+$ . . . . PrO$_2$$^+$ . . . . 39 (SE):   Hb/C1–104
O$_2$Pt . . . . PtO$_2$ . . . . 68 (Pt):   Hb/C1–26/9, 46
–. . . . . PtO$_2$ · H$_2$O . . . . 68 (Pt):   Hb/C1–26, 46/7, 51/2
–. . . . . PtO$_2$ · 2 H$_2$O = Pt(OH)$_4$ . . . . 68 (Pt):   Hb/C1–26, 46/7, 50/1
–. . . . . PtO$_2$ · 3 H$_2$O . . . . 68 (Pt):   Hb/C1–26, 46/7, 49/50
–. . . . . PtO$_2$ · 4 H$_2$O . . . . 68 (Pt):   Hb/C1–26, 46/9

O$_2$Pu . . . . PuO$_2$:
      Behavior as target . . . . Erg. W. 8–287
      Chemical reactions . . . . Erg. W. 4–26/7
      Compatibility with
        construction materials . . . . Erg. W. 4–25/6
      Criticality . . . . Erg. W. 8–368/9, 372
      Physical properties . . . . Erg. W. 4–16/25
                                     Erg. W. 8–298/300
      Preparation . . . . Erg. W. 4–14/5
      Pressing and sintering behavior . . . . Erg. W. 4–25
      Sols . . . . Erg. W. 8–244/5
      Stoichiometry . . . . Erg. W. 4–15

–. . . . . PuO$_2$ · nH$_2$O . . . . Erg. W. 4–29
–. . . . . PuO$_2$ fuels:
      In-pile behavior . . . . Erg. W. 8–259/77
      Preparation . . . . Erg. W. 8–242/8

–. . . . . PuO$_2$ glasses:
      PuO$_2$–Na$_2$O–SiO$_2$ . . . . Erg. W. 4–70

–. . . . . PuO$_2$ solid solutions:
      PuO$_2$–CeO$_2$ . . . . Erg. W. 4–57
      PuO$_2$–CmO$_2$. . . . . Erg. W. 4–58
      PuO$_2$–ThO$_2$ . . . . Erg. W. 4–58
      (Pu, U)O$_2$ . . . . Erg. W. 4–60/70
                                       Erg. W. 8–192
      PuO$_2$–UO$_2$–ZrO$_2$ . . . . Erg. W. 4–71

–. . . . . PuO$_2$ systems:
      PuO$_2$–Al$_2$O$_3$ . . . . Erg. W. 4–56
      PuO$_2$–BeO. . . . . Erg. W. 4–55
      PuO$_2$–EuO$_{1.5}$ . . . . Erg. W. 4–57
      PuO$_2$–Fe$_2$O$_3$. . . . . Erg. W. 4–72
      PuO$_2$–HoO$_{1.5}$ . . . . Erg. W. 4–57
      PuO$_2$–LuO$_{1.5}$ . . . . Erg. W. 4–57
      PuO$_2$–MgO . . . . Erg. W. 4–55/6

See the "Instructions for the user"
at the beginning of the volume

O_2Pu	PuO_2 systems:		
	$PuO_2-PaO_{2.5}$	Erg. W. 4–58	
	$PuO_2-TmO_{1.5}$	Erg. W. 4–57	
	PuO_2-UO_2	Erg. W. 4–59	
	$PuO_2-U_3O_7$	Erg. W. 4–60	
	$PuO_2-YO_{1.5}$	Erg. W. 4–57	
	PuO_2-ZrO_2	Erg. W. 4–70/1	
O_2Pu^+	PuO_2^+	Erg. W. 8–69/70, 72	
O_2Pu^{2+}	PuO_2^{2+}	Erg. W. 4–99	
		Erg. W. 8–69/70, 74/5	
–.	$[Pu_2O_4]^{4+}$	Erg. W. 4–28	
O_2Pu_2S	Pu_2O_2S	Erg. W. 4–172/3	
$O_2Pu_2S_3$	$Pu_2O_2S_3$	Erg. W. 4–172/3	
O_2Pu_2Se	Pu_2O_2Se	Erg. W. 4–189, 191	
O_2Pu_2Te	Pu_2O_2Te	Erg. W. 4–193	
O_2RbSc	$RbScO_2$	39 (SE):	Hb/C2–140/1
O_2RbTm	$RbTmO_2$	39 (SE):	Hb/C2–140
O_2RbY	$RbYO_2$	39 (SE):	Hb/C2–140/1
O_2RbYb	$RbYbO_2$	39 (SE):	Hb/C2–140
O_2Rb_2	Rb_2O_2	24 (Rb):	Hb–110
O_2Re	ReO_2	70 (Re):	Hb–89/90
–.	$ReO_2 \cdot xH_2O$	70 (Re):	Hb–97
O_2Rh	RhO_2	64 (Rh):	Hb–48
–.	$RhO_2 \cdot 2 H_2O$	64 (Rh):	Hb–48/50
O_2Ru	RuO_2	63 (Ru):	Hb–25/7
			Eb–232/3, 234/8
–.	$RuO_2 \cdot H_2O$	63 (Ru):	Eb–238/9
–.	$RuO_2 \cdot nH_2O$	63 (Ru):	Eb–233, 238/9
O_2Ru^+	RuO_2^+	63 (Ru):	Eb–233
O_2Ru^-	RuO_2^-	63 (Ru):	Eb–233
O_2S	SO_2:		
	Acoustic properties	9 (S):	Hb/B1–247
	Behavior as solvent	9 (S):	Hb/B1–310/23
			Hb/B2–1122/30
	Chemical reactions	9 (S):	Hb/B1–262/96
	– in electrical discharges	9 (S):	Hb/B1–263
	– on heating	9 (S):	Hb/B1–262/3
	– with light	9 (S):	Hb/B1–263
	– with metal compounds	9 (S):	Hb/B1–288/94
	– with metals	9 (S):	Hb/B1–270/7
	– with nonmetal compounds	9 (S):	Hb/B1–277/88
	– with nonmetals	9 (S):	Hb/B1–263/70, 323/31, 336/7
	– with organic compounds	9 (S):	Hb/B1–294/6
	Crystallographic properties	9 (S):	Hb/B1–210/1
	Determination	9 (S):	Hb/A2–319, 423, 447
			Hb/B3–1145/9, 1222
	Electrical properties	9 (S):	Hb/B1–259/62

Hb = Main Series Volume; GD = Gmelin-Durrer;   Ab = Appended Volume to Main Series;
Eb = Supplement Volume to Main Series;   Erg. W. = New Supplement Series Volume

O₂S . . . . SO₂:

Enrichment	9 (S):	Hb/A2–307/17
		Hb/B3–1279/82
Formation	9 (S):	Hb/B1–181/204
History	9 (S):	Hb/A1–40/3
Industrial hazards	9 (S):	Hb/A2–481/3
Liquefaction	9 (S):	Hb/A2–318/9
Magnetic properties	9 (S):	Hb/B1–259
Mechanical properties	9 (S):	Hb/B1–211/6, 233/47
Molecule	9 (S):	Hb/B1–204/10
Optical properties	9 (S):	Hb/B1–247/59
Oxidation	9 (S):	Hb/B1–323/31, 336/7
see also H₂O₃S ... H₂SO₃		
–, chamber process	9 (S):	Hb/A2–426/34
–, contact process	9 (S):	Hb/A2–324/51
Preparation	9 (S):	Hb/B1–202/3
–, industrial	9 (S):	Hb/A2–286/317
–, –, concentration and purification . .	9 (S):	Hb/A2–307/18
–, –, preparation from		
hydrogen sulfide	9 (S):	Hb/A2–288
–, –, – from raw sulfur	9 (S):	Hb/A2–286/8
–, –, – from sulfates	9 (S):	Hb/A2–300/7
–, –, – from sulfidic ores	9 (S):	Hb/A2–288/300
–, –, – from sulfites	9 (S):	Hb/A2–300
–, –, – from waste gases	9 (S):	Hb/A2–307
		Hb/B3–1279/82, 1418/9
Purification	9 (S):	Hb/A2–317/8
		Hb/B1–203
Purity testing	9 (S):	Hb/A2–319
Removal from waste gases	9 (S):	Hb/B3–1294, 1519
Solubility in acid solutions	9 (S):	Hb/B3–1220/46
– in organic materials	9 (S):	Hb/B1–298/310
– in salt solutions	9 (S):	Hb/B3–1246/79
– in technical absorbents	9 (S):	Hb/B3–1279/84
– in water	9 (S):	Hb/B3–1131/219
– in other inorganic substances	9 (S):	Hb/B1–296/8
Sorption	9 (S):	Hb/B1–237/47
		Hb/B3–1131/2, 1219,
		1279/84
Spectra	9 (S):	Hb/B1–251/9
Storage	9 (S):	Hb/A2–319
Thermal properties	9 (S):	Hb/B1–214/33
Toxicity	9 (S):	Hb/A2–482, 506/9
Uses	9 (S):	Hb/A2–319
		Hb/B3–1294/5
–, ... ³²S¹⁶O₂	9 (S):	Hb/B1–206/7
–, ... (SO₂)₂	9 (S):	Hb/B1–209
–, ... SO₂ · H₂O	9 (S):	Hb/B2–1121/2

See the "Instructions for the user"
at the beginning of the volume

O_2S . . . . SO$_2$ · 6 H$_2$O:
          Composition . . . . 9 (S):   Hb/B2−1094/9
          Formation and preparation . . . . . . 9 (S):   Hb/B2−1099/108
          Mixed hydrates and
            hydrate solid solutions . . . . 9 (S):   Hb/B2−1118/22
          Properties and behavior . . . . 9 (S):   Hb/B2−1090/4, 1109/18

−. . . . . SO$_2$ mixtures:
          SO$_2$−CO$_2$ . . . . 14 (C):   Hb/C1−546

−. . . . . SO$_2$ solid solutions:
          SO$_2$−Rn . . . . 9 (S):   Hb/B1−210
          SO$_2$ · 6 H$_2$O−Rn · 6 H$_2$O . . . . 9 (S):   Hb/B2−1120/1

−. . . . . SO$_2$ solutions:
          SO$_2$−HCN . . . . 9 (S):   Hb/B1−309
          SO$_2$−H$_2$O
            see H$_2$O$_3$S … H$_2$SO$_3$

          SO$_2$−N . . . . 9 (S):   Hb/B1−296/7
          SO$_2$−O . . . . 9 (S):   Hb/B1−296
          SO$_2$−organic solvents . . . . 9 (S):   Hb/B1−298/310
          SO$_2$−S . . . . 9 (S):   Hb/B1−297
          SO$_2$−SO$_3$ . . . . 9 (S):   Hb/B1−366

−. . . . . SO$_2$ systems:
          SO$_2$−Al$_2$(SO$_4$)$_3$−H$_2$O . . . . 9 (S):   Hb/B3−1282
          SO$_2$−BCl$_3$ . . . . 13 (B):   Eb−200
          SO$_2$−BF$_3$ . . . . 9 (S):   Hb/B1−298
                                     13 (B):   Eb−186
          SO$_2$−BaO−BaS−BaSO$_4$−S . . . . . . 30 (Ba):   Eb−448
          SO$_2$−BaS−BaSO$_4$−S . . . . 30 (Ba):   Eb−448
          SO$_2$−Br . . . . 9 (S):   Hb/B1−297
          SO$_2$−CCl$_4$ . . . . 9 (S):   Hb/B1−307
           SO$_2$−CO(NH$_2$)$_2$−H$_2$O . . . . 14 (C):   Hb/D1−421
          SO$_2$−CH$_3$OH . . . . 9 (S):   Hb/B1−302
          SO$_2$−CO$_2$ . . . . 14 (C):   Hb/C1−543/6
          SO$_2$−CH$_3$COOH−H$_2$O . . . . 9 (S):   Hb/B3−1245/6
          SO$_2$−(CH$_3$)$_2$O . . . . 9 (S):   Hb/B1−305
          SO$_2$−C$_2$H$_5$OH . . . . 9 (S):   Hb/B1−302/3
          SO$_2$−CH$_3$COCH$_3$ . . . . 9 (S):   Hb/B1−304
          SO$_2$−(C$_2$H$_5$)$_2$O . . . . 9 (S):   Hb/B1−305
          SO$_2$−(CH$_3$)$_4$NI−H$_2$O . . . . 9 (S):   Hb/B3−1251/3
          SO$_2$−C$_6$H$_4$(OH)$_2$−H$_2$O . . . . 9 (S):   Hb/B3−1251/4
          SO$_2$−Ca(HSO$_3$)$_2$−H$_2$O . . . . 9 (S):   Hb/B3−1262/79
          SO$_2$−CaO . . . . 28 (Ca):   Hb/B3−657/8
          SO$_2$−CaO−H$_2$O . . . . 28 (Ca):   Hb/B3−665/74
          SO$_2$−CaS . . . . 28 (Ca):   Hb/B3−658
          SO$_2$−CdBr$_2$−H$_2$O . . . . 9 (S):   Hb/B3−1248/50
          SO$_2$−CdCl$_2$−H$_2$O . . . . 9 (S):   Hb/B3−1248/50
          SO$_2$−CdI$_2$−H$_2$O . . . . 9 (S):   Hb/B3−1248/50
          SO$_2$−CdSO$_4$−H$_2$O . . . . 9 (S):   Hb/B3−1256/8

Hb = Main Series Volume; GD = Gmelin-Durrer;   *Ab = Appended Volume to Main Series;*
Eb = Supplement Volume to Main Series;   *Erg. W. = New Supplement Series Volume*

O₂S . . . . SO₂ systems:

SO₂–Cl	9 (S):	Hb/B1–297
SO₂–Cl–SO₂Cl₂.	9 (S):	Hb/B3–1820
SO₂–GeCl₄	45 (Ge):	Eb–540
SO₂–HBr	9 (S):	Hb/B1–298
SO₂–HCl.	9 (S):	Hb/B1–297/8
SO₂–HSCN–H₂O	9 (S):	Hb/B3–1245/6
SO₂–H₂O	9 (S):	Hb/B2–1046/88
		Hb/B3–1132/219
SO₂–H₂S	9 (S):	Hb/B1–298
SO₂–H₂SO₄–H₂O	9 (S):	Hb/B3–1220/45
SO₂–HgBr₂–HgI₂	34 (Hg):	Hb/B2–951
SO₂–I	9 (S):	Hb/B1–297
SO₂–KBr–H₂O	9 (S):	Hb/B3–1248/54
SO₂–KCl–H₂O	9 (S):	Hb/B3–1248/56
SO₂–KI	22 (K):	Hb/3–650/2
SO₂–KI–H₂O	9 (S):	Hb/B3–1248/54
SO₂–KNO₃–H₂O	9 (S):	Hb/B3–1256/8
SO₂–KSCN	22 (K):	Hb/4–912/3
SO₂–KSCN–H₂O	9 (S):	Hb/B3–1248/54
SO₂–K₂SO₄–H₂O	9 (S):	Hb/B3–1256/8
SO₂–LiI	20 (Li):	Hb–201
SO₂–Mg(HSO₃)₂–H₂O	9 (S):	Hb/B3–1262/4, 1273
SO₂–MgSO₃–H₂O	27 (Mg):	Hb/B2–209/10
SO₂–N.	9 (S):	Hb/B1–297
SO₂–NH₃	9 (S):	Hb/B1–297
SO₂–NH₃–H₂O	23 (NH₄):	Hb/2–255/6
SO₂–NH₄Br–H₂O	9 (S):	Hb/B3–1248/50
SO₂–NH₄Cl–H₂O	9 (S):	Hb/B3–1248/50
SO₂–NH₄HSO₃–H₂O	9 (S):	Hb/B3–1273, 1276/9
SO₂–NH₄I	23 (NH₄):	Hb/1–231, 235
SO₂–NH₄NO₃–H₂O.	9 (S):	Hb/B3–1256/8
SO₂–NH₄SCN.	23 (NH₄):	Hb/2–387
SO₂–NH₄SCN–H₂O.	9 (S):	Hb/B3–1248/50
	23 (NH₄):	Hb/2–387/8
SO₂–(NH₄)₂SO₄–H₂O.	9 (S):	Hb/B3–1256/8
SO₂–NaBr–H₂O	9 (S):	Hb/B3–1248/50
SO₂–NaCl–H₂O.	9 (S):	Hb/B3–1248/54
SO₂–NaHSO₄–H₂O.	9 (S):	Hb/B3–1261, 1278/9
SO₂–NaH₂PO₄–H₂O	9 (S):	Hb/B3–1283/4
SO₂–NaH₂PO₄–Na₂HPO₄–H₂O	21 (Na):	Eb/5–2010
SO₂–NaSCN–H₂O	9 (S):	Hb/B3–1248/50
SO₂–Na₂HPO₄–H₂O	21 (Na):	Eb/5–2005
SO₂–Na₂SO₄–H₂O	9 (S):	Hb/B3–1256/61
SO₂–PbO–H₂O	47 (Pb):	Hb/C2–535
SO₂–RbI–H₂O	9 (S):	Hb/B3–1251/3
SO₂–S–SrO–SrS–SrSO₄	29 (Sr):	Eb–257
SO₂–S–SrS–SrSO₄	29 (Sr):	Eb–257
SO₂–SO₂Cl₂.	9 (S):	Hb/B3–1820

See the "Instructions for the user"
at the beginning of the volume

O_2S	SO_2 systems:		
	SO_2-SiCl_4	15 (Si):	Hb/B−692
	SO_2-SnBr_4	46 (Sn):	Hb/C1−412
	SO_2-TiBr_4	41 (Ti):	Hb−328
	SO_2-TiCl_4	41 (Ti):	Hb−314/5
	$SO_2-ZnSO_3-H_2O$	32 (Zn):	Eb−935
O_2S^+	SO_2^+	9 (S):	Hb/B1−262
O_2SSb	SbO_2S	18 (Sb):	Hb/B3−541
O_2SSi	SiO_2S	15 (Si):	Hb/B−752
O_2SSr	$SrSO_2$	29 (Sr):	Eb−249
O_2STl_2	Tl_2SO_2	38 (Tl):	Hb/3−346
O_2SU	UO_2S	55 (U):	Hb−143
O_2SW	WO_2S	54 (W):	Hb−186
O_2SZn.	$ZnSO_2$	32 (Zn):	Hb−219
O_2S_2.	$S_2O_2 = (SO)_2$	9 (S):	Hb/B1−58/9, 167/78
$O_2S_2^{2-}$	$S_2O_2^{2-}$	9 (S):	Hb/B2−376
$O_2S_4U_3$	$UO_2 \cdot 2\, US_2$.	55 (U):	Hb−143
$O_2S_5V_2^{4-}$	$[V_2O_2S_5]^{4-}$	48 (V):	Hb/B1−281
O_2S_x.	S_nO_2 (n = 8 to 16)	9 (S):	Hb/B1−167
O_2Sb^-	SbO_2^-	18 (Sb):	Hb/B2−368/9
O_2Sb_3	Sb_3O_2.	18 (Sb):	Hb/B2−348
O_2Se	SeO_2:		
	Aerosols	10 (Se):	Hb/B−27
	Chemical reactions	10 (Se):	Hb/B−34/40
	Colloidal forms	10 (Se):	Hb/B−27
	Crystallographic properties	10 (Se):	Hb/B−29/31
	Electrical properties	10 (Se):	Hb/B−34
	Formation	10 (Se):	Hb/B−25/6, 27
	Mechanical properties	10 (Se):	Hb/B−31
	Molecule.	10 (Se):	Hb/B−28/9
	Optical properties.	10 (Se):	Hb/B−31/3
	Preparation	10 (Se):	Hb/B−26/7
	Spectra	10 (Se):	Hb/B−28/9, 32/3
	Thermal properties	10 (Se):	Hb/B−31
	Toxicity	10 (Se):	Hb/A1−259
−.	$SeO_2 \cdot H_2O = H_2SeO_3$	10 (Se):	Hb/A1−264/7, 273/81
			Hb/B−44/51
−.	SeO_2 solutions:		
	$SeO_2-H_2F_2$	10 (Se):	Hb/B−40
	SeO_2-H_2O		
	see H_2O_3Se ... H_2SeO_3 solutions		
	$SeO_2-H_2SO_4$	10 (Se):	Hb/B−35, 40/1
	$SeO_2-H_2SeO_4-H_2O$	10 (Se):	Hb/B−41
	SeO_2-organic solvents.	10 (Se):	Hb/B−41/3
−.	SeO_2 systems:		
	$SeO_2-BaSeO_3-H_2O$	30 (Ba):	Eb−454/5
	SeO_2-HCl	10 (Se):	Hb/B−145/6
	SeO_2-HF	10 (Se):	Hb/B−118

Hb = Main Series Volume; GD = Gmelin-Durrer;   Ab = Appended Volume to Main Series;
Eb = Supplement Volume to Main Series;   Erg. W. = New Supplement Series Volume

O_2Se . . . . SeO_2 systems:

SeO_2–H_2O	10 (Se):	Hb/B–44/6
SeO_2–Na_2O–H_2O	21 (Na):	Eb/5–1906/7
SeO_2–PbO–TeO_2	47 (Pb):	Hb/C2–684

O_2SeU . . . . UO_2Se . . . . 55 (U):   Hb–157

O_2Si . . . . SiO_2:

Absorption of light	15 (Si):	Hb/B–349/51
– of X-rays	15 (Si):	Hb/B–373
Adhesion	15 (Si):	Hb/B–331
Behavior as electrode	15 (Si):	Hb/B–403
Birefringence	15 (Si):	Hb/B–361/8
Boiling point	15 (Si):	Hb/B–339
Catalytic properties	15 (Si):	Hb/B–577/82
Chemical reactions	15 (Si):	Hb/B–539/82
– with acids	15 (Si):	Hb/B–565/8
– with alkaline solutions	15 (Si):	Hb/B–563/5
– with aqueous salt solutions	15 (Si):	Hb/B–571/6
– with metal compounds	15 (Si):	Hb/B–549/63
– with metals	15 (Si):	Hb/B–545/7
– with nonmetal compounds	15 (Si):	Hb/B–547/9
– with nonmetals	15 (Si):	Hb/B–543/5
– with organic substances	15 (Si):	Hb/B–576
– with water	15 (Si):	Hb/B–569/71
Cleavage	15 (Si):	Hb/B–310
Cohesion	15 (Si):	Hb/B–331
Color	15 (Si):	Hb/B–351/5
Compressibility	15 (Si):	Hb/B–328/9
Crystallographic properties	15 (Si):	Hb/B–305/20
Deformation	15 (Si):	Hb/B–320
Density	15 (Si):	Hb/B–321/3
Dielectric properties	15 (Si):	Hb/B–376/80
Diffraction of X-rays	15 (Si):	Hb/B–374/5
Dressing of sands and gravels	15 (Si):	Hb/B–271/7
Elasticity	15 (Si):	Hb/B–323/9
Electrical conductivity	15 (Si):	Hb/B–398/400
Electrical properties	15 (Si):	Hb/B–376/400
Electrochemical behavior	15 (Si):	Hb/B–400/8
Electron emission	15 (Si):	Hb/B–400
Electrostriction	15 (Si):	Hb/B–385
Enthalpy	15 (Si):	Hb/B–341/5
Entropy	15 (Si):	Hb/B–346
Faraday effect	15 (Si):	Hb/B–371/2
Formation	15 (Si):	Hb/B–269/70
– and preparation of cristobalite	15 (Si):	Hb/B–283
– – of fused quartz	15 (Si):	Hb/B–283/4
– – of quartz	15 (Si):	Hb/B–277/80
– – of silica gel	15 (Si):	Hb/B–458/99
– – of silicic acid sol	15 (Si):	Hb/B 423/35

See the "Instructions for the user"
at the beginning of the volume

O_2Si . . . . SiO_2:

Formation and preparation
  of tridymite . . . . 15 (Si):   Hb/B–280/3
– – of vitreous silica . . . . 15 (Si):   Hb/B–284/5
– – of vitreous silica ware . . . . 15 (Si):   Hb/B–286/8
– of special forms . . . . 15 (Si):   Hb/B–270
Forms of state . . . . 15 (Si):   Hb/B–298/306, 340/1
Glass electrode . . . . 15 (Si):   Hb/B–864/920
Hardness . . . . 15 (Si):   Hb/B–323
Heat capacity . . . . 15 (Si):   Hb/B–341/5
Heat of fusion . . . . 15 (Si):   Hb/B–340
– of transformation . . . . 15 (Si):   Hb/B–340/1
Lattice energy . . . . 15 (Si):   Hb/B–320
Lattice structure . . . . 15 (Si):   Hb/B–310/20
Luminescence . . . . 15 (Si):   Hb/B–372/3
Magnetic properties . . . . 15 (Si):   Hb/B–376
Mechanical properties . . . . 15 (Si):   Hb/B–321/34
Melting point . . . . 15 (Si):   Hb/B–304/5
Modifications . . . . 15 (Si):   Hb/B–298/306, 340/1
Molecule . . . . 15 (Si):   Hb/B–307
Optical activity . . . . 15 (Si):   Hb/B–369/71
Optical properties . . . . 15 (Si):   Hb/B–348/76
Permeability . . . . 15 (Si):   Hb/B–332/4
Physiological behavior . . . . 15 (Si):   Hb/B–130/4
Piezoelectricity . . . . 15 (Si):   Hb/B–380/97
Polymorphism . . . . 15 (Si):   Hb/B–298/306, 340/1
Preparation
  see Formation

Purification . . . . 15 (Si):   Hb/B–276/7
Purity testing . . . . 15 (Si):   Hb/B–277
Rayleigh scattering . . . . 15 (Si):   Hb/B–360/1
Reflection of light . . . . 15 (Si):   Hb/B–348/9
– of X-rays . . . . 15 (Si):   Hb/B–373/4
Refraction . . . . 15 (Si):   Hb/B–361/8, 375/6
Scattering of X-rays . . . . 15 (Si):   Hb/B–375
Sound propagation . . . . 15 (Si):   Hb/B–334, 395/6
Specific heat . . . . 15 (Si):   Hb/B–341/5
Spectra . . . . 15 (Si):   Hb/B–355/60
Strength . . . . 15 (Si):   Hb/B–329/31
Surface tension . . . . 15 (Si):   Hb/B–331/2
Susceptibility . . . . 15 (Si):   Hb/B–376
Suspensions . . . . 15 (Si):   Hb/B–288/98
Thermal conductivity . . . . 15 (Si):   Hb/B–346/8
Thermal expansion . . . . 15 (Si):   Hb/B–334/9
Thermal properties . . . . 15 (Si):   Hb/B–334/48
Thermal radiation . . . . 15 (Si):   Hb/B–372
Thermoelectric effects . . . . 15 (Si):   Hb/B–400
Toxicity . . . . 15 (Si):   Hb/B–130/4

Hb = Main Series Volume; GD = Gmelin-Durrer;   *Ab = Appended Volume to Main Series;*
Eb = Supplement Volume to Main Series;   *Erg. W. = New Supplement Series Volume*

O_2Si . . . . SiO_2:

Transformation points . . . . 15 (Si):   Hb/B–300/3
Transmission, optical . . . . 15 (Si):   Hb/B–349/51
Ultrasound absorption . . . . 15 (Si):   Hb/B–395/6
Ultrasound generation . . . . 15 (Si):   Hb/B–397
Uses . . . . 15 (Si):   Hb/B–396/7
Vapor pressure . . . . 15 (Si):   Hb/B–339
Viscosity. . . . . 15 (Si):   Hb/B–332
Wettability. . . . . 15 (Si):   Hb/B–331/2

–. . . . . $SiO_2 \cdot 0.5\, H_2O = H_2Si_2O_5$ . . . . . 15 (Si):   Hb/B–412/3
–. . . . . $[2\, SiO_2 \cdot H_2O]_n$ . . . . 15 (Si):   Hb/B–412/3
–. . . . . $SiO_2 \cdot H_2O = H_2SiO_3$. . . . . . . 15 (Si):   Hb/B–411/2
–. . . . . $[SiO_2 \cdot H_2O]_6$ . . . . 15 (Si):   Hb/B–411/2
–. . . . . $SiO_2 \cdot 1.5\, H_2O = H_6Si_2O_7$ . . . . 15 (Si):   Hb/B–410/1
–. . . . . $SiO_2 \cdot 2\, H_2O = Si(OH)_4$ . . . . 15 (Si):   Hb/B–409/10
–. . . . . $SiO_2 \cdot 2.5\, H_2O$ . . . . 15 (Si):   Hb/B–409
–. . . . . $SiO_2 \cdot xH_2O$:

Aerogels. . . . . 15 (Si):   Hb/B–496/8
Hydrosols:
    Chemical reactions
        see $O_2Si ... SiO_2$

    Colloidchemical behavior . . . . . . 15 (Si):   Hb/B–438/48
    Constitution . . . . 15 (Si):   Hb/B–435/8
    Electrochemical behavior . . . . . . 15 (Si):   Hb/B–453/6
    Formation and preparation . . . . 15 (Si):   Hb/B–423/35
    Physical properties. . . . . 15 (Si):   Hb/B–448/53
    Uses. . . . . 15 (Si):   Hb/B–435

Organogels . . . . 15 (Si):   Hb/B–498/9
Organosols . . . . 15 (Si):   Hb/B–456/7
Silica gel:
    Chemical reactions
        see $O_2Si ... SiO_2$

    Colloidchemical behavior . . . . . . 15 (Si):   Hb/B–539/43
    Constitution . . . . 15 (Si):   Hb/B–414/7
    Drying. . . . . 15 (Si):   Hb/B–473/7, 486/7
    Electrochemical behavior . . . . . . 15 (Si):   Hb/B–537/9
    Formation. . . . . 15 (Si):   Hb/B–414, 458
    Physical properties. . . . . 15 (Si):   Hb/B–500/36
    Preparation . . . . 15 (Si):   Hb/B–458/77
    –, industrial. . . . . 15 (Si):   Hb/B–477/89
    – of special forms . . . . 15 (Si):   Hb/B–489/96
    Purification . . . . 15 (Si):   Hb/B–473
    Regeneration . . . . 15 (Si):   Hb/B–499
    Sorption . . . . 15 (Si):   Hb/B–512/31

–. . . . . SiO_2 glasses:
$SiO_2–Ag_2O–B_2O_3$. . . . . 61 (Ag):   Hb/B3–389
$SiO_2–Ag_2O–CaO–Na_2O$ . . . . 61 (Ag):   Hb/B3–384/5

See the "Instructions for the user"
at the beginning of the volume

O_2Si . . . . SiO_2 glasses:

$SiO_2-Al_2O_3-CaO-Na_2O$ . . . . 35 (Al):   Hb/B2−592/3
$SiO_2-Al_2O_3-PbO$ . . . . 47 (Pb):   Hb/C4−1297/300
$SiO_2-B_2O_3$. . . . . 15 (Si):   Hb/B−759/61
$SiO_2-B_2O_3-K_2O-PbO$ . . . . 47 (Pb):   Hb/C3−1109/10
$SiO_2-B_2O_3-Li_2O$ . . . . 20 (Li):   Eb−519
$SiO_2-B_2O_3-Li_2O-Na_2O$ . . . . 21 (Na):   Eb/5−2242
$SiO_2-B_2O_3-Na_2O$. . . . . 21 (Na):   Eb/5−1984/98
$SiO_2-B_2O_3-Na_2O-PbO$ . . . . 47 (Pb):   Hb/C3−1047
$SiO_2-B_2O_3-PbO$ . . . . 47 (Pb):   Hb/C3−858/60
$SiO_2-BaO-PbO$. . . . . 47 (Pb):   Hb/C3−1208/10
$SiO_2-BeO-Na_2O$ . . . . 26 (Be):   Hb−171
$SiO_2-Bi_2O_3$ . . . . 19 (Bi):   Eb−844
$SiO_2-Bi_2O_3-PbO$ . . . . 47 (Pb):   Hb/C3−978
$SiO_2-CaO-K_2O$. . . . . 28 (Ca):   Hb/B3−1308/9
$SiO_2-CaO-K_2O-MgO-Na_2O$ . . . . 28 (Ca):   Hb/B3−1384
$SiO_2-CaO-K_2O-Na_2O$ . . . . 28 (Ca):   Hb/B3−1318
$SiO_2-CaO-MgO$ . . . . 28 (Ca):   Hb/B3−1370
$SiO_2-CaO-Na_2O$ . . . . 28 (Ca):   Hb/B3−1275/8
$SiO_2-CaO-PbO$. . . . . 47 (Pb):   Hb/C3−1191/2
$SiO_2-CdO-PbO$ . . . . 47 (Pb):   Hb/C4−1257
$SiO_2-GeO_2-Na_2O$ . . . . 45 (Ge):   Eb−501
$SiO_2-K_2O-Li_2O-PbO$ . . . . 47 (Pb):   Hb/C3−1116
$SiO_2-K_2O-Na_2O-PbO$. . . . . . . 47 (Pb):   Hb/C3−1120/2
SiO_2-K_2O-PbO. . . . . 47 (Pb):   Hb/C3−1102/6
$SiO_2-La_2O_3-PbO$. . . . . 47 (Pb):   Hb/C4−1376
SiO_2-Li_2O . . . . 20 (Li):   Eb−518/9
$SiO_2-Li_2O-Na_2O$ . . . . 21 (Na):   Hb−991
                                                    Eb/5−2240/2
$SiO_2-Li_2O-Na_2O-PbO$. . . . . 47 (Pb):   Hb/C3−1052
SiO_2-Li_2O-PbO . . . . 47 (Pb):   Hb/C3−993
SiO_2-NaBO_2 . . . . 21 (Na):   Eb/5−1990, 1992
SiO_2-Na_2O:
    Chemical reactions. . . . . 21 (Na):   Eb/4−1505/9
    Crystallographic properties . . . . . 21 (Na):   Eb/4−1486/90
    Devitrification . . . . 21 (Na):   Eb/4−1484/5
    Diffusion . . . . 21 (Na):   Eb/4−1488
    Electrical properties . . . . 21 (Na):   Eb/4−1501/2
    Electrochemical behavior . . . . . . 21 (Na):   Eb/4−1503/5
    Formation and preparation . . . . 21 (Na):   Eb/4−1482/4
    Mechanical properties . . . . 21 (Na):   Eb/4−1490/6
    Optical properties . . . . 21 (Na):   Eb/4−1498/501
    Structure . . . . 21 (Na):   Eb/4−1486/90
    Thermal properties. . . . . 21 (Na):   Eb/4−1496/9
$SiO_2-Na_2O-P_2O_5$. . . . . 21 (Na):   Eb/5−2016
SiO_2-Na_2O-PbO . . . . 47 (Pb):   Hb/C3−1039/46
$SiO_2-Na_2O-PuO_2$ . . . . Erg. W. 4−70
SiO_2-Na_2O-ZnO . . . . 32 (Zn):   Eb−1008

Hb = Main Series Volume; GD = Gmelin-Durrer;   Ab = Appended Volume to Main Series;
Eb = Supplement Volume to Main Series;   Erg. W. = New Supplement Series Volume

O_2Si. . . . . SiO_2 glasses:

SiO_2-PbO:

    Chemical reactions. . . . . 47 (Pb):   Hb/C3−835/6

    Enamels. . . . . 47 (Pb):   Hb/C3−836

    Formation and preparation . . . . . 47 (Pb):   Hb/C3−824/5

    Physical properties. . . . . 47 (Pb):   Hb/C3−827/34

    Solder glasses . . . . 47 (Pb):   Hb/C3−836

    Structure . . . . 47 (Pb):   Hb/C3−825/6

$SiO_2-PbO-Pr_2O_3$ . . . . 47 (Pb):   Hb/C4−1376

$SiO_2-PbO-Rb_2O$ . . . . 47 (Pb):   Hb/C3−1144/5

$SiO_2-PbO-Sb_2O_3$ . . . . 47 (Pb):   Hb/C3−941

$SiO_2-PbO-SrO$ . . . . 47 (Pb):   Hb/C3−1199

$SiO_2-PbO-ThO_2$ . . . . 47 (Pb):   Hb/C4−1471

$SiO_2-PbO-TiO_2$ . . . . 47 (Pb):   Hb/C4−1392/4

$SiO_2-PbO-ZnO$. . . . . 47 (Pb):   Hb/C4−1231

−. . . . . SiO_2 melts:

$SiO_2-B_2O_3-Na_2O-PbO$ . . . . 47 (Pb):   Hb/C3−1047

$SiO_2-Li_2O-Na_2O$ . . . . 21 (Na):   Eb/5−2240/2

SiO_2-Na_2O . . . . 21 (Na):   Eb/4−1482/509

SiO_2-PbO:

    Activity . . . . 47 (Pb):   Hb/C3−812

    Chemical reactions. . . . . 47 (Pb):   Hb/C3−823

    Constitution . . . . 47 (Pb):   Hb/C3−810/2

    Electrochemical behavior . . . . . 47 (Pb):   Hb/C3−820/3

    Physical properties. . . . . 47 (Pb):   Hb/C3−813/20

$SiO_2-PbO-Tl_2O$ . . . . 47 (Pb):   Hb/C4−1354

−. . . . . SiO_2 phosphors:

$SiO_2-PbO-ZnO$. . . . . 47 (Pb):   Hb/C4−1231

−. . . . . SiO_2 solid solutions:

$SiO_2-Ta_2O_5-ZrO_2$ . . . . 50 (Ta):   Hb/B2−349

−. . . . . SiO_2 solutions:

$SiO_2 \cdot xH_2O-H_2O$ . . . . 15 (Si):   Hb/B−417/22

−. . . . . SiO_2 suspensions . . . . 15 (Si):   Hb/B−288/98

−. . . . . SiO_2 systems:

SiO_2-Ag_2O . . . . 61 (Ag):   Hb/B3−379

$SiO_2-Al_2O_3$ . . . . 35 (Al):   Hb/B2−309/10

$SiO_2-Al_2O_3-BaO$. . . . . 35 (Al):   Hb/B2−607

$SiO_2-Al_2O_3-BeO$ . . . . 35 (Al):   Hb/B2−533

$SiO_2-Al_2O_3-CO_2-CaO-Na_2O$ . . . . . 35 (Al):   Hb/B2−593

$SiO_2-Al_2O_3-CO_2-K_2O$ . . . . 35 (Al):   Hb/B2−504

$SiO_2-Al_2O_3-CO_2-Na_2O$ . . . . 35 (Al):   Hb/B2−426

$SiO_2-Al_2O_3-CaO$. . . . . 35 (Al):   Hb/B2−576/81

$SiO_2-Al_2O_3-CaO-Fe_2O_3-MgO$ . . . . 35 (Al):   Hb/B2−613

$SiO_2-Al_2O_3-CaO-Fe_2O_3$

   $-MgO-Na_2O$. . . . . 35 (Al):   Hb/B2−613

$SiO_2-Al_2O_3-CaO-K_2O$ . . . . 35 (Al):   Hb/B2−594

See the "Instructions for the user"
at the beginning of the volume

O_2Si . . . . SiO_2 systems:

$SiO_2-Al_2O_3-CaO-K_2O-Na_2O$	35 (Al):	Hb/B2−594/5
$SiO_2-Al_2O_3-CaO-MgO$	35 (Al):	Hb/B2−596/600
$SiO_2-Al_2O_3-CaO-MgO-Na_2O$	35 (Al):	Hb/B2−600
$SiO_2-Al_2O_3-CaO-Na_2O$	35 (Al):	Hb/B2−589/92
$SiO_2-Al_2O_3-Co_3O_4$	58 (Co):	Hb/A2−470
$SiO_2-Al_2O_3-Cr_2O_3-MgO$	52 (Cr):	Hb/B−897
$SiO_2-Al_2O_3-FeO-Fe_2O_3$	35 (Al):	Hb/B2−613
$SiO_2-Al_2O_3-K_2O$	35 (Al):	Hb/B2−489
$SiO_2-Al_2O_3-K_2O-H_2O$	35 (Al):	Hb/B2−503
$SiO_2-Al_2O_3-Li_2O$	35 (Al):	Hb/B2−356
$SiO_2-Al_2O_3-MgO$	35 (Al):	Hb/B2−540/1
$SiO_2-Al_2O_3-Na_2O$	35 (Al):	Hb/B2−388/90
$SiO_2-Al_2O_3-Na_2O-H_2O$	35 (Al):	Hb/B2−402
$SiO_2-Al_2O_3-Na_2O-ZnO$	35 (Al):	Hb/B2−611
$SiO_2-Al_2O_3-PbO$	47 (Pb):	Hb/C4−1296/7
$SiO_2-Al_2O_3-SrO$	35 (Al):	Hb/B2−603
$SiO_2-Al_2O_3-TiO_2$	41 (Ti):	Hb−464
$SiO_2-Al_2O_3-ZnO$	35 (Al):	Hb/B2−611
$SiO_2-Al_2O_3-ZrO_2$	42 (Zr):	Hb−444
$SiO_2-As_2O_5-NaF$	21 (Na):	Eb/5−2021
$SiO_2-As_2O_5-Na_2O$	21 (Na):	Eb/5−2021
SiO_2-BPO_4	16 (P):	Hb/C−642
$SiO_2-B_2O_3$	15 (Si):	Hb/B−759/61
$SiO_2-B_2O_3-BaO$	30 (Ba):	Eb−490
$SiO_2-B_2O_3-CaO$	13 (B):	Eb−135
	28 (Ca):	Hb/B3−1109/11
$SiO_2-B_2O_3-CaO-MgO-Na_2O$	28 (Ca):	Hb/B3−1382
$SiO_2-B_2O_3-CaO-Na_2O$	13 (B):	Eb−135
	28 (Ca):	Hb/B3−1282
$SiO_2-B_2O_3-K_2O$	22 (K):	Hb/5−979
$SiO_2-B_2O_3-Li_2O$	13 (B):	Eb−135
	20 (Li):	Eb−519
$SiO_2-B_2O_3-Na_2O$	13 (B):	Eb−135
	21 (Na):	Eb/5−1982/4
$SiO_2-B_2O_3-P_2O_5$	16 (P):	Hb/C−642
$SiO_2-B_2O_3-PbO$	47 (Pb):	Hb/C3−857/60
$SiO_2-B_2O_3-ZnO$	32 (Zn):	Eb−994
SiO_2-BaCO_3	30 (Ba):	Hb−331
SiO_2-BaO	30 (Ba):	Hb−331/2
		Eb−486
$SiO_2-BaO-CO_2-Na_2O$	30 (Ba):	Eb−517
$SiO_2-BaO-CaO-MgO-Na_2O$	30 (Ba):	Eb−531
$SiO_2-BaO-CaO-Na_2O$	30 (Ba):	Eb−531
$SiO_2-BaO-H_2O$	30 (Ba):	Eb−488
$SiO_2-BaO-K_2O$	30 (Ba):	Hb−376
		Eb−522/3
$SiO_2-BaO-K_2O-Na_2O$	30 (Ba):	Eb−523
$SiO_2-BaO-MgO$	30 (Ba):	Eb−527

O_2Si . . . . SiO_2 systems:

$SiO_2-BaO-MgO-PbO$	47 (Pb):	Hb/C3−1211
$SiO_2-BaO-Na_2O$	30 (Ba):	Eb−515/7
$SiO_2-BaO-PbO$	47 (Pb):	Hb/C3−1207/8
$SiO_2-BeO-Na_2O$	26 (Be):	Hb−171
$SiO_2-Bi_2O_3$	19 (Bi):	Eb−844
SiO_2-C	15 (Si):	Hb/B−857
SiO_2-CO_2-CaO	28 (Ca):	Hb/B3−1112/4
$SiO_2-CO_2-CaO-K_2O$	28 (Ca):	Hb/B3−1312/3
$SiO_2-CO_2-CaO-MgO$	28 (Ca):	Hb/B3−1379
$SiO_2-CO_2-CaO-Na_2O$	28 (Ca):	Hb/B3−1282/5
$SiO_2-CO_2-K_2O$	22 (K):	Hb/5−979/80
$SiO_2-CO_2-Li_2CO_3-Li_2SiO_3$	20 (Li):	Eb−520
$SiO_2-CO_2-Li_2O$	20 (Li):	Hb−243
		Eb−519/20
$SiO_2-CO_2-Na_2O$	21 (Na):	Hb−885/7
		Eb/5−1982
$SiO_2-CaAl_2Si_2O_8$	35 (Al):	Hb/B2−579
$SiO_2-CaAl_2Si_2O_8-KAlSi_2O_6$	35 (Al):	Hb/B2−594
$SiO_2-CaO \cdot Al_2O_3 \cdot 2\,SiO_2$		
$\quad -K_2O \cdot Al_2O_3 \cdot 2\,SiO_2$		
$\quad\quad -Na_2O \cdot Al_2O_3 \cdot 2\,SiO_2$	35 (Al):	Hb/B2−594
$SiO_2-CaO \cdot Al_2O_3 \cdot 2\,SiO_2-MgO$ . . .	35 (Al):	Hb/B2−596
$SiO_2-CaAl_2Si_2O_8-Mg_2SiO_4$	35 (Al):	Hb/B2−597/8
SiO_2-CaC_2	28 (Ca):	Hb/B3−1112
SiO_2-CaCl_2-CaO	28 (Ca):	Hb/B3−1105/6
SiO_2-CaF_2	28 (Ca):	Hb/B3−1100
SiO_2-CaF_2-CaO	28 (Ca):	Hb/B3−1100/2
SiO_2-CaF_2-KF	28 (Ca):	Hb/B3−1310
SiO_2-CaF_2-MgO	28 (Ca):	Hb/B3−1378/9
SiO_2-CaF_2-NaF	28 (Ca):	Hb/B3−1281
$SiO_2-CaF_2-Na_2O$	28 (Ca):	Hb/B3−1281
$SiO_2-CaO \cdot MgO \cdot SiO_2$		
$\quad -2\,MgO \cdot SiO_2$	28 (Ca):	Hb/B3−1364
$SiO_2-CaO \cdot Nb_2O_5-CaO \cdot TiO_2$	49 (Nb):	Hb/B4−360
SiO_2-CaO	28 (Ca):	Hb/B3−1036/43
$SiO_2-CaO-CaS$	28 (Ca):	Hb/B3−1107
$SiO_2-CaO-CoO$	58 (Co):	Eb/A−844
$SiO_2-CaO-Cr_2O_3$	52 (Cr):	Hb/B−817/8
$SiO_2-CaO-FeO$	59 (Fe):	Hb/B5−1083
$SiO_2-CaO-Fe_2O_3$	59 (Fe):	Hb/B5−1086/7
$SiO_2-CaO-H_2O$	28 (Ca):	Hb/B3−1075/84
$SiO_2-CaO-K_2O$	28 (Ca):	Hb/B3−1307/9
$SiO_2-CaO-Li_2O$	28 (Ca):	Hb/B3−1247/8
$SiO_2-CaO-MgO$	28 (Ca):	Hb/B3−1362/70
$SiO_2-CaO-MgO-H_2O$	28 (Ca):	Hb/B3−1377
$SiO_2-CaO-MgO-Na_2O$	28 (Ca):	Hb/B3−1381/2
$SiO_2-CaO-MgO-NiO$	57 (Ni):	Hb/B3−1122
$SiO_2-CaO-NaCl-Na_2O$	28 (Ca):	Hb/B3−1281

See the "Instructions for the user"
at the beginning of the volume

O$_2$Si . . . . SiO$_2$ systems:

SiO$_2$–CaO–Na$_2$O	28 (Ca):	Hb/B3–1272/8
SiO$_2$–CaO–Na$_2$O–H$_2$O	28 (Ca):	Hb/B3–1280/1
SiO$_2$–CaO–Na$_2$O–TiO$_2$.	41 (Ti):	Hb–431
SiO$_2$–CaO–Nb$_2$O$_5$	49 (Nb):	Hb/B4–357/8
SiO$_2$–CaO–Nb$_2$O$_5$–TiO$_2$	49 (Nb):	Hb/B4–360
SiO$_2$–CaO–P$_2$O$_5$	28 (Ca):	Hb/B1–204/6
		Hb/B3–1215/9
SiO$_2$–CaO–PbO.	47 (Pb):	Hb/C3–1191/2
SiO$_2$–CaO–PbO–ZnO	47 (Pb):	Hb/C4–1231
SiO$_2$–CaO–SO$_3$.	28 (Ca):	Hb/B3–1107/8
SiO$_2$–CaO–SrO	29 (Sr):	Eb–306
SiO$_2$–CaO–TiO$_2$	41 (Ti):	Hb–430
SiO$_2$–CaTiO$_3$	41 (Ti):	Hb–430
SiO$_2$–8 CaO · 7 Nb$_2$O$_5$ · 6 TiO$_2$	49 (Nb):	Hb/B4–360
SiO$_2$–CoO	58 (Co):	Hb/A2–386/7
		Eb/A–735
SiO$_2$–CoO–MgO	58 (Co):	Eb/A–840
SiO$_2$–Co$_3$O$_4$–Fe$_2$O$_3$.	59 (Fe):	Hb/B5–1166
SiO$_2$–Cr$_2$O$_3$	52 (Cr):	Hb/B–435/6
SiO$_2$–Cr$_2$O$_3$–MgO	52 (Cr):	Hb/B–791
SiO$_2$–Cu$_2$O	60 (Cu):	Hb/B2–899
SiO$_2$–Fe–FeO–FeSi$_2$	59 (Fe):	Hb/A8–1759/60
SiO$_2$–Fe–FeO–Ni–NiO	59 (Fe):	Hb/B5–1160
SiO$_2$–Fe–FeO–Ni–NiO–S.	59 (Fe):	Hb/B5–1160
SiO$_2$–FeO	59 (Fe):	Hb/A5–876/81
		Hb/B4–761/2
SiO$_2$–FeO–Fe$_2$O$_3$	59 (Fe):	Hb/B4–762
SiO$_2$–FeO–MgO	59 (Fe):	Hb/B5–1066
SiO$_2$–FeO–MnO	56 (Mn):	Hb/C1–68
	59 (Fe):	Hb/B5–1153
SiO$_2$–Fe$_2$O$_3$–Na$_2$O	59 (Fe):	Hb/B5–908/9
SiO$_2$–KAlSi$_3$O$_8$–NaAlSi$_3$O$_8$	35 (Al):	Hb/B2–506
SiO$_2$–K$_2$O–Na$_2$O	22 (K):	Hb/6–1222
SiO$_2$–K$_2$O–Nb$_2$O$_5$.	49 (Nb):	Hb/B4–356/7
SiO$_2$–K$_2$O–PbO.	47 (Pb):	Hb/C3–1100/1
SiO$_2$–K$_2$O–Ta$_2$O$_5$	50 (Ta):	Hb/B2–348
SiO$_2$–K$_2$O–ZnO	32 (Zn):	Eb–1015
SiO$_2$–K$_2$SiO$_3$	22 (K):	Hb/5–973/4
SiO$_2$–K$_2$SiO$_3$–H$_2$O	22 (K):	Hb/5–974/5
SiO$_2$–K$_2$SiO$_3$–Na$_2$SiO$_3$	22 (K):	Hb/6–1222/4
SiO$_2$–Li$_2$O	20 (Li):	Hb–236/7
		Eb–514/5
SiO$_2$–Li$_2$O–Na$_2$O	21 (Na):	Eb/5–2238/40
SiO$_2$–Li$_2$SiO$_3$–Na$_2$SiO$_3$	21 (Na):	Eb/5–2239
SiO$_2$–MgO	27 (Mg):	Hb/B3–346/8
SiO$_2$–MgO–PbO	47 (Pb):	Hb/C3–1177
SiO$_2$–NaCl–Na$_2$O.	21 (Na):	Eb/5–1981
SiO$_2$–NaF–Na$_2$O	21 (Na):	Eb/5–1980/1

Hb = Main Series Volume; GD = Gmelin-Durrer;   *Ab = Appended Volume to Main Series;*
Eb = Supplement Volume to Main Series;   *Erg. W. = New Supplement Series Volume*

O_2Si . . . . SiO_2 systems:

$SiO_2-Na_2O \cdot 4\,B_2O_3$	21 (Na):	Eb/5−1983/4
$SiO_2-Na_2CO_3-H_2O$	21 (Na):	Eb/5−1982
SiO_2-Na_2O	21 (Na):	Hb−861
		Eb/4−1465/7
$SiO_2-Na_2O-H_2O$	21 (Na):	Hb−873
		Eb/4−1509/15
$SiO_2-Na_2O-Na_2S$	21 (Na):	Eb/5−1981
$SiO_2-Na_2O-Nb_2O_5$	49 (Nb):	Hb/B4−356
$SiO_2-Na_2O-P_2O_5$	21 (Na):	Eb/5−2013/6
SiO_2-Na_2O-PbO	47 (Pb):	Hb/C3−1038/46
$SiO_2-Na_2O-PbO-TiO_2$	47 (Pb):	Hb/C4−1392/3
$SiO_2-Na_2O-ZrO_2$	42 (Zr):	Hb−407/8
$SiO_2-Na_2SO_4-Na_2SiO_3$	21 (Na):	Eb/5−1981/2
$SiO_2-Na_2SiO_3$	21 (Na):	Hb−879
$SiO_2-Na_2SiO_3-PbSiO_3$	47 (Pb):	Hb/C3−1038
$SiO_2-Nb_2O_5$	49 (Nb):	Hb/B1−401/2
$SiO_2-Nb_2O_5-TiO_2$	49 (Nb):	Hb/B4−360
$SiO_2-Nb_2O_5-Y_2O_3$	49 (Nb):	Hb/B4−359
$SiO_2-Nb_2O_5-ZnO$	49 (Nb):	Hb/B4−359
SiO_2-NiO	57 (Ni):	Hb/B3−936
$SiO_2-NiO-H_2O$	57 (Ni):	Hb/B3−940/1
$SiO_2-P_2O_5$	16 (P):	Hb/C−638
$SiO_2-P_2O_5-H_2O$	16 (P):	Hb/C−639/40
$SiO_2-P_2O_5-PbO$	47 (Pb):	Hb/C3−885/6
SiO_2-PbO	47 (Pb):	Hb/C3−808/10
$SiO_2-PbO-PbSO_4$	47 (Pb):	Hb/C3−856
$SiO_2-PbO-ZrO_2$	47 (Pb):	Hb/C4−1431
SiO_2-SrO	29 (Sr):	Hb−210
		Eb−280/1
$SiO_2-SrO-H_2O$	29 (Sr):	Eb−282
$SiO_2-Ta_2O_5$	50 (Ta):	Hb/B1−263
$SiO_2-Ta_2O_5-ZnO$	50 (Ta):	Hb/B2−349
SiO_2-TiO_2-	41 (Ti):	Hb−376/7
SiO_2-ZnO	32 (Zn):	Eb−989/90
SiO_2-ZrO_2	42 (Zr):	Hb−381/2

O_2Si_3 . . . . Si_3O_2 . . . . 15 (Si): Hb/B−257

O_2Sn . . . . SnO_2:

Chemical reactions	46 (Sn):	Hb/C1−97/113
Colloidal forms	46 (Sn):	Hb/C1−60/1
Crystallographic properties	46 (Sn):	Hb/C1−66/71
Electrical properties	46 (Sn):	Hb/C1−73/89
Electrochemical behavior	46 (Sn):	Hb/C1−95/7
Formation and preparation	46 (Sn):	Hb/C1−59/65
History	46 (Sn):	Hb/A−164/7
Magnetic properties	46 (Sn):	Hb/C1−73/4
Mechanical properties	46 (Sn):	Hb/C1−71/3
Molecule	46 (Sn):	Hb/C1−66
Optical properties	46 (Sn):	Hb/C1−89/95

See the "Instructions for the user"
at the beginning of the volume

O_2Sn . . . . SnO_2:

Spectra . . . . 46 (Sn):   Hb/C1–89/94
Thermal properties . . . . 46 (Sn):   Hb/C1–71/3
Toxicity . . . . 46 (Sn):   Hb/B–412

–. . . . . 2 $SnO_2 \cdot H_2O$ = $Sn_2O_5H_2$ . . . . 46 (Sn):   Hb/C1–115, 117/8, 121
–. . . . . $SnO_2 \cdot H_2O$ = SnO_3H_2 . . . . 46 (Sn):   Hb/C1–115/8, 121
–. . . . . $SnO_2 \cdot 1.25 H_2O$ . . . . 46 (Sn):   Hb/C1–117
–. . . . . $SnO_2 \cdot 3 H_2O$ . . . . 46 (Sn):   Hb/C1–116
–. . . . . $SnO_2 \cdot 3.45 H_2O$ . . . . 46 (Sn):   Hb/C1–121
–. . . . . $SnO_2 \cdot xH_2O$ . . . . 46 (Sn):   Hb/C1–113/27
–. . . . . $SnO_2 \cdot xH_2O$, Sols . . . . 46 (Sn):   Hb/C1–127/36
–. . . . . SnO_2 solid solutions:
SnO_2–PbO–Sb_2O_5 . . . . 47 (Pb):   Hb/C4–1595

–. . . . . SnO_2 systems:
SnO_2–CuO . . . . 60 (Cu):   Hb/B3–1194
SnO_2–NiO . . . . 57 (Ni):   Hb/B3–1190/1
SnO_2–PbO . . . . 47 (Pb):   Hb/C4–1581
SnO_2–Sn . . . . 46 (Sn):   Hb/C1–24/5
SnO_2–Ta_2O_5 . . . . 50 (Ta):   Hb/B2–231
SnO_2–$ZnTa_2O_6$ . . . . 50 (Ta):   Hb/B2–232

O_2Sn^{2-} . . . . $[SnO_2]^{2-}$ . . . . 46 (Sn):   Hb/C1–57
O_2Sn_2 . . . . Sn_2O_2 . . . . 46 (Sn):   Hb/C1–34/5, 104
O_2Sr . . . . SrO_2 . . . . 29 (Sr):   Hb–86/7
                                                                     Eb–85, 165/8
–. . . . . $SrO_2 \cdot 8 H_2O$ . . . . 29 (Sr):   Hb–238/9
                                                                     Eb–168
–. . . . . $SrO_2 \cdot 9 H_2O$ . . . . 29 (Sr):   Hb–238
O_2Ta . . . . TaO_2 . . . . 50 (Ta):   Hb/A2–498, 529
                                                                     Hb/B1–17, 21, 25,
                                                                         28/30, 46
–. . . . . $TaO_2 \cdot xH_2O$ . . . . 50 (Ta):   Hb/B1–54
–. . . . . TaO_2 solid solutions:
$(Ta,Nb)O_2$ . . . . 50 (Ta):   Hb/B2–246
–. . . . . TaO_2 systems:
TaO_2–TiO_2 . . . . 50 (Ta):   Hb/B2–221/2
TaO_2–VO_2 . . . . 50 (Ta):   Hb/B2–244
O_2Ta^+ . . . . $[TaO_2]^+$ . . . . 50 (Ta):   Hb/A2–407
                                                                     Hb/B1–59, 135
–. . . . . $[Ta_2O_4]^{2+}$ . . . . 50 (Ta):   Hb/B1–178
O_2Ta^- . . . . $[TaO_2]^-$ . . . . 50 (Ta):   Hb/A2–407
                                                                     Hb/B1–59
O_2Tb . . . . TbO_2 . . . . 39 (SE):   Hb/C1–103/4, 398/9
–. . . . . TbO_2 solid solutions:
TbO_2–PrO_2 . . . . 39 (SE):   Hb/C1–404
O_2Tc . . . . TcO_2 . . . . 14 (C):   Hb/C2–168
O_2Te . . . . TeO_2 . . . . 11 (Te):   Hb–239, 275/82

Hb = Main Series Volume; GD = Gmelin-Durrer;   *Ab = Appended Volume to Main Series;*
Eb = Supplement Volume to Main Series;   *Erg. W. = New Supplement Series Volume*

O_2Te . . . . TeO$_2$ glasses:
                      TeO$_2$–Li$_2$O–PbO . . . . 47 (Pb):   Hb/C3–992/3
                      TeO$_2$–Na$_2$O–PbO. . . . . 47 (Pb):   Hb/C3–1034
                      TeO$_2$–PbO. . . . . 47 (Pb):   Hb/C2–678/9

–. . . . . TeO$_2$ melts:
                      TeO$_2$–V$_2$O$_5$ . . . . 48 (V):   Hb/B1–320

–. . . . . TeO$_2$ systems:
                      TeO$_2$–B$_2$O$_3$–PbO . . . . 47 (Pb):   Hb/C2–697
                      TeO$_2$–Bi$_2$O$_3$ . . . . 19 (Bi):   Eb–817
                      TeO$_2$–HF . . . . 11 (Te):   Hb–322
                      TeO$_2$–HF–H$_2$O . . . . 11 (Te):   Hb–322/3
                      TeO$_2$–Na$_2$O–H$_2$O . . . . 21 (Na):   Eb/3–1205
                                                                   Eb/5–1909/10
                      TeO$_2$–Nb$_2$O$_5$–PbO . . . . 49 (Nb):   Hb/B4–307
                      TeO$_2$–PbO–SO$_3$ . . . . 47 (Pb):   Hb/C2–681
                      TeO$_2$–PbO–SeO$_2$. . . . . 47 (Pb):   Hb/C2–684

O_2Th . . . . ThO$_2$:
                      Chemical reactions . . . . 44 (Th):   Hb–220/6
                      Crystallographic properties . . . . . 44 (Th):   Hb–205/6
                      Electrical properties . . . . 44 (Th):   Hb–217/20
                      Formation and preparation . . . . . 44 (Th):   Hb–202/5
                      Magnetic properties . . . . 44 (Th):   Hb–217
                      Mechanical properties . . . . 44 (Th):   Hb–207/8
                      Optical properties. . . . . 44 (Th):   Hb–211/7
                      Sols . . . . 44 (Th):   Hb–238/9
                      Solubility . . . . 44 (Th):   Hb–225/6
                      Sorption . . . . 44 (Th):   Hb–208
                      Thermal properties . . . . 44 (Th):   Hb–209/11
                      Toxicity . . . . 44 (Th):   Hb–197
                      Uses . . . . 44 (Th):   Hb–77/81

–. . . . . ThO$_2$ · 0.5 H$_2$O . . . . 44 (Th):   Hb–227
–. . . . . ThO$_2$ · 2 H$_2$O . . . . 44 (Th):   Hb–228
–. . . . . ThO$_2$ · xH$_2$O . . . . 44 (Th):   Hb–226/38
–. . . . . ThO$_2$ glasses:
                      ThO$_2$–P$_2$O$_5$–PbO . . . . 47 (Pb):   Hb/C4–1471
                      ThO$_2$–PbO–SiO$_2$ . . . . 47 (Pb):   Hb/C4–1471

–. . . . . ThO$_2$ solid solutions:
                      ThO$_2$–AmO$_2$ . . . . Erg. W. 4–79
                      ThO$_2$–NpO$_2$. . . . . Erg. W. 4–49
                      ThO$_2$–PuO$_2$. . . . . Erg. W. 4–58
                      ThO$_2$–UO$_2$. . . . . 55 (U):   Hb–239

–. . . . . ThO$_2$ systems:
                      ThO$_2$–Al$_2$O$_3$ . . . . 44 (Th):   Hb–351
                      ThO$_2$–Al$_2$O$_3$–BeO . . . . 44 (Th):   Hb–351
                      ThO$_2$–AmO$_{1.5}$. . . . . Erg. W. 4–79
                      ThO$_2$–BeAl$_2$O$_4$–BeO . . . . 44 (Th):   Hb–351

See the "Instructions for the user"
at the beginning of the volume

O_2Th . . . . ThO_2 systems:

ThO_2-BeO	44 (Th):	Hb−346
$ThO_2-BeO-MgO$	44 (Th):	Hb−347
ThO_2-CaO	44 (Th):	Hb−347
$ThO_2-CaO-ZrO_2$	44 (Th):	Hb−354
ThO_2-CeO_2	44 (Th):	Hb−352/3
$ThO_2-CrO_3-H_2O$	52 (Cr):	Hb/B−907
ThO_2-MgO	44 (Th):	Hb−346
$ThO_2-MgO-ZrO_2$	44 (Th):	Hb−354
ThO_2-NbO_2	49 (Nb):	Hb/B3−219
$ThO_2-Nb_2O_5$	49 (Nb):	Hb/B3−219/20
$ThO_2-Ta_2O_5$	50 (Ta):	Hb/B2−229
ThO_2-TiO_2	44 (Th):	Hb−353/4
$ThO_2-TiO_2-ZrO_2$	44 (Th):	Hb−354
$ThO_2-V_2O_5$	48 (V):	Hb/B2−564
ThO_2-ZrO_2	44 (Th):	Hb−354

O_2Ti . . . . TiO_2:

Chemical reactions	41 (Ti):	Hb−252/5, 473/5
Crystallographic properties	41 (Ti):	Hb−232/40
Dressing and treatment of ores	41 (Ti):	Hb−81/117
Electrical properties	41 (Ti):	Hb−247/51
Formation	41 (Ti):	Hb−226/32, 478
Hydrosols	41 (Ti):	Hb−263/8
Magnetic properties	41 (Ti):	Hb−247
Mechanical properties	41 (Ti):	Hb−240/2
Molecule.	41 (Ti):	Hb−232
Optical properties.	41 (Ti):	Hb−243/7
Organosols	41 (Ti):	Hb−268
Physiological action	41 (Ti):	Hb−199
Preparation, industrial	41 (Ti):	Hb−80/117
– of anatase	41 (Ti):	Hb−229/30
– of brookite.	41 (Ti):	Hb−230/1
– of films	41 (Ti):	Hb−231
– of pigments	41 (Ti):	Hb−114/6
– of rutile	41 (Ti):	Hb−103, 111, 226/9
– of sols	41 (Ti):	Hb−263/5, 268
Purity testing	41 (Ti):	Hb−231
Reduction, industrial	41 (Ti):	Hb−117/21
Sorption	41 (Ti):	Hb−242
Thermal properties	41 (Ti):	Hb−242/3
Toxicity	41 (Ti):	Hb−199
Uses	41 (Ti):	Hb−127/8, 467

−. . . . . $TiO_2 \cdot nH_2O$. . . . . 41 (Ti): Hb−256/8, 471/2

−. . . . . TiO_2 glasses:

$TiO_2-PbO-SiO_2$	47 (Pb):	Hb/C4−1392/4

−. . . . . TiO_2 solid solutions:

$Fe_{2-x}Ti_xO_3$.	59 (Fe):	D/2. Eb−430/2
$Fe_{3-x}Ti_xO_4$.	59 (Fe):	D/2. Eb−430/2

Hb = Main Series Volume; GD = Gmelin-Durrer; Ab = Appended Volume to Main Series;
Eb = Supplement Volume to Main Series; Erg. W. = New Supplement Series Volume

O₂Ti . . . . TiO₂ solid solutions:

TiO₂–Ga₂O₃–Nb₂O₅ . . . . 49 (Nb): Hb/B3–205
TiO₂–In₂O₃–Nb₂O₅ . . . . 49 (Nb): Hb/B3–205
(Ti,Nb)O₂ . . . . 49 (Nb): Hb/B3–179/81
TiO₂–Nb₂O₅ . . . . 49 (Nb): Hb/B3–187/8
TiO₂–Nb₂O₅–Y₂O₃ . . . . 49 (Nb): Hb/B3–209
TiO₂–NiF₂ . . . . 57 (Ni): Hb/B3–1181
NiₙTi₄₋ₙO₈₋ₙ . . . . 57 (Ni): Hb/B3–1181
TiO₂–PbNb₂O₆ . . . . 49 (Nb): Hb/B3–285
TiO₂–Pb₃Nb₄O₁₃ . . . . 49 (Nb): Hb/B3–287

–. . . . . TiO₂ systems:

TiO₂–Al(PO₃)₃–NaF . . . . 41 (Ti): Hb–464
TiO₂–Al₂O₃ . . . . 41 (Ti): Hb–463
TiO₂–Al₂O₃–Cr₂O₃ . . . . 52 (Cr): Hb/B–905
TiO₂–Al₂O₃–SiO₂ . . . . 41 (Ti): Hb–464
TiO₂–Al₂TiO₅–CaTiO₃–MgTiO₃ . . . . 41 (Ti): Hb–464
TiO₂–B₂O₃–BaO–PbO . . . . 47 (Pb): Hb/C4–1412/3
TiO₂–B₂O₃–PbO . . . . 47 (Pb): Hb/C4–1391
TiO₂–BaO . . . . 41 (Ti): Hb–434
TiO₂–BaO–CaO . . . . 41 (Ti): Hb–456/7
TiO₂–BaO–MgO . . . . 41 (Ti): Hb–456
TiO₂–BaO–P₂O₅–PbO . . . . 47 (Pb): Hb/C4–1413
TiO₂–BaO–SrO . . . . 41 (Ti): Hb–457
TiO₂–BeO . . . . 41 (Ti): Hb–416/7
TiO₂–Bi₂O₃–PbO . . . . 47 (Pb): Hb/C4–1394
TiO₂–CO₂–CaO . . . . 41 (Ti): Hb–430
TiO₂–CO₂–CaO–K₂O . . . . 41 (Ti): Hb–432
TiO₂–CO₂–CaO–Na₂O . . . . 41 (Ti): Hb–431
TiO₂–CO₂–K₂O . . . . 41 (Ti): Hb–393/4
TiO₂–CO₂–Na₂O . . . . 41 (Ti): Hb–384
TiO₂–CaO · MgO · 2 SiO₂ . . . . . . 41 (Ti): Hb–433
TiO₂–CaO . . . . 41 (Ti): Hb–424/6
TiO₂–CaO–La₂O₃ . . . . 41 (Ti): Hb–468
TiO₂–CaO–Na₂O–SiO₂ . . . . 41 (Ti): Hb–431
TiO₂–CaO–Nb₂O₅ . . . . 49 (Nb): Hb/B3–191/3
TiO₂–CaO–Nb₂O₅–SiO₂ . . . . 49 (Nb): Hb/B4–360
TiO₂–CaO–SiO₂ . . . . 41 (Ti): Hb–430
TiO₂–CaO–ZrO₂ . . . . 42 (Zr): Hb–448
TiO₂–CaSiO₃ . . . . 41 (Ti): Hb–430
TiO₂–CaTiO₃–MgTiO₃ . . . . 41 (Ti): Hb–432
TiO₂–CoO–CuO . . . . 60 (Cu): Hb/B3–1264
TiO₂–Cr₂O₃ . . . . 52 (Cr): Hb/B–904
TiO₂–CuO–Fe₂O₃ . . . . 60 (Cu): Hb/B3–1327
TiO₂–La₂O₃–Nb₂O₅–PbO . . . . . . 49 (Nb): Hb/B3–302
TiO₂–MgO . . . . 41 (Ti): Hb–418/9
TiO₂–MgO–Ti₂O₃ . . . . 41 (Ti): Hb–423
TiO₂–MgO–ZrO₂ . . . . 42 (Zr): Hb–448
TiO₂–Na₂O . . . . 41 (Ti): Hb–383/4
TiO₂–Na₂O–H₂O . . . . 41 (Ti): Hb–384

See the "Instructions for the user"
at the beginning of the volume

O_2Ti . . . . TiO_2 systems:

      $TiO_2-Na_2O-PbO-SiO_2$ . . . . 47 (Pb):   Hb/C4−1392/3

      TiO_2-NbO_2 . . . . 49 (Nb):   Hb/B3−179/81

      $TiO_2-NbO_2-Nb_2O_5$ . . . . 49 (Nb):   Hb/B3−179

      $TiO_2-Nb_2O_5$ . . . . 49 (Nb):   Hb/B3−181/3

      $TiO_2-Nb_2O_5-PbO$ . . . . 49 (Nb):   Hb/B3−284

      $TiO_2-Nb_2O_5-SiO_2$ . . . . 49 (Nb):   Hb/B4−360

      TiO_2-Ni . . . . 57 (Ni):   Hb/B1−98

      TiO_2-NiO . . . . 57 (Ni):   Hb/B3−1179

      $TiO_2-NiO-ZnO$ . . . . 57 (Ni):   Hb/B3−1184/5

      TiO_2-PbO . . . . 47 (Pb):   Hb/C4−1378/9

      $TiO_2-PbO-V_2O_5$ . . . . 48 (V):   Hb/B2−574

      $TiO_2-PbO-ZrO_2$ . . . . 47 (Pb):   Hb/C4−1439/40

      $TiO_2-SO_3-H_2O$ . . . . 41 (Ti):   Hb−345/7

      TiO_2-SiO_2 . . . . 41 (Ti):   Hb−376/7

      TiO_2-TaO_2 . . . . 50 (Ta):   Hb/B2−221/2

      $TiO_2-Ta_2O_5$ . . . . 50 (Ta):   Hb/B2−222

      $TiO_2-Ta_2O_5-ZrO_2$ . . . . 50 (Ta):   Hb/B2−229

      TiO_2-ThO_2 . . . . 44 (Th):   Hb−353/4

      $TiO_2-ThO_2-ZrO_2$ . . . . 44 (Th):   Hb−354

      TiO_2-TiS_2 . . . . 41 (Ti):   Hb−343

      $TiO_2-VO-ZrO_2$ . . . . 48 (V):   Hb/B2−564

      TiO_2-VO_2 . . . . 48 (V):   Hb/B2−558/9

      TiO_2-ZnO . . . . 41 (Ti):   Hb−460

      $TiO_2-ZnTa_2O_6$ . . . . 50 (Ta):   Hb/B2−226

      TiO_2-ZrO_2 . . . . 42 (Zr):   Hb−446/7

O_2Ti^{2+} . . . . $Ti(O \cdot O)^{2+}$ . . . . 41 (Ti):   Hb−270/1

−. . . . . $(TiO \cdot O \cdot xH_2O)^{2+}$ . . . . . . 41 (Ti):   Hb−471

O_2Tl . . . . $TlO_2 = Tl_2O_4$ . . . . 38 (Tl):   Hb/2−233

O_2U . . . . UO_2 . . . . 55 (U):   Hb−83/8, 100

−. . . . . UO_2, Sols. . . . . Erg. W. 8−244/5

−. . . . . $UO_2 \cdot xH_2O$ . . . . 55 (U):   Hb−99/100

−. . . . . UO_2 fuels:

      In-pile behavior . . . . Erg. W. 8−259/77

      Preparation . . . . Erg. W. 8−242/8

−. . . . . UO_2 solid solutions:

      UO_2-NpO_2 . . . . Erg. W. 4−49

      UO_2-PuO_2. . . . . Erg. W. 4−60/70

                                   Erg. W. 8−192

      $UO_2-PuO_2-ZrO_2$ . . . . Erg. W. 4−71

      UO_2-ThO_2. . . . . 55 (U):   Hb−239

−. . . . . UO_2 systems:

      UO_2-PuO_2. . . . . Erg. W. 4−59

O_2U^{2+} . . . . UO_2^{2+}:

      Chemical reactions . . . . 55 (U):   Hb−257/70

      Electrochemical behavior . . . . 55 (U):   Hb−255/7

      Ionic susceptibility . . . . 55 (U):   Hb−154, 255

Hb = Main Series Volume; GD = Gmelin-Durrer;   *Ab = Appended Volume to Main Series;*
Eb = Supplement Volume to Main Series;   *Erg. W. = New Supplement Series Volume*

O_2U^{2+} . . . . UO_2^{2+}:

    Spectra . . . . 21 (Na):   Eb/6−354/6

    55 (U):   Hb−246/54

−. . . . . $^{239}UO_2^{2+}$, Recoil reactions . . . . Erg. W. 8−318

O_2V . . . . $VO_2 = V_2O_4$:

    Chemical reactions . . . . 48 (V):   Hb/B1−60/1

    Determination. . . . . 48 (V):   Hb/A2−714

    Electrochemical behavior . . . . . . 48 (V):   Hb/A2−467/9, 498, 502

    Formation and preparation . . . . . 48 (V):   Hb/B1−51/3

    Physical properties . . . . 48 (V):   Hb/A2−376, 380

    Hb/B1−30, 54/60

−. . . . . V_4O_8 . . . . 48 (V):   Hb/B1−84

−. . . . . V_6O_{12} . . . . 48 (V):   Hb/B1−84

−. . . . . $VO_2 \cdot H_2O = VO(OH)_2$ . . . . 48 (V):   Hb/A2−501, 535/6

    Hb/B1−65/70, 346

−. . . . . $V_2O_4 \cdot H_2O$ . . . . 48 (V):   Hb/B1−65, 68

−. . . . . $3\,V_2O_4 \cdot 4\,H_2O$ . . . . 48 (V):   Hb/B1−66

−. . . . . $V_2O_4 \cdot 7\,H_2O$ . . . . 48 (V):   Hb/B1−67

−. . . . . VO_2 solid solutions:

    VO_2−$VNbO_4$ . . . . 49 (Nb):   Hb/B3−322/3

    VO_2−$VTaO_4$ . . . . 50 (Ta):   Hb/B2−244

    V_2O_4−$ZrSiO_4$ . . . . 48 (V):   Hb/B2−563

−. . . . . VO_2 systems:

    VO_2−Al_2O_3 . . . . 48 (V):   Hb/B2−544

    VO_2−BeO . . . . 48 (V):   Hb/B2−500

    VO_2−CaO . . . . 48 (V):   Hb/B2−511

    VO_2−CeO_2 . . . . 48 (V):   Hb/B2−552

    VO_2−CoO . . . . 58 (Co):   Eb/A−873

    VO_2−HF−H_2O . . . . 48 (V):   Hb/B1−191/2

    VO_2−Li_2O−V_2O_5 . . . . 48 (V):   Hb/B2−370

    VO_2−MgO . . . . 48 (V):   Hb/B2−503

    VO_2−MgO−V_2O_5 . . . . 48 (V):   Hb/B2−504

    VO_2−NbO_2 . . . . 49 (Nb):   Hb/B3−322/4

    VO_2−Nb_2O_5 . . . . 49 (Nb):   Hb/B3−324

    VO_2−TaO_2 . . . . 50 (Ta):   Hb/B2−244

    VO_2−TiO_2 . . . . 48 (V):   Hb/B2−558/9

    VO_2−V_2O_5−ZnO . . . . 48 (V):   Hb/B2−534

    VO_2−ZnO . . . . 48 (V):   Hb/B2−533

O_2V^+ . . . . $[VO_2]^+$:

    Chemical reactions . . . . 48 (V):   Hb/B1−158/60

    Detection . . . . 48 (V):   Hb/A2−647/52

    Determination. . . . . 48 (V):   Hb/A2−676

    Electrochemical behavior . . . . . . 48 (V):   Hb/A2−468, 478, 498, 500/2, 505/6

    Properties of the ion . . . . 48 (V):   Hb/B1−157/8

    Thermodynamic data of formation . . 48 (V):   Hb/B1−157

    Transition reactions . . . . 48 (V):   Hb/A2−550/640

See the "Instructions for the user"
at the beginning of the volume

O_2V^{3+} . . . . $V(O_2)^{3+}$ . . . . 48 (V):   Hb/B1–162/3, 166

O_2V_2 . . . . V_2O_2

                see OV ... VO

$O_2V_2^{4+}$ . . . . $(VO)_2^{4+}$ . . . . 48 (V):   Hb/A2–601

O_2W . . . . WO_2 . . . . 54 (W):   Hb–108/15

O_2Xe . . . . $XeO_2 \cdot 2H_2O$ . . . . Erg. W. 1–101

O_2Xe^+ . . . . XeO_2^+ . . . . Erg. W. 1–101

O_2Y_2 . . . . Y_2O_2 . . . . 39 (SE):   Hb/C1–91, 100

O_2Zn . . . . ZnO_2 . . . . 32 (Zn):   Eb–817/8

–. . . . . $ZnO_2 \cdot 0.5\ H_2O$ . . . . 32 (Zn):   Hb–146

                                                                        Eb–817/8

–. . . . . $ZnO_2 \cdot H_2O$ . . . . 32 (Zn):   Eb–817

O_2Zn^{2-} . . . . $[ZnO_2]^{2-}$ . . . . 32 (Zn):   Hb–112

                                                                        Eb–403, 511

O_2Zr . . . . ZrO_2:

               Chemical reactions . . . . 42 (Zr):   Hb–238/51

               Crystallographic properties . . . . . . 42 (Zr):   Hb–217/22

               Electrical properties . . . . 42 (Zr):   Hb–237/8

               Electrochemical behavior . . . . 42 (Zr):   Hb–238

               Formation . . . . 42 (Zr):   Hb–215/6

               Magnetic properties . . . . 42 (Zr):   Hb–236/7

               Mechanical properties . . . . 42 (Zr):   Hb–222/6

               Optical properties. . . . . 42 (Zr):   Hb–232/6

               Preparation . . . . 42 (Zr):   Hb–215/6

               –, industrial . . . . 42 (Zr):   Hb–76/7

               Solubility in glasses and enamels . . . 42 (Zr):   Hb–248/9

               Thermal properties . . . . 42 (Zr):   Hb–226/32

               Toxicity . . . . 42 (Zr):   Hb–191

               Uses . . . . 42 (Zr):   Hb–80/4

–. . . . . $ZrO_2 \cdot xH_2O$. . . . . 42 (Zr):   Hb–251/61

–. . . . . ZrO_2 solid solutions:

               $ZrO_2–DyNbO_4$. . . . . 49 (Nb):   Hb/B3–217

               $ZrO_2–ErNbO_4$ . . . . 49 (Nb):   Hb/B3–217

               $ZrO_2–GdNbO_4$ . . . . 49 (Nb):   Hb/B3–217

               $ZrO_2–HoNbO_4$. . . . . 49 (Nb):   Hb/B3–217

               $ZrO_2–LuNbO_4$. . . . . 49 (Nb):   Hb/B3–217

               $ZrO_2–PbNb_2O_6$ . . . . 49 (Nb):   Hb/B3–302

               $ZrO_2–PuO_2–UO_2$ . . . . Erg. W. 4–71

               $ZrO_2–Ta_2O_5–SiO_2$ . . . . 50 (Ta):   Hb/B2–349

               $ZrO_2–TbNbO_4$. . . . . 49 (Nb):   Hb/B3–217

               $ZrO_2–TmNbO_4$ . . . . 49 (Nb):   Hb/B3–217

               $ZrO_2–YbNbO_4$ . . . . 49 (Nb):   Hb/B3–217

–. . . . . ZrO_2 systems:

               $ZrO_2–Al_2O_3$ . . . . 42 (Zr):   Hb–443/4

               $ZrO_2–Al_2O_3–BaO$. . . . . 42 (Zr):   Hb–445

               $ZrO_2–Al_2O_3–BeO$. . . . . 42 (Zr):   Hb–444/5

               $ZrO_2–Al_2O_3–MgO$ . . . . 42 (Zr):   Hb–445

               $ZrO_2–Al_2O_3–SiO_2$. . . . . 42 (Zr):   Hb–444

Hb = Main Series Volume; GD = Gmelin-Durrer;   *Ab = Appended Volume to Main Series;*
Eb = Supplement Volume to Main Series;   *Erg. W. = New Supplement Series Volume*

O_2Zr . . . . ZrO_2 systems:

$ZrO_2-AmO_{1.5}$	Erg. W. 4–80	
ZrO_2-AmO_2	Erg. W. 4–80	
$ZrO_2-B_2O_3$	42 (Zr):	Hb–360
ZrO_2-BaO	42 (Zr):	Hb–440
ZrO_2-BeO	42 (Zr):	Hb–430
$ZrO_2-BeO-CaO$	42 (Zr):	Hb–438
$ZrO_2-BeO-Cr_2O_3$	52 (Cr):	Hb/B–906
$ZrO_2-BeO-MgO$	42 (Zr):	Hb–434
ZrO_2-CaCl_2	42 (Zr):	Hb–437
ZrO_2-CaO	42 (Zr):	Hb–434/6
$ZrO_2-CaO-MgO$	42 (Zr):	Hb–438
$ZrO_2-CaO-ThO_2$	44 (Th):	Hb–354
$ZrO_2-CaO-TiO_2$	42 (Zr):	Hb–448
ZrO_2-CoO	58 (Co):	Hb/A2–473
		Eb/A–871
$ZrO_2-Cr_2O_3$	52 (Cr):	Hb/B–905
$ZrO_2-Cr_2O_3-MgO$	52 (Cr):	Hb/B–907
$ZrO_2-Cr_2O_3-NiO$	57 (Ni):	Hb/B3–1217
ZrO_2-CuO	60 (Cu):	Hb/B3–1191
$ZrO_2-HNO_3-H_2O$	42 (Zr):	Hb–272
$ZrO_2-HfO_2-Ta_2O_5$	50 (Ta):	Hb/B2–229
$ZrO_2-MgO \cdot Cr_2O_3$	52 (Cr):	Hb/B–907
ZrO_2-MgO	42 (Zr):	Hb–431/3
$ZrO_2-MgO-ThO_2$	44 (Th):	Hb–354
$ZrO_2-MgO-TiO_2$	42 (Zr):	Hb–448
$ZrO_2-Na_2O-SiO_2$	42 (Zr):	Hb–407/8
ZrO_2-NbO	49 (Nb):	Hb/B3–209/10
ZrO_2-NbO_2	49 (Nb):	Hb/B3–209/10
$ZrO_2-Nb_2O_5$	49 (Nb):	Hb/B3–210/1
ZrO_2-NiO	57 (Ni):	Hb/B3–1185
$ZrO_2-P_2O_5-PbO$	47 (Pb):	Hb/C4–1431
ZrO_2-PbO	47 (Pb):	Hb/C4–1423
$ZrO_2-PbO-SiO_2$	47 (Pb):	Hb/C4–1431
$ZrO_2-PbO-TiO_2$	47 (Pb):	Hb/C4–1439/40
ZrO_2-PuO_2	Erg. W. 4–70/1	
$ZrO_2-SO_3-H_2O$	42 (Zr):	Hb–335/7
ZrO_2-SiO_2	42 (Zr):	Hb–381/2
ZrO_2-SrO	42 (Zr):	Hb–439
$ZrO_2-Ta_2O_5$	50 (Ta):	Hb/B2–228
$ZrO_2-Ta_2O_5-TiO_2$	50 (Ta):	Hb/B2–229
$ZrO_2-Ta_2O_5-Y_2O_3$	50 (Ta):	Hb/B2–229
ZrO_2-ThO_2	44 (Th):	Hb–354
$ZrO_2-ThO_2-TiO_2$	44 (Th):	Hb–354
ZrO_2-TiO_2	42 (Zr):	Hb–446/7
ZrO_2-TiO_2-VO	48 (V):	Hb/B2–564
$ZrO_2-V_2O_5$	48 (V):	Hb/B2–562
ZrO_2-ZnO	42 (Zr):	Hb–442

$O_{2.17}V$ . . . . $VO_{2.17} = V_6O_{13}$ . . . . 48 (V): Hb/B1–17, 61/3

See the "Instructions for the user"
at the beginning of the volume

$O_{2.33}V$	. . . .	$VO_{2.33} \cdot 0.65\ H_2O = V_6O_{14} \cdot 4\ H_2O$		
		$= V_3O_5(OH)_4$	48 (V):	Hb/B1–70
$O_{2.5}SrV$	. . . .	$SrVO_{2.5}$	48 (V):	Hb/B2–518
–.	. . . .	$SrVO_{2.5}$ systems:		
		$SrVO_{2.5}$–$LaVO_3$	48 (V):	Hb/B2–557
		$SrVO_{2.5}$–$SrVO_3$	48 (V):	Hb/B2–519
$O_{2.75}SrV$	. . . .	$SrVO_{2.75}$	48 (V):	Hb/B2–519
O_3	. . . .	O_3		
		see O … Ozone		
–.	. . . .	$^{16}O^{16}O^{18}O$	3 (O):	Hb/4–1111
–.	. . . .	$^{16}O^{18}O^{16}O$	3 (O):	Hb/4–1111
–.	. . . .	$^{16}O^{18}O^{18}O$	3 (O):	Hb/4–1111
–.	. . . .	$^{18}O_3$.	3 (O):	Hb/4–1111
–.	. . . .	$O_3 \cdot H_2O = H_2O_4$	3 (O):	Hb/8–2626/7
O_3^+	. . . .	O_3^+	3 (O):	Hb/4–1071, 1080
O_3^-	. . . .	O_3^-	3 (O):	Hb/3–354
				Hb/4–1071
O_3^{2-}	. . . .	O_3^{2-}	3 (O):	Hb/8–2625
O_3Os	. . . . . .	OsO_3	66 (Os):	Hb–30
–.	. . . .	$OsO_3 \cdot H_2O$	66 (Os):	Hb–30/1
O_3OsS	. . . . . .	$OsSO_3$	66 (Os):	Hb–60
–.	. . . .	$2\ OsO_3S \cdot 3\ H_2O$	66 (Os):	Hb–60
O_3Os_2	. . . . . .	Os_2O_3.	66 (Os):	Hb–25
–.	. . . .	$Os_2O_3 \cdot 3\ H_2O$	66 (Os):	Hb–25
O_3PRb	. . . . . .	$RbPO_3$	24 (Rb):	Hb–234
O_3PTl	. . . . . .	$TlPO_3$	38 (Tl):	Hb/3–431/2
–.	. . . .	$(TlPO_3)_n$	38 (Tl):	Hb/3–431/2
O_3PTl_2.	. . . . . .	$Tl_2PO_3 = Tl_4P_2O_6$	38 (Tl):	Hb/3–427
O_3P_2.	. . . .	P_2O_3	16 (P):	Hb/C–68/75
–.	. . . .	P_4O_6	16 (P):	Hb/C–70/1
–.	. . . .	$(P_2O_3)_x$	16 (P):	Hb/C–67/8
O_3Pb^{2-}	. . . .	$[PbO_3]^{2-}$	47 (Pb):	Hb/B2–11
				Hb/C1–198/9
O_3PbRb_2	. . . . . .	Rb_2PbO_3	47 (Pb):	Hb/C3–1137
O_3PbS	. . . . . .	$PbSO_3$	47 (Pb):	Hb/C2–535/6
–.	. . . .	$PbSO_3 \cdot H_2O$	47 (Pb):	Hb/C2–536/7
O_3PbSSe	. . . . . .	$PbSeSO_3$.	47 (Pb):	Hb/C2–640
O_3PbS_2	. . . . . .	PbS_2O_3	47 (Pb):	Hb/C2–595/6
–.	. . . .	PbS_2O_3 solutions:		
		PbS_2O_3–H_2O	47 (Pb):	Hb/C2–596/8
		PbS_2O_3–$Na_2S_2O_3$–H_2O.	47 (Pb):	Hb/C3–1031
O_3PbSe	. . . . . .	$PbSeO_3$.	47 (Pb):	Hb/C2–632/4
–.	. . . .	$PbSeO_3$ systems:		
		$PbSeO_3$–PbO	47 (Pb):	Hb/C2–634/5
O_3PbSi	. . . . . .	$PbSiO_3 = PbO \cdot SiO_2$	47 (Pb):	Hb/B2–403, 413/4
				Hb/C3–848/54
–.	. . . .	$PbSiO_3 \cdot xH_2O$	47 (Pb):	Hb/C3–854

Hb = Main Series Volume; GD = Gmelin-Durrer;   Ab = Appended Volume to Main Series;
Eb = Supplement Volume to Main Series;   Erg. W. = New Supplement Series Volume

O₃PbSi	PbSiO₃ systems:		
	PbSiO₃–CaO · TiO₂ · SiO₂	47 (Pb):	Hb/C4–1401
	PbSiO₃–Cu₂S	60 (Cu):	Hb/B3–1203
	PbSiO₃–FeS–FeSiO₃–PbS.	59 (Fe):	Hb/B5–1137
	PbSiO₃–Na₂SiO₃–SiO₂.	47 (Pb):	Hb/C3–1038
	PbSiO₃–PbZrO₃.	47 (Pb):	Hb/C4–1431
O₃PbSn	PbSnO₃.	47 (Pb):	Hb/C4–1581/3
–.	PbSnO₃ · nH₂O.	47 (Pb):	Hb/C4–1583
–.	PbSnO₃ solid solutions:		
	PbSnO₃–BaSnO₃–BaTiO₃–PbTiO₃ . .	47 (Pb):	Hb/C4–1600/1
	Pb(Hf, Sn, Ti)O₃		
	(modified by Nb₂O₅)	49 (Nb):	Hb/B3–321
	PbSnO₃–PbNb₂O₆	49 (Nb):	Hb/B3–320
	Pb(Sn, Ti, Zr)O₃		
	(modified by Nb₂O₅)	49 (Nb):	Hb/B3–320/1
–.	PbSnO₃ systems:		
	PbSnO₃–BaSnO₃.	47 (Pb):	Hb/C4–1595/6
	PbSnO₃–BaTiO₃	47 (Pb):	Hb/C4–1601/2
	PbSnO₃–PbTiO₃	47 (Pb):	Hb/C4–1597/9
	PbSnO₃–PbTiO₃–PbZrO₃	47 (Pb):	Hb/C4–1605/7
	PbSnO₃–PbZrO₃	47 (Pb):	Hb/C4–1603/5
O₃PbSr	SrPbO₃	47 (Pb):	Hb/C3–1196
O₃PbTe	3 PbTeO₃ · 2 H₂O.	47 (Pb):	Hb/C2–679
O₃PbTh	PbThO₃.	47 (Pb):	Hb/C4–1471
–.	PbThO₃ solid solutions:		
	Pb(Th, Zr)O₃.	47 (Pb):	Hb/C4–1472
O₃PbTi	PbTiO₃	47 (Pb):	Hb/C4–1379/90
–.	PbTiO₃ solid solutions:		
	(Pb, Ba)TiO₃–(Ba, Pb)Nb₂O₆	49 (Nb):	Hb/B3–202
	PbTiO₃–BaSnO₃–BaTiO₃.	47 (Pb):	Hb/C4–1603
	PbTiO₃–BaSnO₃–BaTiO₃–PbSnO₃ . .	47 (Pb):	Hb/C4–1600/1
	PbTiO₃–BaTiO₃	47 (Pb):	Hb/C4–1405/12
	(Pb, Ba) (Ti, Zr)O₃		
	(modified by Nb₂O₅)	49 (Nb):	Hb/B3–318
	PbTiO₃–BaTiO₃–CaTiO₃	47 (Pb):	Hb/C4–1414/6
	PbTiO₃–BaTiO₃–SrTiO₃	47 (Pb):	Hb/C4–1416
	PbTiO₃–Ba₅Nb₄O₁₅	49 (Nb):	Hb/B3–297/8
	PbTiO₃–Bi₀.₆₇TiO₃–SrTiO₃	47 (Pb):	Hb/C4–1404/5
	(1−x)Pb(ZrᵧTi₁₋ᵧ)O₃ · xBiAlO₃	47 (Pb):	Hb/C4–1462
	PbTiO₃–CaSnO₃	47 (Pb):	Hb/C4–1599
	PbTiO₃–CaTiO₃	47 (Pb):	Hb/C4–1400/1
	(Pb, Ca) (Zr, Ti)O₃	47 (Pb):	Hb/C4–1455
	PbTiO₃–CdTiO₃	47 (Pb):	Hb/C4–1417
	PbTiO₃–K₀.₅Bi₀.₅TiO₃	47 (Pb):	Hb/C4–1398/9
	PbTiO₃–KNbO₃	49 (Nb):	Hb/B3–290/1
	Pb(Ti, Zr)O₃–KNbO₃	49 (Nb):	Hb/B3–314/5
	(1−x)Pb(ZrᵧTi₁₋ᵧ)O₃ · xKSbO₃	47 (Pb):	Hb/C4–1455
	PbTiO₃–La₀.₆₇TiO₃	47 (Pb):	Hb/C4–1417/8

See the "Instructions for the user"
at the beginning of the volume

O_3PbTi . . . . PbTiO$_3$ solid solutions:

PbTiO$_3$–LaAlO$_3$ . . . . 47 (Pb):   Hb/C4–1418/9

$(1-x)$Pb$(Zr_yTi_{1-y})O_3 \cdot x$LaAlO$_3$ . . . . . 47 (Pb):   Hb/C4–1462/3

PbTiO$_3$–MgTiO$_3$ . . . . 47 (Pb):   Hb/C4–1400

PbTiO$_3$–Na$_{0.5}$Bi$_{0.5}$TiO$_3$ . . . . 47 (Pb):   Hb/C4–1396/7

$(1-x)$Pb$(Zr_yTi_{1-y})O_3 \cdot x$NaBiO$_3$ . . . . . 47 (Pb):   Hb/C4–1455

PbTiO$_3$–NaNbO$_3$ . . . . 49 (Nb):   Hb/B3–289/90

Pb(Ti, Zr)O$_3$–NaNbO$_3$. . . . . 49 (Nb):   Hb/B3–314

$(1-x)$Pb$(Zr_yTi_{1-y})O_3 \cdot x$NaSbO$_3$ . . . . 47 (Pb):   Hb/C4–1455

Pb(Hf, Ti)O$_3$ . . . . 47 (Pb):   Hb/C4–1467

PbTiO$_3$–PbHfO$_3$–PbNb$_2$O$_6$ . . . . . . 49 (Nb):   Hb/B3–320

Pb(Hf, Sn, Ti)O$_3$
(modified by Nb$_2$O$_5$) . . . . 49 (Nb):   Hb/B3–321

Pb(Hf, Zr, Ti)O$_3$ . . . . 47 (Pb):   Hb/C4–1468

PbTiO$_3$–PbNb$_2$O$_6$ . . . . 49 (Nb):   Hb/B3–285/7
                                                                   Hb/B4–313/4

Pb(Ti, Sn, Zr)O$_3$
(modified by Nb$_2$O$_5$) . . . . 49 (Nb):   Hb/B3–320/1

Pb(Ti, Zr)O$_3$ . . . . 47 (Pb):   Hb/C4–1442/54

– (modified by Nb$_2$O$_5$) . . . . 49 (Nb):   Hb/B3–309/14

– (modified by Ta$_2$O$_5$) . . . . 50 (Ta):   Hb/B2–242/3

Pb(Ti, Zr)O$_3$–Pb$_2$SbNbO$_6$. . . . . . . 49 (Nb):   Hb/B3–313/4

Pb(Ti, Zr)O$_3$–Pb$_3$MgNb$_2$O$_9$ . . . . . . 49 (Nb):   Hb/B3–315/7

(Pb, Sr) (Zr, Ti)O$_3$ . . . . 47 (Pb):   Hb/C4–1455/8

– (modified by Nb$_2$O$_5$) . . . . 49 (Nb):   Hb/B3–317/8

PbTiO$_3$–Pb$_2$ScNbO$_6$ . . . . 49 (Nb):   Hb/B3–300/1

PbTiO$_3$–Pb$_2$YbNbO$_6$ . . . . 49 (Nb):   Hb/B3–302

PbTiO$_3$–Pb$_3$MgNb$_2$O$_9$ . . . . 49 (Nb):   Hb/B3–291/2

PbTiO$_3$–Pb$_3$ZnNb$_2$O$_9$ . . . . 49 (Nb):   Hb/B3–298/9

PbTiO$_3$–Pb$_5$Nb$_4$O$_{15}$ . . . . 49 (Nb):   Hb/B3–285

PbTiO$_3$–SrSnO$_3$. . . . . 47 (Pb):   Hb/C4–1600

PbTiO$_3$–SrTiO$_3$ . . . . 47 (Pb):   Hb/C4–1401/4

PbTiO$_3$–SrZrO$_3$ . . . . 47 (Pb):   Hb/C4–1458

PbTiO$_3$–Sr$_5$Nb$_4$O$_{15}$ . . . . 49 (Nb):   Hb/B3–294/5

PbTiO$_3$–ZnBi$_2$Nb$_2$O$_9$ . . . . 49 (Nb):   Hb/B3–299/300

PbTiO$_3$–ZnTiO$_3$ . . . . 47 (Pb):   Hb/C4–1416/7

–. . . . . PbTiO$_3$ systems:

PbTiO$_3$–B$_2$O$_3$–PbO–PbZrO$_3$ . . . . . . 47 (Pb):   Hb/C4–1454

PbTiO$_3$–BaSnO$_3$ . . . . 47 (Pb):   Hb/C4–1602/3

PbTiO$_3$–BaTiO$_3$–BaZrO$_3$–PbZrO$_3$ . . . 47 (Pb):   Hb/C4–1459/60

PbTiO$_3$–BaTiO$_3$–Na$_2$SiO$_3$ . . . . 47 (Pb):   Hb/C4–1413/4

PbTiO$_3$–BaZrO$_3$ . . . . 47 (Pb):   Hb/C4–1461/2

PbTiO$_3$–Ba$_3$(PO$_4$)$_2$ . . . . 47 (Pb):   Hb/C4–1413

PbTiO$_3$–K$_2$CO$_3$ . . . . 47 (Pb):   Hb/C4–1398

PbTiO$_3$–K$_2$F$_2$ . . . . 47 (Pb):   Hb/C4–1398

PbTiO$_3$–K$_2$SiO$_3$ . . . . 47 (Pb):   Hb/C4–1398

PbTiO$_3$–K$_2$SiO$_3$–Na$_2$SiO$_3$ . . . . . . 47 (Pb):   Hb/C4–1399/400

PbTiO$_3$–K$_2$TiO$_3$ . . . . 47 (Pb):   Hb/C4–1397/8

(PbTiO$_3$)$_2$–K$_4$P$_2$O$_7$. . . . . 47 (Pb):   Hb/C4–1398

Hb = Main Series Volume; GD = Gmelin-Durrer;   Ab = Appended Volume to Main Series;
Eb = Supplement Volume to Main Series;   Erg. W. = New Supplement Series Volume

O_3PbTi . . . . PbTiO$_3$ systems:

PbTiO$_3$–LaNbO$_4$ . . . . 49 (Nb):   Hb/B3–302

PbTiO$_3$–Na$_2$CO$_3$. . . . . 47 (Pb):   Hb/C4–1396

PbTiO$_3$–Na$_2$F$_2$. . . . . 47 (Pb):   Hb/C4–1396

PbTiO$_3$–Na$_2$SiO$_3$ . . . . 47 (Pb):   Hb/C4–1396

(PbTiO$_3$)$_2$–Na$_4$P$_2$O$_7$ . . . . 47 (Pb):   Hb/C4–1396

PbTiO$_3$–PbNb$_2$O$_6$–PbZrO$_3$ . . . . . . 49 (Nb):   Hb/B3–308/9

PbTiO$_3$–PbSnO$_3$ . . . . 47 (Pb):   Hb/C4–1597/9

PbTiO$_3$–PbSnO$_3$–PbZrO$_3$ . . . . . . 47 (Pb):   Hb/C4–1605/7

PbTiO$_3$–PbTa$_2$O$_6$ . . . . 50 (Ta):   Hb/B2–242

PbTiO$_3$–PbZrO$_3$. . . . . 47 (Pb):   Hb/C4–1440/2

PbTiO$_3$–PbZrO$_3$–SrTiO$_3$–SrZrO$_3$ . . . 47 (Pb):   Hb/C4–1455

PbTiO$_3$–Pb$_2$ScTaO$_6$ . . . . 50 (Ta):   Hb/B2–243

O_3PbZr . . . . PbZrO$_3$ . . . . 47 (Pb):   Hb/C4–1423/30

–. . . . . PbZrO$_3$ solid solutions:

PbZrO$_3$–BaTiO$_3$ . . . . 47 (Pb):   Hb/C4–1460

(Pb, Ba) (Zr, Ti)O$_3$

(modified by Nb$_2$O$_5$) . . . . 49 (Nb):   Hb/B3–318

(1–x)Pb(Zr$_y$Ti$_{1-y}$)O$_3$ · xBiAlO$_3$ . . . . . 47 (Pb):   Hb/C4–1462

(Pb, Ca) (Zr, Ti)O$_3$ . . . . 47 (Pb):   Hb/C4–1455

PbZrO$_3$–CdZrO$_3$. . . . . 47 (Pb):   Hb/C4–1439

K$_x$Pb$_{1-2x}$Bi$_x$ZrO$_3$ . . . . 47 (Pb):   Hb/C4–1433

Pb(Zr, Ti)O$_3$–KNbO$_3$ . . . . 49 (Nb):   Hb/B3–314/5

(1–x)Pb(Zr$_y$Ti$_{1-y}$)O$_3$ · xKSbO$_3$ . . . . . 47 (Pb):   Hb/C4–1455

(1–x)Pb(Zr$_y$Ti$_{1-y}$)O$_3$ · xLaAlO$_3$ . . . . . 47 (Pb):   Hb/C4–1462/3

(1–x)Pb(Zr$_y$Ti$_{1-y}$)O$_3$ · xNaBiO$_3$ . . . . . 47 (Pb):   Hb/C4–1455

Na$_x$Pb$_{1-2x}$Bi$_x$ZrO$_3$. . . . . 47 (Pb):   Hb/C4–1432/3

PbZrO$_3$–NaNbO$_3$ . . . . 49 (Nb):   Hb/B3–304/5

Pb(Zr, Ti)O$_3$–NaNbO$_3$ . . . . 49 (Nb):   Hb/B3–314

(1–x)Pb(Zr$_y$Ti$_{1-y}$)O$_3$ · xNaSbO$_3$ . . . . 47 (Pb):   Hb/C4–1455

PbZrO$_3$–PbCeO$_3$ . . . . 47 (Pb):   Hb/C4–1439

Pb(Hf, Zr)O$_3$. . . . . 47 (Pb):   Hb/C4–1467/8

Pb(Hf, Zr, Ti)O$_3$ . . . . 47 (Pb):   Hb/C4–1468

PbZrO$_3$–PbNb$_2$O$_6$. . . . . 49 (Nb):   Hb/B3–232, 302/4

Pb(Zr, Sn, Ti)O$_3$

(modified by Nb$_2$O$_5$) . . . . 49 (Nb):   Hb/B3–320/1

Pb(Th, Zr)O$_3$ . . . . 47 (Pb):   Hb/C4–1472

Pb(Zr, Ti)O$_3$ . . . . 47 (Pb):   Hb/C4–1442/54

– (modified by Nb$_2$O$_5$) . . . . 49 (Nb):   Hb/B3–309/14

– (modified by Ta$_2$O$_5$) . . . . 50 (Ta):   Hb/B2–242/3

Pb(Zr,Ti)O$_3$–Pb$_2$SbNbO$_6$ . . . . 49 (Nb):   Hb/B3–313/4

Pb(Zr,Ti)O$_3$–Pb$_3$MgNb$_2$O$_9$ . . . . . . 49 (Nb):   Hb/B3–315/7

Pb(Zr,Ti)O$_3$–Pb$_3$ZnNb$_2$O$_9$. . . . . . . 49 (Nb):   Hb/B3–314

(Pb,Sr) (Zr,Ti)O$_3$. . . . . 47 (Pb):   Hb/C4–1455/8

– (modified by Nb$_2$O$_5$) . . . . 49 (Nb):   Hb/B3–317/8

PbZrO$_3$–Pb$_2$ScNbO$_6$ . . . . 49 (Nb):   Hb/B3–307/8

PbZrO$_3$–Pb$_3$MgNb$_2$O$_9$ . . . . 49 (Nb):   Hb/B3–305

PbZrO$_3$–SrZrO$_3$. . . . . 47 (Pb):   Hb/C4–1435/7

PbZrO$_3$–ZnZrO$_3$. . . . . 47 (Pb):   Hb/C4–1438/9

See the "Instructions for the user"
at the beginning of the volume

O_3PbZr . . . . PbZrO$_3$ systems:

PbZrO$_3$–B$_2$O$_3$–PbO–PbTiO$_3$ . . . . . . 47 (Pb):   Hb/C4–1454

PbZrO$_3$–BaNb$_2$O$_6$–BaZrO$_3$

–PbNb$_2$O$_6$ . . . . 49 (Nb):   Hb/B3–305/7

PbZrO$_3$–BaTiO$_3$–BaZrO$_3$–PbTiO$_3$ . . . 47 (Pb):   Hb/C4–1459/60

PbZrO$_3$–BaZrO$_3$ . . . . 47 (Pb):   Hb/C4–1437/8

PbZrO$_3$–CaZrO$_3$ . . . . 47 (Pb):   Hb/C4–1433/4

PbZrO$_3$–K$_2$B$_2$O$_4$ . . . . 47 (Pb):   Hb/C4–1432

PbZrO$_3$–K$_2$CO$_3$ . . . . 47 (Pb):   Hb/C4–1432

PbZrO$_3$–K$_2$Cl$_2$ . . . . 47 (Pb):   Hb/C4–1432

PbZrO$_3$–K$_2$F$_2$ . . . . 47 (Pb):   Hb/C4–1432

PbZrO$_3$–K$_2$SO$_4$ . . . . 47 (Pb):   Hb/C4–1432

PbZrO$_3$–K$_2$TiO$_3$ . . . . 47 (Pb):   Hb/C4–1432

PbZrO$_3$–Li$_2$B$_2$O$_4$ . . . . 47 (Pb):   Hb/C4–1432

PbZrO$_3$–Li$_2$Cl$_2$ . . . . 47 (Pb):   Hb/C4–1432

PbZrO$_3$–Li$_2$F$_2$ . . . . 47 (Pb):   Hb/C4–1432

PbZrO$_3$–Li$_2$SO$_4$ . . . . 47 (Pb):   Hb/C4–1432

PbZrO$_3$–Na$_2$B$_2$O$_4$ . . . . 47 (Pb):   Hb/C4–1432

PbZrO$_3$–Na$_2$CO$_3$ . . . . 47 (Pb):   Hb/C4–1432

PbZrO$_3$–Na$_2$Cl$_2$ . . . . 47 (Pb):   Hb/C4–1432

PbZrO$_3$–Na$_2$F$_2$ . . . . 47 (Pb):   Hb/C4–1432

PbZrO$_3$–Na$_2$SO$_4$ . . . . 47 (Pb):   Hb/C4–1432

PbZrO$_3$–Na$_2$SiO$_3$ . . . . 47 (Pb):   Hb/C4–1432

PbZrO$_3$–Na$_2$TiO$_3$ . . . . 47 (Pb):   Hb/C4–1432

PbZrO$_3$–Na$_4$P$_2$O$_7$ . . . . 47 (Pb):   Hb/C4–1432

PbZrO$_3$–Pb(BO$_2$)$_2$ . . . . 47 (Pb):   Hb/C4–1431

PbZrO$_3$–PbF$_2$ . . . . 47 (Pb):   Hb/C4–1431

PbZrO$_3$–PbNb$_2$O$_6$–PbTiO$_3$ . . . . . . 49 (Nb):   Hb/B3–308/9

PbZrO$_3$–Pb(PO$_3$)$_2$ . . . . 47 (Pb):   Hb/C4–1431

PbZrO$_3$–PbSiO$_3$ . . . . 47 (Pb):   Hb/C4–1431

PbZrO$_3$–PbSnO$_3$ . . . . 47 (Pb):   Hb/C4–1603/5

PbZrO$_3$–PbSnO$_3$–PbTiO$_3$ . . . . . . 47 (Pb):   Hb/C4–1605/7

PbZrO$_3$–PbTa$_2$O$_6$ . . . . 50 (Ta):   Hb/B2–242

PbZrO$_3$–PbTiO$_3$ . . . . 47 (Pb):   Hb/C4–1440/2

PbZrO$_3$–PbTiO$_3$–SrTiO$_3$–SrZrO$_3$ . . . 47 (Pb):   Hb/C4–1455

PbZrO$_3$–Pb$_2$ScTaO$_6$ . . . . 50 (Ta):   Hb/B2–243

PbZrO$_3$–Pb$_3$(PO$_4$)$_2$ . . . . 47 (Pb):   Hb/C4–1431

PbZrO$_3$–SrZrO$_3$ . . . . 47 (Pb):   Hb/C4–1434

O_3Pb_2 . . . . Pb$_2$O$_3$ . . . . 47 (Pb):   Hb/A1–111

                                                                                 Hb/C1–139/44

– . . . . Pb$_2$O$_3$ · 3 H$_2$O . . . . 47 (Pb):   Hb/C1–144

– . . . . Pb$_2$O$_3$ systems:

Pb$_2$O$_3$–Pb$_3$O$_4$ . . . . 47 (Pb):   Hb/C1–54/5

O_3Pd_2 . . . . Pd$_2$O$_3$ · xH$_2$O . . . . 65 (Pd):   Hb/2–265/6

O_3PmSc . . . . PmScO$_3$ = Sc$_2$O$_3$ · Pm$_2$O$_3$ . . . . 39 (SE):   Hb/C1–113, 124, 313

O_3Pm_2 . . . . Pm$_2$O$_3$:

Chemical reactions . . . . 39 (SE):   Hb/C1–313

Crystallographic properties . . . . . . 39 (SE):   Hb/C1–104/6, 119/21, 312

O₃Pm₂ . . . . Pm₂O₃:
    Formation and preparation . . . . . . 39 (SE):   Hb/C1–132, 312
    Magnetic properties . . . . 39 (SE):   Hb/C1–142/3, 312/3
    Mechanical properties . . . . 39 (SE):   Hb/C1–127/8
    Optical properties. . . . . 39 (SE):   Hb/C1–150/1
    Thermal properties . . . . 39 (SE):   Hb/C1–130, 139, 312
–. . . . . Pm₂O₃ · 3 H₂O = Pm(OH)₃ . . . . 39 (SE):   Hb/C2–9, 17, 56/7
–. . . . . Pm₂O₃ solid solutions:
    Pm₂O₃–Sm₂O₃ . . . . 39 (SE):   Hb/C1–322
O₃Po . . . . PoO₃ . . . . 12 (Po):   Hb–133
O₃Po²⁻ . . . . [PoO₃]²⁻ . . . . 12 (Po):   Hb–102
O₃Po₂ . . . . Po₂O₃. . . . . 12 (Po):   Hb–133
O₃PrSc . . . . PrScO₃ = Sc₂O₃ · Pr₂O₃ . . . . 39 (SE):   Hb/C1–113, 124, 127
O₃PrV . . . . PrVO₃. . . . . 48 (V):   Hb/B2–551/2
O₃Pr₂ . . . . Pr₂O₃:
    Catalytic properties. . . . . 39 (SE):   Hb/C1–275
    Chemical reactions . . . . 39 (SE):   Hb/C1–156/75, 273/5
    Crystallographic properties . . . . . . 39 (SE):   Hb/C1–104/6, 113/23,
                                                             270/2
    Electrical properties . . . . 39 (SE):   Hb/C1–144/6, 149, 273
    Electrochemical behavior . . . . 39 (SE):   Hb/C1–155
    Formation and preparation . . . . . . 39 (SE):   Hb/C1–106, 131/3,
                                                             269/70
    Magnetic properties . . . . 39 (SE):   Hb/C1–142/3, 272/3
    Mechanical properties . . . . 39 (SE):   Hb/C1–127/8, 272
    Optical properties. . . . . 39 (SE):   Hb/C1–151, 273
    Thermal properties . . . . 39 (SE):   Hb/C1–130/41, 272
–. . . . . Pr₂O₃ · H₂O = PrO(OH) . . . . 39 (SE):   Hb/C2–15, 93, 95, 102
–. . . . . Pr₂O₃ · 2 H₂O . . . . 39 (SE):   Hb/C2–15
–. . . . . Pr₂O₃ · 3 H₂O
    see H₃O₃Pr ... Pr(OH)₃
–. . . . . Pr₂O₃ glasses:
    Pr₂O₃–PbO–SiO₂ . . . . 47 (Pb):   Hb/C4–1376
–. . . . . Pr₂O₃ solid solutions:
    Pr₂O₃–Nd₂O₃ . . . . 39 (SE):   Hb/C1–308
    Pr₂O₃–Tb₂O₃ . . . . 39 (SE):   Hb/C1–400
–. . . . . Pr₂O₃ systems:
    Pr₂O₃–H₂O . . . . 39 (SE):   Hb/C2–50, 102
    Pr₂O₃–Lu₂O₃ . . . . 39 (SE):   Hb/C1–436
    Pr₂O₃–Nb₂O₅ . . . . 49 (Nb):   Hb/B3–155
    Pr₂O₃–Nb₂O₅–H₂O . . . . 49 (Nb):   Hb/B3–155
    PrO₁.₅–NpO₂ . . . . Erg. W. 4–46/8
    Pr₂O₃–Pr(NO₃)₃–Pr₂(CO₃)₃–H₂O. . . . 39 (SE):   Hb/C2–50
    Pr₂O₃–Pr₂(CO₃)₃–H₂O . . . . 39 (SE):   Hb/C2–50
    Pr₂O₃–Y₂O₃ . . . . 39 (SE):   Hb/C1–294
    Pr₂O₃–Yb₂O₃ . . . . 39 (SE):   Hb/C1–433
O₃Pt . . . . PtO₃ · H₂O . . . . 68 (Pt):   Hb/C1–53/4

See the "Instructions for the user"
at the beginning of the volume

O_3Pt_2	$Pt_2O_3 \cdot 2\,H_2O$	68 (Pt):	Hb/C1−26, 53
$O_3Pt_2Sn_3$	$Pt_2Sn_3O_3$	68 (Pt):	Hb/C3−310
O_3Pu	$PuO_3 \cdot 0.8\,H_2O$	Erg. W. 4−29	
−.	$PuO_3 \cdot H_2O$	Erg. W. 4−29	
−.	$PuO_3 \cdot 2\,H_2O$	Erg. W. 4−29	
O_3PuSc	$PuScO_3$	Erg. W. 4−56	
O_3PuSr	$SrPuO_3$	Erg. W. 4−53/5	
O_3PuV	$PuVO_3$	Erg. W. 4−56, 71	
O_3Pu_2	Pu_2O_3	Erg. W. 4−12/4	
		Erg. W. 8−192/3, 372	
O_3RbReS	$RbReO_3S$	70 (Re):	Hb−144
O_3RbTa	$RbTaO_3$	50 (Ta):	Hb/B2−170/1
−.	$RbTaO_3$ solid solutions:		
	$Rb(Ta,Nb)O_3$	50 (Ta):	Hb/B2−260
O_3RbV	$RbVO_3 = Rb_2O \cdot V_2O_5$	48 (V):	Hb/A2−368
			Hb/B2−486/7
−.	$RbVO_3$ systems:		
	$RbVO_3–RbCl$	48 (V):	Hb/B2−489
	$RbVO_3–RbF$	48 (V):	Hb/B2−488
	$RbVO_3–Rb_3[VO_3F_2]$	48 (V):	Hb/B2−488
O_3Rb_2	Rb_2O_3	24 (Rb):	Hb−111
O_3Rb_2S	Rb_2SO_3	24 (Rb):	Hb−207
O_3Rb_2SSe	Rb_2SeSO_3	24 (Rb):	Hb−218
$O_3Rb_2S_2$	$Rb_2S_2O_3$	24 (Rb):	Hb−214/5
−.	$Rb_2S_2O_3 \cdot 2\,H_2O$	24 (Rb):	Hb−215
O_3Rb_2Se	$Rb_2SeO_3 \cdot H_2O$	24 (Rb):	Hb−216
O_3Rb_2Si	Rb_2SiO_3	24 (Rb):	Hb−232
O_3Rb_2Tb	Rb_2TbO_3	39 (SE):	Hb/C2−141
O_3Rb_2Ti	Rb_2TiO_3	41 (Ti):	Hb−411
O_3Re	ReO_3	70 (Re):	Hb−90/3
O_3ReSTl	$TlReO_3S$	70 (Re):	Hb−148/9
O_3Re_2	$Re_2O_3 \cdot xH_2O$	70 (Re):	Hb−96/7
O_3Rh_2	Rh_2O_3	64 (Rh):	Hb−46/7
−.	$Rh_2O_3 \cdot H_2O$	64 (Rh):	Hb−47
−.	$Rh_2O_3 \cdot 3\,H_2O$	64 (Rh):	Hb−48
−.	$Rh_2O_3 \cdot 5\,H_2O$	64 (Rh):	Hb−47
O_3Rn	RnO_3	Erg. W. 1−130	
O_3Ru	RuO_3	63 (Ru):	Eb−232/3, 240
O_3Ru^+	RuO_3^+	63 (Ru):	Eb−233
O_3Ru^-	RuO_3^-	63 (Ru):	Eb−233
O_3Ru^{2-}	RuO_3^{2-}	63 (Ru):	Eb−249
O_3RuSr	$SrRuO_3$	63 (Ru):	Eb−254/5
−.	$SrRuO_3$ solid solutions:		
	$(Ba_{1-x}Sr_x)RuO_3$	63 (Ru):	Eb−256/8
O_3Ru_2	Ru_2O_3	63 (Ru):	Hb−25
O_3S	SO_3:		
	Absorption by H_2SO_4	9 (S):	Hb/A2−408/11

Hb = Main Series Volume; GD = Gmelin-Durrer;   Ab = Appended Volume to Main Series;
Eb = Supplement Volume to Main Series;   Erg. W. = New Supplement Series Volume

O_3S . . . . SO$_3$:

	Behavior as solvent	9 (S):	Hb/B1$-$366/7
	Chemical reactions	9 (S):	Hb/B1$-$353/66
	$-$ with metal compounds	9 (S):	Hb/B1$-$360/5
	$-$ with metals	9 (S):	Hb/B1$-$356
	$-$ with nonmetal compounds	9 (S):	Hb/B1$-$356/60
	$-$ with nonmetals	9 (S):	Hb/B1$-$354/5
	$-$ with organic substances	9 (S):	Hb/B1$-$365/6
	Colloidal forms	9 (S):	Hb/B1$-$334
	Crystallographic properties	9 (S):	Hb/B1$-$341/3
	Decomposition	9 (S):	Hb/B1$-$353/4
	Electrical properties	9 (S):	Hb/B1$-$353
	Formation	9 (S):	Hb/B1$-$323/31, 335/7
	History	9 (S):	Hb/A1$-$59/60
	Magnetic properties	9 (S):	Hb/B1$-$352/3
	Mechanical properties	9 (S):	Hb/B1$-$343/4, 349/50
	Molecule	9 (S):	Hb/B1$-$337/41
	Optical properties	9 (S):	Hb/B1$-$350/2
	Preparation	9 (S):	Hb/B1$-$332/5
	$-$, industrial	9 (S):	Hb/A2$-$320/426
	$-$, $-$, contact process	9 (S):	Hb/A2$-$364/408
	$-$, $-$, further treatment	9 (S):	Hb/A2$-$408/22
	$-$, $-$, process control	9 (S):	Hb/A2$-$422/5
	$-$, $-$, purification of gases	9 (S):	Hb/A2$-$351/64
	$-$, $-$, theoretical	9 (S):	Hb/A2$-$324/51
	$-$, $-$, waste gas utilization	9 (S):	Hb/A2$-$425/6
	$-$ in pure state	9 (S):	Hb/B1$-$332/3
	$-$ of modifications	9 (S):	Hb/B1$-$333/4
	$-$ of special forms	9 (S):	Hb/B1$-$334/5
	Spectra	9 (S):	Hb/B1$-$351/2
	Thermal properties	9 (S):	Hb/B1$-$344/9
$-$	S$_2$O$_6$	9 (S):	Hb/B1$-$338, 351/2
$-$	S$_3$O$_9$	9 (S):	Hb/B1$-$337/8
$-$	(SO$_3$)$_m$	9 (S):	Hb/B1$-$337
$-$	11 SO$_3 \cdot$ 2 H$_2$O = 2 H$_2$SO$_4 \cdot$ 9 SO$_3$	9 (S):	Hb/B2$-$626
$-$	4 SO$_3 \cdot$ H$_2$O = H$_2$SO$_4 \cdot$ 3 SO$_3$	9 (S):	Hb/B2$-$625/6
$-$	3 SO$_3 \cdot$ H$_2$O = H$_2$SO$_4 \cdot$ 2 SO$_3$	9 (S):	Hb/B2$-$625
$-$	2 SO$_3 \cdot$ H$_2$O = H$_2$SO$_4 \cdot$ SO$_3$ = H$_2$S$_2$O$_7$ . .	9 (S):	Hb/B2$-$624/5, 795/8
$-$	4 SO$_3 \cdot$ 3 H$_2$O	9 (S):	Hb/B2$-$622, 624
$-$	7 SO$_3 \cdot$ 6 H$_2$O	9 (S):	Hb/B2$-$622
$-$	SO$_3 \cdot$ H$_2$O		
	see H$_2$O$_4$S ... H$_2$SO$_4$		
$-$	SO$_3 \cdot$ 2 H$_2$O = H$_2$SO$_4 \cdot$ H$_2$O	9 (S):	Hb/B2$-$614/5, 621/2
$-$	SO$_3 \cdot$ 3 H$_2$O = H$_2$SO$_4 \cdot$ 2 H$_2$O	9 (S):	Hb/B2$-$614/5, 621
$-$	SO$_3 \cdot$ 4 H$_2$O	9 (S):	Hb/B2$-$614/5, 620/1
$-$	SO$_3 \cdot$ 5 H$_2$O	9 (S):	Hb/B2$-$614/5, 620
$-$	SO$_3 \cdot$ 7 H$_2$O	9 (S):	Hb/B2$-$614/5, 620
$-$	SO$_3 \cdot$ 9 H$_2$O	9 (S):	Hb/B2$-$620

See the "Instructions for the user"
at the beginning of the volume

O₃S . . . . SO₃ solutions:

        SO₃−HCN . . . . 9 (S):   Hb/B1−366

        SO₃−HNO₃−H₂SO₄ . . . . 9 (S):   Hb/B3−1678/80

        SO₃−H₂O

          see H₂O₄S ... H₂SO₄

        SO₃−H₂SO₄:

          Chemical reactions. . . . . 9 (S):   Hb/B2−796/8

          Electrochemical behavior . . . . . . 9 (S):   Hb/B2−795/6

          Physical properties. . . . . 9 (S):   Hb/B2−787/95

          Thermodynamic data

           of formation and solution. . . . . 9 (S):   Hb/B2−784/7

        SO₃−N₂O . . . . 9 (S):   Hb/B1−366

        SO₃−organic solvents . . . . 9 (S):   Hb/B1−366

        SO₃−SO₂ . . . . 9 (S):   Hb/B1−366

−. . . . . SO₃ systems:

        SO₃−B₂O₃−H₂O . . . . 13 (B):   Eb−210

        SO₃−B₂O₃−PbO. . . . . 47 (Pb):   Hb/C2−697

        SO₃−Bi₂O₃−H₂O . . . . 19 (Bi):   Eb−747

        SO₃−CaO . . . . 28 (Ca):   Hb/B3−658, 675/9

        SO₃−CaO−SiO₂. . . . . 28 (Ca):   Hb/B3−1107/8

        SO₃−CrO₃−H₂O. . . . . 52 (Cr):   Hb/B−331/2

        SO₃−CrO₃−K₂O−H₂O. . . . . 52 (Cr):   Hb/B−646/7

        SO₃−CuO−Fe₂O₃−H₂O . . . . 60 (Cu):   Hb/B3−1297

        SO₃−Fe₂O₃−H₂O . . . . 59 (Fe):   Hb/B2−446/7

        SO₃−HCl. . . . . 9 (S):   Hb/B3−1842

        SO₃−HCl−H₂O . . . . 9 (S):   Hb/B3−1830

        SO₃−HNO₃ . . . . 9 (S):   Hb/B3−1667/9,

                                       1678/9

        SO₃−HNO₃−H₂SO₄ . . . . 9 (S):   Hb/B3−1667/9, 1680

        SO₃−HSO₃Cl . . . . 9 (S):   Hb/B3−1842/4

        SO₃−H₂O . . . . 9 (S):   Hb/B2−613/9

        SO₃−H₂S−HgO−H₂O . . . . 34 (Hg):   Hb/B3−1021/2

        SO₃−H₂SO₄−K₂SO₄. . . . . 22 (K):   Hb/3−744

        SO₃−H₂SO₄−(NH₄)₂SO₄ . . . . 23 (NH₄):   Hb/2−293

        SO₃−H₂SeO₄ . . . . 10 (Se):   Hb/B−185

        SO₃−HgO−H₂O . . . . 34 (Hg):   Hb/B3−1021

        SO₃−K₂O−P₂O₅−H₂O. . . . . 22 (K):   Hb/5−1002

        SO₃−NH₃−P₂O₅−H₂O. . . . . 23 (NH₄):   Hb/2−430/1

        SO₃−(NH₄)₂SO₄−Nb₂O₅−H₂O . . . . . 49 (Nb):   Hb/B4−303

        SO₃−(NH₄)₂SO₄−Ta₂O₅−H₂O . . . . . 50 (Ta):   Hb/B2−316

        SO₃−N₂O₃−H₂O. . . . . 9 (S):   Hb/B3−1650

        SO₃−N₂O₅−H₂O. . . . . 9 (S):   Hb/B3−1665/96

        SO₃−Na₂O. . . . . 21 (Na):   Eb/3−1146

        SO₃−Na₂O−V₂O₅ . . . . 48 (V):   Hb/B2−405

        SO₃−Na₂O−V₂O₅−H₂O . . . . 48 (V):   Hb/B2−406

        SO₃−Nb₂O₅−H₂O . . . . 49 (Nb):   Hb/B1−269

        SO₃−NiO−UO₃−H₂O . . . . 57 (Ni):   Hb/B3−1235

        SO₃−NiO−UO₃−D₂O . . . . 57 (Ni):   Hb/B3−1235

Hb = Main Series Volume; GD = Gmelin-Durrer;   *Ab = Appended Volume to Main Series;*

Eb = Supplement Volume to Main Series;   *Erg. W. = New Supplement Series Volume*

O_3S . . . . SO_3 systems:

SO_3-POCl_3 . . . . 9 (S): Hb/B1−360

16 (P): Hb/C−597

$SO_3-PbO-TeO_2$ . . . . 47 (Pb): Hb/C2−681

$SO_3-SO_2Cl_2$ . . . . 9 (S): Hb/B3−1820

$SO_3-Ta_2O_5-H_2O$ . . . . 50 (Ta): Hb/B1−176

$SO_3-TiO_2-H_2O$ . . . . 41 (Ti): Hb−345/7

$SO_3-Tl_2O_3-H_2O$ . . . . 38 (Tl): Hb/3−364/5

$SO_3-ZrO_2-H_2O$ . . . . 42 (Zr): Hb−335/7

O_3S^{2-} . . . . $[SO_3]^{2-}$

see $H_2O_3S \ldots H_2SO_3$

O_3SSe . . . . $SSeO_3$ . . . . 10 (Se): Hb/B−176/7, 181

−. . . . . $SeSO_3$ . . . . 10 (Se): Hb/B−176/7

−. . . . . $(SeSO_3)_2$ . . . . 10 (Se): Hb/B−179

−. . . . . $SeSO_3$ solutions:

$SeSO_3-H_2SO_4$ . . . . 10 (Se): Hb/B−177/81

O_3SSe^{2-} . . . . $[SeSO_3]^{2-}$ . . . . 10 (Se): Hb/B−181/5

O_3SSr . . . . $SrSO_3$ . . . . 29 (Sr): Hb−171/2

Eb−249

−. . . . . $SrSO_3 \cdot 0.5 \, H_2O$ . . . . 29 (Sr): Hb−172

O_3STe . . . . $TeSO_3$ . . . . 11 (Te): Hb−355/7

−. . . . . $TeSO_3$ solutions:

$TeSO_3-H_2SO_4$ . . . . 11 (Te): Hb−357

$TeSO_3-H_2SO_4-H_2O$ . . . . 11 (Te): Hb−357/9

O_3STl_2 . . . . Tl_2SO_3 . . . . 38 (Tl): Hb/3−349

O_3SV^{3-} . . . . $[VO_3S]^{3-}$ . . . . 48 (V): Hb/B1−281

O_3SZn . . . . $ZnSO_3 \cdot 2 \, H_2O$ . . . . 32 (Zn): Hb−219/20

Eb−935

−. . . . . $ZnSO_3 \cdot 2.5 \, H_2O$ . . . . 32 (Zn): Hb−219/20

Eb−935

−. . . . . $ZnSO_3$ systems:

$ZnSO_3-Na_2SO_3-H_2O$ . . . . 32 (Zn): Eb−1006

$ZnSO_3-SO_2-H_2O$ . . . . 32 (Zn): Eb−935

O_3S_2 . . . . S_2O_3 . . . . 9 (S): Hb/B1−179/81

$O_3S_2^{2-}$ . . . . $[S_2O_3]^{2-}$

see $H_2O_3S_2 \ldots H_2S_2O_3$ solutions

O_3S_2Sr . . . . SrS_2O_3 . . . . 29 (Sr): Hb−181

Eb−258/9

−. . . . . $SrS_2O_3 \cdot H_2O$ . . . . 29 (Sr): Hb−182

Eb−258/9

−. . . . . $SrS_2O_3 \cdot 3 \, H_2O$ . . . . 29 (Sr): Eb−258

−. . . . . $SrS_2O_3 \cdot 5 \, H_2O$ . . . . 29 (Sr): Hb−181/2

Eb−258

−. . . . . SrS_2O_3 systems:

$SrS_2O_3-H_2O$ . . . . 29 (Sr): Eb−258/9

$O_3S_2Tl^-$ . . . . $[Tl_2(S_2O_3)_2]^{2-}$ . . . . 38 (Tl): Hb/3−368

$O_3S_2Tl_2$ . . . . $Tl_2S_2O_3$ . . . . 38 (Tl): Hb/3−367/8

See the "Instructions for the user"
at the beginning of the volume

O₃S₂Zn . . . . ZnS₂O₃ . . . . 32 (Zn):   Hb−241/2
                                                                            Eb−965/6
O₃S₄Sb₄ . . . . . . Sb₄O₃S₄ . . . . 18 (Sb):   Hb/B3−541
O₃S₆Sb₆ . . . . . . 2 Sb₂S₃ · Sb₂O₃ = Sb₂OS₂ . . . . 18 (Sb):   Hb/A3−305
                                                                            Hb/B3−540
O₃S₁₅Sb₁₂. . . . . . 5 Sb₂S₃ · Sb₂O₃ = Sb₄OS₅ . . . . 18 (Sb):   Hb/B3−540
O₃Sb³⁻ . . . . SbO₃³⁻ . . . . 18 (Sb):   Hb/B2−368/9
O₃Sb₂ . . . . Sb₂O₃:
                    Chemical reactions . . . . 18 (Sb):   Hb/B2−359/65
                    Formation and preparation . . . . . . 18 (Sb):   Hb/B2−349/52
                    History. . . . . 18 (Sb):   Hb/A1−23/4, 29/30,
                                                                            36/9
                    Physical properties . . . . 18 (Sb):   Hb/B2−352/9
                    Toxicity . . . . 18 (Sb):   Hb/B2−276
                    Uses . . . . 18 (Sb):   Hb/A3−305
−. . . . . Sb₂O₃ · xH₂O . . . . 18 (Sb):   Hb/B2−365/7
−. . . . . Sb₂O₃ glasses:
                    Sb₂O₃−PbO−SiO₂ . . . . 47 (Pb):   Hb/C3−941
−. . . . . Sb₂O₃ solid solutions:
                    Sb₂O₃−PbNb₂O₆ . . . . 49 (Nb):   Hb/B3−241
                    Sb₂O₃−Pb₂Nb₂O₇ . . . . 49 (Nb):   Hb/B3−241
−. . . . . Sb₂O₃ systems:
                    Sb₂O₃−Ag₂O−Sb₂O₅ . . . . 61 (Ag):   Hb/B4−89
                    Sb₂O₃−Bi₂O₃ . . . . 19 (Bi):   Eb−858
                    Sb₂O₃−CaO . . . . 28 (Ca):   Hb/B3−1239
                    Sb₂O₃−HCl−SbCl₃−H₂O . . . . 18 (Sb):   Hb/B2−462/5
                    Sb₂O₃−NaOH−H₂O . . . . 21 (Na):   Eb−4/1710
                    Sb₂O₃−Na₂O−P₂O₅ . . . . 21 (Na):   Eb−5/2026
                    Sb₂O₃−(NaPO₃)ₓ . . . . 21 (Na):   Eb−5/2026
                    Sb₂O₃−PbO . . . . 47 (Pb):   Hb/C3−927/8
                    Sb₂O₃−PbO−Sb₂S₃ . . . . 47 (Pb):   Hb/C3−939
                    Sb₂O₃−Sb₂S₃ . . . . 18 (Sb):   Hb/B3−539/40
O₃ScSm . . . . . . SmScO₃ = Sc₂O₃ · Sm₂O₃ . . . . 39 (SE):   Hb/C1−113, 124, 321/2
O₃ScY. . . . . YScO₃ = Sc₂O₃ · Y₂O₃. . . . . 39 (SE):   Hb/C1−113, 124/5,
                                                                            127, 205
O₃Sc₂ . . . . Sc₂O₃:
                    Chemical reactions . . . . 39 (SE):   Hb/C1−157, 162/4,
                                                                            168/70, 187/8
                    Crystallographic properties . . . . . . 39 (SE):   Hb/C1−104/6, 121/2, 183
                    Diffusion. . . . . 39 (SE):   Hb/C1−128
                    Electrical properties . . . . 39 (SE):   Hb/C1−144/50, 185
                    Electrochemical behavior . . . . . . 39 (SE):   Hb/C1−155
                    Formation . . . . 39 (SE):   Hb/C1−106/10, 131/3,
                                                                            181/3
                    Mechanical properties . . . . 39 (SE):   Hb/C1−127, 184
                    Optical properties. . . . . 39 (SE):   Hb/C1−150/1, 154/5,
                                                                            186/7
                    Preparation . . . . 39 (SE):   Hb/C1−106/12, 181/3

Hb = Main Series Volume; GD = Gmelin-Durrer;   Ab = Appended Volume to Main Series;
Eb = Supplement Volume to Main Series;   Erg. W. = New Supplement Series Volume

O_3Sc_2 . . . . Sc_2O_3:

              Preparation of special forms . . . . . . 39 (SE):   Hb/C1–106/12, 182

              Spectra . . . . 39 (SE):   Hb/C1–154/5, 186

              Thermal properties . . . . 39 (SE):   Hb/C1–129/41, 184/5

–. . . . . Sc_2O_{3-x} . . . . 39 (SE):   Hb/C1–177/8

–. . . . . $Sc_2O_3 \cdot H_2O = ScO(OH)$ . . . . 39 (SE):   Hb/C2–23/4, 93, 98/9

–. . . . . $Sc_2O_3 \cdot 3\ H_2O$

              see H_3O_3Sc ... $Sc(OH)_3$

–. . . . . $Sc_2O_3 \cdot nH_2O = Sc(OH)_3 \cdot nH_2O$. . . . . 39 (SE):   Hb/C2–19/26

–. . . . . Sc_2O_3 systems:

              $ScO_{1.5}–AmO_2$ . . . . Erg. W. 4–79

              $Sc_2O_3–CeO_2$. . . . . 39 (SE):   Hb/C1–262

              $Sc_2O_3–Er_2O_3$ . . . . 39 (SE):   Hb/C1–420

              $Sc_2O_3–HNO_3–H_2O$ . . . . 39 (SE):   Hb/C2–25

              $Sc_2O_3–Ho_2O_3$ . . . . 39 (SE):   Hb/C1–413/4

              $Sc_2O_3–La_2O_3$ . . . . 39 (SE):   Hb/C1–217/8

              $Sc_2O_3–Lu_2O_3$ . . . . 39 (SE):   Hb/C1–435

              $Sc_2O_3–N_2O_5–H_2O$ . . . . 39 (SE):   Hb/C2–225/6

              $Sc_2O_3–Nb_2O_5$ . . . . 49 (Nb):   Hb/B3–136/7

              $Sc_2O_3–Nd_2O_3$ . . . . 39 (SE):   Hb/C1–306/7

              $ScO_{1.5}–NpO_2$ . . . . Erg. W. 4–45

              $Sc_2O_3–Ta_2O_5$ . . . . 50 (Ta):   Hb/B2–204

              $Sc_2O_3–Y_2O_3$ . . . . 39 (SE):   Hb/C1–205

               $Sc_2O_3–Yb_2O_3$ . . . . 39 (SE):   Hb/C1–430

O_3Se . . . . SeO_3 . . . . 10 (Se):   Hb/B–43/4

–. . . . . SeO_3 systems:

              $SeO_3–CrO_3–H_2O$ . . . . 52 (Cr):   Hb/B–341

              $SeO_3–Nb_2O_5–H_2O$ . . . . 49 (Nb):   Hb/B1–81

              $SeO_3–Tl_2O_3–H_2O$ . . . . 38 (Tl):   Hb/3–376

O_3Se^{2-} . . . . $[SeO_3]^{2-}$

              see H_2O_3Se ... H_2SeO_3

O_3SeSr . . . . $SrSeO_3$ . . . . 29 (Sr):   Hb–186

                                                         Eb–261

–. . . . . $SrSeO_3 \cdot 3.5\ H_2O$ . . . . 29 (Sr):   Hb–186

O_3SeTe . . . . $TeSeO_3$ . . . . 11 (Te):   Hb–362/3

O_3SeTl_2 . . . . Tl_2SeO_3. . . . . 38 (Tl):   Hb/3–373

O_3SeZn . . . . $ZnSeO_3$ . . . . 32 (Zn):   Hb–246

                                                         Eb–968

–. . . . . $ZnSeO_3 \cdot H_2O$ . . . . 32 (Zn):   Hb–247

–. . . . . $ZnSeO_3 \cdot 2\ H_2O$. . . . . 32 (Zn):   Hb–247

O_3Se_2 . . . . Se_2O_3. . . . . 10 (Se):   Hb/B–25

O_3Si^{2-} . . . . $[SiO_3]^{2-}$ . . . . 15 (Si):   Hb/B–421/2

O_3SiSr . . . . $SrO \cdot SiO_2 = SrSiO_3$. . . . . 29 (Sr):   Hb–210/1

                                                         Eb–282

–. . . . . $SrO \cdot SiO_2 \cdot 0.6\ H_2O$ . . . . 29 (Sr):   Eb–282

–. . . . . $SrSiO_3$ systems:

              $SrSiO_3–CaSiO_3$ . . . . 29 (Sr):   Hb–237

See the "Instructions for the user"
at the beginning of the volume

O$_3$SiSr	SrSiO$_3$ systems:		
	SrSiO$_3$–Li$_2$SiO$_3$	29 (Sr):	Hb–223
	SrSiO$_3$–Na$_2$SiO$_3$	29 (Sr):	Hb–226
O$_3$SiTl$_2$	Tl$_2$SiO$_3$	38 (Tl):	Hb/3–424
O$_3$SiZn	ZnSiO$_3$	32 (Zn):	Hb–273/4
			Eb–989
–.	ZnSiO$_3$ systems:		
	ZnSiO$_3$–Li$_2$SiO$_3$	32 (Zn):	Hb–294
O$_3$Si$_2$	Si$_2$O$_3$	15 (Si):	Hb/B–257
O$_3$SmV	SmVO$_3$	48 (V):	Hb/B2–551/2
O$_3$Sm$_2$	Sm$_2$O$_3$:		
	Catalytic properties	39 (SE):	Hb/C1–321
	Chemical reactions	39 (SE):	Hb/C1–156/76, 319/21
	Crystallographic properties	39 (SE):	Hb/C1–104/6, 115/23,
			316/7
	Diffusion.	39 (SE):	Hb/C1–317
	Electrical properties	39 (SE):	Hb/C1–144/50, 319
	Electrochemical behavior	39 (SE):	Hb/C1–155
	Formation	39 (SE):	Hb/C1–106/10,
			131/3, 315
	Magnetic properties	39 (SE):	Hb/C1–142/3, 318
	Mechanical properties	39 (SE):	Hb/C1–127/8, 317
	Optical properties.	39 (SE):	Hb/C1–150/5, 319
	Preparation	39 (SE):	Hb/C1–106/12, 315
	– of special forms.	39 (SE):	Hb/C1–110/2, 315
	Spectra	39 (SE):	Hb/C1–153/5, 319
	Thermal properties	39 (SE):	Hb/C1–129/41, 317
–.	Sm$_2$O$_3$ · H$_2$O = SmO(OH).	39 (SE):	Hb/C2–14, 15, 93/7,
			103
–.	Sm$_2$O$_3$ · 3 H$_2$O		
	see H$_3$O$_3$Sm … Sm(OH)$_3$		
–.	Sm$_2$O$_3$ · nH$_2$O	39 (SE):	Hb/C2–15
–.	Sm$_2$O$_3$ solid solutions:		
	Sm$_2$O$_3$–Gd$_2$O$_3$	39 (SE):	Hb/C1–386
	Sm$_2$O$_3$–La$_2$O$_3$.	39 (SE):	Hb/C1–322
	Sm$_2$O$_3$–Nd$_2$O$_3$	39 (SE):	Hb/C1–322
	Sm$_2$O$_3$–Pm$_2$O$_3$	39 (SE):	Hb/C1–322
–.	Sm$_2$O$_3$ systems:		
	Sm$_2$O$_3$–BaO–Nb$_2$O$_5$	49 (Nb):	Hb/B3–162
	Sm$_2$O$_3$–CeO$_2$	39 (SE):	Hb/C1–323
	Sm$_2$O$_3$–Dy$_2$O$_3$.	39 (SE):	Hb/C1–409
	Sm$_2$O$_3$–Er$_2$O$_3$	39 (SE):	Hb/C1–421
	Sm$_2$O$_3$–Eu$_2$O$_3$.	39 (SE):	Hb/C1–377
	Sm$_2$O$_3$–Ho$_2$O$_3$	39 (SE):	Hb/C1–414
	Sm$_2$O$_3$–La$_2$O$_3$–Lu$_2$O$_3$	39 (SE):	Hb/C1–436
	Sm$_2$O$_3$–Lu$_2$O$_3$.	39 (SE):	Hb/C1–436
	Sm$_2$O$_3$–Nb$_2$O$_5$	49 (Nb):	Hb/B3–160

O_3Sm_2	Sm_2O_3 systems:		
	$SmO_{1.5}-NpO_2$	Erg. W. 4-46/8	
	$Sm_2O_3-PrO_2$	39 (SE):	Hb/C1-295
	$Sm_2O_3-PrO_x$	39 (SE):	Hb/C1-294/5
	Sm_2O_3-Sm	39 (SE):	Hb/C1-313/4
	$Sm_2O_3-Ta_2O_5$	50 (Ta):	Hb/B2-214
	$Sm_2O_3-Tm_2O_3$	39 (SE):	Hb/C1-425
	$Sm_2O_3-Yb_2O_3$	39 (SE):	Hb/C1-433
O_3Sn	$2 SnO_3 \cdot H_2O$	46 (Sn):	Hb/C1-139
-.	$SnO_3 \cdot 2 H_2O$	46 (Sn):	Hb/C1-139
O_3SnSr	$SrSnO_3$ solid solutions:		
	$SrSnO_3-PbTiO_3$	47 (Pb):	Hb/C4-1600
	$SrSnO_3-SrNb_2O_6$	49 (Nb):	Hb/B3-223
O_3SnTa	$SnTaO_3$	50 (Ta):	Hb/B2-231
O_3Sn_2	Sn_2O_3	46 (Sn):	Hb/C1-57/9
O_3Sn_3	Sn_3O_3	46 (Sn):	Hb/C1-34/5, 104
O_3SrTe	$SrTeO_3$	29 (Sr):	Hb-187
			Eb-262
O_3SrTh	$SrThO_3$	44 (Th):	Hb-348
O_3SrTi	$SrTiO_3$	41 (Ti):	Hb-433/4
-.	$SrTiO_3$ solid solutions:		
	$SrTiO_3-BaTiO_3-NaNbO_3$	49 (Nb):	Hb/B3-199
	$SrTiO_3-BaTiO_3-PbTiO_3$	47 (Pb):	Hb/C4-1416
	$SrTiO_3-Bi_{0.67}TiO_3-PbTiO_3$	47 (Pb):	Hb/C4-1404/5
	$SrTiO_3-Bi_2Ti_3O_9-PbNb_2O_6$	49 (Nb):	Hb/B3-295/6
	$SrTiO_3-NaNbO_3$	49 (Nb):	Hb/B3-195
	$SrTiO_3-PbNb_2O_6$	49 (Nb):	Hb/B3-292/4
	$SrTiO_3-PbTiO_3$	47 (Pb):	Hb/C4-1401/4
	$(Pb,Sr) (Zr,Ti)O_3$	47 (Pb):	Hb/C4-1455/8
	– (modified by Nb_2O_5)	49 (Nb):	Hb/B3-317/8
	$SrTiO_3-SrNb_2O_6$	49 (Nb):	Hb/B3-194
-.	$SrTiO_3$ systems:		
	$SrTiO_3-BaTiO_3$	41 (Ti):	Hb-457/60
	$SrTiO_3-PbTiO_3-PbZrO_3-SrZrO_3$ . . .	47 (Pb):	Hb/C4-1455
O_3SrV	$SrVO_3$	48 (V):	Hb/B2-519
-.	$SrVO_3$ solid solutions:		
	$Sr_xLa_{1-x}VO_3$	48 (V):	Hb/B2-557
-.	$SrVO_3$ systems:		
	$SrVO_3-SrVO_{2.5}$	48 (V):	Hb/B2-519
O_3SrZr	$SrZrO_3$	42 (Zr):	Hb-439
-.	$SrZrO_3$ solid solutions:		
	$SrZrO_3-PbTiO_3$	47 (Pb):	Hb/C4-1458
	$(Pb,Sr) (Zr,Ti)O_3$	47 (Pb):	Hb/C4-1455/8
	– (modified by Nb_2O_5)	49 (Nb):	Hb/B3-317/8
	$SrZrO_3-PbZrO_3$	47 (Pb):	Hb/C4-1435/7
	$SrZrO_3-SrNb_2O_6$	49 (Nb):	Hb/B3-214

See the "Instructions for the user"
at the beginning of the volume

O_3SrZr . . . . SrZrO₃ systems:
                    $SrZrO_3-PbTiO_3-PbZrO_3-SrTiO_3$ . . . 47 (Pb):   Hb/C4−1455
                    $SrZrO_3-PbZrO_3$ . . . . 47 (Pb):   Hb/C4−1434
                    $SrZrO_3-SrTa_2O_6$ . . . . 50 (Ta):   Hb/B2−228/9
O_3Ta . . . . TaO₃ . . . . 50 (Ta):   Hb/A1−8
                                                                        Hb/B1−30
O_3Ta^- . . . . $[TaO_3]^-$ . . . . 50 (Ta):   Hb/A2−407
                                                                        Hb/B1−57, 59/60
O_3Ta_2 . . . . Ta₂O₃ . . . . 50 (Ta):   Hb/A1−8
                                                                        Hb/A2−316
                                                                        Hb/B1−30
$O_3Ta_3^-$ . . . . $[Ta_3O_3]^-$ . . . . 50 (Ta):   Hb/B1−59
O_3Tb_2 . . . . Tb₂O₃:
                    Chemical reactions . . . . 39 (SE):   Hb/C1−156/75, 391
                    Crystallographic properties . . . . . . 39 (SE):   Hb/C1−104/6, 115/23,
                                                                        388/9
                    Electrical properties . . . . 39 (SE):   Hb/C1−144/50, 390
                    Formation and preparation . . . . . . 39 (SE):   Hb/C1−109, 111,
                                                                        131/3, 387/8
                    Magnetic properties . . . . 39 (SE):   Hb/C1−142/3, 389/90
                    Mechanical properties . . . . 39 (SE):   Hb/C1−127/8
                    Optical properties. . . . . 39 (SE):   Hb/C1−150/2, 390
                    Thermal properties . . . . 39 (SE):   Hb/C1−130/9
−. . . . . Tb₂O₃ · H₂O = TbO(OH) . . . . 39 (SE):   Hb/C2−14, 15, 93/7,
                                                                        103/4
−. . . . . Tb₂O₃ · 3 H₂O
                    see H_3O_3Tb ... Tb(OH)₃
−. . . . . Tb₂O₃ · nH₂O . . . . 39 (SE):   Hb/C2−65/6, 68
−. . . . . Tb₂O₃ solid solutions:
                    Tb₂O₃−Pr₂O₃ . . . . 39 (SE):   Hb/C1−400
                    Tb₂O₃−Y₂O₃ . . . . 39 (SE):   Hb/C1−400
−. . . . . Tb₂O₃ systems:
                    Tb₂O₃−Er₂O₃ . . . . 39 (SE):   Hb/C1−421
                    TbO₁.₅−NpO₂ . . . . Erg. W. 4−46/8
                    Tb₂O₃−Tb(NO₃)₃−H₂O . . . . 39 (SE):   Hb/C2−272
O_3Te . . . . TeO₃ . . . . 11 (Te):   Hb−282/4
−. . . . . TeO₃ · 3 H₂O
                    see H_6O_6Te ... H_6TeO_6
−. . . . . TeO₃ systems:
                    TeO₃−NaOH−H₂O . . . . 21 (Na):   Eb/5−1910
O_3Te^{2-} . . . . $[TeO_3]^{2-}$ . . . . 11 (Te):   Hb−286/7
O_3Ti . . . . TiO₃ · 2 H₂O = Ti(OH)₃(OOH) . . . . . . 41 (Ti):   Hb−268/71
−. . . . . TiO₃ · 3 H₂O . . . . 41 (Ti):   Hb−269
O_3Ti^{2-} . . . . $[TiO_3]^{2-}$ . . . . 41 (Ti):   Hb−468
O_3TiV . . . . TiVO₃ . . . . 48 (V):   Hb/B2−558
O_3TiZn . . . . ZnTiO₃ = Zn₂TiO₄ · TiO₂ . . . . 41 (Ti):   Hb−460, 461/2

Hb = Main Series Volume; GD = Gmelin-Durrer;   Ab = Appended Volume to Main Series;
Eb = Supplement Volume to Main Series;   Erg. W. = New Supplement Series Volume

O_3TiZn . . . . ZnTiO$_3$ solid solutions:
                    $ZnTiO_3 - BaTiO_3 - NiZrO_3$ . . . . 57 (Ni):   Hb/B3 − 1186
                    $ZnTiO_3 - PbTiO_3$ . . . . 47 (Pb):   Hb/C4 − 1416/7

O_3Ti_2 . . . . Ti_2O_3 . . . . 41 (Ti):   Hb − 221/4, 478
−. . . . . Ti_2O_3 systems:
                    $Ti_2O_3 - Al_2O_3$ . . . . 41 (Ti):   Hb − 463
                    $Ti_2O_3 - Fe - FeO - FeO \cdot TiO_2 - Fe_2Ti$ . . . 59 (Fe):   Hb/A9 − 1873
                    $Ti_2O_3 - MgO - TiO_2$ . . . . 41 (Ti):   Hb − 423
                    $Ti_2O_3 - TiC$ . . . . 41 (Ti):   Hb − 368
                    $Ti_2O_3 - V_2O_3$ . . . . 48 (V):   Hb/B2 − 558

O_3TlV . . . . $TlVO_3 = Tl_2O \cdot V_2O_5$ . . . . 48 (V):   Hb/B2 − 548
O_3Tl_2 . . . . Tl_2O_3 . . . . 38 (Tl):   Hb/2 − 222/9
−. . . . . $Tl_2O_3 \cdot H_2O = TlO(OH)$ . . . . 38 (Tl):   Hb/2 − 232
−. . . . . $Tl_2O_3 \cdot 3 H_2O = Tl(OH)_3$ . . . . 38 (Tl):   Hb/2 − 232/3
−. . . . . Tl_2O_3 systems:
                    $Tl_2O_3 - Nb_2O_5$ . . . . 49 (Nb):   Hb/B3 − 127
                    $Tl_2O_3 - SO_3 - H_2O$ . . . . 38 (Tl):   Hb/3 − 364/5
                    $Tl_2O_3 - SeO_3 - H_2O$ . . . . 38 (Tl):   Hb/3 − 376
                    $Tl_2O_3 - Ta_2O_5$ . . . . 50 (Ta):   Hb/B2 − 201

O_3Tm_2 . . . . Tm_2O_3:
                    Chemical reactions . . . . 39 (SE):   Hb/C1 − 156/75, 424/5
                    Crystallographic properties . . . . . . 39 (SE):   Hb/C1 − 104/6, 115/23,
                                                                 422/3
                    Diffusion. . . . . 39 (SE):   Hb/C1 − 128
                    Electrical properties . . . . 39 (SE):   Hb/C1 − 145/8
                    Formation . . . . 39 (SE):   Hb/C1 − 106/10,
                                                              131/3, 422
                    Mechanical properties . . . . 39 (SE):   Hb/C1 − 127/8, 423/4
                    Optical properties. . . . . 39 (SE):   Hb/C1 − 150/4, 424
                    Preparation . . . . 39 (SE):   Hb/C1 − 106/12, 422
                    − of special forms. . . . . 39 (SE):   Hb/C1 − 110/2, 422
                    Spectra . . . . 39 (SE):   Hb/C1 − 154/5
                    Thermal properties . . . . 39 (SE):   Hb/C1 − 129/41, 424
−. . . . . $Tm_2O_3 \cdot H_2O = TmO(OH)$ . . . . 39 (SE):   Hb/C2 − 14, 15,
                                                                93/7, 104

−. . . . . $Tm_2O_3 \cdot 3 H_2O$
                see H_3O_3Tm ... $Tm(OH)_3$
−. . . . . $Tm_2O_3 \cdot nH_2O$ . . . . 39 (SE):   Hb/C2 − 75
−. . . . . Tm_2O_3 solid solutions:
                    $Tm_2O_3 - Yb_2O_3$. . . . . 39 (SE):   Hb/C1 − 433/4

−. . . . . Tm_2O_3 systems:
                    $Tm_2O_3 - Dy_2O_3$. . . . . 39 (SE):   Hb/C1 − 425
                    $Tm_2O_3 - Er_2O_3$ . . . . 39 (SE):   Hb/C1 − 425
                    $Tm_2O_3 - Eu_2O_3$. . . . . 39 (SE):   Hb/C1 − 425
                    $Tm_2O_3 - Gd_2O_3$ . . . . 39 (SE):   Hb/C1 − 425
                    $Tm_2O_3 - Ho_2O_3$. . . . . 39 (SE):   Hb/C1 − 425
                    $Tm_2O_3 - La_2O_3$ . . . . 39 (SE):   Hb/C1 − 425

See the "Instructions for the user"
at the beginning of the volume

O_3Tm_2 . . . . Tm_2O_3 systems:

$Tm_2O_3-Lu_2O_3$ . . . . 39 (SE):   Hb/C1−436

$Tm_2O_3-Nd_2O_3$ . . . . 39 (SE):   Hb/C1−425

$TmO_{1.5}-NpO_2$ . . . . Erg. W. 4−46/8

$TmO_{1.5}-PuO_2$ . . . . Erg. W. 4−57

$Tm_2O_3-Sm_2O_3$ . . . . 39 (SE):   Hb/C1−425

O_3U . . . . UO_3 . . . . 55 (U):   Hb−95/9

−. . . . . $UO_3 \cdot 0.5\, H_2O = H_2U_2O_7$ . . . . 55 (U):   Hb−104

−. . . . . $UO_3 \cdot H_2O$ . . . . 55 (U):   Hb−102/4

−. . . . . $UO_3 \cdot 1.5\, H_2O$. . . . . 55 (U):   Hb−101

−. . . . . $UO_3 \cdot 2\, H_2O$ . . . . 55 (U):   Hb−101/2

−. . . . . UO_3 systems:

$UO_3-NiO-SO_3-H_2O$ . . . . 57 (Ni):   Hb/B3−1235

$UO_3-NiO-SO_3-D_2O$ . . . . 57 (Ni):   Hb/B3−1235

O_3V^+ . . . . $[VO_3]^+ = [V(O)\,(O)_2]^+$ . . . . 48 (V):   Hb/A2−588/9, 638/40

                                                               Hb/B1−162/3, 165/6

O_3V^- . . . . $[VO_3]^-$:

Chemical reactions . . . . 48 (V):   Hb/B1−132/45, 149

Detection . . . . 48 (V):   Hb/A2−647/52, 655

Electrochemical behavior . . . . . . 48 (V):   Hb/A2−474/6, 493/4,

                                                                  506/9

Spectra . . . . 48 (V):   Hb/A2−368/9

                                                                 Hb/B1−149

Transition reactions . . . . 48 (V):   Hb/A2−550/640

O_3V^{2-} . . . . $[VO_3]^{2-}$ . . . . 48 (V):   Hb/A2−555/6

O_3VY . . . . YVO_3 . . . . 48 (V):   Hb/B2−551/2

O_3V_2 . . . . V_2O_3:

Chemical reactions . . . . 48 (V):   Hb/B1−45/9

Crystallographic properties . . . . . 48 (V):   Hb/B1−36/7

Determination. . . . . 48 (V):   Hb/A2−714

Electrical properties . . . . 48 (V):   Hb/B1−30, 43/5

Electrochemical behavior . . . . . . 48 (V):   Hb/A1−306

                                                                  Hb/A2−465/8, 498/9

Formation and preparation . . . . . 48 (V):   Hb/B1−32/6

History. . . . . 48 (V):   Hb/A1−7

Magnetic properties . . . . 48 (V):   Hb/B1−16/7, 40/3

Mechanical properties . . . . 48 (V):   Hb/B1−37/9

Spectra . . . . 48 (V):   Hb/A2−376/7, 379/80,

                                                                  387

                                                                  Hb/B1−40

Thermal properties . . . . 48 (V):   Hb/B1−37/40

Toxicity . . . . 48 (V):   Hb/A2−725/8

−. . . . . V_2O_3 systems:

$V_2O_3-Al_2O_3$ . . . . 48 (V):   Hb/B2−543/4

V_2O_3-C-CO. . . . . 48 (V):   Hb/B1−47

$V_2O_3-Cr_2O_3$ . . . . 52 (Cr):   Hb/B−939

$V_2O_3-Ti_2O_3$ . . . . 48 (V):   Hb/B2−558

$V_2O_3-V_2O_5$ . . . . 48 (V):   Hb/B1−13

Hb = Main Series Volume; GD = Gmelin-Durrer;   *Ab = Appended Volume to Main Series;*
Eb = Supplement Volume to Main Series;   *Erg. W. = New Supplement Series Volume*

$O_3V_2^+$	$[V_2O_3]^+$	48 (V):	Hb/B1−84
$O_3V_2^{2+}$	$[(VO)_2O]^{2+}$	48 (V):	Hb/A2−601
O_3W	WO_3:		
	Chemical reactions	54 (W):	Hb−108/10, 123/7
	Colloidal forms	54 (W):	Hb−132/7
	Formation and preparation	54 (W):	Hb−117/21
	Physical properties	54 (W):	Hb−121/2
	Preparation	54 (W):	Hb−19/23
−.	$WO_3 \cdot H_2O = H_2WO_4$	54 (W):	Hb−127/32
−.	$WO_3 \cdot 2 H_2O$	54 (W):	Hb−127/31
−.	$WO_3 \cdot xH_2O$	54 (W):	Hb−127/8
−.	WO_3 systems:		
	$WO_3−Ag_2WO_4$	61 (Ag):	Hb/B4−349
	$WO_3−K_2WO_4$	54 (W):	Hb−246/7
	$WO_3−Li_2WO_4$	54 (W):	Hb−209/10
	$WO_3−NaPO_3$	54 (W):	Hb−241
	$WO_3−Na_2WO_4$	54 (W):	Hb−219/20
	$WO_3−PbO$	54 (W):	Hb−317
O_3W_2	W_2O_3	54 (W):	Hb−111
O_3W_4	W_4O_3	54 (W):	Hb−111
O_3Xe	XeO_3	Erg. W. 1−101/12	
O_3Xe^+	XeO_3^+	Erg. W. 1−101	
O_3Y_2	Y_2O_3:		
	Catalytic properties	39 (SE):	Hb/C1−204
	Chemical reactions	39 (SE):	Hb/C1−157/76, 201/4
	Crystallographic properties	39 (SE):	Hb/C1−104/6, 115/7, 120/3, 192/3
	Diffusion	39 (SE):	Hb/C1−128, 193
	Electrical properties	39 (SE):	Hb/C1−144/50, 195/9
	Electrochemical behavior	39 (SE):	Hb/C1−155
	Formation	39 (SE):	Hb/C1−106/9, 131/3, 189/90
	Magnetic properties	39 (SE):	Hb/C1−142/3, 195
	Mechanical properties	39 (SE):	Hb/C1−127/8, 193/4
	Optical properties	39 (SE):	Hb/C1−150/5, 199/201
	Preparation	39 (SE):	Hb/C1−106/12, 189/92
	− of special forms	39 (SE):	Hb/C1−110/2, 190/2
	Spectra	39 (SE):	Hb/C1−154/5
	Thermal properties	39 (SE):	Hb/C1−129/41, 194/5
−.	Y_2O_{3-x}	39 (SE):	Hb/C1−177/9
−.	$Y_2O_3 \cdot H_2O = YO(OH)$	39 (SE):	Hb/C2−15, 94/7, 99/100
−.	$Y_2O_3 \cdot 2 H_2O$	39 (SE):	Hb/C2−15
−.	$Y_2O_3 \cdot 3 H_2O$ see H_3O_3Y ... $Y(OH)_3$		
−.	$Y_2O_3 \cdot 4.2 H_2O$	39 (SE):	Hb/C2−15
−.	Y_2O_3 solid solutions:		
	$Y_2O_3−CeO_2$	39 (SE):	Hb/C1−263/5

See the "Instructions for the user"
at the beginning of the volume

O_3Y_2 . . . . Y_2O_3 solid solutions:

$Y_2O_3-Dy_2O_3$ . . . . 39 (SE):   Hb/C1−409
$Y_2O_3-Ho_2O_3$ . . . . 39 (SE):   Hb/C1−414
$Y_2O_3-La_2O_3$ . . . . 39 (SE):   Hb/C1−218/9
$Y_2O_3-Nb_2O_5-TiO_2$ . . . . 49 (Nb):   Hb/B3−209
$Y_2O_3-Nd_2O_3$ . . . . 39 (SE):   Hb/C1−307/8
$Y_2O_3-TbO_x$ . . . . 39 (SE):   Hb/C1−400/1
$Y_2O_3-Tb_2O_3$ . . . . 39 (SE):   Hb/C1−400

−. . . . . Y_2O_3 systems:

$Y_2O_3-Al_2O_3-B_2O_3-PbO$ . . . . 47 (Pb):   Hb/C4−1375/6
$Y_2O_3-CeO_2$ . . . . 39 (SE):   Hb/C1−262/4
Y_2O_3-CuO . . . . 60 (Cu):   Hb/B3−1186
$Y_2O_3-Eu_2O_3$ . . . . 39 (SE):   Hb/C1−377
$Y_2O_3-La_2O_3$ . . . . 39 (SE):   Hb/C1−218/9
$Y_2O_3-Nb_2O_5$ . . . . 49 (Nb):   Hb/B3−138/9
$Y_2O_3-Nb_2O_5-SiO_2$ . . . . 49 (Nb):   Hb/B4−359
$Y_2O_3-Nd_2O_3$ . . . . 39 (SE):   Hb/C1−306/8
$Y_2O_3-PrO_x$ . . . . 39 (SE):   Hb/C1−294/5
$Y_2O_3-Pr_2O_3$ . . . . 39 (SE):   Hb/C1−294
$YO_{1.5}-PuO_2$ . . . . Erg. W. 4−57
$Y_2O_3-Sc_2O_3$ . . . . 39 (SE):   Hb/C1−205
Y_2O_3-Ta . . . . 50 (Ta):   Hb/B2−205
$Y_2O_3-Ta_2O_5$ . . . . 50 (Ta):   Hb/B2−205
$Y_2O_3-Ta_2O_5-ZrO_2$ . . . . 50 (Ta):   Hb/B2−229

O_3Yb_2 . . . . Yb_2O_3:

Chemical reactions . . . . 39 (SE):   Hb/C1−156/75, 429/30
Crystallographic properties . . . . . . 39 (SE):   Hb/C1−104/6,
                                                                        115/23, 427
Electrical properties . . . . 39 (SE):   Hb/C1−144/50, 429
Formation . . . . 39 (SE):   Hb/C1−106/10, 131/3,
                                                                        426/7
Magnetic properties . . . . 39 (SE):   Hb/C1−142/3, 428/9
Mechanical properties . . . . 39 (SE):   Hb/C1−127/8, 428
Optical properties . . . . 39 (SE):   Hb/C1−150/5, 429
Preparation . . . . 39 (SE):   Hb/C1−106/12, 426/7
− of special forms . . . . 39 (SE):   Hb/C1−110/2
Spectra . . . . 39 (SE):   Hb/C1−154/5, 429
Thermal properties . . . . 39 (SE):   Hb/C1−129/41, 428

−. . . . . $Yb_2O_3 \cdot H_2O = YbO(OH)$ . . . . 39 (SE):   Hb/C2−14, 93/7, 104/5
−. . . . . $Yb_2O_3 \cdot nH_2O$ . . . . 39 (SE):   Hb/C2−76
−. . . . . Yb_2O_3 solid solutions:

$Yb_2O_3-Dy_2O_3$ . . . . 39 (SE):   Hb/C1−433/4
$Yb_2O_3-Er_2O_3$ . . . . 39 (SE):   Hb/C1−433/4
$Yb_2O_3-Ho_2O_3$ . . . . 39 (SE):   Hb/C1−433/4
$Yb_2O_3-Tm_2O_3$ . . . . 39 (SE):   Hb/C1−433/4

−. . . . . Yb_2O_3 systems:

$Yb_2O_3-CeO_2$ . . . . 39 (SE):   Hb/C1−434
$Yb_2O_3-Eu_2O_3$ . . . . 39 (SE):   Hb/C1−433

Hb = Main Series Volume; GD = Gmelin-Durrer;   *Ab = Appended Volume to Main Series;*
Eb = Supplement Volume to Main Series;   *Erg. W. = New Supplement Series Volume*

O_3Yb_2 . . . . Yb_2O_3 systems:
$\qquad$ Yb_2O_3–Gd_2O_3 . . . . 39 (SE): Hb/C1–433
$\qquad$ Yb_2O_3–La_2O_3 . . . . 39 (SE): Hb/C1–430/2
$\qquad$ Yb_2O_3–Lu_2O_3 . . . . 39 (SE): Hb/C1–436
$\qquad$ Yb_2O_3–Nd_2O_3 . . . . 39 (SE): Hb/C1–433
$\qquad$ $YbO_{1.5}$–NpO_2 . . . . Erg. W. 4–46/8
$\qquad$ Yb_2O_3–Pr_2O_3 . . . . 39 (SE): Hb/C1–433
$\qquad$ Yb_2O_3–Sc_2O_3 . . . . 39 (SE): Hb/C1–430
$\qquad$ Yb_2O_3–Sm_2O_3 . . . . 39 (SE): Hb/C1–433
O_3ZnZr . . . . $ZnZrO_3$ . . . . 42 (Zr): Hb–442
–. . . . . $ZnZrO_3$ solid solutions:
$\qquad$ $ZnZrO_3$–$PbZrO_3$ . . . . 47 (Pb): Hb/C4–1438/9
O_3Zn_2 . . . . Zn_2O_3 . . . . 32 (Zn): Hb–145
–. . . . . $Zn_2O_3 \cdot H_2O$ . . . . 32 (Zn): Eb–817
O_3Zr . . . . $ZrO_3 \cdot xH_2O$ . . . . 42 (Zr): Hb–261/3
O_3Zr_2 . . . . Zr_2O_3 . . . . 42 (Zr): Hb–215
$O_{3.5}S_{0.5}Zr$ . . . . . . $ZrO_2 \cdot 0.5\,SO_3$ . . . . 42 (Zr): Hb–340
–. . . . . $ZrO_2 \cdot 0.5\,SO_3 \cdot xH_2O$ . . . . 42 (Zr): Hb–351
$O_{3.5}U$ . . . . $UO_{3.5} \cdot 2\,H_2O$ . . . . 55 (U): Hb–105
–. . . . . $UO_{3.5} \cdot 3\,H_2O$ . . . . 55 (U): Hb–105
O_4 . . . . O_4 . . . . 3 (O): Hb/3–352/4, 362,
$\qquad$ 367, 575/8
–. . . . . $(O_2)_2$ . . . . 3 (O): Hb/3–362
O_4^+ . . . . $(O_4)^+$ . . . . 3 (O): Hb/4–1071
O_4^- . . . . O_4^- . . . . 3 (O): Hb/3–354
O_4^{2-} . . . . O_4^{2-} . . . . 3 (O): Hb/8–2627
O_4Os . . . . OsO_4:
$\qquad$ Chemical reactions . . . . 66 (Os): Hb–41/2
$\qquad$ Formation . . . . 66 (Os): Hb–31/33, 36
$\qquad$ Physical properties . . . . 66 (Os): Hb–36/41
$\qquad$ Physiological action . . . . 66 (Os): Hb–42
$\qquad$ Preparation . . . . 66 (Os): Hb–33/6
$\qquad$ Solubility . . . . 66 (Os): Hb–43
–. . . . . $(OsO_4)_4$ . . . . 66 (Os): Hb–48
–. . . . . OsO_4 solutions:
$\qquad$ OsO_4–CCl_4 . . . . 66 (Os): Hb–48
$\qquad$ OsO_4–H_2O . . . . 66 (Os): Hb–43/8
$\qquad$ OsO_4–H_2SO_4 . . . . 66 (Os): Hb–48
$\qquad$ OsO_4–organic solvents. . . . . 66 (Os): Hb–48/9
$\qquad$ OsO_4–$POCl_3$ . . . . 66 (Os): Hb–48
O_4P^{3-} . . . . $[PO_4]^{3-}$ . . . . 16 (P): Hb/B–343/8, 364/77
$\qquad$ Hb/C–170/2

$\qquad$ *see also H_3O_4P ... H_3PO_4 solutions*

$O_4PPb_{0.5}Th_{0.5}$ . . . . . $Pb_{0.5}Th_{0.5}PO_4$ . . . . 47 (Pb): Hb/C4–1471
O_4PPu . . . . $PuPO_4$ . . . . Erg. W. 4–252
–. . . . . $PuPO_4 \cdot 0.5\,H_2O$ . . . . Erg. W. 4–251/2
O_4PRb_3 . . . . $Rb_3PO_4 \cdot 4\,H_2O$ . . . . 24 (Rb): Hb–233

*See the "Instructions for the user"
at the beginning of the volume*

O_4PRh.	$RhPO_4$	64 (Rh):	Hb–72
O_4PSb	$5\,SbPO_4 \cdot 2\,H_2O$	18 (Sb):	Hb/B3–561
–.	$SbPO_4 \cdot 2\,H_2O$	18 (Sb):	Hb/B3–560/1
O_4PTl	$TlPO_4 \cdot 2\,H_2O$.	38 (Tl):	Hb/3–430
O_4PTl_3	Tl_3PO_4	38 (Tl):	Hb/3–427/8
O_4P_2.	P_2O_4	16 (P):	Hb/C–75/7
–.	P_2O_4 systems:		
	P_2O_4–Na_2O–H_2O	21 (Na):	Hb–890
$O_4P_2Pb_3S_4$	$Pb_3(PO_2S_2)_2 \cdot 4\,H_2O$	47 (Pb):	Hb/C3–884
$O_4P_2S_4Zn_3$	$Zn_3P_2S_4O_4 \cdot H_2O$	32 (Zn):	Eb–997
$O_4P_4S_3$	$P_4S_3O_4$	16 (P):	Hb/C–582
$O_4P_{10}Zn_5$	$Zn_5P_{10}O_4$	32 (Zn):	Hb–276
$O_4Pb_{0.5}Th_{0.5}V$. . . .	$Th_{0.5}Pb_{0.5}VO_4$	48 (V):	Hb/B2–574
O_4Pb^{4-}	$[PbO_4]^{4-}$	47 (Pb):	Hb/C1–198/9
O_4PbS.	$PbSO_4$:		
	Chemical reactions	47 (Pb):	Hb/C2–560/9
	Colloidal forms	47 (Pb):	Hb/C2–546
	Electrochemical behavior	47 (Pb):	Hb/C2–554/60
	Formation	47 (Pb):	Hb/C2–537/42, 544/5
	History.	47 (Pb):	Hb/A1–140/3
	Physical properties	47 (Pb):	Hb/C2–546/54
	Preparation	47 (Pb):	Hb/C2–537/46
	–, industrial	47 (Pb):	Hb/C2–542/4
	Solubility	47 (Pb):	Hb/C2–569/80
			Hb/C3–1136
	Special forms	47 (Pb):	Hb/C2–545/6
–.	$PbSO_4$ solid solutions:		
	$PbSO_4$–$BaSO_4$	30 (Ba):	Hb–267
			Eb–424
		47 (Pb):	Hb/C3–1206
	$PbSO_4$–$RaSO_4$	47 (Pb):	Hb/C3–1212
	$PbSO_4$–$SrSO_4$.	47 (Pb):	Hb/C3–1198
–.	$PbSO_4$ solutions:		
	$PbSO_4$–acids	47 (Pb):	Hb/B2–283/4
			Hb/C2–570/3
	$PbSO_4$–H_2O	47 (Pb):	Hb/C2–569/70
	$PbSO_4$–organic solvents	47 (Pb):	Hb/C2–578/80
	$PbSO_4$–salt solutions.	47 (Pb):	Hb/C2–573
			Hb/C3–1136
–.	$PbSO_4$ systems:		
	$PbSO_4$–$AgCl$–Ag_2SO_4–$PbCl_2$	61 (Ag):	Hb/B4–267
	$PbSO_4$–$Al(NO_3)_3$–H_2O	47 (Pb):	Hb/C4–1295
	$PbSO_4$–$BaSO_4$–H_2O	47 (Pb):	Hb/C3–1207
	$PbSO_4$–$Ca(NO_3)_2$–H_2O	47 (Pb):	Hb/C3–1190
	$PbSO_4$–$CsCl$–Cs_2SO_4–$PbCl_2$.	47 (Pb):	Hb/C3–1156
	$PbSO_4$–Cs_2SO_4	47 (Pb):	Hb/C3–1155
	$PbSO_4$–Cs_2SO_4–K_2SO_4.	47 (Pb):	Hb/C3–1158
	$PbSO_4$–Cs_2SO_4–Li_2SO_4	47 (Pb):	Hb/C3–1157

Hb = Main Series Volume; GD = Gmelin-Durrer;   *Ab = Appended Volume to Main Series;*
Eb = Supplement Volume to Main Series;   *Erg. W. = New Supplement Series Volume*

O$_4$PbS . . . . PbSO$_4$ systems:

PbSO$_4$–Cs$_2$SO$_4$–Rb$_2$SO$_4$ . . . . 47 (Pb):   Hb/C3–1158

PbSO$_4$–H$_2$SO$_4$–H$_2$O . . . . 47 (Pb):   Hb/C2–572

PbSO$_4$–KCl–K$_2$SO$_4$–PbCl$_2$ . . . . . . 47 (Pb):   Hb/C3–1096/7

PbSO$_4$–KF–K$_2$SO$_4$–PbF$_2$ . . . . 47 (Pb):   Hb/C3–1096

PbSO$_4$–K$_2$SO$_4$ . . . . 47 (Pb):   Hb/C3–1093/4

PbSO$_4$–K$_2$SO$_4$–H$_2$O . . . . 47 (Pb):   Hb/C3–1094/5

PbSO$_4$–K$_2$SO$_4$–Li$_2$SO$_4$ . . . . 47 (Pb):   Hb/C3–1115

PbSO$_4$–K$_2$SO$_4$–Na$_2$SO$_4$ . . . . 47 (Pb):   Hb/C3–1119

PbSO$_4$–K$_2$SO$_4$–Rb$_2$SO$_4$ . . . . 47 (Pb):   Hb/C3–1146

PbSO$_4$–K$_2$SO$_4$–Tl$_2$SO$_4$ . . . . 47 (Pb):   Hb/C4–1357

PbSO$_4$–LiCl–Li$_2$SO$_4$–PbCl$_2$ . . . . . . 47 (Pb):   Hb/C3–992

PbSO$_4$–LiF–Li$_2$SO$_4$–PbF$_2$ . . . . . . 47 (Pb):   Hb/C3–992

PbSO$_4$–Li$_2$SO$_4$ . . . . 47 (Pb):   Hb/C3–992

PbSO$_4$–Li$_2$SO$_4$–H$_2$O . . . . 47 (Pb):   Hb/C3–992

PbSO$_4$–Li$_2$SO$_4$–Mg(NO$_3$)$_2$–H$_2$O . . . . 47 (Pb):   Hb/C3–1178

PbSO$_4$–Li$_2$SO$_4$–Na$_2$SO$_4$ . . . . 47 (Pb):   Hb/C3–1051

PbSO$_4$–Li$_2$SO$_4$–Rb$_2$SO$_4$ . . . . 47 (Pb):   Hb/C3–1145

PbSO$_4$–Li$_2$SO$_4$–Tl$_2$SO$_4$ . . . . 47 (Pb):   Hb/C4–1355

PbSO$_4$–NaCl–H$_2$O . . . . 47 (Pb):   Hb/C3–1033

PbSO$_4$–NaCl–Na$_2$SO$_4$–PbCl$_2$ . . . . . 47 (Pb):   Hb/C3–1033

PbSO$_4$–NaF–Na$_2$SO$_4$–PbF$_2$ . . . . . 47 (Pb):   Hb/C3–1032

PbSO$_4$–Na$_2$SO$_4$ . . . . 47 (Pb):   Hb/C3–1030

PbSO$_4$–Na$_2$SO$_4$–H$_2$O . . . . 47 (Pb):   Hb/C3–1030/1

PbSO$_4$–Na$_2$SO$_4$–Tl$_2$SO$_4$ . . . . 47 (Pb):   Hb/C4–1356

PbSO$_4$–PbCl$_2$ . . . . 47 (Pb):   Hb/C2–606

PbSO$_4$–PbCrO$_4$ . . . . 52 (Cr):   Hb/B–933/4

PbSO$_4$–PbF$_2$ . . . . 47 (Pb):   Hb/C2–605

PbSO$_4$–PbMoO$_4$ . . . . 53 (Mo):   Hb–310/1

PbSO$_4$–PbO . . . . 47 (Pb):   Hb/C2–582/3

PbSO$_4$–PbO–SiO$_2$ . . . . 47 (Pb):   Hb/C3–856

PbSO$_4$–PbWO$_4$ . . . . 54 (W):   Hb–321

PbSO$_4$–Rb$_2$SO$_4$ . . . . 47 (Pb):   Hb/C3–1143

PbSO$_4$–Tl$_2$SO$_4$ . . . . 47 (Pb):   Hb/C4–1353

O$_4$PbS$_2$ . . . . PbS$_2$O$_4$ · xH$_2$O . . . . 47 (Pb):   Hb/C2–535

O$_4$PbSb$_2$ . . . . Pb(SbO$_2$)$_2$ = PbO · Sb$_2$O$_3$ . . . . . . 47 (Pb):   Hb/C3–930

O$_4$PbSe . . . . PbSeO$_4$ . . . . 47 (Pb):   Hb/C2–636/8

O$_4$PbSiZn . . . . . . PbZnSiO$_4$ . . . . 47 (Pb):   Hb/C4–1229

– . . . . PbZnSiO$_4$ solid solutions:

(Pb,Ca)$_4$Zn$_4$Si$_4$O$_{16}$ . . . . 47 (Pb):   Hb/C4–1231

O$_4$PbSr$_2$ . . . . Sr$_2$PbO$_4$ . . . . 47 (Pb):   Hb/C3–1195

O$_4$PbU . . . . PbUO$_4$ . . . . 55 (U):   Hb–239

O$_4$PbW . . . . PbWO$_4$ . . . . 54 (W):   Hb–317/9

– . . . . PbWO$_4$ systems:

PbWO$_4$–Bi$_2$(WO$_4$)$_3$ . . . . 54 (W):   Hb–321

PbWO$_4$–PbCrO$_4$ . . . . 54 (W):   Hb–321

PbWO$_4$–PbMoO$_4$ . . . . 54 (W):   Hb–321

PbWO$_4$–PbSO$_4$ . . . . 54 (W):   Hb–321

See the "Instructions for the user"
at the beginning of the volume

O_4Pb_2S	Pb_2SO_4	47 (Pb):	Hb/C2–537
$O_4Pb_2S^{2+}$	$[Pb_2SO_4]^{2+}$	47 (Pb):	Hb/C2–580
$O_4Pb_2S_2$	$PbS_2O_3 \cdot PbO \cdot 2\ H_2O$	47 (Pb):	Hb/C2–600
–.	$PbS_2O_3 \cdot PbO \cdot xH_2O$	47 (Pb):	Hb/C2–598
O_4Pb_2Se	$PbSeO_3 \cdot PbO$	47 (Pb):	Hb/C2–635
O_4Pb_2Si	$Pb_2SiO_4 = 2\ PbO \cdot SiO_2$	47 (Pb):	Hb/B2–403, 413/4
			Hb/C3–842/7
–.	$2\ PbO \cdot SiO_2 \cdot H_2O$	47 (Pb):	Hb/C3–838
–.	Pb_2SiO_4 solid solutions:		
	$(Pb,Ba)_2SiO_4$	47 (Pb):	Hb/C3–1210
	Pb_2SiO_4–$KPbAsO_4$	47 (Pb):	Hb/C3–1112
	Pb_2SiO_4–$NaPbAsO_4$	47 (Pb):	Hb/C3–1050/1
	Pb_2SiO_4–$NaPbPO_4$	47 (Pb):	Hb/C3–1050
	Pb_2SiO_4–$Pb_2MgSi_2O_7$	47 (Pb):	Hb/C3–1178
O_4Pb_2Sn	Pb_2SnO_4	47 (Pb):	Hb/C4–1581/2
O_4Pb_2Ti	Pb_2TiO_4	47 (Pb):	Hb/C4–1379
O_4Pb_2Zr	Pb_2ZrO_4	47 (Pb):	Hb/C4–1423
O_4Pb_3	Pb_3O_4:		
	Adulterating	47 (Pb):	Hb/C1–127
	Aftertreatment	47 (Pb):	Hb/C1–126
	Chemical reactions	47 (Pb):	Hb/C1–135/9
	Electrochemical behavior	47 (Pb):	Hb/C1–133/5
	Extending	47 (Pb):	Hb/C1–127
	Formation	47 (Pb):	Hb/C1–122/3
	History	47 (Pb):	Hb/A1–111/23
	Performance tests	47 (Pb):	Hb/C1–127
	Physical properties	47 (Pb):	Hb/C1–129/33
			Hb/C4–1745
	Preparation	47 (Pb):	Hb/C1–122/9
	–, industrial	47 (Pb):	Hb/C1–124/9
	Purification	47 (Pb):	Hb/C1–126
	Purity testing	47 (Pb):	Hb/C1–128/9
	Standards	47 (Pb):	Hb/C1–127
–.	Pb_3O_4 systems:		
	Pb_3O_4–KNO_3–$Pb(NO_3)_2$	47 (Pb):	Hb/C3–1064
	Pb_3O_4–PbO	47 (Pb):	Hb/C1–54
	Pb_3O_4–PbO_2	47 (Pb):	Hb/C1–55/6
	Pb_3O_4–Pb_2O_3	47 (Pb):	Hb/C1–54/5
	Pb_3O_4–$Pb_{12}O_{19}$	47 (Pb):	Hb/C1–54/5
O_4PdS	$PdSO_4 \cdot xH_2O$	65 (Pd):	Hb/2–295
O_4PdSe	$PdSeO_4$	65 (Pd):	Hb/2–296/7
O_4PmTa	$PmTaO_4$	50 (Ta):	Hb/B2–202/3, 213
O_4PrTa	$PrTaO_4$	50 (Ta):	Hb/B2–202/3, 211
–.	$PrTaO_4$ solid solutions:		
	$(Pr,Nd)TaO_4$	50 (Ta):	Hb/B2–213
O_4PrV	$PrVO_4$	48 (V):	Hb/B2–553/5
O_4Pr_2	$Pr_2O_4 \cdot nH_2O$	39 (SE):	Hb/C2–106/8, 112

O_4Pt^{2-}	$[PtO_4]^{2-}$	68 (Pt):	Hb/B4–265
$O_4Pt_2Sn_3$	$Pt_2Sn_3O_4$	68 (Pt):	Hb/C3–310
O_4Pt_3	$Pt_3O_4 \cdot 9\ H_2O$	68 (Pt):	Hb/C1–26, 52
O_4Pu	$PuO_4 \cdot nH_2O$	Erg. W. 4–28	
O_4PuRb_2	Rb_2PuO_4	Erg. W. 4–52/3	
O_4PuSi	$PuSiO_4$	Erg. W. 4–56, 70	
O_4PuSr	$SrPuO_4$	Erg. W. 4–53/5	
O_4PuTa	$PuTaO_4$	Erg. W. 4–71	
$O_4Pu_2^{4+}$	$[Pu_2O_4]^{4+}$	Erg. W. 4–28	
$O_4Pu_2S_3U_2$	$U_2Pu_2O_4S_3$	Erg. W. 4–173/4	
$O_4Pu_4S_3$	$Pu_4O_4S_3$	Erg. W. 4–172/3	
O_4RaS	$RaSO_4$	31 (Ra):	Hb–44/5, 59, 62/3
–.	$RaSO_4$ solid solutions:		
	$RaSO_4$–$BaSO_4$	30 (Ba):	Eb–424
	$RaSO_4$–$CsClO_4$	25 (Cs):	Hb/2–169
	$RaSO_4$–$PbSO_4$	47 (Pb):	Hb/C3–1212
	$RaSO_4$–$RbClO_4$	24 (Rb):	Hb–160
O_4RbRe	$RbReO_4$	70 (Re):	Hb–143
O_4RbS^-	$RbSO_4^-$	24 (Rb):	Hb–213
O_4Rb_2	Rb_2O_4	24 (Rb):	Hb–111/2
O_4Rb_2Ru	Rb_2RuO_4	63 (Ru):	Hb–116
O_4Rb_2S	Rb_2SO_4	24 (Rb):	Hb–207/11
–.	Rb_2SO_4 solutions:		
	Rb_2SO_4–H_2O	24 (Rb):	Hb–211/3
	Rb_2SO_4–NH_3–H_2O	24 (Rb):	Hb–213
	Rb_2SO_4–organic solvents	24 (Rb):	Hb–213
–.	Rb_2SO_4 systems:		
	Rb_2SO_4–Ag_2SO_4	61 (Ag):	Hb/B3–106
	Rb_2SO_4–Ag_2SO_4–H_2O	61 (Ag):	Hb/B3–106
	Rb_2SO_4–$CaSO_4$	28 (Ca):	Hb/B3–1335
	Rb_2SO_4–$CaSO_4$–H_2O	28 (Ca):	Hb/B3–1335/6
	Rb_2SO_4–$CdSO_4$–H_2O	33 (Cd):	Eb–702/3
	Rb_2SO_4–$CoCl_2$–$CoSO_4$–$RbCl$	58 (Co):	Eb/A–831
	Rb_2SO_4–$CoSO_4$–H_2O	58 (Co):	Eb/A–829/30
	Rb_2SO_4–Cs_2SO_4–$PbSO_4$	47 (Pb):	Hb/C3–1158
	Rb_2SO_4–KCl–K_2SO_4–$RbCl$	24 (Rb):	Hb–249
	Rb_2SO_4–K_2SO_4	24 (Rb):	Hb–249
	Rb_2SO_4–K_2SO_4–$PbSO_4$	47 (Pb):	Hb/C3–1146
	Rb_2SO_4–Li_2SO_4–$PbSO_4$	47 (Pb):	Hb/C3–1145
	Rb_2SO_4–$NiSO_4$–H_2O	57 (Ni):	Hb/B3–1096
	Rb_2SO_4–$PbSO_4$	47 (Pb):	Hb/C3–1143
	Rb_2SO_4–$RbCl$	24 (Rb):	Hb–213
$O_4Rb_2S_2$	$Rb_2S_2O_4$	24 (Rb):	Hb–206
O_4Rb_2Se	Rb_2SeO_4	24 (Rb):	Hb–216/7
O_4Rb_2Te	$Rb_2TeO_4 \cdot 3\ H_2O = Rb_2H_4[TeO_6] \cdot H_2O$	24 (Rb):	Hb–219
O_4Rb_2U	Rb_2UO_4	55 (U):	Hb–226
O_4Rb_2W	Rb_2WO_4	54 (W):	Hb–282

See the "Instructions for the user"
at the beginning of the volume

O_4Rb_2W . . . . Rb_2WO_4 systems:

$Rb_2WO_4-AgI-Ag_2WO_4-RbI$ . . . . . . 61 (Ag):   Hb/B4−353/4

$Rb_2WO_4-Ag_2WO_4$ . . . . 61 (Ag):   Hb/B4−351/3

O_4Rb_3V . . . . $Rb_3VO_4 = 3 Rb_2O \cdot V_2O_5$ . . . . . 48 (V):   Hb/B2−486

O_4Re^- . . . . $[ReO_4]^-$ . . . . 70 (Re):   Hb−61, 99

O_4ReTl . . . . $TlReO_4$ . . . . 70 (Re):   Hb−147/8

O_4Ru . . . . RuO_4 . . . . 63 (Ru):   Hb−27/31
                                                                    Eb−232/3, 240/7

O_4Ru^+ . . . . $[RuO_4]^+$ . . . . 63 (Ru):   Eb−233

O_4Ru^- . . . . $[RuO_4]^-$ . . . . 63 (Ru):   Hb−91
                                                                    Eb−233, 251/2

O_4Ru^{2-} . . . . $[RuO_4]^{2-}$ . . . . 63 (Ru):   Hb−90
                                                                    Eb−233, 249/51

O_4RuS . . . . $RuSO_4$ . . . . 63 (Ru):   Eb−316

O_4RuSr_2 . . . . Sr_2RuO_4 . . . . 63 (Ru):   Eb−255

O_4S . . . . SO_4 . . . . 9 (S):   Hb/B1−371/2

O_4S^{2-} . . . . $[SO_4]^{2-}$

see $H_2O_4S \ldots H_2SO_4$

$H_2O_4S \ldots H_2SO_4$ solutions

O_4SSb_2 . . . . Sb_2O_4S . . . . 18 (Sb):   Hb/B3−541

O_4SSn . . . . $SnSO_4$ solutions:

$SnSO_4-H_2O$ . . . . 46 (Sn):   Hb/B−306/24, 413

O_4SSn^{2+} . . . . . . $[SnSO_4]^{2+}$ . . . . 46 (Sn):   Hb/B−393

O_4SSr . . . . $SrSO_4$:

Chemical reactions . . . . 29 (Sr):   Hb−176/8
                                                                    Eb−255/6

Electrochemical behavior . . . . . . 29 (Sr):   Eb−254/5

Formation . . . . 29 (Sr):   Hb−172/4
                                                                    Eb−249/50

Physical properties . . . . 29 (Sr):   Hb−174/6
                                                                    Eb−250/4

Preparation . . . . 29 (Sr):   Hb−172/4

−, industrial . . . . 29 (Sr):   Hb−16
                                                                    Eb−87

Solubility . . . . 29 (Sr):   Hb−178/80
                                                                    Eb−256/7

Uses . . . . 29 (Sr):   Eb−87

−. . . . . $SrSO_4 \cdot 2 H_2O$ . . . . 29 (Sr):   Eb−256

−. . . . . $SrSO_4 \cdot xH_2O$ . . . . 29 (Sr):   Hb−178

−. . . . . $SrSO_4$ solid solutions:

$SrSO_4-BaSO_4$ . . . . 30 (Ba):   Hb−267
                                                                    Eb−424, 533/4

$SrSO_4-PbSO_4$ . . . . 47 (Pb):   Hb/C3−1198

−. . . . . $SrSO_4$ solutions:

$SrSO_4-C_2H_5OH-H_2O$ . . . . 29 (Sr):   Eb−257

$SrSO_4-H_2O$ . . . . 29 (Sr):   Hb−180/1
                                                                    Eb−256/7

$SrSO_4-H_2SO_4$ . . . . 29 (Sr):   Hb−181

Hb = Main Series Volume; GD = Gmelin-Durrer;   *Ab = Appended Volume to Main Series;*
Eb = Supplement Volume to Main Series;   *Erg. W. = New Supplement Series Volume*

O$_4$SSr . . . . SrSO$_4$ systems:

                 SrSO$_4$–BaSO$_4$ . . . . 30 (Ba):   Hb–388

                 SrSO$_4$–BaSO$_4$–CaSO$_4$ . . . . 30 (Ba):   Hb–390

                 SrSO$_4$–CaF$_2$ . . . . 29 (Sr):   Eb–304

                 SrSO$_4$–CaSO$_4$ . . . . 29 (Sr):   Hb–237

                 SrSO$_4$–K$_2$SO$_4$ . . . . 29 (Sr):   Hb–229/30

                 SrSO$_4$–K$_2$SO$_4$–H$_2$O . . . . 29 (Sr):   Hb–230/1

                 SrSO$_4$–LiF–Li$_2$SO$_4$–SrF$_2$ . . . . . . 29 (Sr):   Eb–299

                 SrSO$_4$–Li$_2$SO$_4$ . . . . 29 (Sr):   Hb–223

                 SrSO$_4$–(NH$_4$)$_2$SO$_4$–H$_2$O . . . . . . 29 (Sr):   Hb–233

                 SrSO$_4$–NaCl . . . . 29 (Sr):   Hb–224

                 SrSO$_4$–NaCl–Na$_2$CO$_3$ . . . . 29 (Sr):   Hb–225

                 SrSO$_4$–Na$_2$CO$_3$–Na$_2$SO$_4$

                   –SrCO$_3$–H$_2$O . . . . 29 (Sr):   Hb–225

                 SrSO$_4$–Na$_2$SO$_4$ . . . . 29 (Sr):   Hb–224

                 SrSO$_4$–Na$_2$SO$_4$–H$_2$O . . . . 29 (Sr):   Hb–224

                 SrSO$_4$–S–SO$_2$–SrO–SrS . . . . . . 29 (Sr):   Eb–257

                 SrSO$_4$–S–SO$_2$–SrS . . . . 29 (Sr):   Eb–257

O$_4$STl$^-$ . . . . [TlSO$_4$]$^-$ . . . . 38 (Tl):   Hb/3–358

O$_4$STl$_2$ . . . . Tl$_2$SO$_4$ . . . . 38 (Tl):   Hb/1–23, 29

                                              Hb/3–350/60

–. . . . . Tl$_2$SO$_4$ systems:

                 Tl$_2$SO$_4$–AgBr–Ag$_2$SO$_4$–TlBr . . . . . 61 (Ag):   Hb/B4–226/7

                 Tl$_2$SO$_4$–AgCl–Ag$_2$SO$_4$–TlCl . . . . . 61 (Ag):   Hb/B4–226

                 Tl$_2$SO$_4$–AgI–Ag$_2$SO$_4$–TlI . . . . 61 (Ag):   Hb/B4–227

                 Tl$_2$SO$_4$–AgNO$_3$–Ag$_2$SO$_4$–TlNO$_3$. . . 61 (Ag):   Hb/B4–226

                 Tl$_2$SO$_4$–Ag$_2$SO$_4$ . . . . 61 (Ag):   Hb/B4–225

                 Tl$_2$SO$_4$–Am$_2$(SO$_4$)$_3$–H$_2$O . . . . Erg. W. 4–186

                 Tl$_2$SO$_4$–CdSO$_4$–H$_2$O . . . . 38 (Tl):   Hb/3–486

                 Tl$_2$SO$_4$–CoSO$_4$–H$_2$O . . . . 58 (Co):   Eb/A–868

                 Tl$_2$SO$_4$–CuSO$_4$–H$_2$O . . . . 60 (Cu):   Hb/B3–1184

                 Tl$_2$SO$_4$–CuSO$_4$–K$_2$SO$_4$–H$_2$O . . . . . 60 (Cu):   Hb/B3–1186

                 Tl$_2$SO$_4$–CuSO$_4$–(NH$_4$)$_2$SO$_4$–H$_2$O . . . 60 (Cu):   Hb/B3–1186

                 Tl$_2$SO$_4$–FeSO$_4$–H$_2$O . . . . 59 (Fe):   Hb/B5–1123/4

                 Tl$_2$SO$_4$–H$_2$SO$_4$–H$_2$O . . . . 38 (Tl):   Hb/3–360

                 Tl$_2$SO$_4$–H$_2$SO$_4$–Tl$_2$(SO$_4$)$_3$–H$_2$O . . . . 38 (Tl):   Hb/3–362

                 Tl$_2$SO$_2$–HgBr$_2$–HgSO$_4$–TlBr . . . . . . 38 (Tl):   Hb/3–490

                 Tl$_2$SO$_4$–HgCl$_2$–HgSO$_4$–TlCl . . . . . . 38 (Tl):   Hb/3–489/90

                 Tl$_2$SO$_4$–HgI$_2$–HgSO$_4$–TlI . . . . 38 (Tl):   Hb/3–490

                 Tl$_2$SO$_4$–K$_2$SO$_4$–H$_2$O . . . . 38 (Tl):   Hb/3–458

                 Tl$_2$SO$_4$–K$_2$SO$_4$–PbSO$_4$ . . . . 47 (Pb):   Hb/C4–1357

                 Tl$_2$SO$_4$–Li$_2$SO$_4$–PbSO$_4$ . . . . 47 (Pb):   Hb/C4–1355

                 Tl$_2$SO$_4$–MgSO$_4$–H$_2$O . . . . 38 (Tl):   Hb/3–474

                 Tl$_2$SO$_4$–NaBr–Na$_2$SO$_4$–TlBr . . . . . . 38 (Tl):   Hb/3–444

                 Tl$_2$SO$_4$–NaCl–NaNO$_3$–Na$_2$SO$_4$

                   –TlCl–TlNO$_3$ . . . . 38 (Tl):   Hb/3–444

                 Tl$_2$SO$_4$–NaCl–Na$_2$SO$_4$–TlCl . . . . . . 38 (Tl):   Hb/3–444

                 Tl$_2$SO$_4$–Na$_2$SO$_4$–PbSO$_4$ . . . . 47 (Pb):   Hb/C4–1356

                 Tl$_2$SO$_4$–NiSO$_4$–H$_2$O . . . . 57 (Ni):   Hb/B3–1166/7

See the "Instructions for the user"
at the beginning of the volume

O_4STl_2 . . . . Tl_2SO_4 systems:

                  $Tl_2SO_4–PbSO_4$ . . . . 47 (Pb):   Hb/C4–1353

                  $Tl_2SO_4–TlCl$ . . . . 38 (Tl):   Hb/3–360

                  $Tl_2SO_4–ZnSO_4–H_2O$ . . . . 38 (Tl):   Hb/3–480

O_4SU . . . . $UOSO_3 \cdot 2 H_2O = U(OH)_2SO_3 \cdot H_2O$ . . . 55 (U):   Hb–145

O_4SV . . . . VSO_4

                 see also V^{2+} ... V^{2+}

–. . . . . $VSO_4 \cdot 6 H_2O$ . . . . 48 (V):   Hb/B1–281/2

–. . . . . $VSO_4 \cdot 7 H_2O = [V(H_2O)_6]SO_4 \cdot H_2O$ . . . 48 (V):   Hb/A2–337, 359/60

                                                     Hb/B1–281/2

–. . . . . VSO_4 solid solutions:

                  $(V,Fe)SO_4 \cdot 7 H_2O$ . . . . 59 (Fe):   Hb/B5–1138

                  $(V,Mg)SO_4 \cdot 7 H_2O$. . . . . 48 (V):   Hb/B2–508

–. . . . . VSO_4 solutions:

                  $VSO_4–H_2O$. . . . . 48 (V):   Hb/A2–502/3, 511, 553

                                                     Hb/B1–282/4

O_4SZn. . . . . $ZnSO_4$ . . . . 32 (Zn):   Hb–220/2

                                                     Eb–936/41

–. . . . . $ZnSO_4 \cdot H_2O$ . . . . 32 (Zn):   Hb–227

                                                     Eb–946/7

–. . . . . $ZnSO_4 \cdot 2 H_2O$ . . . . 32 (Zn):   Hb–227

                                                     Eb–946

–. . . . . $ZnSO_4 \cdot 3 H_2O$ . . . . 32 (Zn):   Hb–227

                                                     Eb–946

–. . . . . $ZnSO_4 \cdot 4 H_2O$ . . . . 32 (Zn):   Eb–946

–. . . . . $ZnSO_4 \cdot 5 H_2O$ . . . . 32 (Zn):   Hb–227

–. . . . . $ZnSO_4 \cdot 6 H_2O$ . . . . 32 (Zn):   Hb–226/7

                                                     Eb–946

–. . . . . $ZnSO_4 \cdot 7 H_2O = [Zn(H_2O)_6] [SO_4 \cdot H_2O]$

                  $= Zn(OH)HSO_4 \cdot 6 H_2O$ . . . . 32 (Zn):   Hb–222/6

                                                    Eb–944/6

–. . . . . $ZnSO_4 \cdot 13 H_2O$. . . . . 32 (Zn):   Hb–222

–. . . . . $ZnSO_4 \cdot 20 H_2O$ . . . . 32 (Zn):   Hb–222

–. . . . . $ZnSO_4$ solid solutions:

                  $(Zn,Cu)SO_4 \cdot H_2O$ . . . . 60 (Cu):   Hb/B3–1149

                  $(Zn,Cu)SO_4 \cdot 5 H_2O$ . . . . 60 (Cu):   Hb/B3–1148/9

                  $(Zn,Cu)SO_4 \cdot 7 H_2O$ . . . . 60 (Cu):   Hb/B3–1148/9

                  $(Zn,Cu,Fe)SO_4$. . . . . 60 (Cu):   Hb/B3–1326

                  $ZnSO_4 \cdot H_2O–FeSO_4 \cdot H_2O$ . . . . 59 (Fe):   Hb/B5–1102

                  $ZnSO_4 \cdot 7 H_2O–FeSO_4 \cdot 7 H_2O$ . . . . . 59 (Fe):   Hb/B5–1102

                  $(Zn,Mg)SO_4 \cdot 7 H_2O$ . . . . 32 (Zn):   Hb–324

                                                    Eb–1024

                  $(Ni,Zn,Mg)SO_4 \cdot 7 H_2O$ . . . . 57 (Ni):   Hb/B3–1140

                  $ZnSO_4–(NH_4)_2PO_3F$. . . . . 32 (Zn):   Eb–1020

–. . . . . $ZnSO_4$ solutions:

                  $ZnSO_4–HF$. . . . . 32 (Zn):   Eb–958

                  $ZnSO_4–H_2O$ . . . . 32 (Zn):   Hb–228/39

                                                    Eb–231/3, 947/58

Hb = Main Series Volume; GD = Gmelin-Durrer;   *Ab = Appended Volume to Main Series;*
Eb = Supplement Volume to Main Series;   *Erg. W. = New Supplement Series Volume*

O_4SZn . . . . ZnSO$_4$ solutions:
        ZnSO$_4$–H$_2$SO$_4$ . . . . 32 (Zn): Eb–958
        ZnSO$_4$–organic solvents . . . . 32 (Zn): Hb–239/40
        Eb–958/9
        ZnSO$_4$–salt solutions. . . . . 32 (Zn): Eb–958

−. . . . . ZnSO$_4$ systems:
        ZnSO$_4$–AgNO$_3$ . . . . 61 (Ag): Hb/B4–141
        ZnSO$_4$–BeSO$_4$–H$_2$O . . . . 32 (Zn): Eb–1022/3
        ZnSO$_4$–(CH$_2$OH)$_2$CHOH–H$_2$O . . . . . 32 (Zn): Eb–959
        ZnSO$_4$–(CH$_3$)$_3$COH–H$_2$O . . . . 32 (Zn): Eb–958
        ZnSO$_4$–CuSO$_4$–FeSO$_4$–H$_2$SO$_4$–H$_2$O . 60 (Cu): Hb/B3–1326
        ZnSO$_4$–CuSO$_4$–H$_2$O . . . . 60 (Cu): Hb/B3–1148/50
        ZnSO$_4$–H$_2$O . . . . 32 (Zn): Hb–225/6
        Eb–941/3
        ZnSO$_4$–H$_2$SO$_4$–H$_2$O . . . . 32 (Zn): Eb–962/3
        ZnSO$_4$–KCl–K$_2$SO$_4$–NaCl
          –Na$_2$SO$_4$–ZnCl$_2$ . . . . 32 (Zn): Eb–1016
        ZnSO$_4$–K$_2$SO$_4$–H$_2$O . . . . 32 (Zn): Eb–1011/2
        ZnSO$_4$–MgSO$_4$–H$_2$O . . . . 32 (Zn): Eb–1024
        ZnSO$_4$–NH$_3$–H$_2$O. . . . . 32 (Zn): Eb–959/60
        ZnSO$_4$–(NH$_4$)$_2$SO$_4$–H$_2$O . . . . . . 32 (Zn): Eb–1018/9
        ZnSO$_4$–Na$_2$SO$_4$–H$_2$O. . . . . 32 (Zn): Hb–297/8
        Eb–1006/7
        ZnSO$_4$–NiSO$_4$. . . . . 57 (Ni): Hb/B3–1136
        ZnSO$_4$–Tl$_2$SO$_4$–H$_2$O . . . . 38 (Tl): Hb/3–480
        ZnSO$_4$–ZnO–H$_2$O. . . . . 32 (Zn): Eb–963/4
O_4SZn_2 . . . . ZnSO$_3$ · ZnO · 2 H$_2$O . . . . 32 (Zn): Hb–220
$O_4S_2{}^{2-}$ . . . . [S$_2$O$_4$]$^{2-}$ . . . . 9 (S): Hb/B2–390/1
O_4S_2Sr . . . . . . SrS$_2$O$_4$ . . . . 29 (Sr): Hb–171
−. . . . . SrS$_2$O$_4$ · 4.5 H$_2$O . . . . 29 (Sr): Eb–249
$O_4S_2Th_3$ . . . . Th$_3$O$_4$S$_2$ . . . . 44 (Th): Hb–280
O_4S_2Zn . . . . . . ZnS$_2$O$_4$ . . . . 32 (Zn): Hb–242/3
        Eb–934/5
O_4Sb^{3-} . . . . [SbO$_4$]$^{3-}$ . . . . 18 (Sb): Hb/B2–388/9
O_4SbTa . . . . SbTaO$_4$. . . . . 50 (Ta): Hb/B1–273/4
−. . . . . SbTaO$_4$ systems:
        SbTaO$_4$–SbNbO$_4$ . . . . 50 (Ta): Hb/B2–248
O_4SbV . . . . SbVO$_4$ . . . . 48 (V): Hb/B1–367
O_4Sb_2 . . . . Sb$_2$O$_4$. . . . . 18 (Sb): Hb/A1–23/4, 29/30
        Hb/B2–371/5
−. . . . . Sb$_2$O$_4$ · H$_2$O = H$_2$Sb$_2$O$_5$ . . . . 18 (Sb): Hb/B2–375
−. . . . . Sb$_2$O$_4$ · xH$_2$O . . . . 18 (Sb): Hb/B2–375
−. . . . . Sb$_2$O$_4$ systems:
        Sb$_2$O$_4$–PbO . . . . 47 (Pb): Hb/C3–928/9
O_4Sb_2Zn . . . . . . ZnSb$_2$O$_4$ . . . . 32 (Zn): Eb–999
O_4ScTa . . . . ScTaO$_4$. . . . . 50 (Ta): Hb/B2–202/3, 205
O_4ScV. . . . . ScVO$_4$ . . . . 48 (V): Hb/B2–553/5
O_4Sc_2 . . . . Sc$_2$O$_4$ · nH$_2$O . . . . 39 (SE): Hb/C2–106

See the "Instructions for the user"
at the beginning of the volume

O_4Sc_3 . . . . Sc_3O_4 . . . . 39 (SE):   Hb/C1−181

O_4Se^{2-} . . . . $[SeO_4]^{2-}$

                      see H_2O_4Se ... H_2SeO_4

O_4SeSr . . . . $SrSeO_4$ . . . . 29 (Sr):   Hb−186/7
                                                        Eb−261

O_4SeTi . . . . $TiO_2 \cdot SeO_2 \cdot H_2O$ . . . . 41 (Ti):   Hb−353

O_4SeTl_2 . . . . Tl_2SeO_4 . . . . 38 (Tl):   Hb/3−373/5

O_4SeV . . . . $VOSeO_3 \cdot 2 H_2O$ . . . . 48 (V):   Hb/B1−313

−. . . . . $VOSeO_3 \cdot 3 H_2O$ . . . . 48 (V):   Hb/B1−313

O_4SeZn . . . . $ZnSeO_4$ . . . . 32 (Zn):   Eb−968

−. . . . . $ZnSeO_4 \cdot H_2O$ . . . . 32 (Zn):   Eb−968

−. . . . . $ZnSeO_4 \cdot 2 H_2O$ . . . . 32 (Zn):   Hb−247

−. . . . . $ZnSeO_4 \cdot 5 H_2O$ . . . . 32 (Zn):   Hb−247
                                                        Eb−968

−. . . . . $ZnSeO_4 \cdot 6 H_2O$ . . . . 32 (Zn):   Hb−247
                                                        Eb−968

−. . . . . $ZnSeO_4 \cdot 7 H_2O$ . . . . 32 (Zn):   Hb−247

−. . . . . $ZnSeO_4$ systems:
                              $ZnSeO_4–H_2O$ . . . . 32 (Zn):   Eb−968

O_4SeZr . . . . $ZrO \cdot SeO_3 \cdot 2 H_2O$
                              = $ZrO_2 \cdot SeO_2 \cdot 2 H_2O$ . . . . 42 (Zr):   Hb−355

O_4Se_3 . . . . Se_3O_4 . . . . 10 (Se):   Hb/B−25

O_4Si . . . . $Si(O_2)_2$ . . . . 15 (Si):   Hb/B−582

O_4Si^{4-} . . . . $[SiO_4]^{4-}$ . . . . 15 (Si):   Hb/B−421/2

O_4SiSr_2 . . . . $2 SrO \cdot SiO_2 = Sr_2SiO_4$ . . . . 29 (Sr):   Hb−211
                                                         Eb−281/2

−. . . . . $2 SrO \cdot SiO_2$ systems:
                              $2 SrO \cdot SiO_2–2 BaO \cdot SiO_2$ . . . . . . 30 (Ba):   Eb−535

O_4SiTh . . . . $ThSiO_4$ . . . . 44 (Th):   Hb−313

−. . . . . $ThSiO_4 \cdot xH_2O$ . . . . 44 (Th):   Hb−313

O_4SiTl_4 . . . . Tl_4SiO_4 . . . . 38 (Tl):   Hb/3−424

O_4SiZn_2 . . . . Zn_2SiO_4 . . . . 32 (Zn):   Hb−272/3
                                                        Eb−989/92

−. . . . . $Zn_2SiO_4 \cdot H_2O$ . . . . 32 (Zn):   Hb−273
                                                        Eb−992/3

−. . . . . Zn_2SiO_4 systems:
                              $Zn_2SiO_4–Ni_2SiO_4$ . . . . 57 (Ni):   Hb/B3−1138
                              $Zn_2SiO_4–Zn_2GeO_4$ . . . . 45 (Ge):   Eb−574

O_4SiZr . . . . $ZrSiO_4$ . . . . 42 (Zr):   Hb−81/3, 382/9

−. . . . . $ZrSiO_4$ solid solutions:
                              $ZrSiO_4–NbSiO_4$ . . . . 49 (Nb):   Hb/B4−360
                              $(Ta,Zr)SiO_4$ . . . . 50 (Ta):   Hb/B2−349
                              $ZrSiO_4–V_2O_4$ . . . . 48 (V):   Hb/B2−563

−. . . . . $ZrSiO_4$ systems:
                              $ZrSiO_4–Li_4SiO_4$ . . . . 42 (Zr):   Hb−401

O_4Si_3 . . . . Si_3O_4 . . . . 15 (Si):   Hb/B−257

O_4SmTa	$SmTaO_4$	50 (Ta):	Hb/B2–202/3, 214
O_4SmV	$SmVO_4$	48 (V):	Hb/B2–553/5
O_4Sm_2	$Sm_2O_4 \cdot nH_2O$	39 (SE):	Hb/C2–106/7
O_4SnW	$SnWO_4 \cdot 6 H_2O$	54 (W):	Hb–316
O_4SnZn_2	Zn_2SnO_4 solid solutions:		
	$Ni_xZn_{2-x}SnO_4$	57 (Ni):	Hb/B3–1193
–.	Zn_2SnO_4 systems:		
	$Zn_2SnO_4–Co_2SnO_4$	58 (Co):	Eb/A–872
O_4Sn_3	Sn_3O_4	46 (Sn):	Hb/C1–58/9
O_4Sn_4	Sn_4O_4	46 (Sn):	Hb/C1–34/5, 104
$O_4Sr_{0.5}Th_{0.5}V$	$Sr_{0.5}Th_{0.5}VO_4$	48 (V):	Hb/B2–566
O_4SrTe	$SrTeO_4$	29 (Sr):	Hb–187
			Eb–262
O_4SrU	$SrUO_4$	29 (Sr):	Hb–74
		55 (U):	Hb–232
O_4SrV_2	SrV_2O_4	48 (V):	Hb/B2–518
O_4SrW	$SrWO_4$	54 (W):	Hb–295/6
O_4Sr_2Ti	Sr_2TiO_4	41 (Ti):	Hb–433
O_4Sr_2V	Sr_2VO_4	48 (V):	Hb/B2–519
O_4Sr_2Zr	Sr_2ZrO_4	42 (Zr):	Hb–439
O_4TaTb	$TbTaO_4$	50 (Ta):	Hb/B2–202/3, 217
O_4TaTi	$TiTaO_4$	50 (Ta):	Hb/B2–224
O_4TaTl	$TlTaO_4$	50 (Ta):	Hb/B2–201
O_4TaTm	$TmTaO_4$	50 (Ta):	Hb/B2–202/3, 220
O_4TaV	$VTaO_4$	50 (Ta):	Hb/B2–245
–.	$VTaO_4$ solid solutions:		
	$VTaO_4–VO_2$	50 (Ta):	Hb/B2–244
O_4TaY	$YTaO_4$	50 (Ta):	Hb/B2–202/3, 206
–.	$YTaO_4$ solid solutions:		
	$Y(Ta,Nb)O_4$	50 (Ta):	Hb/B2–264
–.	$YTaO_4$ systems:		
	$YTaO_4–B_2O_3–BaO$	50 (Ta):	Hb/B2–187, 188
	$YTaO_4–BaO$	50 (Ta):	Hb/B2–187
O_4TaYb	$YbTaO_4$	50 (Ta):	Hb/B2–202/3, 220
$O_4Ta_2^{2+}$	$[Ta_2O_4]^{2+}$	50 (Ta):	Hb/B1–178
O_4TbV	$TbVO_4$	48 (V):	Hb/B2–553/5
O_4Te^{2-}	$[TeO_4]^{2-}$	11 (Te):	Hb–292, 301
O_4TeTl_2	Tl_2TeO_4	38 (Tl):	Hb/3–377
–.	$Tl_2TeO_4 \cdot 0.5 H_2O$	38 (Tl):	Hb/3–377
O_4TeZn	$ZnTeO_4$	32 (Zn):	Eb–934, 970
O_4TeZr	$ZrO_2 \cdot TeO_2 \cdot H_2O$		
	$= Zr(TeO_3)_2 \cdot Zr(OH)_4$	42 (Zr):	Hb–355/6
O_4ThZr	$ZrO_2 \cdot ThO_2$	44 (Th):	Hb–354
O_4TiV	$TiVO_4 = Ti_2O_3 \cdot V_2O_5$	48 (V):	Hb/B2–559/60
O_4TiZn_2	Zn_2TiO_4	41 (Ti):	Hb–460/1
–.	Zn_2TiO_4 solid solutions:		
	$Zn_2TiO_4–LiZnNbO_4$	49 (Nb):	Hb/B3–204

See the "Instructions for the user"
at the beginning of the volume

O₄TiZn₂ . . . . Zn₂TiO₄ systems:

O_4TiZn_2 . . . . Zn_2TiO_4 systems:
  $Zn_2TiO_4-Co_2TiO_4$ . . . . 58 (Co):   Eb/A-870
O_4TiZr . . . . $ZrTiO_4$ solid solutions:
  $ZrTiO_4-PbNb_2O_6$ . . . . 49 (Nb):   Hb/B3-308
  $ZrTiO_4-Zr_6Ta_2O_{17}$ . . . . 50 (Ta):   Hb/B2-229
O_4Ti_3 . . . . Ti_3O_4 . . . . 41 (Ti):   Hb-211
O_4TlV . . . . $TlVO_4 = Tl_2O_3 \cdot V_2O_5$ . . . . 48 (V):   Hb/B2-548
O_4Tl_2 . . . . $Tl_2O_4 = TlO_2$ . . . . 38 (Tl):   Hb/2-233
O_4Tl_2W . . . . Tl_2WO_4 . . . . 54 (W):   Hb-314
O_4Tl_3V . . . . $Tl_3VO_4 = 3\ Tl_2O \cdot V_2O_5$ . . . . 48 (V):   Hb/B2-547
O_4Tl_4 . . . . $Tl_4O_4 = Tl_2O \cdot Tl_2O_3$ . . . . 38 (Tl):   Hb/2-221
O_4TmV . . . . $TmVO_4$ . . . . 48 (V):   Hb/B2-553/5
O_4U . . . . $UO_4 \cdot 2\ H_2O$. . . . . . . 55 (U):   Hb-105/7
−. . . . . $UO_4 \cdot 3\ H_2O$ . . . . 55 (U):   Hb-106
−. . . . . $UO_4 \cdot 4.5\ H_2O$. . . . . . . 55 (U):   Hb-105
O_4UV . . . . $VO_2 \cdot UO_2$ . . . . 48 (V):   Hb/B1-91
O_4UZn . . . . $ZnUO_4$ . . . . 55 (U):   Hb-236
O_4V^- . . . . $[VO_4]^-$ . . . . 48 (V):   Hb/A2-469, 639
  Hb/B1-165
O_4V^{3-} . . . . $[VO_4]^{3-}$ . . . . 48 (V):   Hb/A2-391, 468, 478,
  498, 550/640
  Hb/B1-132/47
−. . . . . $[VO_4(H_2O)_6]^{3-}$ . . . . 48 (V):   Hb/B1-146
O_4VY . . . . YVO_4 . . . . 48 (V):   Hb/A2-343
  Hb/B2-553/5
O_4VYb . . . . $YbVO_4$ . . . . 48 (V):   Hb/B2-553/5
O_4VZn_2 . . . . . . Zn_2VO_4 . . . . 48 (V):   Hb/B2-533
O_4V_2 . . . . V_2O_4
  see $O_2V \dots VO_2$
O_4V_2Zn . . . . . . ZnV_2O_4 . . . . 48 (V):   Hb/B2-532/3
−. . . . . ZnV_2O_4 solid solutions:
  $(Mg,Zn)\ (V,Al)_2O_4$ . . . . 48 (V):   Hb/B2-547
O_4W^{2-} . . . . $[WO_4]^{2-}$ . . . . 54 (W):   Hb-144
O_4WZn . . . . . . $ZnWO_4$ . . . . 54 (W):   Hb-304/6
−. . . . . $ZnWO_4 \cdot H_2O$ . . . . 54 (W):   Hb-306
O_4Xe . . . . XeO_4 . . . . . . Erg. W. 1-112/3
O_4Xe^+ . . . . $[XeO_4]^+$ . . . . Erg. W. 1-101
O_4Xe^{2-} . . . . $[XeO_4]^{2-}$ . . . . Erg. W. 1-109
O_4Zn_2Zr . . . . Zn_2ZrO_4. . . . . . . 42 (Zr):   Hb-442
$O_{4.5}SZn_{1.5}$ . . . . . $ZnSO_4 \cdot 0.5\ ZnO$ . . . . 32 (Zn):   Hb-240
  see also $O_9S_2Zn_3 \dots ZnO \cdot 2\ ZnSO_4$
$O_{4.9}Pr_2$ . . . . . . $Pr_2O_{4.9} \cdot 5\ H_2O$ . . . . . 39 (SE):   Hb/C2-112
$O_{4.98}Rb_{2.98}S$ . . . . . $PbSO_3 \cdot 1.98\ PbO \cdot 0.61\ H_2O$ . . . . . . 47 (Pb):   Hb/C2-537
O_5^+ . . . . O_5^+ . . . . 3 (O):   Hb/3-354
O_5Os_2 . . . . Os_2O_5. . . . . . . 66 (Os):   Hb-30
$O_5Os_3S_7$ . . . . . . $Os_3O_5S_7 \cdot 2\ H_2O$ . . . . . . 66 (Os):   Hb-60
O_5P^{3-} . . . . $[PO_5]^{3-}$ . . . . 16 (P):   Hb/C-300

Hb = Main Series Volume; GD = Gmelin-Durrer;   *Ab = Appended Volume to Main Series;*
Eb = Supplement Volume to Main Series;   *Erg. W. = New Supplement Series Volume*

O$_5$PTa . . . . TaOPO$_4$
        see O$_{10}$P$_2$Ta$_2$ ... Ta$_2$O$_5$ · P$_2$O$_5$

O$_5$PV . . . . VOPO$_4$
        see O$_{10}$P$_2$V$_2$ ... V$_2$O$_5$ · P$_2$O$_5$

O$_5$P$_2$ . . . . P$_2$O$_5$:
Chemical reactions	16 (P):	Hb/C−87/93
Formation	16 (P):	Hb/C−79/80, 82
Handling	16 (P):	Hb/B−74/6
Modifications	16 (P):	Hb/C−77/9
Physical properties	16 (P):	Hb/C−83/7
Preparation	16 (P):	Hb/C−80/2
−, industrial		
see also P ... Phosphorus		
−, −, oxidation of phosphorus	16 (P):	Hb/B−50/8
− of special forms	16 (P):	Hb/B−70/2
Purity testing	16 (P):	Hb/B−73
Shipment	16 (P):	Hb/B−76
Standards	16 (P):	Hb/B−72/3
Toxicity	16 (P):	Hb/B−93

−. . . . . P$_4$O$_{10}$ . . . . 16 (P): Hb/C−77/9, 82/3

−. . . . . P$_2$O$_5$ glasses:
P$_2$O$_5$−Ag$_2$O−MoO$_3$	61 (Ag):	Hb/B4−343
P$_2$O$_5$−B$_2$O$_3$−Na$_2$O	21 (Na):	Eb/5−2013
P$_2$O$_5$−Bi$_2$O$_3$	19 (Bi):	Eb−847
P$_2$O$_5$−Bi$_2$O$_3$−PbO	47 (Pb):	Hb/C3−978
P$_2$O$_5$−CdO	33 (Cd):	Eb−662
P$_2$O$_5$−Li$_2$O	20 (Li):	Eb−521
P$_2$O$_5$−Na$_2$O−SiO$_2$	21 (Na):	Eb/5−2016
P$_2$O$_5$−PbO	47 (Pb):	Hb/C3−864/5
P$_2$O$_5$−PbO−ThO$_2$	47 (Pb):	Hb/C4−1471

−. . . . . P$_2$O$_5$ melts:
P$_2$O$_5$−BaO−Na$_2$O−V$_2$O$_5$	48 (V):	Hb/B2−532
P$_2$O$_5$−BaO−V$_2$O$_5$	48 (V):	Hb/B2−531

−. . . . . P$_2$O$_5$ systems:
P$_2$O$_5$−Ag$_2$O−H$_2$O	61 (Ag):	Hb/B4−7
P$_2$O$_5$−As$_2$O$_5$−Li$_2$O−Na$_2$O	21 (Na):	Eb/5−2244
P$_2$O$_5$−As$_2$O$_5$−Na$_2$O	21 (Na):	Eb/5−2021
P$_2$O$_5$−As$_2$O$_5$−Na$_2$O−H$_2$O	21 (Na):	Eb/5−2021
P$_2$O$_5$−B$_2$O$_3$−H$_2$O	16 (P):	Hb/C−613/4
P$_2$O$_5$−B$_2$O$_3$−K$_2$O	13 (B):	Eb−135
P$_2$O$_5$−B$_2$O$_3$−SiO$_2$	16 (P):	Hb/C−642
P$_2$O$_5$−BaO−H$_2$O	30 (Ba):	Hb−339
P$_2$O$_5$−BaO−PbO−TiO$_2$	47 (Pb):	Hb/C4−1413
P$_2$O$_5$−Bi$_2$O$_3$	19 (Bi):	Eb−847
P$_2$O$_5$−Bi$_2$O$_3$−Na$_2$O	21 (Na):	Eb/5−2026
P$_2$O$_5$−CaO	28 (Ca):	Hb/B3−1122/6
P$_2$O$_5$−CaO−HF−H$_2$O	28 (Ca):	Hb/B3−1199

See the "Instructions for the user"
at the beginning of the volume

O_5P_2 . . . . P_2O_5 systems:

$P_2O_5-CaO-H_2O$ . . . . 28 (Ca):   Hb/B3−1126/39

$P_2O_5-CaO-H_2SiF_6-N_2O_5-H_2O$ . . . . 28 (Ca):   Hb/B3−1222

$P_2O_5-CaO-MgO-H_2O$ . . . . 28 (Ca):   Hb/B3−1379

$P_2O_5-CaO-N_2O_5-H_2O$ . . . . 28 (Ca):   Hb/B1−243

Hb/B3−1189/92

$P_2O_5-CaO-SiO_2$ . . . . 28 (Ca):   Hb/B1−204/6

Hb/B3−1215/9

$P_2O_5-Ca_2P_2O_7$ . . . . 28 (Ca):   Hb/B3−1123/5

$P_2O_5-[Co(H_2O) (NH_3)_5]_2O_3-H_2O$ . . . . 58 (Co):   Eb/B2−388

$P_2O_5-[Co(NH_3)_6]_2O_3$ . . . . 58 (Co):   Eb/B2−341

$P_2O_5-Cr_2O_3-H_2O$ . . . . 52 (Cr):   Hb/B−440

$P_2O_5-FeO-H_2O$ . . . . 59 (Fe):   Hb/B4−771

$P_2O_5-Fe_2O_3-H_2O$ . . . . 59 (Fe):   Hb/B4−773

$P_2O_5-HF-H_2O$ . . . . 16 (P):   Hb/C−404

$P_2O_5-H_2O$ . . . . 16 (P):   Hb/C−155/60

$P_2O_5-K_2O-H_2O$ . . . . 22 (K):   Hb/5−984/8

$P_2O_5-K_2O-NH_3-H_2O$ . . . . 23 (NH$_4$):   Hb/2−517/8

$P_2O_5-K_2O-SO_3-H_2O$ . . . . 22 (K):   Hb/5−1002

$P_2O_5-Li_2O$ . . . . 20 (Li):   Eb−521

$P_2O_5-Li_2O-H_2O$ . . . . 20 (Li):   Eb−521/2

$P_2O_5-NH_3-H_2O$ . . . . 23 (NH$_4$):   Hb/2−418/9

$P_2O_5-NH_3-SO_3-H_2O$ . . . . 23 (NH$_4$):   Hb/2−430/1

$P_2O_5-N_2O_5-H_2O$ . . . . 16 (P):   Hb/C−332/5

$P_2O_5-Na_2O$ . . . . 21 (Na):   Eb/4−1561/2

$P_2O_5-Na_2O-H_2O$ . . . . 21 (Na):   Eb/4−1562/7

$P_2O_5-Na_2O-Sb_2O_3$ . . . . 21 (Na):   Eb/5−2026

$P_2O_5-Na_2O-SiO_2$ . . . . 21 (Na):   Eb/5−2013/6

$P_2O_5-Nb_2O_5-PbO$ . . . . 49 (Nb):   Hb/B4−361

$P_2O_5-Nb_2O_5-ZnO$ . . . . 49 (Nb):   Hb/B4−361

$P_2O_5-POCl_3-PSCl_3-P_2S_5$ . . . . . . 16 (P):   Hb/C−597/8

P_2O_5-PbO . . . . 47 (Pb):   Hb/C3−863/5

$P_2O_5-PbO-H_2O$ . . . . 47 (Pb):   Hb/C3−870

$P_2O_5-PbO-SiO_2$ . . . . 47 (Pb):   Hb/C3−885/6

$P_2O_5-PbO-V_2O_5$ . . . . 48 (V):   Hb/B2−572

$P_2O_5-PbO-ZrO_2$ . . . . 47 (Pb):   Hb/C4−1431

$P_2O_5-SiO_2$ . . . . 16 (P):   Hb/C−638

$P_2O_5-SiO_2-H_2O$ . . . . 16 (P):   Hb/C−639/40

$P_2O_5-SrO-H_2O$ . . . . 29 (Sr):   Hb−213

$P_2O_5-Ta_2O_5-ZnO$ . . . . 50 (Ta):   Hb/B2−349

$P_2O_5-V_2O_5$ . . . . 48 (V):   Hb/B1−361

$P_2O_5-V_2O_5-H_2O$ . . . . 48 (V):   Hb/B1−362

$P_2O_5-ZnO-H_2O$ . . . . 32 (Zn):   Hb−277

$O_5P_{14}Zn_7$ . . . . $Zn_7P_{14}O_5$ . . . . 32 (Zn):   Hb−276

O_5Pa_2 . . . . Pa_2O_5 . . . . 51 (Pa):   Hb−67

−. . . . . Pa_2O_5 systems:

$PaO_{2.5}-PuO_2$ . . . . Erg. W. 4−58

O_5PbS . . . . $PbOSO_4 \cdot H_2O$ . . . . 47 (Pb):   Hb/C2−594

Hb = Main Series Volume; GD = Gmelin-Durrer;   *Ab = Appended Volume to Main Series;*
Eb = Supplement Volume to Main Series;   *Erg. W. = New Supplement Series Volume*

O_5PbSb_2	$PbSb_2O_5 = PbO \cdot Sb_2O_4$	47 (Pb):	Hb/C3−930
O_5PbSe_2	$PbSe_2O_5$	47 (Pb):	Hb/C2−636
$O_5PbSiZr$	$PbZrSiO_5$	42 (Zr):	Hb−438
		47 (Pb):	Hb/C4−1431
O_5PbSi_2	$PbSi_2O_5$	47 (Pb):	Hb/C3−854
O_5Pb_2	Pb_2O_5	47 (Pb):	Hb/C1−196
O_5Pb_2S	$PbSO_4 \cdot PbO$	47 (Pb):	Hb/C2−584/8
$O_5Pb_2S_2V_2$	$Pb_2V_2O_5S_2$	48 (V):	Hb/B2−572
O_5Pb_2Se	$PbSeO_4 \cdot PbO$	47 (Pb):	Hb/C2−638/9
O_5Pb_2W	$2\,PbO \cdot WO_3$	54 (W):	Hb−317
O_5Pb_3Se	$PbSeO_3 \cdot 2\,PbO$	47 (Pb):	Hb/C2−635
O_5Pb_3Si	Pb_3SiO_5	47 (Pb):	Hb/C3−842
O_5Pr_2	$Pr_2O_5 \cdot nH_2O$	39 (SE):	Hb/C2−106/8, 112
O_5Pu^{3-}	$[PuO_5]^{3-}$		Erg. W. 8−76
O_5PuRb_3	Rb_3PuO_5		Erg. W. 4−52/3
$O_5Rb_2S_2$	$Rb_2S_2O_5$	24 (Rb):	Hb−207
$O_5Rb_2S_3$	$Rb_2S_2O_3 \cdot SO_2$	24 (Rb):	Hb−215
$O_5Rb_2Ti_2$	$Rb_2Ti_2O_5$	41 (Ti):	Hb−411
−.	$Rb_2Ti_2O_5$ systems:		
	$Rb_2Ti_2O_5–RbF$	41 (Ti):	Hb−412
O_5Re_2	$Re_2O_5 \cdot xH_2O$	70 (Re):	Hb−97
O_5RuS	$RuOSO_4$	63 (Ru):	Eb−314/5
O_5Ru_2	Ru_2O_5	63 (Ru):	Hb−27
−.	$Ru_2O_5 \cdot 2\,H_2O$	63 (Ru):	Hb−27
			Eb−240
O_5SSb_2	Sb_2O_5S	18 (Sb):	Hb/B3−541
O_5SSe	$SeO_2 \cdot SO_3$	10 (Se):	Hb/B−181
O_5STh	$ThOSO_4 \cdot H_2O$	44 (Th):	Hb−293
−.	$ThOSO_4 \cdot 2\,H_2O$	44 (Th):	Hb−293
−.	$ThOSO_4 \cdot 3\,H_2O$	44 (Th):	Hb−292
−.	$ThOSO_4 \cdot 5\,H_2O$	44 (Th):	Hb−292
O_5STi	$TiOSO_4 = TiO_2 \cdot SO_3$	41 (Ti):	Hb−347, 348
−.	$TiOSO_4 \cdot H_2O$	41 (Ti):	Hb−346/7, 349
−.	$TiOSO_4 \cdot 2\,H_2O$	41 (Ti):	Hb−345/7, 348/9
−.	$TiOSO_4 \cdot 5\,H_2O$	41 (Ti):	Hb−348
O_5SU	$UOSO_4 \cdot 2\,H_2O$	55 (U):	Hb−151
−.	$UO_2SO_3 \cdot 2.75\,H_2O$		Hb−146
	$= 4\,UO_3 \cdot 4\,SO_2 \cdot 11\,H_2O$	55 (U):	
−.	$UO_2SO_3 \cdot 4\,H_2O = UO_3 \cdot SO_2 \cdot 4\,H_2O$ . . .	55 (U):	Hb−145/6
O_5SV	$VOSO_4$	48 (V):	Hb/B1−92, 288/91
	see also OV^{2+} ... VO^{2+}		
−.	$VOSO_4 \cdot H_2O$	48 (V):	Hb/B1−295/6
−.	$VOSO_4 \cdot 1.5\,H_2O$	48 (V):	Hb/B1−291, 295, 299
−.	$VOSO_4 \cdot 2\,H_2O$	48 (V):	Hb/B1−72, 295
−.	$VOSO_4 \cdot 2.5\,H_2O$	48 (V):	Hb/B1−295
−.	$VOSO_4 \cdot 3\,H_2O$	48 (V):	Hb/A2−376, 621
			Hb/B1−291, 294/5, 299
−.	$VOSO_4 \cdot 3.5\,H_2O$	48 (V):	Hb/B1−293/4

See the "Instructions for the user"
at the beginning of the volume

O_5SV . . . . VOSO$_4$ · 4 H$_2$O = [VO(H$_2$O)$_4$SO$_4$] . . . . 48 (V): Hb/B1–293, 296
–. . . . . VOSO$_4$ · 5 H$_2$O . . . . 48 (V): Hb/B1–71/2, 292/3, 299
–. . . . . VOSO$_4$ · 6 H$_2$O . . . . 48 (V): Hb/B1–291/2, 299
–. . . . . VOSO$_4$ · 6.5 H$_2$O . . . . 48 (V): Hb/B1–291/2
–. . . . . VOSO$_4$ solutions:
               VOSO$_4$–H$_2$O. . . . . 48 (V): Hb/A2–480/1
               Hb/B1–296/8
               VOSO$_4$–H$_2$SO$_4$–H$_2$O . . . . 48 (V): Hb/A2–476, 510/1
               Hb/B1–301/2
               VOSO$_4$–LiCl–H$_2$O . . . . 48 (V): Hb/B2–376
               VOSO$_4$–organic solvents. . . . . . . 48 (V): Hb/B1–298
–. . . . . VOSO$_4$ systems:
               VOSO$_4$–C$_4$H$_6$O$_6$–H$_2$O. . . . . 48 (V): Hb/B1–350
               VOSO$_4$–H$_2$C$_2$O$_4$–H$_2$O. . . . . 48 (V): Hb/B1–348
               VOSO$_4$–H$_2$O. . . . . 48 (V): Hb/B1–291
               VOSO$_4$–H$_2$SO$_4$–H$_2$O . . . . 48 (V): Hb/B1–298/300
               VOSO$_4$–NaC$_4$H$_5$O$_6$–H$_2$O . . . . 48 (V): Hb/B1–350
               VOSO$_4$–NaKC$_4$H$_4$O$_6$–H$_2$O . . . . . . 48 (V): Hb/B1–350
               VOSO$_4$–Na$_2$C$_2$O$_4$–H$_2$O . . . . 48 (V): Hb/B1–348
O_5SV^+ . . . . [VO(H$_2$O)SO$_4$]$^+$ . . . . 48 (V): Hb/B1–305
O_5SZn_2 . . . . ZnSO$_4$ · ZnO . . . . 32 (Zn): Hb–240
–. . . . . ZnSO$_4$ · ZnO · H$_2$O . . . . 32 (Zn): Hb–241
–. . . . . ZnSO$_4$ · ZnO · xH$_2$O . . . . 32 (Zn): Hb–240
O_5SZr . . . . ZrOSO$_4$ · 4 H$_2$O . . . . 42 (Zr): Hb–346
–. . . . . ZrOSO$_4$ · xH$_2$O . . . . 42 (Zr): Hb–345/6
$O_5S_2^{2-}$ . . . . [S$_2$O$_5$]$^{2-}$ . . . . 9 (S): Hb/B2–613
O_5S_2U . . . . UO$_2$S$_2$O$_3$ . . . . 55 (U): Hb–156
O_5Sb_2 . . . . Sb$_2$O$_5$. . . . . 18 (Sb): Hb/B2–276, 376/80
–. . . . . Sb$_2$O$_5$ · H$_2$O = HSbO$_3$ . . . . 18 (Sb): Hb/B2–381
–. . . . . 3 Sb$_2$O$_5$ · 5 H$_2$O = H$_5$Sb$_3$O$_{10}$ . . . . . . 18 (Sb): Hb/B2–384/5
–. . . . . Sb$_2$O$_5$ · 2 H$_2$O = H$_4$Sb$_2$O$_7$ . . . . 18 (Sb): Hb/B2–381
–. . . . . Sb$_2$O$_5$ · 3 H$_2$O = H$_3$SbO$_4$ . . . . 18 (Sb): Hb/B2–381
–. . . . . Sb$_2$O$_5$ · xH$_2$O . . . . 18 (Sb): Hb/B2–380/7
–. . . . . Sb$_2$O$_5$ solid solutions:
               Sb$_2$O$_5$–PbO–SnO$_2$ . . . . 47 (Pb): Hb/C4–1595
–. . . . . Sb$_2$O$_5$ systems:
               Sb$_2$O$_5$–Ag$_2$O–Sb$_2$O$_3$ . . . . 61 (Ag): Hb/B4–89
               Sb$_2$O$_5$–Na$_2$O–H$_2$O . . . . 21 (Na): Eb/4–1711/2
               Sb$_2$O$_5$–PbO . . . . 47 (Pb): Hb/C3–929
O_5SeTi . . . . TiO$_2$ · SeO$_3$ · H$_2$O. . . . . 41 (Ti): Hb–353
O_5SeU . . . . UO$_2$SeO$_3$ . . . . 55 (U): Hb–157
–. . . . . UO$_2$SeO$_3$ · 2 H$_2$O. . . . . 55 (U): Hb–158
$O_5SiSrTi$ . . . . . . SrTiSiO$_5$ . . . . 41 (Ti): Hb–434
$O_5SiSrZr$ . . . . . . SrZrSiO$_5$ . . . . 42 (Zr): Hb–438
O_5SiSr_3 . . . . 3 SrO · SiO$_2$ . . . . 29 (Sr): Eb–281
$O_5SiZnZr$ . . . . . . ZnZrSiO$_5$ . . . . 42 (Zr): Hb–438
O_5Si_2 . . . . Si$_2$O$_5$ . . . . 15 (Si): Hb/B–257

Hb = Main Series Volume; GD = Gmelin-Durrer;   *Ab = Appended Volume to Main Series;*
Eb = Supplement Volume to Main Series;   *Erg. W. = New Supplement Series Volume*

$O_5Si_2^{2-}$	$[Si_2O_5]^{2-}$	15 (Si):	Hb/B−421/2
O_5Sm_2	$Sm_2O_5 \cdot nH_2O$	39 (SE):	Hb/C2−106/7, 113
O_5SrTa_2	$SrTa_2O_5$.	50 (Ta):	Hb/B2−181
−.	$SrTa_2O_5$ solid solutions:		
	$(Sr,Ba)Ta_2O_5$	50 (Ta):	Hb/B2−192
O_5SrV_2	SrV_2O_5	48 (V):	Hb/B2−519
$O_5Sr_2V_2$	$Sr_2V_2O_5$		
	see $O_{2.5}SrV \ldots SrVO_{2.5}$		
O_5TaV.	$VTaO_5$	50 (Ta):	Hb/B2−245
O_5Ta_2	Ta_2O_5:		
	Chemical reactions	50 (Ta):	Hb/B1−45/53
	Electrochemical behavior	50 (Ta):	Hb/A2−412, 490/3
			Hb/B1−45
	Formation and preparation	50 (Ta):	Hb/B1−17, 31/4
	History.	50 (Ta):	Hb/A1−8
	Physical properties	50 (Ta):	Hb/A2−322/3
			Hb/B1−34/45
	Purity testing	49 (Nb):	Hb/A−352/3
	Toxicity	50 (Ta):	Hb/A1−276
	Uses	50 (Ta):	Hb/A1−276
−.	$Ta_2O_5 \cdot xH_2O$	50 (Ta):	Hb/A1−247
			Hb/B1−54/8, 136
−.	Ta_2O_5 solid solutions:		
	$Ta_2O_5-BaTiO_3$.	50 (Ta):	Hb/B2−224/6
	$Ta_2O_5-K_2O-Nb_2O_5-PbO$	50 (Ta):	Hb/B2−265
	$Ta_2O_5-Na_2O-Nb_2O_5-PbO$.	50 (Ta):	Hb/B2−265
	$Ta_2O_5-SiO_2-ZrO_2$	50 (Ta):	Hb/B2−349
−.	Ta_2O_5 systems:		
	$Ta_2O_5-Ag_2O$	61 (Ag):	Hb/B4−285
	$Ta_2O_5-Al_2O_3$	50 (Ta):	Hb/B2−197
	Ta_2O_5-BaO	50 (Ta):	Hb/B2−186
	$Ta_2O_5-Bi_2O_3$	50 (Ta):	Hb/B1−275
	Ta_2O_5-CaO	50 (Ta):	Hb/B2−176/7
	Ta_2O_5-CdO	50 (Ta):	Hb/B2−195
	$Ta_2O_5-CeO_2$.	50 (Ta):	Hb/B2−210
	$Ta_2O_5-Ce_2O_3$	50 (Ta):	Hb/B2−210
	Ta_2O_5-CuO	60 (Cu):	Hb/B3−1214
	$Ta_2O_5-Dy_2O_3$	50 (Ta):	Hb/B2−217
	$Ta_2O_5-Eu_2O_3$	50 (Ta):	Hb/B2−215
	$Ta_2O_5-Ga_2O_3$	50 (Ta):	Hb/B2−199
	$Ta_2O_5-Gd_2O_3$	50 (Ta):	Hb/B2−216
	$Ta_2O_5-GeO_2-K_2O$	50 (Ta):	Hb/B2−230
	$Ta_2O_5-GeO_2-PbO$	50 (Ta):	Hb/B2−243
	$Ta_2O_5-HF-H_2O$	50 (Ta):	Hb/B1−88
	$Ta_2O_5-HfO_2-ZrO_2$	50 (Ta):	Hb/B2−229
	$Ta_2O_5-In_2O_3$.	50 (Ta):	Hb/B2−200
	Ta_2O_5-KCl.	50 (Ta):	Hb/B2−287

See the "Instructions for the user"
at the beginning of the volume

O_5Ta_2 . . . . Ta_2O_5 systems:

$Ta_2O_5-KCl-K_2TaF_7$ . . . . 50 (Ta):	Hb/B2−289/90
Ta_2O_5-KF . . . . 50 (Ta):	Hb/B2−275
$Ta_2O_5-KF-K_2TaF_7$ . . . . 50 (Ta):	Hb/B2−275/6
$Ta_2O_5-K_2CO_3$ . . . . 50 (Ta):	Hb/B2−151/2
$Ta_2O_5-K_2O$ . . . . 50 (Ta):	Hb/B2−151/2
$Ta_2O_5-K_2O-C_2H_5OH-H_2O$ . . . . . . 50 (Ta):	Hb/B2−165
$Ta_2O_5-K_2O-H_2O$ . . . . 50 (Ta):	Hb/B2−163/5
$Ta_2O_5-K_2O-SiO_2$ . . . . 50 (Ta):	Hb/B2−348
$Ta_2O_5-K_2TaF_7$ . . . . 50 (Ta):	Hb/B2−275
$Ta_2O_5-La_2O_3$ . . . . 50 (Ta):	Hb/B2−207/8
$Ta_2O_5-La_2O_3-Li_2O$ . . . . 50 (Ta):	Hb/B2−209
$Ta_2O_5-Li_2O$ . . . . 50 (Ta):	Hb/B2−126
Ta_2O_5-MgO . . . . 50 (Ta):	Hb/B2−173
$Ta_2O_5-(NH_4)_2SO_4-SO_3-H_2O$ . . . . . 50 (Ta):	Hb/B2−316
$Ta_2O_5-Na_2O$ . . . . 50 (Ta):	Hb/B2−141
$Ta_2O_5-Na_2O-H_2O$ . . . . 50 (Ta):	Hb/B2−147
$Ta_2O_5-Nb_2O_5$ . . . . 50 (Ta):	Hb/B2−246/8
$Ta_2O_5-Nd_2O_3$ . . . . 50 (Ta):	Hb/B2−212
$Ta_2O_5-P_2O_5-ZnO$ . . . . 50 (Ta):	Hb/B2−349
Ta_2O_5-PbO . . . . 50 (Ta):	Hb/B2−232/3
$Ta_2O_5-SO_3-H_2O$ . . . . 50 (Ta):	Hb/B1−176
$Ta_2O_5-Sc_2O_3$ . . . . 50 (Ta):	Hb/B2−204
$Ta_2O_5-SiO_2$ . . . . 50 (Ta):	Hb/B1−263
$Ta_2O_5-SiO_2-ZnO$ . . . . 50 (Ta):	Hb/B2−349
$Ta_2O_5-Sm_2O_3$ . . . . 50 (Ta):	Hb/B2−214
Ta_2O_5-SnO . . . . 50 (Ta):	Hb/B2−231
$Ta_2O_5-SnO_2$ . . . . 50 (Ta):	Hb/B2−231
Ta_2O_5-SrO . . . . 50 (Ta):	Hb/B2−181
$Ta_2O_5-ThO_2$ . . . . 50 (Ta):	Hb/B2−229
$Ta_2O_5-TiO_2$ . . . . 50 (Ta):	Hb/B2−222
$Ta_2O_5-TiO_2-ZrO_2$ . . . . 50 (Ta):	Hb/B2−229
$Ta_2O_5-Tl_2O_3$ . . . . 50 (Ta):	Hb/B2−201
$Ta_2O_5-V_2O_5$ . . . . 50 (Ta):	Hb/B2−244
$Ta_2O_5-Y_2O_3$ . . . . 50 (Ta):	Hb/B2−205
$Ta_2O_5-Y_2O_3-ZrO_2$ . . . . 50 (Ta):	Hb/B2−229
Ta_2O_5-ZnO . . . . 50 (Ta):	Hb/B2−193
$Ta_2O_5-ZrO_2$ . . . . 50 (Ta):	Hb/B2−228

$O_5Ta_2^-$	$[Ta_2O_5]^-$ . . . . 50 (Ta):	Hb/B1−59
O_5TeTh	$ThOTeO_4 \cdot 4 H_2O$ . . . . 44 (Th):	Hb−295
−.	$ThOTeO_4 \cdot 8 H_2O$ . . . . 44 (Th):	Hb−295
O_5TiV_2	V_2TiO_5 . . . . 48 (V):	Hb/B2−558
O_5Ti_2	$Ti_2O_5 \cdot xH_2O$ . . . . 41 (Ti):	Hb−269
O_5Ti_2V	Ti_2VO_5 . . . . 48 (V):	Hb/B2−558
O_5Ti_3	Ti_3O_5 . . . . 41 (Ti):	Hb−224/6
−.	Ti_3O_5 systems:	
	$Ti_3O_5-V_3O_5$ . . . . 48 (V):	Hb/B2−558
O_5Tl_3	$Tl_3O_5 = Tl_2O_3 \cdot TlO_2$ . . . . 38 (Tl):	Hb/2−233

O_5UV . . . . UVO$_5$ . . . . 48 (V):   Hb/B1−91
O_5U_2 . . . . U$_2$O$_5$ . . . . 55 (U):   Hb−88/9
O_5V^- . . . . [V(O) (O$_2$)$_2$]$^-$ . . . . 48 (V):   Hb/B1−162/3, 165

O_5V_2 . . . . V$_2$O$_5$:
                          Catalytic properties . . . . 48 (V):   Hb/B1−84
                          Chemical reactions . . . . 48 (V):   Hb/B1−84/95
                          Determination . . . . 48 (V):   Hb/A2−714
                          Diffusion of gases . . . . 48 (V):   Hb/B1−95/6
                          Electrochemical behavior . . . . 48 (V):   Hb/A2−468/9, 488/9,
                                                                                        496/8
                                                                                Hb/B1−84
                          Formation . . . . 48 (V):   Hb/B1−73
                          Gels . . . . 48 (V):   Hb/B1−130/1
                          History . . . . 48 (V):   Hb/A1−6/10
                          Physical properties . . . . 48 (V):   Hb/A2−376/80, 387
                                                                                Hb/B1−30, 73/83
                          Preparation . . . . 48 (V):   Hb/A1−280/5
                          Purification . . . . 48 (V):   Hb/A1−283/4
                          Purity testing . . . . 48 (V):   Hb/A1−284/5
                          Sols . . . . 48 (V):   Hb/B1−98/130
                          Solubility . . . . 48 (V):   Hb/B1−97
                          Sorption of gases . . . . 48 (V):   Hb/B1−95/7
                          Special forms . . . . 48 (V):   Hb/A1−270, 284
                                                                                Hb/B1−72, 98/131
                          Toxicity . . . . 48 (V):   Hb/A2−725/8
−. . . . . V$_4$O$_{10}$ . . . . 48 (V):   Hb/B1−84
−. . . . . V$_2$O$_5$ · 0.5 H$_2$O . . . . 48 (V):   Hb/B1−98
−. . . . . V$_2$O$_5$ · H$_2$O . . . . 48 (V):   Hb/B1−97/8
−. . . . . V$_2$O$_5$ · 2 H$_2$O . . . . 48 (V):   Hb/B1−97
−. . . . . V$_2$O$_5$ · nH$_2$O . . . . 48 (V):   Hb/B1−97
−. . . . . V$_2$O$_5$ melts:
                          V$_2$O$_5$−As$_2$O$_3$ . . . . 48 (V):   Hb/B1−366
                          V$_2$O$_5$−B$_2$O$_3$−ZnO . . . . 48 (V):   Hb/B2−536
                          V$_2$O$_5$−BaO−Na$_2$O−P$_2$O$_5$ . . . . 48 (V):   Hb/B2−532
                          V$_2$O$_5$−BaO−P$_2$O$_5$ . . . . 48 (V):   Hb/B2−531
                          V$_2$O$_5$−Cs$_2$O . . . . 48 (V):   Hb/B2−492
                          V$_2$O$_5$−Li$_2$SO$_4$ . . . . 48 (V):   Hb/B2−376
                          V$_2$O$_5$−Na$_2$SO$_4$ . . . . 48 (V):   Hb/B2−405/6
                          V$_2$O$_5$−Na$_2$S$_2$O$_7$ . . . . 48 (V):   Hb/B2−406
                          V$_2$O$_5$−Rb$_2$O . . . . 48 (V):   Hb/B2−486
                          V$_2$O$_5$−TeO$_2$ . . . . 48 (V):   Hb/B1−320
−. . . . . V$_2$O$_5$ solutions:
                          V$_2$O$_5$−H$_2$O . . . . 48 (V):   Hb/B1−132/45
−. . . . . V$_2$O$_5$ systems:
                          V$_2$O$_5$−Ag$_2$O . . . . 61 (Ag):   Hb/B4−277
                          V$_2$O$_5$−Al . . . . 48 (V):   Hb/B2−544/5
                          V$_2$O$_5$−Al$_2$O$_3$ . . . . 48 (V):   Hb/B2−545
                          V$_2$O$_5$−B$_2$O$_3$ . . . . 48 (V):   Hb/B1−325

See the "Instructions for the user"
at the beginning of the volume

O_5V_2 . . . . V_2O_5 systems:

V_2O_5-BaO	48 (V):	Hb/B2–525
V_2O_5-BeO	48 (V):	Hb/B2–500
$V_2O_5-Bi_2O_3$	48 (V):	Hb/B1–368
$V_2O_5-Bi_2O_3-PbO$	48 (V):	Hb/B2–573
V_2O_5-CaO	48 (V):	Hb/B2–511/2
V_2O_5-CoO	58 (Co):	Hb/A2–478
		Eb/A–873
$V_2O_5-Cr_2O_3$	52 (Cr):	Hb/B–939
$V_2O_5-Cr_2O_3-NiO$	57 (Ni):	Hb/B3–1217
$V_2O_5-Cs_2O$	48 (V):	Hb/B2–492
$V_2O_5-Cs_2SO_4$	48 (V):	Hb/B2–499
V_2O_5-CuO	60 (Cu):	Hb/B3–1209
$V_2O_5-CuO-Fe_2O_3$	60 (Cu):	Hb/B3–1330
$V_2O_5-HF-H_2O$	48 (V):	Hb/B1–194
$V_2O_5-H_2SO_4-H_2O$	48 (V):	Hb/B1–303/5
$V_2O_5-KPO_3$	48 (V):	Hb/B2–444
$V_2O_5-KVO_3$	48 (V):	Hb/B2–412
$V_2O_5-K_2O$	48 (V):	Hb/B2–412/3
$V_2O_5-K_2SO_4$	48 (V):	Hb/B2–438/9
$V_2O_5-K_2S_2O_7$	48 (V):	Hb/B2–439
$V_2O_5-Li_2O$	48 (V):	Hb/B2–371
$V_2O_5-Li_2O-VO_2$	48 (V):	Hb/B2–370
V_2O_5-MgO	48 (V):	Hb/B2–504/5
$V_2O_5-MgO-VO_2$	48 (V):	Hb/B2–504
$V_2O_5-NH_3-H_2O$	48 (V):	Hb/B2–449/50
$V_2O_5-Na_2O$	48 (V):	Hb/B2–380/1
$V_2O_5-Na_2O-H_2O$	48 (V):	Hb/B2–382
$V_2O_5-Na_2O-SO_3$	48 (V):	Hb/B2–405
$V_2O_5-Na_2O-SO_3-H_2O$	48 (V):	Hb/B2–406
$V_2O_5-Nb_2O_5$	49 (Nb):	Hb/B3–324/6
$V_2O_5-Nb_2O_5-ZnO$	49 (Nb):	Hb/B3–329
V_2O_5-NiO	57 (Ni):	Hb/B3–1198/9
$V_2O_5-P_2O_5$	48 (V):	Hb/B1–361
$V_2O_5-P_2O_5-H_2O$	48 (V):	Hb/B1–362
$V_2O_5-P_2O_5-PbO$	48 (V):	Hb/B2–572
V_2O_5-PbO	48 (V):	Hb/B2–567/8
$V_2O_5-PbO-TiO_2$	48 (V):	Hb/B2–574
$V_2O_5-Ta_2O_5$	50 (Ta):	Hb/B2–244
$V_2O_5-ThO_2$	48 (V):	Hb/B2–564
V_2O_5-VO	48 (V):	Hb/B1–13
$V_2O_5-VO_2-ZnO$	48 (V):	Hb/B2–534
$V_2O_5-V_2O_3$	48 (V):	Hb/B1–13
V_2O_5-ZnO	48 (V):	Hb/B2–534
$V_2O_5-ZrO_2$	48 (V):	Hb/B2–562

$O_5V_2^{2-}$ . . . . $[V_2O_5]^{2-}$ . . . . 48 (V):   Hb/A2–477

$O_5V_2Zn_x$ . . . . $Zn_xV_2O_5$ . . . . 48 (V):   Hb/B2–534

O_5V_3 . . . . $V_3O_5 = VO_{1.67}$ . . . . 48 (V):   Hb/B1–13, 17,
                                                          49/50, 86

Hb = Main Series Volume; GD = Gmelin-Durrer;   *Ab = Appended Volume to Main Series;*
Eb = Supplement Volume to Main Series;   *Erg. W. = New Supplement Series Volume*

O_5V_3 . . . . V$_3$O$_5$ systems:

    V$_3$O$_5$–Ti$_3$O$_5$ . . . . 48 (V):   Hb/B2–558

O_5W_2 . . . . W$_2$O$_5$ . . . . 54 (W):   Hb–107/11, 115/7

O_5Zn_3 . . . . Zn$_3$O$_5$ . . . . 32 (Zn):   Hb–145

–. . . . . Zn$_3$O$_5$ · H$_2$O . . . . 32 (Zn):   Hb–146
                                                                                   Eb–817

O_5Zr_2 . . . . Zr$_2$O$_5$ · xH$_2$O . . . . 42 (Zr):   Hb–261

O_6. . . . . O$_6$. 3 (O):   Hb/4–1041

O_6^+ . . . . (2 O$_3$)$^+$ . . . . 3 (O):   Hb/4–1071

O_6P_2. . . . . P$_2$O$_6$ . . . . 16 (P):   Hb/C–93/4

O_6P_2Pb . . . . [Pb(PO$_3$)$_2$]$_x$ . . . . 47 (Pb):   Hb/C3–875/6

–. . . . . Pb(PO$_3$)$_2$ systems:

    Pb(PO$_3$)$_2$–BaTiO$_3$ . . . . 47 (Pb):   Hb/C4–1413

    Pb(PO$_3$)$_2$–PbZrO$_3$ . . . . 47 (Pb):   Hb/C4–1431

$O_6P_2Pb_2$. . . . . . . Pb$_2$P$_2$O$_6$ . . . . 47 (Pb):   Hb/C3–863

$O_6P_2Pb_3S_2$ . . . . . . Pb$_3$(PO$_3$S)$_2$ . . . . 47 (Pb):   Hb/C3–884

$O_6P_2Rb_4$. . . . . . . Rb$_4$P$_2$O$_6$ · 10 H$_2$O. . . . . 24 (Rb):   Hb–233

O_6P_2Sr . . . . Sr(PO$_3$)$_2$ . . . . 29 (Sr):   Hb–216
                                                                                   Eb–286

–. . . . . Sr(PO$_3$)$_2$ · 2 H$_2$O . . . . 29 (Sr):   Hb–216

$O_6P_2Sr_2$ . . . . Sr$_2$P$_2$O$_6$. . . . . 29 (Sr):   Eb–283

–. . . . . Sr$_2$P$_2$O$_6$ · 2 H$_2$O . . . . 29 (Sr):   Eb–283

O_6P_2Th . . . . ThP$_2$O$_6$ · 11 H$_2$O . . . . 44 (Th):   Hb–316

$O_6P_2Tl_4$ . . . . Tl$_4$P$_2$O$_6$ = Tl$_2$PO$_3$ . . . . 38 (Tl):   Hb/3–427

O_6P_2U. . . . . UP$_2$O$_6$. . . . . 55 (U):   Hb–180

O_6P_2Zn . . . . Zn(PO$_3$)$_2$ . . . . 32 (Zn):   Hb–279

–. . . . . Zn$_2$(PO$_3$)$_4$ . . . . 32 (Zn):   Hb–280

–. . . . . Zn$_3$(PO$_3$)$_6$ . . . . 32 (Zn):   Eb–997

–. . . . . Zn(PO$_3$)$_2$ · 4 H$_2$O . . . . 32 (Zn):   Hb–279/80

–. . . . . Zn$_2$(PO$_3$)$_4$ · 10 H$_2$O . . . . 32 (Zn):   Hb–280

–. . . . . Zn$_3$(PO$_3$)$_6$ · 9 H$_2$O . . . . 32 (Zn):   Hb–280

$O_6P_2Zn_2$. . . . . . . Zn$_2$P$_2$O$_6$ · 2 H$_2$O . . . . 32 (Zn):   Hb–277

O_6P_2Zr . . . . ZrP$_2$O$_6$ · H$_2$O . . . . 42 (Zr):   Hb–392

O_6P_4. . . . . P$_4$O$_6$ . . . . 16 (P):   Hb/C–70/1

$O_6P_4S_4$ . . . . P$_4$S$_4$O$_6$ . . . . 16 (P):   Hb/C–582/3

$O_6P_4Se_4$. . . . . . . P$_4$O$_6$Se$_4$. . . . . 16 (P):   Hb/C–75

O_6PbS_2 . . . . PbS$_2$O$_6$ . . . . 47 (Pb):   Hb/B2–284/6

–. . . . . PbS$_2$O$_6$ · H$_2$O. . . . . 47 (Pb):   Hb/C2–598

–. . . . . PbS$_2$O$_6$ · 4 H$_2$O . . . . 47 (Pb):   Hb/C2–598/9

–. . . . . PbS$_2$O$_6$ solid solutions:

    (Pb,Ba)S$_2$O$_6$ · 4 H$_2$O . . . . 30 (Ba):   Hb–284
                                                                    47 (Pb):   Hb/C3–1207

–. . . . . PbS$_2$O$_6$ systems:

    PbS$_2$O$_6$–SrS$_2$O$_6$–H$_2$O . . . . 47 (Pb):   Hb/C3–1198

O_6PbS_3 . . . . PbS$_3$O$_6$. . . . . 47 (Pb):   Hb/C2–600

O_6PbS_4 . . . . PbS$_4$O$_6$ . . . . 47 (Pb):   Hb/B2–286

–. . . . . PbS$_4$O$_6$ · 2 H$_2$O. . . . . 47 (Pb):   Hb/C2–600

See the "Instructions for the user"
at the beginning of the volume

$O_6PbS_4{}^{2-}$ . . . . . . $[Pb(S_2O_3)_2]^{2-}$ . . . . 47 (Pb): Hb/B1–365
Hb/C2–597
O_6PbS_5 . . . . PbS_5O_6 . . . . 47 (Pb): Hb/C2–600
O_6PbSb_2 . . . . . . $Pb(SbO_3)_2$ . . . . 47 (Pb): Hb/C3–932/3
–. . . . . $Pb(SbO_3)_2 \cdot nH_2O$ . . . . 47 (Pb): Hb/C3–933
O_6PbTa_2 . . . . . . $PbTa_2O_6$ . . . . 50 (Ta): Hb/B2–233/6
–. . . . . $PbTa_2O_6$ solid solutions:
$(Pb,Ba,Ca)Ta_2O_6$. . . . . . . 50 (Ta): Hb/B2–239
$(Pb,Ba,Sr)Ta_2O_6$. . . . . . . 50 (Ta): Hb/B2–239
$Pb(Ta,Nb)_2O_6$ . . . . 50 (Ta): Hb/B2–265/6
$(Pb,Sr)Ta_2O_6$. . . . . 50 (Ta): Hb/B2–239
–. . . . . $PbTa_2O_6$ systems:
$PbTa_2O_6–PbTiO_3$ . . . . 50 (Ta): Hb/B2–242
$PbTa_2O_6–PbZrO_3$ . . . . 50 (Ta): Hb/B2–242
O_6PbV_2 . . . . $PbO \cdot V_2O_5 = Pb(VO_3)_2$ . . . . . 48 (V): Hb/B2–570
O_6Pb_2PrTa . . . . . . Pb_2PrTaO_6 . . . . 50 (Ta): Hb/B2–240
O_6Pb_2SbTa . . . . . Pb_2SbTaO_6 . . . . 50 (Ta): Hb/B2–238
O_6Pb_2ScTa . . . . . Pb_2ScTaO_6 . . . . 50 (Ta): Hb/B2–240
–. . . . . Pb_2ScTaO_6 systems:
$Pb_2ScTaO_6–PbHfO_3$ . . . . 50 (Ta): Hb/B2–243
$Pb_2ScTaO_6–PbTiO_3$. . . . . 50 (Ta): Hb/B2–243
$Pb_2ScTaO_6–PbZrO_3$. . . . . 50 (Ta): Hb/B2–243
O_6Pb_2SmTa . . . . . Pb_2SmTaO_6 . . . . 50 (Ta): Hb/B2–240
O_6Pb_2TaY. . . . . . Pb_2YTaO_6 . . . . 50 (Ta): Hb/B2–240
O_6Pb_2TaYb . . . . . Pb_2YbTaO_6. . . . . 50 (Ta): Hb/B2–240/1
O_6Pb_3S . . . . $PbSO_4 \cdot 2 PbO$ . . . . 47 (Pb): Hb/C2–588/9
$O_6Pb_3Sb_2$ . . . . . . $Pb_3(SbO_3)_2 = 3 PbO \cdot Sb_2O_3$ . . . . 47 (Pb): Hb/C3–930
O_6Pb_4S . . . . $PbSO_3 \cdot 3 PbO \cdot 0.6 H_2O$ . . . . 47 (Pb): Hb/C2–537
O_6Pb_4Si . . . . Pb_4SiO_6. . . . . 47 (Pb): Hb/C3–840/2
O_6Pd_5 . . . . Pd_5O_6 . . . . 65 (Pd): Hb/2–259
$O_6Pm_2Sc_2$. . . . . . $Sc_2O_3 \cdot Pm_2O_3 = PmScO_3$. . . . . 39 (SE): Hb/C1–113, 124, 313
O_6PrSr_2Ta . . . . . Sr_2PrTaO_6 . . . . 50 (Ta): Hb/B2–212
$O_6Pr_2Sc_2$ . . . . . . $Sc_2O_3 \cdot Pr_2O_3 = PrScO_3$ . . . . 39 (SE): Hb/C1–113, 124, 127
O_6PtS_2 . . . . $Pt(SO_3)_2$ . . . . 68 (Pt): Hb/C1–109
$O_6PtS_2{}^{2-}$ . . . . . . $[Pt(SO_3)_2]^{2-}$ . . . . 68 (Pt): Hb/C2–158
O_6PuSe_2 . . . . . . $Pu(SeO_3)_2$ . . . . Erg. W. 4–191
O_6PuSr_3. . . . . . . Sr_3PuO_6 . . . . Erg. W. 4–54/5
$O_6Rb_2S_2$ . . . . $Rb_2S_2O_6$ . . . . 24 (Rb): Hb–215/6
$O_6Rb_2S_2Se$ . . . . . $Rb_2SeS_2O_6$ . . . . 24 (Rb): Hb–218
$O_6Rb_2S_3$ . . . . $Rb_2S_3O_6$ . . . . 24 (Rb): Hb–216
$O_6Rb_2S_4$. . . . . . . $Rb_2S_4O_6$ . . . . 24 (Rb): Hb–216
O_6Rb_2U . . . . $Rb_2UO_6 \cdot 3 H_2O$. . . . . 55 (U): Hb–226
$O_6Rb_2V_2$ . . . . $Rb_2O \cdot V_2O_5$
see $O_3RbV \ldots RbVO_3$
O_6Rb_4Xe . . . . . . Rb_4XeO_6 . . . . Erg. W. 1–116
O_6RuS_2 . . . . RuS_2O_6 . . . . 63 (Ru): Hb–77
O_6SSb_2 . . . . $Sb_2O_3 \cdot SO_3 = (SbO)_2SO_4$. . . . . 18 (Sb): Hb/B3–545
–. . . . . $Sb_2O_3 \cdot SO_3 \cdot H_2O$ . . . . 18 (Sb): Hb/B3–545

Hb = Main Series Volume; GD = Gmelin-Durrer; Ab = Appended Volume to Main Series;
Eb = Supplement Volume to Main Series; Erg. W. = New Supplement Series Volume

O_6STh.	$ThO_2SO_4 \cdot 3\,H_2O$	44 (Th):	Hb–293
O_6SU	UO_2SO_4	55 (U):	Hb–151/2, 246/54
–.	$UO_2SO_4 \cdot H_2O$	55 (U):	Hb–154
–.	$UO_2SO_4 \cdot 2\,H_2O$	55 (U):	Hb–152
–.	$UO_2SO_4 \cdot 3\,H_2O$	55 (U):	Hb–153/4
–.	$UO_2SO_4 \cdot 10\,H_2O$	55 (U):	Hb–153
–.	UO_2SO_4 solutions:		
	UO_2SO_4–H_2O	55 (U):	Hb–154/5, 246/57
–.	UO_2SO_4 systems:		
	UO_2SO_4–Ag_2SO_4–H_2O	61 (Ag):	Hb/B4–358
	UO_2SO_4–$CuSO_4$–H_2SO_4–H_2O	60 (Cu):	Hb/B3–1241
	UO_2SO_4–$CuSO_4$–H_2SO_4–D_2O	60 (Cu):	Hb/B3–1241
	UO_2SO_4–$CuSO_4$–H_2SO_4		
	–$NiSO_4$–H_2O	60 (Cu):	Hb/B3–1262
	UO_2SO_4–H_2SO_4–H_2O	55 (U):	Hb–152/3
	UO_2SO_4–K_2SO_4–H_2O.	55 (U):	Hb–205
	UO_2SO_4–$(NH_4)_2SO_4$–H_2O	55 (U):	Hb–216
	UO_2SO_4–Na_2SO_4–H_2O	55 (U):	Hb–193
O_6SV_2.	$2\,VO_2 \cdot SO_2 \cdot n\,H_2O$	48 (V):	Hb/B1–281
O_6SW.	$WO_3 \cdot SO_3$	54 (W):	Hb–186
O_6S_2.	S_2O_6	9 (S):	Hb/B1–338, 351/2
$O_6S_2^{2-}$	$[S_2O_6]^{2-}$		
	see $H_2O_6S_2 \dots H_2S_2O_6$		
O_6S_2Se	$(SeSO_3)SO_3$	10 (Se):	Hb/B–179
$O_6S_2Se^{2-}$.	$[SeS_2O_6]^{2-}$	10 (Se):	Hb/B–186/9
$O_6S_2Se_2$.	$(SeSO_3)_2$	10 (Se):	Hb/B–179
$O_6S_2Se_2^{2-}$	$[Se_2S_2O_6]^{2-}$	10 (Se):	Hb/B–187, 190/1
$O_6S_2Se_3^{2-}$	$[Se_3S_2O_6]^{2-}$	10 (Se):	Hb/B–187, 190
O_6S_2Sr	SrS_2O_6	29 (Sr):	Hb–183/4
			Eb–259
–.	$SrS_2O_6 \cdot H_2O$	29 (Sr):	Eb–259
–.	$SrS_2O_6 \cdot 4\,H_2O$	29 (Sr):	Hb–183/4
			Eb–259
–.	SrS_2O_6 solid solutions:		
	SrS_2O_6–$K_2S_2O_6$–$Tl_2S_2O_6$.	38 (Tl):	Hb/3–477
	SrS_2O_6–$Tl_2S_2O_6$.	38 (Tl):	Hb/3–477
–.	SrS_2O_6 systems:		
	SrS_2O_6–H_2O.	29 (Sr):	Eb–259
	SrS_2O_6–$(NH_4)_2S_2O_6$–H_2O	29 (Sr):	Hb–234
	SrS_2O_6–$Na_2S_2O_6$–H_2O	29 (Sr):	Hb–224
	SrS_2O_6–PbS_2O_6–H_2O	47 (Pb):	Hb/C3–1198
O_6S_2Th	$Th(SO_3)_2 \cdot H_2O$	44 (Th):	Hb–283
–.	$Th(SO_3)_2 \cdot 4\,H_2O$	44 (Th):	Hb–283
$O_6S_2Tl_2$	$Tl_2S_2O_6$	38 (Tl):	Hb/3–368
–.	$Tl_2S_2O_6$ solid solutions:		
	$Tl_2S_2O_6 \cdot 2\,H_2O$–$BaS_2O_6 \cdot 2\,H_2O$	30 (Ba):	Hb–284
	$Tl_2S_2O_6$–$K_2S_2O_6$–SrS_2O_6.	38 (Tl):	Hb/3–477
	$Tl_2S_2O_6$–SrS_2O_6.	38 (Tl):	Hb/3–477

See the "Instructions for the user"
at the beginning of the volume

$O_6S_2Tl_2$ . . . . $Tl_2S_2O_6$ systems:

$Tl_2S_2O_6$–$K_2S_2O_6$–H_2O . . . . 38 (Tl):   Hb/3–459

$O_6S_2Tl_2Zn$ . . . . . $Tl_2[Zn(SO_3)_2]$ = $Tl_2SO_3 \cdot ZnSO_3$ . . . . . 38 (Tl):   Hb/3–479

O_6S_2U . . . . $U(SO_3)_2$ . . . . 55 (U):   Hb–145

O_6S_2Zn . . . . $ZnS_2O_6 \cdot 6 H_2O$ . . . . 32 (Zn):   Hb–243

Eb–966

–. . . . . $ZnS_2O_6 \cdot nH_2O$ . . . . 32 (Zn):   Eb–966

O_6S_2Zr . . . . $Zr(SO_3)_2 \cdot 7 H_2O$ . . . . 42 (Zr):   Hb–333

$O_6S_3{}^{2-}$ . . . . $[S_3O_6]^{2-}$ . . . . 9 (S):   Hb/B2–856, 979/92,

1031/2, 1041

$O_6S_3Se^{2-}$ . . . . $[SeS_3O_6]^{2-}$ . . . . 10 (Se):   Hb/B–189/91

$O_6S_3Tl_2$ . . . . $Tl_2S_3O_6$ . . . . 38 (Tl):   Hb/3–369

O_6S_3Zn . . . . ZnS_3O_6 . . . . 32 (Zn):   Hb–243

Eb–966

$O_6S_4{}^{2-}$ . . . . $[S_4O_6]^{2-}$ . . . . 9 (S):   Hb/B2–992/1011, 1032

$O_6S_4Se^{2-}$ . . . . $[SeS_4O_6]^{2-}$ . . . . 10 (Se):   Hb/B–190/1

O_6S_4Sr . . . . $SrS_4O_6 \cdot 2 H_2O$ . . . . 29 (Sr):   Hb–184

–. . . . . $SrS_4O_6 \cdot 6 H_2O$ . . . . 29 (Sr):   Hb–184

O_6S_4Th . . . . $Th(S_2O_3)_2$ . . . . 44 (Th):   Hb–293

$O_6S_4Tl_2{}^{2-}$ . . . . $[Tl_2(S_2O_3)_2]^{2-}$ . . . . 38 (Tl):   Hb/3–368

O_6S_4Zn . . . . ZnS_4O_6 . . . . 32 (Zn):   Hb–243

Eb–966

O_6S_4Zr . . . . $Zr(S_2O_3)_2$ . . . . 42 (Zr):   Hb–353

$O_6S_5{}^{2-}$ . . . . $[S_5O_6]^{2-}$ . . . . 9 (S):   Hb/B2–1011/24,

1032/3, 1041/2

O_6S_5Zn . . . . ZnS_5O_6 . . . . 32 (Zn):   Hb–243

Eb–966

$O_6S_6{}^{2-}$ . . . . $[S_6O_6]^{2-}$ . . . . 9 (S):   Hb/B2–1024/8, 1033,

1036/46

O_6S_6Sr . . . . $SrS_6O_6 \cdot 2 H_2O$ . . . . 29 (Sr):   Eb–260

O_6Sb_2Sr . . . . $Sr(SbO_3)_2$ . . . . 29 (Sr):   Hb–220

Eb–294/5

–. . . . . – (doped with Nb_2O_5) . . . . 49 (Nb):   Hb/B3–31

–. . . . . $Sr(SbO_3)_2 \cdot 6 H_2O$ . . . . 29 (Sr):   Hb–220

O_6Sb_2Zn . . . . $ZnSb_2O_6$ = $Zn(SbO_3)_2$ . . . . 32 (Zn):   Hb–290/1

Eb–999/1000

–. . . . . $Zn(SbO_3)_2 \cdot nH_2O$ . . . . 32 (Zn):   Hb–290/1

$O_6ScTaTi$ . . . . $ScTiTaO_6$ . . . . 50 (Ta):   Hb/B2–227

$O_6Sc_2Sm_2$ . . . . . . $Sc_2O_3 \cdot Sm_2O_3$ = $SmScO_3$ . . . . 39 (SE):   Hb/C1–113, 124, 321/2

$O_6Sc_2Y_2$ . . . . $Sc_2O_3 \cdot Y_2O_3$ = $YScO_3$ . . . . 39 (SE):   Hb/C1–113, 124/5,

127, 205

O_6SeTi_2 . . . . $2 TiO_2 \cdot SeO_2 \cdot H_2O$ . . . . 41 (Ti):   Hb–353

O_6SeU . . . . $UO_2SeO_4 \cdot xH_2O$ . . . . 55 (U):   Hb–158

O_6Se_2Th . . . . $Th(SeO_3)_2 \cdot H_2O$ . . . . 44 (Th):   Hb–294

–. . . . . $Th(SeO_3)_2 \cdot 8 H_2O$ . . . . 44 (Th):   Hb–294

O_6Se_2U . . . . $U(SeO_3)_2$ . . . . 55 (U):   Hb–157

O_6Se_2Zr . . . . $Zr(SeO_3)_2$ . . . . 42 (Zr):   Hb–354

–. . . . . $Zr(SeO_3)_2 \cdot H_2O$ . . . . 42 (Zr):   Hb–354

O_6SiZn_4 . . . . $ZnSiO_3 \cdot 3 ZnO$ . . . . 32 (Zn):   Eb–993

Hb = Main Series Volume; GD = Gmelin-Durrer;   *Ab = Appended Volume to Main Series;*

Eb = Supplement Volume to Main Series;   *Erg. W. = New Supplement Series Volume*

$O_6Si_2^{4-}$ . . . . $[Si_2O_6]^{4-}$ . . . . 15 (Si): Hb/B–421/2

O_6Si_2Th . . . . $ThSi_2O_6$. . . . . 44 (Th): Hb–313

O_6SmSr_2Ta . . . . . Sr_2SmTaO_6. . . . . 50 (Ta): Hb/B2–214

$O_6SmTaTi$ . . . . . . $SmTiTaO_6$ . . . . 50 (Ta): Hb/B2–227/8

$O_6Sn_{0.4}Sr_{1.2}Ta_{1.6}$ . . . $Sr_{1.2}(Ta_{0.8}Sn_{0.2})_2O_6$
                            $= 3 SrO \cdot SnO_2 \cdot 2 Ta_2O_5$ . . . . 50 (Ta): Hb/B2–232

O_6SnTa_2 . . . . $SnTa_2O_6$ . . . . 50 (Ta): Hb/B2–231

O_6Sn_5 . . . . Sn_5O_6 . . . . 46 (Sn): Hb/C1–58/9

O_6SrTa_2 . . . . $SrTa_2O_6$ . . . . 50 (Ta): Hb/B2–181, 182

–. . . . . $SrTa_2O_6$ solid solutions:
                            $(Sr,Ba,Pb)Ta_2O_6$ . . . . 50 (Ta): Hb/B2–239
                            $(Sr,Pb)Ta_2O_6$ . . . . 50 (Ta): Hb/B2–239

–. . . . . $SrTa_2O_6$ systems:
                            $SrTa_2O_6–SrZrO_3$ . . . . 50 (Ta): Hb/B2–228/9

O_6SrV_2 . . . . $Sr(VO_3)_2 = SrO \cdot V_2O_5$ . . . . . . 48 (V): Hb/B2–520

–. . . . . $Sr(VO_3)_2 \cdot 4 H_2O$ . . . . 48 (V): Hb/B2–520/1

$O_6Sr_{1.2}Ta_{1.6}Zr_{0.4}$ . . . $Sr_{1.2}(Zr_{0.2}Ta_{0.8})_2O_6$
                            $= 3 SrO \cdot ZrO_2 \cdot 2 Ta_2O_5$ . . . . . . 50 (Ta): Hb/B2–228/9

O_6Sr_2TaTb . . . . . Sr_2TbTaO_6 . . . . 50 (Ta): Hb/B2–217

O_6Sr_2TaTm . . . . . Sr_2TmTaO_6 . . . . 50 (Ta): Hb/B2–220

O_6Sr_2TaY . . . . . . Sr_2YTaO_6 . . . . 50 (Ta): Hb/B2–207

O_6Sr_2TaYb . . . . . Sr_2YbTaO_6 . . . . 50 (Ta): Hb/B2–221

$O_6TaTbTi$ . . . . $TbTiTaO_6$ solid solutions:
                            $TbTi(Ta,Nb)O_6$. . . . . 50 (Ta): Hb/B2–264

O_6TaTiY . . . . . . $YTiTaO_6$ . . . . 50 (Ta): Hb/B2–227

$O_6TaTiYb$ . . . . . . $YbTiTaO_6$. . . . . 50 (Ta): Hb/B2–227/8

$O_6Ta_2^-$ . . . . $[Ta_2O_6]^-$ . . . . 50 (Ta): Hb/B1–59

O_6Ta_2V . . . . VTa_2O_6 . . . . 50 (Ta): Hb/B2–244

O_6Ta_2Zn . . . . $ZnTa_2O_6$ . . . . 50 (Ta): Hb/B2–193/4, 349

–. . . . . $ZnTa_2O_6$ solid solutions:
                            $ZnTa_2O_6–BaTiO_3$ . . . . 50 (Ta): Hb/B2–226

–. . . . . $ZnTa_2O_6$ systems:
                            $ZnTa_2O_6–GeO_2$ . . . . 50 (Ta): Hb/B2–230
                            $ZnTa_2O_6–SnO_2$ . . . . 50 (Ta): Hb/B2–232
                            $ZnTa_2O_6–TiO_2$. . . . . 50 (Ta): Hb/B2–226

O_6TeZn_3 . . . . . . Zn_3TeO_6 . . . . 32 (Zn): Hb–248

O_6ThTi_2 . . . . $ThO_2 \cdot 2 TiO_2$ . . . . 44 (Th): Hb–353/4

$O_6Ti_2Zn_2$ . . . . . . $Zn_2TiO_4 \cdot TiO_2 = ZnTiO_3$ . . . . 41 (Ti): Hb–460, 461/2

$O_6Tl_2V_2$ . . . . . . $Tl_2O \cdot V_2O_5 = TlVO_3$ . . . . 48 (V): Hb/B2–548

O_6UW . . . . $UO_3 \cdot WO_3 \cdot 2 H_2O$ . . . . 55 (U): Hb–244

O_6V^{3-} . . . . $[V(O)_2(O_2)_2]^{3-}$ . . . . 48 (V): Hb/A2–588/9
                                                                                        Hb/B1–162/3, 165
                                                                                        Hb/B2–424

O_6V_2Zn . . . . $ZnO \cdot V_2O_5 = Zn(VO_3)_2$ . . . . 48 (V): Hb/B2–534, 535

–. . . . . $Zn(VO_3)_2 \cdot 2 H_2O$ . . . . 48 (V): Hb/B2–535

$O_6V_3^+$ . . . . $[V_3O_6]^+$ . . . . 48 (V): Hb/B1–84

O_6Xe^{4-} . . . . $[XeO_6]^{4-}$ . . . . Erg. W. 1–109

See the "Instructions for the user"
at the beginning of the volume

O_6Y_2	$Y_2(O_2)_3$	39 (SE):	Hb/C1−204/5
$O_{6.3}Ta_2Tl_2$.	$Tl_2Ta_2O_{6.3}$	50 (Ta):	Hb/B2−201
O_7^+	O_7^+	3 (O):	Hb/3−354
O_7^-	O_7^-	3 (O):	Hb/3−354
$O_7P_2^{4-}$	$[P_2O_7]^{4-}$	16 (P):	Hb/B−348/9, 377/9
			Hb/C−224, 228/39
O_7P_2Pb	PbP_2O_7	47 (Pb):	Hb/C3−877/8
$O_7P_2Pb^{2-}$	$[PbP_2O_7]^{2-}$	16 (P):	Hb/C−238/9
		47 (Pb):	Hb/B1−367
			Hb/C3−874
$O_7P_2Pb_2$.	$Pb_2P_2O_7$.	47 (Pb):	Hb/C3−873
−.	$Pb_2P_2O_7$ solutions:		
	$Pb_2P_2O_7$–H_2O	47 (Pb):	Hb/B2−371/80
			Hb/C3−874
O_7P_2Pt	PtP_2O_7	68 (Pt):	Hb/C1−137
O_7P_2Pu	PuP_2O_7	Erg. W. 4−252/3	
$O_7P_2Rb_4$.	$Rb_4P_2O_7$.	24 (Rb):	Hb−234
O_7P_2Si	$P_2O_5 \cdot SiO_2$	16 (P):	Hb/C−638/9
−.	$P_2O_5 \cdot SiO_2 \cdot H_2O$	16 (P):	Hb/C−640
−.	$P_2O5 \cdot SiO_2 \cdot 2 H_2O$.	16 (P):	Hb/C−640
$O_7P_2Sn^{2-}$	$[SnP_2O_7]^{2-}$	16 (P):	Hb/C−239
		46 (Sn):	Hb/B−387/8
$O_7P_2Sn_2$.	$Sn_2P_2O_7$	46 (Sn):	Hb/B−329/31
$O_7P_2Sr^{2-}$	$[SrP_2O_7]^{2-}$	16 (P):	Hb/C−238
$O_7P_2Sr_2$	$Sr_2P_2O_7$	29 (Sr):	Hb−215
			Eb−285/6
−.	$Sr_2P_2O_7 \cdot H_2O$	29 (Sr):	Hb−215
−.	$Sr_2P_2O_7 \cdot 2.5 H_2O$	29 (Sr):	Hb−215
O_7P_2Th	$ThP_2O_7 \cdot 2 H_2O$	44 (Th):	Hb−318/9
O_7P_2Ti.	$TiO_2 \cdot P_2O_5$	41 (Ti):	Hb−380
$O_7P_2Tl^{3-}$	$[TlP_2O_7]^{3-}$	16 (P):	Hb/C−238
$O_7P_2Tl_4$	$Tl_4P_2O_7$	38 (Tl):	Hb/3−430/1
−.	$Tl_4P_2O_7 \cdot 2 H_2O$	38 (Tl):	Hb/3−431
O_7P_2U	UP_2O_7	55 (U):	Hb−182
−.	$UP_2O_7 \cdot 3 H_2O$	55 (U):	Hb−181
$O_7P_2Zn^{2-}$	$[ZnP_2O_7]^{2-}$	16 (P):	Hb/C−238, 239
$O_7P_2Zn_2$	$Zn_2P_2O_7$	32 (Zn):	Hb−279
			Eb−996
−.	$Zn_2P_2O_7 \cdot 5 H_2O$	32 (Zn):	Eb−996
O_7P_2Zr	$ZrP_2O_7 = ZrO_2 \cdot P_2O_5$	42 (Zr):	Hb−396/8
−.	$ZrO_2 \cdot P_2O_5 \cdot 1.5 H_2O$	42 (Zr):	Hb−393
−.	$ZrO_2 \cdot P_2O_5 \cdot 2 H_2O$.	42 (Zr):	Hb−393
−.	$ZrO_2 \cdot P_2O_5 \cdot nH_2O$	42 (Zr):	Hb−393
O_7PbS_2	PbS_2O_7	47 (Pb):	Hb/C2−600/1
O_7PbTi_3	$PbTi_3O_7$.	47 (Pb):	Hb/C4−1390
O_7PbU_2	$PbO \cdot 2 UO_3$	55 (U):	Hb−239/40
−.	$PbO \cdot 2 UO_3 \cdot 2.5 H_2O$	55 (U):	Hb−240
O_7PbW_2	$PbO \cdot 2 WO_3 \cdot 2 H_2O$	54 (W):	Hb−319

Hb = Main Series Volume; GD = Gmelin-Durrer;   *Ab = Appended Volume to Main Series;*
Eb = Supplement Volume to Main Series;   *Erg. W. = New Supplement Series Volume*

$O_7Pb_2Pr_2$	$Pr_2Pb_2O_7$	47 (Pb):	Hb/C4–1373
$O_7Pb_2Rb_2Si_2$	$Rb_2Pb_2Si_2O_7$	47 (Pb):	Hb/C3–1145
$O_7Pb_2Ru_2$	$Pb_2Ru_2O_7$	63 (Ru):	Eb–259
$O_7Pb_2Sb_2$	$Pb_2Sb_2O_7$	47 (Pb):	Hb/C3–932
–.	$Pb_2Sb_2O_7 \cdot nH_2O$	47 (Pb):	Hb/C3–932
$O_7Pb_2Se_2$	$Pb_2Se_2O_7$	47 (Pb):	Hb/C2–636
$O_7Pb_2Si_2Zn$	$Pb_2ZnSi_2O_7$	47 (Pb):	Hb/C4–1229
$O_7Pb_2Sm_2$	$Sm_2Pb_2O_7$	47 (Pb):	Hb/C4–1373
$O_7Pb_2Ta_2$	$Pb_2Ta_2O_7$	50 (Ta):	Hb/B2–232/3, 237/8
$O_7Pb_2Tb_2$	$Tb_2Pb_2O_7$	47 (Pb):	Hb/C4–1373
$O_7Pb_2V_2$	$2\,PbO \cdot V_2O_5 = Pb_2V_2O_7$	48 (V):	Hb/B2–567/8, 569
–.	$2\,PbO \cdot V_2O_5 \cdot xH_2O = Pb_2V_2O_7 \cdot xH_2O$ . .	48 (V):	Hb/B2–569/70
$O_7Pb_2Y_2$	$Y_2Pb_2O_7$	47 (Pb):	Hb/C4–1373
$O_7Pb_3Sb_2$	$Pb_3Sb_2O_7 = 3\,PbO \cdot Sb_2O_4$	47 (Pb):	Hb/C3–930
$O_7Pb_3Si_2$	$Pb_3Si_2O_7$	47 (Pb):	Hb/C3–847/8
–.	$Pb_3Si_2O_7$ solid solutions:		
	$Pb_9(Si_2O_7)_3 – Pb_5(SiO_4)_2SO_4$	47 (Pb):	Hb/C3–857
O_7Pb_4S	$PbSO_4 \cdot 3\,PbO \cdot H_2O$	47 (Pb):	Hb/C2–589/91
O_7Pb_5Se	$PbSeO_3 \cdot 4\,PbO$	47 (Pb):	Hb/C2–635/6
$O_7Pr_2Ru_2$	$Pr_2Ru_2O_7$	63 (Ru):	Eb–258
$O_7Pu_2Zr_2$	$Pu_2Zr_2O_7$	Erg. W. 4–70	
$O_7Rb_2S_2$	$Rb_2S_2O_7 = Rb_2SO_4 \cdot SO_3$	24 (Rb):	Hb–214
$O_7Rb_4V_2$	$2\,Rb_2O \cdot V_2O_5 = Rb_4V_2O_7$	48 (V):	Hb/B2–486
O_7Re_2	Re_2O_7	70 (Re):	Hb–93/5
$O_7Ru_2Sm_2$	$Sm_2Ru_2O_7$	63 (Ru):	Eb–258
$O_7Ru_2Tb_2$	$Tb_2Ru_2O_7$	63 (Ru):	Eb–258
$O_7Ru_2Tm_2$	$Tm_2Ru_2O_7$	63 (Ru):	Eb–258
$O_7Ru_2Y_2$	$Y_2Ru_2O_7$	63 (Ru):	Eb–258
$O_7Ru_2Yb_2$	$Yb_2Ru_2O_7$	63 (Ru):	Eb–258
O_7SSe^{2-}	$[SeSO_7]^{2-}$	10 (Se):	Hb/B–185
O_7STe_2	$2\,TeO_2 \cdot SO_3$	11 (Te):	Hb–359/60
O_7STi_2	$2\,TiO_2 \cdot SO_3$	41 (Ti):	Hb–349, 351
–.	$2\,TiO_2 \cdot SO_3 \cdot xH_2O$	41 (Ti):	Hb–349
O_7SZn_4	$ZnSO_4 \cdot 3\,ZnO \cdot H_2O$	32 (Zn):	Hb–241
–.	$ZnSO_4 \cdot 3\,ZnO \cdot 2\,H_2O$	32 (Zn):	Hb–241
–.	$ZnSO_4 \cdot 3\,ZnO \cdot 3\,H_2O$	32 (Zn):	Hb–241
–.	$ZnSO_4 \cdot 3\,ZnO \cdot 5\,H_2O$	32 (Zn):	Hb–241
–.	$ZnSO_4 \cdot 3\,ZnO \cdot 6\,H_2O$	32 (Zn):	Hb–241
–.	$ZnSO_4 \cdot 3\,ZnO \cdot 7\,H_2O$	32 (Zn):	Hb–241
–.	$ZnSO_4 \cdot 3\,ZnO \cdot 8\,H_2O$	32 (Zn):	Hb–241
–.	$ZnSO_4 \cdot 3\,ZnO \cdot xH_2O$	32 (Zn):	Eb–956
O_7SZr_2	$ZrOSO_4 \cdot ZrO_2 = Zr(SO_4)_2 \cdot 3\,ZrO_2$	42 (Zr):	Hb–350
–.	$ZrOSO_4 \cdot ZrO_2 \cdot xH_2O$	42 (Zr):	Hb–350/1
$O_7S_{1.5}V_2$	$2\,VO_2 \cdot 1.5\,SO_2 \cdot nH_2O$	48 (V):	Hb/B1–281
O_7S_2	S_2O_7	9 (S):	Hb/B1–367/70
–.	$(S_2O_7)_x$	9 (S):	Hb/B1–367/70
O_7S_2Sr	SrS_2O_7	29 (Sr):	Hb–181
$O_7S_2Tl_2$	$Tl_2S_2O_7 = Tl_2SO_4 \cdot SO_3$	38 (Tl):	Hb/3–367

See the "Instructions for the user"
at the beginning of the volume

O_7S_2V . . . . $VOS_2O_6 \cdot xH_2O$ . . . . 48 (V): Hb/B1−307
O_7S_2Zn . . . . ZnS_2O_7 . . . . 32 (Zn): Hb−241
$O_7S_2Zn_3$ . . . . $ZnO \cdot 2 ZnSO_3 \cdot 3 H_2O$ . . . . 32 (Zn): Eb−935
$O_7Sb_2Se_2$ . . . . $Sb_2O_3 \cdot 2 SeO_2 \cdot 0.5 H_2O$ . . . . 18 (Sb): Hb/B3−553
$O_7Sb_2Sr_2$ . . . . $Sr_2Sb_2O_7$ . . . . 29 (Sr): Eb−294
O_7Sb_2Zr . . . . $ZrO_2 \cdot Sb_2O_5 \cdot 7.5 H_2O$ . . . . 42 (Zr): Hb−400
O_7Sb_4 . . . . Sb_4O_7 . . . . 18 (Sb): Hb/B2−359/60
O_7Sc_3Ta . . . . Sc_3TaO_7 . . . . 50 (Ta): Hb/B2−205
O_7SeTe_2 . . . . $2 TeO_2 \cdot SeO_3$ . . . . 11 (Te): Hb−363
O_7SeTi_2 . . . . $2 TiO_2 \cdot SeO_3 \cdot H_2O$. . . . . 41 (Ti): Hb−353
$O_7Si_2Tl_6$ . . . . $3 Tl_2O \cdot 2 SiO_2 \cdot H_2O$ . . . . 38 (Tl): Hb/3−424
$O_7Sm_2Ti_2$ . . . . . . $Sm_2Ti_2O_7$ . . . . 41 (Ti): Hb−468
O_7Sm_3Ta . . . . . . Sm_3TaO_7 . . . . 50 (Ta): Hb/B2−214
O_7Sn_2 . . . . $Sn_2O_7 \cdot 3 H_2O$ . . . . 46 (Sn): Hb/C1−139
−. . . . . $Sn_2O_7 \cdot 5 H_2O$ . . . . 46 (Sn): Hb/C1−139
$O_7Sn_2Sr_3$ . . . . . . $Sr_3Sn_2O_7$ . . . . 29 (Sr): Eb−160
$O_7Sn_2Ta_2$ . . . . . . $Sn_2Ta_2O_7$ . . . . 50 (Ta): Hb/B2−231/2
−. . . . . $Sn_2Ta_2O_7$ solid solutions:
                               $Sn_2(Ta,Nb)_2O_7$ . . . . 50 (Ta): Hb/B2−264
O_7SrU_2 . . . . $SrO \cdot 2 UO_3$. . . . . 55 (U): Hb−232
O_7SrW_2 . . . . $SrO \cdot 2 WO_3 \cdot 3 H_2O$ . . . . 54 (W): Hb−296
$O_7Sr_2Ta_2$ . . . . . . $Sr_2Ta_2O_7$ . . . . 50 (Ta): Hb/B2−181, 183
−. . . . . $Sr_2Ta_2O_7$ solid solutions:
                               $(Sr,Ba)_2Ta_2O_7$ . . . . 50 (Ta): Hb/B2−192/3
                               $(Sr,Ca)_2Ta_2O_7$ . . . . 50 (Ta): Hb/B2−185
                               $Sr_2Ta_2O_7 − NaNbO_3$ . . . . 50 (Ta): Hb/B2−262
                               $Sr_2(Ta,Nb)_2O_7$ . . . . 50 (Ta): Hb/B2−262
$O_7Sr_2V_2$ . . . . $2 SrO \cdot V_2O_5 = Sr_2V_2O_7$ . . . . 48 (V): Hb/B2−519/20
−. . . . . $Sr_2V_2O_7 \cdot 10 H_2O$ . . . . 48 (V): Hb/B2−520
O_7TaY_3 . . . . Y_3TaO_7 . . . . 50 (Ta): Hb/B2−205, 206
$O_7Ta_2{}^-$ . . . . $[Ta_2O_7]^-$ . . . . 50 (Ta): Hb/B1−59
O_7Ta_2Ti . . . . $TiTa_2O_7$ . . . . 50 (Ta): Hb/B2−221/4
−. . . . . $TiTa_2O_7$ solid solutions:
                               $Ti(Ta,Nb)_2O_7$ . . . . 50 (Ta): Hb/B2−264
$O_7Ta_2Tl_2$ . . . . . . $Tl_2Ta_2O_7$ . . . . 50 (Ta): Hb/B2−201
$O_7Ta_2Zn_2$ . . . . . . $Zn_2Ta_2O_7$ . . . . 50 (Ta): Hb/B2−194
O_7Tb_4 . . . . $Tb_4O_7 = TbO_{1.75}$ . . . . 39 (SE): Hb/C1−130, 156,
                                                                                                      397/8
−. . . . . $Tb_4O_7 \cdot nH_2O$ . . . . 39 (SE): Hb/C2−2
O_7Tc_2 . . . . Tc_2O_7 . . . . 14 (C): Hb/C2−168
O_7TeZn_4 . . . . . . $4 ZnO \cdot TeO_3$ . . . . 32 (Zn): Eb−970
−. . . . . $ZnTeO_4 \cdot 3 ZnO \cdot 4 H_2O$ . . . . 32 (Zn): Eb−970
O_7Te_3 . . . . $Te_3O_7 = 2 TeO_2 \cdot TeO_3$ . . . . 11 (Te): Hb−282
O_7ThV_2 . . . . $ThO_2 \cdot V_2O_5 = ThV_2O_7$. . . . . 48 (V): Hb/B2−565
−. . . . . $ThV_2O_7 \cdot H_2O$ . . . . 48 (V): Hb/B2−565
O_7Th_2 . . . . $Th_2O_7 \cdot 4 H_2O$. . . . . 44 (Th): Hb−240
$O_7Ti_2Y_2$ . . . . $Y_2Ti_2O_7$ . . . . 41 (Ti): Hb−468
$O_7Ti_2Zn_3$ . . . . . . $3 ZnO \cdot 2 TiO_2$ . . . . 41 (Ti): Hb−462

Hb = Main Series Volume; GD = Gmelin-Durrer;   *Ab = Appended Volume to Main Series;*
Eb = Supplement Volume to Main Series;   *Erg. W. = New Supplement Series Volume*

O_7Ti_3Zn	$ZnO \cdot 3\,TiO_2$	41 (Ti):	Hb–462
$O_7Tl_2W_2$	$Tl_2O \cdot 2\,WO_3 \cdot H_2O$	54 (W):	Hb–314
$O_7Tl_4V_2$	$2\,Tl_2O \cdot V_2O_5 = Tl_4V_2O_7$	48 (V):	Hb/B2–547
O_7U_3	U_3O_7 systems:		
	U_3O_7–PuO_2	Erg. W. 4–60	
O_7V^{3-}	$[V(O)(O_2)_3]^{3-}$	48 (V):	Hb/B1–162/3, 164
$O_7V_2^{4-}$	$[V_2O_7]^{4-}$	48 (V):	Hb/A2–498
			Hb/B1–132/45, 146/8
$O_7V_2Zn_2$	$2\,ZnO \cdot V_2O_5 = Zn_2V_2O_7$	48 (V):	Hb/B2–534, 535
O_7V_2Zr	$ZrO_2 \cdot V_2O_5 = ZrV_2O_7$	48 (V):	Hb/B2–562, 563
O_7V_3	$V_3O_7 \cdot 3\,H_2O$	48 (V):	Hb/B1–70
O_7V_4	$V_4O_7 = VO_{1.75}$	48 (V):	Hb/B1–13, 30, 50/1
O_7W_2Zn	$ZnO \cdot 2\,WO_3 \cdot 3\,H_2O$	54 (W):	Hb–306
O_7Zn_4	Zn_4O_7	32 (Zn):	Hb–145
–.	$Zn_4O_7 \cdot 2\,H_2O$	32 (Zn):	Hb–146
$O_{7.5}SZn_{4.5}$	$ZnSO_4 \cdot 3.5\,ZnO \cdot 3\,H_2O$	32 (Zn):	Eb–963/4
–.	$ZnSO_4 \cdot 3.5\,ZnO \cdot 7\,H_2O$	32 (Zn):	Hb–240
–.	$ZnSO_4 \cdot 3.5\,ZnO \cdot 12\,H_2O$	32 (Zn):	Hb–240
$O_8P_2^{4-}$	$[P_2O_8]^{4-}$	16 (P):	Hb/C–300/2
$O_8P_2PbZn_2$	$Zn_2Pb(PO_4)_2$	47 (Pb):	Hb/C1–117
$O_8P_2Pb_2$	$Pb_2P_2O_8$	47 (Pb):	Hb/C3–879
$O_8P_2Pb_3$	$Pb_3(PO_4)_2$	47 (Pb):	Hb/C3–865/9
–.	$Pb_3(PO_4)_2$ systems:		
	$Pb_3(PO_4)_2$–$BaTiO_3$	47 (Pb):	Hb/C4–1413
	$Pb_3(PO_4)_2$–$PbCl_2$	47 (Pb):	Hb/C3–881
	$Pb_3(PO_4)_2$–PbF_2	47 (Pb):	Hb/C3–880
	$Pb_3(PO_4)_2$–$PbZrO_3$	47 (Pb):	Hb/C4–1431
$O_8P_2Rb_4$	$Rb_4P_2O_8$	24 (Rb):	Hb–234
O_8P_2SrTh	$SrO \cdot ThO_2 \cdot P_2O_5 = Sr[Th(PO_4)_2]$	44 (Th):	Hb–348
O_8P_2SrU	$SrU(PO_4)_2$	55 (U):	Hb–232
$O_8P_2Sr_3$	$Sr_3(PO_4)_2$	29 (Sr):	Hb–212/3
			Eb–283/4
–.	$Sr_3(PO_4)_2$ systems:		
	$Sr_3(PO_4)_2$–$SrCl_2$	29 (Sr):	Hb–216
	$Sr_3(PO_4)_2$–SrF_2	29 (Sr):	Hb–216
O_8P_2U	$UO_2(PO_3)_2$	55 (U):	Hb–184
$O_8P_2Zn_2$	$Zn_2P_2O_8 \cdot 4\,H_2O$	32 (Zn):	Eb–997
$O_8P_2Zn_3$	$Zn_3(PO_4)_2$	32 (Zn):	Hb–277/8
			Eb–995
–.	$Zn_3(PO_4)_2 \cdot 4\,H_2O$	32 (Zn):	Hb–277/8
			Eb–995/6
$O_8PbRb_2S_2$	$Rb_2Pb(SO_4)_2$	47 (Pb):	Hb/C3–1144
$O_8PbRb_2Se_2$	$Rb_2Pb(SeO_4)_2$	47 (Pb):	Hb/C3–1144
O_8PbS_2	$Pb(SO_4)_2$	47 (Pb):	Hb/C2–580/1
–.	$PbS_2O_8 \cdot 3\,H_2O$	47 (Pb):	Hb/C2–601
$O_8PbS_2Tl_2$	$Tl_2Pb(SO_4)_2$	47 (Pb):	Hb/C4–1353
$O_8PbSe_2Tl_2$	$Tl_2Pb(SeO_4)_2$	47 (Pb):	Hb/C4–1353

See the "Instructions for the user"
at the beginning of the volume

O_8PbV_2 . . . . Pb(VO$_4$)$_2$ . . . . 48 (V):   Hb/B2–570/1

$O_8Pb_3S_2^{2+}$ . . . . . . [Pb$_3$(SO$_4$)$_2$]$^{2+}$ . . . . 47 (Pb):   Hb/C2–580

$O_8Pb_3Sb_2$ . . . . . . Pb$_3$(SbO$_4$)$_2$ . . . . 47 (Pb):   Hb/C3–931

$O_8Pb_3Ta_2$ . . . . . . Pb$_3$Ta$_2$O$_8$ . . . . 50 (Ta):   Hb/B2–232/3, 238

$O_8Pb_3V_2$ . . . . 3 PbO · V$_2$O$_5$ = Pb$_3$(VO$_4$)$_2$ . . . . 48 (V):   Hb/B2–568/9

–. . . . . Pb$_3$(VO$_4$)$_2$ solid solutions:
                        (Pb,Pr)$_3$(VO$_4$)$_2$ . . . . 48 (V):   Hb/B2–569

–. . . . . Pb$_3$(VO$_4$)$_2$ systems:
                        Pb$_3$(VO$_4$)$_2$–LiCl–Li$_3$VO$_4$–PbCl$_2$ . . . . . 48 (V):   Hb/B2–573
                        Pb$_3$(VO$_4$)$_2$–PbCl$_2$ . . . . 48 (V):   Hb/B2–571

O_8Pb_5 . . . . Pb$_5$O$_8$ . . . . 47 (Pb):   Hb/C1–144/7

O_8Pb_5S . . . . . . PbSO$_4$ · 4 PbO . . . . 47 (Pb):   Hb/C2–591/3

$O_8Pb_5Sb_2$ . . . . . . Pb$_5$(SbO$_4$)$_2$ = 5 PbO · Sb$_2$O$_3$ . . . . . 47 (Pb):   Hb/C3–929

O_8Pb_5Se . . . . . PbSeO$_4$ · 4 PbO . . . . 47 (Pb):   Hb/C2–638/9

O_8Pb_6Si . . . . . . 6 PbO · SiO$_2$ · 4 H$_2$O . . . . 47 (Pb):   Hb/C3–839

$O_8Pr_3V_2$ . . . . Pr$_3$(VO$_4$)$_2$ solid solutions:
                        (Pb,Pr)$_3$(VO$_4$)$_2$ . . . . 48 (V):   Hb/B2–569

O_8PtS_2 . . . . Pt(SO$_4$)$_2$ . . . . 68 (Pt):   Hb/C1–111

–. . . . . Pt(SO$_4$)$_2$ · 4 H$_2$O
                        = H$_2$[Pt(SO$_4$)$_2$(OH)$_2$] · 2 H$_2$O . . . . . . 68 (Pt):   Hb/C1–111/2

O_8PuRbS_2 . . . . . RbPu(SO$_4$)$_2$ · 4 H$_2$O . . . . Erg. W. 4–176/7

O_8PuS_2 . . . . Pu(SO$_4$)$_2$ . . . . Erg. W. 4–178/9, 184

–. . . . . Pu(SO$_4$)$_2$ · 0.33 H$_2$O . . . . Erg. W. 4–181/2

–. . . . . Pu(SO$_4$)$_2$ · 4 H$_2$O . . . . Erg. W. 4–179/82

–. . . . . Pu(SO$_4$)$_2$ · xH$_2$O . . . . Erg. W. 4–179

O_8PuS_2Tl . . . . . TlPu(SO$_4$)$_2$ · 4 H$_2$O . . . . Erg. W. 4–176, 178

O_8RbS_2Sb . . . . . Rb[Sb(SO$_4$)$_2$] . . . . 24 (Rb):   Hb–239

O_8RbS_2Ti . . . . . . RbTi(SO$_4$)$_2$ · 12 H$_2$O . . . . 41 (Ti):   Hb–413

O_8RbS_2Tl . . . . . . RbTl(SO$_4$)$_2$ = Rb$_2$SO$_4$ · Tl$_2$(SO$_4$)$_3$ . . . . . 38 (Tl):   Hb/3–469

–. . . . . RbTl(SO$_4$)$_2$ · 4 H$_2$O . . . . 38 (Tl):   Hb/3–469

O_8RbS_2V . . . . . . RbV(SO$_4$)$_2$ · 6 H$_2$O . . . . 48 (V):   Hb/B2–491

–. . . . . RbV(SO$_4$)$_2$ · 12 H$_2$O . . . . 48 (V):   Hb/B2–489/91

$O_8Rb_2S_2$ . . . . . . Rb$_2$S$_2$O$_8$ . . . . 24 (Rb):   Hb–214

$O_8Rb_2S_2V$ . . . . . . Rb$_2$V(SO$_4$)$_2$ · 6 H$_2$O . . . . 48 (V):   Hb/B2–489

$O_8Rb_2S_2Zn$ . . . . . Rb$_2$Zn(SO$_4$)$_2$ · 6 H$_2$O
                        = [Zn(H$_2$O)$_6$] (SO$_4$Rb)$_2$ . . . . 32 (Zn):   Hb–111, 318
                                                                  Eb–1020

$O_8Rb_2Se_2Zn$ . . . . ZnSeO$_4$ · Rb$_2$SeO$_4$ · 6 H$_2$O . . . . 32 (Zn):   Hb–319

O_8Rb_2W . . . . Rb$_2$WO$_8$ · 3 H$_2$O . . . . 54 (W):   Hb–283

O_8Rb_3Ta . . . . . . Rb$_3$TaO$_8$ . . . . 50 (Ta):   Hb/B2–171

O_8Rb_4U . . . . . . Rb$_4$UO$_8$ · 8 H$_2$O . . . . 55 (U):   Hb–226

$O_8Rb_6V_2$ . . . . 3 Rb$_2$O · V$_2$O$_5$ = Rb$_3$VO$_4$ . . . . . . 48 (V):   Hb/B2–486

O_8Re_2 . . . . Re$_2$O$_8$ . . . . 70 (Re):   Hb–95/6

O_8Re_2Sr . . . . Sr(ReO$_4$)$_2$ · 2 H$_2$O . . . . 70 (Re):   Hb–146

O_8Re_2Zn . . . . Zn(ReO$_4$)$_2$ · xH$_2$O . . . . 70 (Re):   Hb–147

O_8RuS_2 . . . . Ru(SO$_4$)$_2$ . . . . 63 (Ru):   Hb–76

O_8STa_2 . . . . Ta$_2$O$_4$SO$_4$ = (TaO$_2$)$_2$SO$_4$ . . . . 50 (Ta):   Hb/B1–177

–. . . . . Ta$_2$O$_4$SO$_4$ · 3 H$_2$O . . . . 50 (Ta):   Hb/B1–177

O_8STa_2	$Ta_2O_5SO_3 \cdot nH_2O$	50 (Ta):	Hb/B1−176
O_8SV_2	$(VO_2)_2SO_4$	48 (V):	Hb/A2−505, 581, 637/8
−.	$V_2O_5 \cdot SO_3$	48 (V):	Hb/B1−302
O_8SZn_5	$ZnSO_4 \cdot 4\,ZnO$	32 (Zn):	Hb−240
−.	$ZnSO_4 \cdot 4\,ZnO \cdot 4\,H_2O$	32 (Zn):	Hb−241
$O_8S_2{}^{2-}$	$[S_2O_8]^{2-}$		

see $H_2O_8S_2$ ... $H_2S_2O_8$ solutions

O_8S_2SbTl	$Tl[Sb(SO_4)_2]$	38 (Tl):	Hb/3−437
O_8S_2Sr	$SrS_2O_8 \cdot H_2O$	29 (Sr):	Hb−181
−.	$SrS_2O_8 \cdot 4\,H_2O$	29 (Sr):	Hb−181
O_8S_2Th	$Th(SO_4)_2$	44 (Th):	Hb−283/4
−.	$Th(SO_4)_2 \cdot 2\,H_2O$	44 (Th):	Hb−289
−.	$Th(SO_4)_2 \cdot 4\,H_2O$	44 (Th):	Hb−289, 291
−.	$Th(SO_4)_2 \cdot 6\,H_2O$	44 (Th):	Hb−288
−.	$Th(SO_4)_2 \cdot 8\,H_2O$	44 (Th):	Hb−287/8, 291
−.	$Th(SO_4)_2 \cdot 9\,H_2O$	44 (Th):	Hb−286/7
−.	$Th(SO_4)_2$ solid solutions:		
	$Th(SO_4)_2 \cdot 4\,H_2O−Ce(SO_4)_2 \cdot 4\,H_2O$ . .	44 (Th):	Hb−353
	$Th(SO_4)_2 \cdot 8\,H_2O−U(SO_4)_2 \cdot 8\,H_2O$ . . .	55 (U):	Hb−239
−.	$Th(SO_4)_2$ solutions:		
	$Th(SO_4)_2−H_2O$	44 (Th):	Hb−290
	$Th(SO_4)_2−$organic solvents	44 (Th):	Hb−291
	$Th(SO_4)_2−$salt solutions	44 (Th):	Hb−291
−.	$Th(SO_4)_2$ systems:		
	$Th(SO_4)_2−H_2O$	44 (Th):	Hb−284/6
	$Th(SO_4)_2−H_2SO_4−H_2O$	44 (Th):	Hb−286
	$Th(SO_4)_2−K_2SO_4−H_2O$	44 (Th):	Hb−333
	$Th(SO_4)_2−Li_2SO_4−H_2O$	44 (Th):	Hb−322
	$Th(SO_4)_2−(NH_4)_2SO_4−H_2O$	44 (Th):	Hb−338/9
	$Th(SO_4)_2−Na_2SO_4−H_2O$	44 (Th):	Hb−325
O_8S_2Ti	$Ti(SO_4)_2$	41 (Ti):	Hb−347, 349/50
−.	$Ti(SO_4)_2 \cdot 3\,H_2O$	41 (Ti):	Hb−348
−.	$Ti(SO_4)_2 \cdot 4\,H_2O$	41 (Ti):	Hb−347
−.	$Ti(SO_4)_2 \cdot 9\,H_2O$	41 (Ti):	Hb−347
$O_8S_2Ti^{2-}$	$[Ti(SO_4)_2]^{2-}$	41 (Ti):	Hb−468
O_8S_2TlV	$TlV(SO_4)_2 \cdot 12\,H_2O$	48 (V):	Hb/B2−550/1
$O_8S_2Tl_2$	$Tl_2S_2O_8$	38 (Tl):	Hb/3−367
$O_8S_2Tl_2Zn$	$Tl_2Zn(SO_4)_2 \cdot 6\,H_2O$		
	$= Tl_2SO_4 \cdot ZnSO_4 \cdot 6\,H_2O$	38 (Tl):	Hb/3−480/1
O_8S_2U	$U(SO_4)_2$	55 (U):	Hb−146, 245
−.	$U(SO_4)_2 \cdot 2\,H_2O$	55 (U):	Hb−150
−.	$U(SO_4)_2 \cdot 4\,H_2O$	55 (U):	Hb−149/50
−.	$U(SO_4)_2 \cdot 8\,H_2O$	55 (U):	Hb−148/9
−.	$U(SO_4)_2 \cdot 9\,H_2O$	55 (U):	Hb−147/8
−.	$U(SO_4)_2$ solid solutions:		
	$U(SO_4)_2 \cdot 8\,H_2O−Th(SO_4)_2 \cdot 8\,H_2O$ . . .	55 (U):	Hb−239
	$U(SO_4)_2 \cdot 4\,H_2O−Zr(SO_4)_2 \cdot 4\,H_2O$ . . .	55 (U):	Hb−239

See the "Instructions for the user"
at the beginning of the volume

O_8S_2U . . . . U(SO_4)_2 solutions:

                   $U(SO_4)_2-H_2O$ . . . . 55 (U):   Hb−150/1, 245/6

O_8S_2Zn . . . . ZnS_2O_8 . . . . 32 (Zn):   Hb−241

                                                           Eb−965

$O_8S_2Zn^{2-}$ . . . . . . [Zn_2(SO_4)_4]^{4-} . . . . 32 (Zn):   Eb−1007

O_8S_2Zr . . . . Zr(SO_4)_2 = ZrOSO_4 · SO_3 . . . . 42 (Zr):   Hb−334/5

−. . . . . Zr(SO_4)_2 · H_2O . . . . 42 (Zr):   Hb−338, 340

−. . . . . Zr(SO_4)_2 · 4 H_2O . . . . 42 (Zr):   Hb−334, 337/8

−. . . . . Zr(SO_4)_2 solid solutions:

                   Zr(SO_4)_2 · 4 H_2O−U(SO_4)_2 · 4 H_2O . . . 55 (U):   Hb−239

−. . . . . Zr(SO_4)_2 solutions:

                   Zr(SO_4)_2−CH_3OH . . . . 42 (Zr):   Hb−342

                   Zr(SO_4)_2−H_2O . . . . 42 (Zr):   Hb−339/42

O_8S_3 . . . . SO(S_2O_7) = 2 SO_3 · SO_2 . . . . 9 (S):   Hb/B1−323, 358

O_8Sb_5 . . . . Sb_5O_8 . . . . 18 (Sb):   Hb/B2−359

O_8Se_2Th . . . . . . Th(SeO_4)_2 · 9 H_2O . . . . 44 (Th):   Hb−294

$O_8Se_2Tl_2Zn$ . . . . . Tl_2Zn(SeO_4)_2 · 6 H_2O . . . . 38 (Tl):   Hb/3−481

O_8Se_2Zr . . . . . . Zr(SeO_4)_2 · 4 H_2O . . . . 42 (Zr):   Hb−355

$O_8Si_3^{4-}$ . . . . [Si_3O_8]^{4-} . . . . 15 (Si):   Hb/B−421/2

$O_8Sn_3Ta_2$ . . . . . . Sn_3Ta_2O_8 . . . . 50 (Ta):   Hb/B2−231/2

O_8SrV_2 . . . . Sr(VO_4)_2 . . . . 48 (V):   Hb/B2−522

O_8SrW_2 . . . . SrO · 2 WO_3 · O · 6 H_2O . . . . . . 54 (W):   Hb−297

$O_8Sr_2Ti_3$ . . . . . . 2 SrO · 3 TiO_2 . . . . 41 (Ti):   Hb−434

$O_8Sr_2Zr_3$ . . . . . . 2 SrO · 3 ZrO_2 . . . . 42 (Zr):   Hb−439/40

$O_8Sr_3V_2$ . . . . 3 SrO · V_2O_5 = Sr_3(VO_4)_2 . . . . 48 (V):   Hb/B1−93

                                                           Hb/B2−519

$O_8Ta_2^-$ . . . . [Ta_2O_8]^- . . . . 50 (Ta):   Hb/B1−59

$O_8Ta_2Zn_3$ . . . . . . Zn_3Ta_2O_8 . . . . 50 (Ta):   Hb/B2−193, 194, 349

−. . . . . Zn_3Ta_2O_8 solid solutions:

                   Zn_3(Ta,Nb)_2O_8 . . . . 50 (Ta):   Hb/B2−263

O_8ThTi_2Zr . . . . . . ThO_2 · ZrO_2 · 2 TiO_2 . . . . 44 (Th):   Hb−354

$O_8Ti_2V_2$ . . . . Ti_2O_3 · V_2O_5 = TiVO_4 . . . . 48 (V):   Hb/B2−559/60

$O_8Tl_2V_2$ . . . . Tl_2O_3 · V_2O_5 = TlVO_4 . . . . 48 (V):   Hb/B2−548

$O_8Tl_6V_2$ . . . . 3 Tl_2O · V_2O_5 = Tl_3VO_4 . . . . 48 (V):   Hb/B2−547

O_8UV_2 . . . . UO_3 · V_2O_5 . . . . 48 (V):   Hb/B1−33

−. . . . . UO_3 · V_2O_5 · H_2O . . . . 55 (U):   Hb−241

O_8UV_3 . . . . 3 VO_2 · UO_2 . . . . 48 (V):   Hb/B1−91

O_8U_3 . . . . U_3O_8 . . . . 55 (U):   Hb−89/94

−. . . . . U_3O_8 · xH_2O . . . . 55 (U):   Hb−100/1

−. . . . . U_3O_8 solid solutions:

                   (U,Pu)_3O_{8-x} . . . . Erg. W. 4−59

O_8V^{3-} . . . . [V(O_2)_4]^{3-} . . . . 48 (V):   Hb/B1−162/3, 164

                                                           Hb/B2−424

O_8V_2 . . . . [V(O_2)] [V(O)_2(O_2)_2] . . . . 48 (V):   Hb/B1−163, 166

$O_8V_2Zn_3$ . . . . 3 ZnO · V_2O_5 = Zn_3(VO_4)_2 . . . . 48 (V):   Hb/B2−534, 535

O_8V_4 . . . . V_4O_8 . . . . 48 (V):   Hb/B1−84

O_8W_2Th . . . . Th(WO_4)_2 . . . . 54 (W):   Hb−316

Hb = Main Series Volume; GD = Gmelin-Durrer;   *Ab = Appended Volume to Main Series;*
Eb = Supplement Volume to Main Series;   *Erg. W. = New Supplement Series Volume*

O_8W_2Zr	$Zr(WO_4)_2$	54 (W):	Hb−315
O_8Zn_5	Zn_5O_8	32 (Zn):	Hb−145
$O_9P_2Pb_2$	$2\,PbO_2 \cdot P_2O_5 \cdot nH_2O$	47 (Pb):	Hb/C3−877
$O_9P_2Pb_2Ti$	$Pb_2TiP_2O_9$	47 (Pb):	Hb/C4−1394
$O_9P_2Pb_2Zr$	$Pb_2ZrP_2O_9$	47 (Pb):	Hb/C4−1432
$O_9P_2Pb_4$	$Pb_3(PO_4)_2 \cdot PbO = 4\,PbO \cdot P_2O_5$	47 (Pb):	Hb/C3−869/70, 881
$O_9P_2Si_2$	$P_2O_5 \cdot 2\,SiO_2$	16 (P):	Hb/C−639
$O_9P_2Ti_2$	$2\,TiO_2 \cdot P_2O_5$	41 (Ti):	Hb−380
−.	$2\,TiO_2 \cdot P_2O_5 \cdot H_2O$	41 (Ti):	Hb−380
−.	$2\,TiO_2 \cdot P_2O_5 \cdot 3\,H_2O$	41 (Ti):	Hb−380
$O_9P_2U_2$	$2\,UO_2 \cdot P_2O_5 = UP_2O_7 \cdot UO_2$	55 (U):	Hb−183
$O_9P_2Zn_4$	$4\,ZnO \cdot P_2O_5 \cdot H_2O$		
	$= Zn_3(PO_4)_2 \cdot Zn(OH)_2$	32 (Zn):	Hb−279
			Eb−996
$O_9P_2Zr_2$	$2\,ZrO_2 \cdot P_2O_5 = (ZrO)_2P_2O_7$	42 (Zr):	Hb−398
$O_9P_3{}^{3-}$	$[P_3O_9]^{3-}$	16 (P):	Hb/C−280/7
$O_9P_3Sr^-$	$[SrP_3O_9]^-$	16 (P):	Hb/C−287
O_9P_3Ti	$Ti(PO_3)_3$	41 (Ti):	Hb−379
$O_9PbRb_4S_6$	$Rb_4Pb(S_2O_3)_3$	47 (Pb):	Hb/C3−1144
$O_9PbS_6{}^{4-}$	$[Pb(S_2O_3)_3]^{4-}$	47 (Pb):	Hb/B1−365
			Hb/C2−597
$O_9Pb_2S_6{}^{2-}$	$[Pb_2(S_2O_3)_3]^{2-}$	47 (Pb):	Hb/C2−598
$O_9Pb_3Te_2$	$3\,PbO \cdot 2\,TeO_3 \cdot 2\,H_2O$	47 (Pb):	Hb/C2−679
$O_9Pb_3U_2$	$3\,PbO \cdot 2\,UO_3$	55 (U):	Hb−239
O_9Pb_4SSi	$2\,PbO \cdot PbSiO_3 \cdot PbSO_4$	47 (Pb):	Hb/C3−856
O_9PrTa_3	$PrTa_3O_9$	50 (Ta):	Hb/B2−201/2, 211
O_9PrV_3	$Pr(VO_3)_3 \cdot 4\,H_2O$	48 (V):	Hb/B2−555/6
$O_9Pu_2S_3$	$Pu_2(SO_3)_3 \cdot nH_2O$	Erg. W. 4−175	
$O_9Rb_2SV_2$	$V_2O_5 \cdot Rb_2SO_4$	48 (V):	Hb/B2−491
$O_9Rb_3S_6Sb$	$Rb_3[Sb(S_2O_3)_3]$	24 (Rb):	Hb−239
$O_9Rh_2S_3$	$Rh_2(SO_3)_3 \cdot 6\,H_2O$	64 (Rh):	Hb−66
O_9RuS_2	$RuO(SO_4)_2 = RuO_2S_2O_7$	63 (Ru):	Eb−316
$O_9Ru_2S_3$	$Ru_2(SO_3)_3$	63 (Ru):	Hb−73/4
O_9Ru_4	Ru_4O_9	63 (Ru):	Hb−27
−.	$Ru_4O_9 \cdot xH_2O$	63 (Ru):	Hb−27
O_9SSb_4	$2\,Sb_2O_3 \cdot SO_3$	18 (Sb):	Hb/B3−545
−.	$2\,Sb_2O_3 \cdot SO_3 \cdot nH_2O$	18 (Sb):	Hb/B3−543, 545
O_9STi_3	$3\,TiO_2 \cdot SO_3$	41 (Ti):	Hb−349
O_9SZn_6	$ZnSO_4 \cdot 5\,ZnO$	32 (Zn):	Hb−240
−.	$ZnSO_4 \cdot 5\,ZnO \cdot 10\,H_2O$	32 (Zn):	Hb−240
O_9SZr_3	$3\,ZrO_2 \cdot SO_3 \cdot xH_2O$	42 (Zr):	Hb−352
$O_9S_2Sb_2$	$Sb_2O_3 \cdot 2\,SO_3 = Sb(SbO)(SO_4)_2$	18 (Sb):	Hb/B3−544
−.	$Sb_2O_3 \cdot 2\,SO_3 \cdot H_2O = Sb(OH)SO_4$	18 (Sb):	Hb/B3−544
−.	$Sb_2O_3 \cdot 2\,SO_3 \cdot 3\,H_2O$	18 (Sb):	Hb/B3−544
$O_9S_2Ti^{2-}$	$[TiO(SO_4)_2]^{2-}$	41 (Ti):	Hb−468
$O_9S_2Tl_2$	$Tl_2O(SO_4)_2 \cdot 3\,H_2O$	38 (Tl):	Hb/3−365
−.	$Tl_2O(SO_4)_2 \cdot 5\,H_2O$	38 (Tl):	Hb/3−365
O_9S_2U	$UO_2S_2O_7$	55 (U):	Hb−156

See the "Instructions for the user"
at the beginning of the volume

$O_9S_2V^-$ . . . . [$VO(SO_4)_2$]$^-$ . . . . 48 (V):   Hb/B1−305
$O_9S_2V^{2-}$ . . . . . . [$VO(SO_4)_2$]$^{2-}$ . . . . 48 (V):   Hb/A2−536/7
                                                                                 Hb/B1−298
$O_9S_2Zn_3$ . . . . $ZnO \cdot 2\,ZnSO_4$ . . . . 32 (Zn):   Eb−931
                          see also $O_{4.5}SZn_{1.5} \ldots ZnSO_4 \cdot 0.5\,ZnO$
$O_9S_2Zn_5$ . . . . $2\,ZnSO_3 \cdot 3\,ZnO \cdot 3\,H_2O$ . . . . . 32 (Zn):   Hb−220
$O_9S_2Zr^{2-}$ . . . . . . [$ZrO(SO_4)_2$]$^{2-}$ . . . . 42 (Zr):   Hb−340
O_9S_3 . . . . S_3O_9 . . . . 9 (S):   Hb/B1−337/8
$O_9S_3Sb_2$ . . . . $Sb_2(SO_3)_3$ . . . . 18 (Sb):   Hb/B3−541
$O_9S_6SbTl_3$ . . . . . $Tl_3[Sb(S_2O_3)_3]$ . . . . 38 (Tl):   Hb/3−437
$O_9S_6Zn^{4-}$ . . . . . [$Zn(S_2O_3)_3$]$^{4-}$ . . . . 32 (Zn):   Eb−966
O_9SeU^{6-} . . . . . . [$UO_5(SeO_4)$]$^{6-}$ . . . . 55 (U):   Hb−158
$O_9Se_3Tl_2$ . . . . . . $Tl_2(SeO_3)_3$ . . . . 38 (Tl):   Hb/3−373
O_9SmTa_3 . . . . . . $SmTa_3O_9$ . . . . 50 (Ta):   Hb/B2−201/2, 214
O_9SmV_3 . . . . $Sm(VO_3)_3 \cdot 4\,H_2O$ . . . . 48 (V):   Hb/B2−555/6
$O_9Sr_3Ta_2Zn$ . . . . . $ZnSr_3Ta_2O_9$ . . . . 50 (Ta):   Hb/B2−195
$O_9Sr_4Ta_2$ . . . . . . $Sr_4Ta_2O_9$ . . . . 50 (Ta):   Hb/B2−181, 183
$O_9Ta_2^-$ . . . . [Ta_2O_9]$^-$ . . . . 50 (Ta):   Hb/B1−59
$O_9Ta_2Th_2$ . . . . . . $Th_2Ta_2O_9$ . . . . 50 (Ta):   Hb/B2−229, 230
$O_9Ta_2Zr_2$ . . . . . . $(ZrO)_2Ta_2O_7$ . . . . 50 (Ta):   Hb/B2−228
O_9Ta_3Tb . . . . $TbTa_3O_9$ . . . . 50 (Ta):   Hb/B2−201/2, 217
O_9Ta_3Y . . . . YTa_3O_9 . . . . 50 (Ta):   Hb/B2−201/2, 205/6
O_9Ta_3Yb . . . . $YbTa_3O_9$ . . . . 50 (Ta):   Hb/B2−201/2, 220
O_9Ta_{32} . . . . $Ta_{32}O_9$ . . . . 50 (Ta):   Hb/B1−24/5
O_9Tl_{10} . . . . $3\,Tl_2O \cdot 2\,Tl_2O_3$ . . . . 38 (Tl):   Hb/2−221
$O_9V_3^{3-}$ . . . . [V_3O_9]$^{3-}$ . . . . 48 (V):   Hb/B1−132/45,
                                                                                 149/51
O_9V_3Y . . . . $Y(VO_3)_3 \cdot 4\,H_2O$ . . . . 48 (V):   Hb/B2−555/6
O_9V_4 . . . . V_4O_9 . . . . 48 (V):   Hb/B1−14/5, 18
$O_9V_4^+$ . . . . [V_4O_9]$^+$ . . . . 48 (V):   Hb/B1−84
$O_9V_4^{2-}$ . . . . [V_4O_9]$^{2-}$ . . . . 48 (V):   Hb/A2−477, 498
                                                                                 Hb/B1−160
O_9V_5 . . . . $V_5O_9 = VO_{1.8}$ . . . . 48 (V):   Hb/B1−13, 17, 50/1
O_9W_5 . . . . W_5O_9 . . . . 54 (W):   Hb−111
O_9Y_4 . . . . $Y_4O_9 \cdot 9\,H_2O = Y(OH)_3 \cdot 0.75\,H_2O_2$ . . . . 39 (SE):   Hb/C2−106, 110
O_9Zn_5 . . . . Zn_5O_9 . . . . 32 (Zn):   Hb−145
−. . . . . $4\,ZnO_2 \cdot ZnO \cdot 2\,H_2O$ . . . . . . 32 (Zn):   Eb−817
$O_{10}P_2STh$ . . . . . . $Th(SO_4)(PO_3)_2$ . . . . 44 (Th):   Hb−320
$O_{10}P_2Ta_2$ . . . . . . $Ta_2O_5 \cdot P_2O_5 = TaOPO_4$ . . . . . . 50 (Ta):   Hb/B1−269
                                                                                 Hb/B2−349
−. . . . . $Ta_2O_5 \cdot P_2O_5 \cdot 5\,H_2O$ . . . . 50 (Ta):   Hb/B1−269
−. . . . . $Ta_2O_5 \cdot P_2O_5 \cdot 9\,H_2O$ . . . . 50 (Ta):   Hb/B1−268, 269
−. . . . . $Ta_2O_5 \cdot P_2O_5 \cdot xH_2O$ . . . . 50 (Ta):   Hb/B1−268
$O_{10}P_2V_2$ . . . . $V_2O_5 \cdot P_2O_5 = VPO_5$ . . . . 48 (V):   Hb/B1−361/2
−. . . . . $V_2O_5 \cdot P_2O_5 \cdot 2\,H_2O$ . . . . 48 (V):   Hb/B1−363
−. . . . . $V_2O_5 \cdot P_2O_5 \cdot 4\,H_2O$ . . . . 48 (V):   Hb/B1−362/3
−. . . . . $V_2O_5 \cdot P_2O_5 \cdot nH_2O$
                          $= (VO_2)H_2PO_4 \cdot nH_2O$ . . . . 48 (V):   Hb/B1−362

Hb = Main Series Volume; GD = Gmelin-Durrer;   *Ab = Appended Volume to Main Series;*
Eb = Supplement Volume to Main Series;   *Erg. W. = New Supplement Series Volume*

$O_{10}P_3^{5-}$	$[P_3O_{10}]^{5-}$	16 (P):	Hb/B−380
			Hb/C−239/41
$O_{10}P_3Pb^{3-}$	$[PbP_3O_{10}]^{3-}$	16 (P):	Hb/C−248
		47 (Pb):	Hb/B1−367
$O_{10}P_3Sr^{3-}$	$[SrP_3O_{10}]^{3-}$	16 (P):	Hb/C−247
$O_{10}P_3Zn^{3-}$	$[ZnP_3O_{10}]^{3-}$	16 (P):	Hb/C−247
$O_{10}P_4$	P_4O_{10}	16 (P):	Hb/C−77/9, 82/3
$O_{10}P_6S_5$	$P_6S_5O_{10}$	16 (P):	Hb/C−583
$O_{10}PbU_3$	$PbO \cdot 3\, UO_3$	55 (U):	Hb−240
$O_{10}PbW_3$	$PbO \cdot 3\, WO_3 \cdot 2\, H_2O$	54 (W):	Hb−319
$O_{10}Pb_3SSi_2$	$2\, PbSiO_3 \cdot PbSO_4$	47 (Pb):	Hb/C3−856
$O_{10}Pb_3Sb_4$	$Pb_3Sb_4O_{10} = 3\, PbO \cdot Sb_2O_3 \cdot Sb_2O_4$ . . .	47 (Pb):	Hb/C3−930
$O_{10}Pb_3U_2$	$Pb_3U_2O_{10} = 2\, PbO \cdot UO_4 \cdot PbUO_4$	55 (U):	Hb−240
$O_{10}Pb_6$	$Pb_6O_{10} = PbO_{1.66}$	47 (Pb):	Hb/C1−144/7
$O_{10}Pt_2Sn_6$	$Pt_2O \cdot 3\, Sn_2O_3 = Pt_2Sn_6O_{10}$	68 (Pt):	Hb/C3−310
$O_{10}Rb_2S_2U$	$Rb_2SO_4 \cdot UO_2SO_4 \cdot 2\, H_2O$		
	$= Rb_2UO_2(SO_4)_2 \cdot 2\, H_2O$	55 (U):	Hb−227
$O_{10}RuS_4$	$Ru(S_2O_5)_2$	63 (Ru):	Hb−74
$O_{10}STh_2$	$Th_2O_7SO_3$	44 (Th):	Hb−293
$O_{10}SZr_2$	$Zr_2O_6SO_4 \cdot 8\, H_2O$	42 (Zr):	Hb−353
$O_{10}S_2Ti^{2-}$	$[TiO \cdot O(SO_4)_2]^{2-}$	41 (Ti):	Hb−270, 471/2
$O_{10}S_2Tl_2U$	$Tl_2SO_4 \cdot UO_2SO_4 \cdot 3\, H_2O$		
	$= Tl_2UO_2(SO_4)_2 \cdot 3\, H_2O$	55 (U):	Hb−238
$O_{10}S_2V_3$	$3\, VO_2 \cdot 2\, SO_2 \cdot 4.5\, H_2O$	48 (V):	Hb/B1−281
$O_{10}S_2Zr^{2-}$	$[ZrO_2(SO_4)_2]^{2-}$	42 (Zr):	Hb−263
$O_{10}S_3Zr_2$	$2\, ZrO_2 \cdot 3\, SO_2$	42 (Zr):	Hb−344
$O_{10}Sb_3^{5-}$	$[Sb_3O_{10}]^{5-}$	18 (Sb):	Hb/B2−390
$O_{10}Se_3U_2$	$U(SeO_3)_2 \cdot UO(SeO_3)$	55 (U):	Hb−157
$O_{10}Sn_4^{4-}$	$[Sn_4O_{10}]^{4-}$	46 (Sn):	Hb/C1−138
$O_{10}SrW_3$	$SrO \cdot 3\, WO_3 \cdot 5\, H_2O$	54 (W):	Hb−297
$O_{10}Sr_5Ta_2$	$Sr_5Ta_2O_{10}$	50 (Ta):	Hb/B2−183
−	$Sr_5Ta_2O_{10}$ solid solutions:		
	$(Sr,Ba)_5Ta_2O_{10}$	50 (Ta):	Hb/B2−193, 199
	$(Sr,Ba,Ca)_5Ta_2O_{10}$	50 (Ta):	Hb/B2−193
$O_{10}UV_3$	UV_3O_{10}	48 (V):	Hb/B1−91
$O_{10}U_3$	$U_3O_{10} \cdot 2\, H_2O$	55 (U):	Hb−100
$O_{10}V_3^{5-}$	$[V_3O_{10}]^{5-}$	48 (V):	Hb/B1−150
$O_{10}V_3Zn_4$	$4\, ZnO \cdot 3\, VO_2$	48 (V):	Hb/B2−533
$O_{10}V_4$	V_4O_{10}	48 (V):	Hb/B1−84
$O_{10}W_3Zn$	$2\, ZnO \cdot 6\, WO_3 \cdot 10\, H_2O$	54 (W):	Hb−306
$O_{11}P_2Pu_2$	$(PuO_2)_2P_2O_7$		Erg. W. 4−254
$O_{11}P_2Sb_4$	$2\, Sb_2O_3 \cdot P_2O_5$	18 (Sb):	Hb/B3−561
$O_{11}P_2Ti_2$	$2\, TiO_3 \cdot P_2O_5 \cdot 3\, H_2O$	41 (Ti):	Hb−380
$O_{11}P_2Ti_3$	$3\, TiO_2 \cdot P_2O_5 \cdot 6\, H_2O$	41 (Ti):	Hb−380
$O_{11}P_2Tl_4$	$2\, Tl_2O_3 \cdot P_2O_5 \cdot 5\, H_2O$	38 (Tl):	Hb/3−430
$O_{11}P_2U_2$	$(UO_2)_2P_2O_7$	55 (U):	Hb−183
−	$(UO_2)_2P_2O_7 \cdot 4\, H_2O$	55 (U):	Hb−183
−	$(UO_2)_2P_2O_7 \cdot 5\, H_2O$	55 (U):	Hb−183

See the "Instructions for the user"
at the beginning of the volume

$O_{11}P_2U_2$	$(UO_2)_2P_2O_7 \cdot xH_2O$	55 (U):	Hb−183
$O_{11}P_2V_3$	$(VO)_3(PO_4)_2 \cdot xH_2O$	48 (V):	Hb/B1−361
$O_{11}P_4$	P_4O_{11}	16 (P):	Hb/C−94
$O_{11}P_4Pb$	$[PbP_4O_{11}]_n$ solid solutions:		
	$[(Pb,Sr)P_4O_{11}]_n$	47 (Pb):	Hb/C3−1199
$O_{11}P_4Sr$	$[SrP_4O_{11}]_n$ solid solutions:		
	$[(Sr,Pb)P_4O_{11}]_n$	47 (Pb):	Hb/C3−1199
$O_{11}PbTa_4$	$PbTa_4O_{11}$	50 (Ta):	Hb/B2−232, 233
$O_{11}PbV_4$	$PbO \cdot 2 V_2O_5 = PbV_4O_{11}$	48 (V):	Hb/B2−570
$O_{11}Pb_3Si_4$	$3 PbO \cdot 4 SiO_2 \cdot H_2O$	47 (Pb):	Hb/C3−838
$O_{11}Pb_7$	Pb_7O_{11}	47 (Pb):	Hb/C1−144/7
$O_{11}PdSn_5$	$PdO \cdot 5 SnO_2 \cdot 4 H_2O$	65 (Pd):	Hb/2−339
$O_{11}Pd_8S$	$PdSO_4 \cdot 7 PdO \cdot 6 H_2O$	65 (Pd):	Hb/2−295
−	$PdSO_4 \cdot 7 PdO \cdot 10 H_2O$	65 (Pd):	Hb/2−295
$O_{11}Pr_6$	$Pr_6O_{11} = PrO_{1.833}$:		
	Chemical reactions	39 (SE):	Hb/C1−156, 166/7,
			170, 288/90
	Formation and preparation	39 (SE):	Hb/C1−106/7, 286/7
	Physical properties	39 (SE):	Hb/C1−130, 144/5,
			152/5, 286/8
−	$Pr_6O_{11} \cdot nH_2O$	39 (SE):	Hb/C1−289
			Hb/C2−2, 15
−	Pr_6O_{11} systems:		
	Pr_6O_{11}−H_2O	39 (SE):	Hb/C2−50
$O_{11}PtSn_5$	$PtO \cdot 5 SnO_2 \cdot 4 H_2O$	68 (Pt):	Hb/C3−310
$O_{11}Rb_2S_2Zr_2$	$Rb_2Zr_2O_3(SO_4)_2 \cdot 15 H_2O$	42 (Zr):	Hb−427
$O_{11}Rb_2V_4$	$Rb_2O \cdot 2 V_2O_5$	48 (V):	Hb/B2−487
$O_{11}S_2Ta_2$	$Ta_2O_3(SO_4)_2$	50 (Ta):	Hb/B2−146
$O_{11}S_2V_2$	$V_2O_5 \cdot 2 SO_3$	48 (V):	Hb/B1−302/3
−	$V_2O_5 \cdot 2 SO_3 \cdot H_2O$	48 (V):	Hb/B1−302
−	$V_2O_5 \cdot 2 SO_3 \cdot 3 H_2O$	48 (V):	Hb/B1−304
−	$V_2O_5 \cdot 2 SO_3 \cdot 8 H_2O$	48 (V):	Hb/B1−304
$O_{11}S_2V_3Zn$	$ZnSO_3 \cdot VOSO_3 \cdot 2 VO_2$	48 (V):	Hb/B2−536
$O_{11}S_2Zr_2{}^{2-}$	$[Zr_2O_3(SO_4)_2]^{2-}$	42 (Zr):	Hb−340
$O_{11}S_3$	S_3O_{11}	9 (S):	Hb/B1−370/1
$O_{11}Sb_2Se_4$	$Sb_2O_3 \cdot 4 SeO_2$	18 (Sb):	Hb/B3−553
$O_{11}SnTa_4$	$SnTa_4O_{11}$	50 (Ta):	Hb/B2−231
$O_{11}SrTa_4$	$SrTa_4O_{11}$	50 (Ta):	Hb/B2−182
$O_{11}SrV_4$	$SrO \cdot 2 V_2O_5 = SrV_4O_{11}$	48 (V):	Hb/B2−521
$O_{11}Sr_6Ta_2$	$Sr_6Ta_2O_{11}$	50 (Ta):	Hb/B2−181, 183
$O_{11}Tb_6$	$Tb_6O_{11} = TbO_{1.833}$	39 (SE):	Hb/C1−387
$O_{11}TiV_9$	TiV_9O_{11}	48 (V):	Hb/B2−558
$O_{11}UW_3$	$UO_2 \cdot 3 WO_3 \cdot 6 H_2O$	55 (U):	Hb−244
$O_{11}U_2V_2$	$2 UO_3 \cdot V_2O_5$	55 (U):	Hb−241
$O_{11}V_2{}^{4-}$	$[V_2(O)_3(O_2)_4]^{4-}$	48 (V):	Hb/B2−424
$O_{11}V_2Zn_6$	$6 ZnO \cdot V_2O_5 \cdot xH_2O$	48 (V):	Hb/B2−535

Hb = Main Series Volume; GD = Gmelin-Durrer; Ab = Appended Volume to Main Series;
Eb = Supplement Volume to Main Series; Erg. W. = New Supplement Series Volume

$O_{11}V_4^{2-}$	$[V_4O_{11}]^{2-}$	48 (V):	Hb/B1–144
			Hb/B2–397
$O_{11}V_4Zn$	$ZnO \cdot 2 V_2O_5 = ZnV_4O_{11}$	48 (V):	Hb/B2–536
$O_{11}V_6$	$V_6O_{11} = VO_{1.83}$	48 (V):	Hb/B1–13, 30, 50/1
$O_{11}W_4$	W_4O_{11}	54 (W):	Hb–107/11, 115/7
O_{12}	O_{12}	3 (O):	Hb/4–1116
$O_{12}PPb_5SiV$	$Pb_5SiO_4PO_4VO_4$	48 (V):	Hb/B2–573
$O_{12}P_2Pb_3SSr$	$Pb_3SrSO_4(PO_4)_2$	47 (Pb):	Hb/C3–1199
$O_{12}P_2Pb_3SZn$	$Pb_3ZnSO_4(PO_4)_2$	47 (Pb):	Hb/C4–1229
$O_{12}P_2Pb_4S$	$Pb_3(PO_4)_2 \cdot PbSO_4$	47 (Pb):	Hb/C3–884
$O_{12}P_2Pb_5Si$	$Pb_5(PO_4)_2SiO_4$	47 (Pb):	Hb/C3–886/7
$O_{12}P_2SZr_2$	$2 ZrO_2 \cdot P_2O_5 \cdot SO_3$	42 (Zr):	Hb–398/9
$O_{12}P_3Pb_4Rb$	$RbPb_4(PO_4)_3$	47 (Pb):	Hb/C3–1145
$O_{12}P_3Pb_4Tl$	$TlPb_4(PO_4)_3$	47 (Pb):	Hb/C4–1355
$O_{12}P_4^{4-}$	$[P_4O_{12}]^{4-}$	16 (P):	Hb/C–287/93
$O_{12}P_4Pb_2$	$Pb_2P_4O_{12}$	47 (Pb):	Hb/C3–877
–	$Pb_2P_4O_{12} \cdot 4 H_2O$	47 (Pb):	Hb/C3–877
$O_{12}P_4Pu$	$Pu(PO_3)_4$	Erg. W. 4–252/3	
$O_{12}P_4Si$	$2 P_2O_5 \cdot SiO_2 \cdot 2 H_2O$		
	$= Si(PO_3)_4 \cdot 2 H_2O$	16 (P):	Hb/C–640
–	$2 P_2O_5 \cdot SiO_2 \cdot 4 H_2O = H_8[Si(PO_4)_4]$ . .	16 (P):	Hb/C–640
$O_{12}P_4Sr^{2-}$	$[SrP_4O_{12}]^{2-}$	16 (P):	Hb/C–293
$O_{12}P_4Sr_2$	$Sr_2(PO_3)_4$	29 (Sr):	Hb–216
–	$Sr_2(PO_3)_4 \cdot 8 H_2O$	29 (Sr):	Hb–216
$O_{12}P_4Th$	$Th(PO_3)_4$	44 (Th):	Hb–319
$O_{12}P_4U$	$U(PO_3)_4$	55 (U):	Hb–183
–	$U(PO_3)_4 \cdot 6 H_2O$	55 (U):	Hb–184
$O_{12}P_4Zn_2$	$Zn_2(PO_3)_4$	32 (Zn):	Hb–280
–	$Zn_2(PO_3)_4 \cdot 10 H_2O$	32 (Zn):	Hb–280
$O_{12}PbRb_2S_3$	$Rb_2Pb(SO_4)_3$	47 (Pb):	Hb/C2–582
$O_{12}PbS_3Sn$	$PbSn(SO_4)_3 \cdot 3 H_2O$	47 (Pb):	Hb/C4–1586
$O_{12}PbS_8^{6-}$	$[Pb(S_2O_3)_4]^{6-}$	47 (Pb):	Hb/B1–365
			Hb/C2–597
$O_{12}Pb_2Rb_2S_3$	$Rb_2Pb_2(SO_4)_3$	47 (Pb):	Hb/C3–1143
$O_{12}Pb_2S_3Tl_2$	$Tl_2Pb_2(SO_4)_3$	47 (Pb):	Hb/C4–1353
$O_{12}Pb_3S_8^{2-}$	$[Pb_3(S_2O_3)_4]^{2-}$	47 (Pb):	Hb/C2–598
$O_{12}Pb_4TlV_3$	$TlPb_4(VO_4)_3$	48 (V):	Hb/B2–574
$O_{12}Pb_5SSi_2$	$Pb_5(SiO_4)_2SO_4$	47 (Pb):	Hb/C3–856
–	$Pb_5(SiO_4)_2SO_4$ solid solutions:		
	$Pb_5(SiO_4)_2SO_4–Pb_9(Si_2O_7)_3$	47 (Pb):	Hb/C3–857
$O_{12}Pb_5Sb_4$	$Pb_5Sb_4O_{12} = 5 PbO \cdot Sb_2O_3 \cdot Sb_2O_4$ . . .	47 (Pb):	Hb/C3–930
$O_{12}Pb_5SiV_2$	$Pb_5SiO_4(VO_4)_2$	48 (V):	Hb/B2–572
$O_{12}Pr_6$	$Pr_6O_{12} \cdot nH_2O$	39 (SE):	Hb/C2–106
$O_{12}Pr_7$	$Pr_7O_{12} = PrO_{1.714}$	39 (SE):	Hb/C1–284/5
$O_{12}PtS_4^{6-}$	$[Pt(SO_3)_4]^{6-}$	68 (Pt):	Hb/C2–158
$O_{12}PuS_3Tl_3$	$Tl_3Pu(SO_4)_3$	Erg. W. 4–178	
$O_{12}PuTa_4$	$PuO_2 \cdot 2 Ta_2O_5$	Erg. W. 4–56, 72	
$O_{12}Pu_2S_3$	$Pu_2(SO_4)_3$	Erg. W. 4–175/6	

See the "Instructions for the user"
at the beginning of the volume

$O_{12}Pu_2S_3$ . . . . $Pu_2(SO_4)_3 \cdot 5\,H_2O$ . . . . Erg. W. 4–176

–. . . . . $Pu_2(SO_4)_3 \cdot 7\,H_2O$ . . . . Erg. W. 4–175/6

–. . . . . $Pu_2(SO_4)_3 \cdot xH_2O$ . . . . Erg. W. 4–179

–. . . . . $Pu_2(SO_4)_3$ systems:

                    $Pu_2(SO_4)_3–K_2SO_4–H_2O$ . . . . Erg. W. 4–177

$O_{12}Rb_2S_3Th$ . . . . . . $Rb_2SO_4 \cdot Th(SO_4)_2 \cdot 2\,H_2O$ . . . . 44 (Th):   Hb–344/5

$O_{12}Rb_2W_4$ . . . . $Rb_2W_4O_{12}$ . . . . 54 (W):   Hb–283

$O_{12}Rb_4S_3V_2$ . . . . . . $2\,Rb_2SO_3 \cdot VOSO_3 \cdot VO_2 \cdot H_2O$ . . . . . . 48 (V):   Hb/B2–489

$O_{12}Rh_2S_3$ . . . . $Rh_2(SO_4)_3$. . . . . 64 (Rh):   Hb–67

–. . . . . $Rh_2(SO_4)_3 \cdot 4\,H_2O$ . . . . 64 (Rh):   Hb–68

–. . . . . $Rh_2(SO_4)_3 \cdot 6\,H_2O$ . . . . 64 (Rh):   Hb–68

–. . . . . $Rh_2(SO_4)_3 \cdot 9\,H_2O$ . . . . 64 (Rh):   Hb–68

–. . . . . $Rh_2(SO_4)_3 \cdot 12\,H_2O$. . . . . 64 (Rh):   Hb–68

–. . . . . $Rh_2(SO_4)_3 \cdot 15\,H_2O$. . . . . 64 (Rh):   Hb–67/8

–. . . . . $Rh_2(SO_4)_3$ systems:

                    $Rh_2(SO_4)_3–H_2O$ . . . . 64 (Rh):   Hb–67

$O_{12}Ru_2S_3$ . . . . $Ru_2(SO_4)_3$. . . . . 63 (Ru):   Hb–74

$O_{12}SSb_6$ . . . . $3\,Sb_2O_3 \cdot SO_3$ . . . . 18 (Sb):   Hb/B3–545

–. . . . . $3\,Sb_2O_3 \cdot SO_3 \cdot 1.67\,H_2O$. . . . . . . 18 (Sb):   Hb/B3–545

$O_{12}SU_3$ . . . . $3\,UO_3 \cdot SO_3 \cdot 2\,H_2O$ . . . . 55 (U):   Hb–156

$O_{12}S_2V_2{}^{2-}$ . . . . . $[V_2O_4(SO_4)_2]^{2-}$ . . . . 48 (V):   Hb/B2–440

$O_{12}S_2Zr_3$ . . . . $Zr(SO_4)_2 \cdot 2\,ZrO_2$ . . . . 42 (Zr):   Hb–349

–. . . . . $3\,ZrO_2 \cdot 2\,SO_3 \cdot 6\,H_2O$ . . . . 42 (Zr):   Hb–349

–. . . . . $2\,ZrOSO_4 \cdot ZrO_2 \cdot 8\,H_2O$ . . . . 42 (Zr):   Hb–349

–. . . . . $2\,ZrOSO_4 \cdot ZrO_2 \cdot 8.5\,H_2O$ . . . . 42 (Zr):   Hb–349

–. . . . . $3\,ZrO_2 \cdot 2\,SO_3 \cdot 12\,H_2O$ . . . . 42 (Zr):   Hb–349

–. . . . . $3\,ZrO_2 \cdot 2\,SO_3 \cdot xH_2O$ . . . . 42 (Zr):   Hb–349

$O_{12}S_3Sb_2$ . . . . $Sb_2(SO_4)_3$. . . . . 18 (Sb):   Hb/B3–542/3

$O_{12}S_3SrTi$ . . . . $SrTi(SO_4)_3$ . . . . 41 (Ti):   Hb–434

$O_{12}S_3ThTl_2$ . . . . . . $Tl_2SO_4 \cdot Th(SO_4)_2 \cdot 4\,H_2O$ . . . . 44 (Th):   Hb–351

$O_{12}S_3Ti_2$ . . . . $Ti_2(SO_4)_3$ . . . . 41 (Ti):   Hb–343/5, 469/70,

                                                      478/9

–. . . . . $Ti_2(SO_4)_3 \cdot 8\,H_2O$ . . . . 41 (Ti):   Hb–345

$O_{12}S_3Tl_2$ . . . . $Tl_2(SO_4)_3$ . . . . 38 (Tl):   Hb/3–365, 366/7

–. . . . . $Tl_2(SO_4)_3 \cdot H_2O$ . . . . 38 (Tl):   Hb/3–365

–. . . . . $Tl_2(SO_4)_3 \cdot 7\,H_2O$ . . . . 38 (Tl):   Hb/3–365

–. . . . . $Tl_2(SO_4)_3$ systems:

                    $Tl_2(SO_4)_3–H_2SO_4–Tl_2SO_4–H_2O$ . . . . 38 (Tl):   Hb/3–362

$O_{12}S_3Tl_2Zr_2$ . . . . . . $Tl_2SO_4 \cdot 2\,Zr(SO_4) \cdot 8\,H_2O$ . . . . 42 (Zr):   Hb–446

$O_{12}S_3Tl_4$ . . . . $Tl_4(SO_4)_3$

               see $O_{24}S_6Tl_8$ … $3\,Tl_2SO_4 \cdot Tl_2(SO_4)_3$

$O_{12}S_3Tl_4V_2$ . . . . . . $2\,Tl_2SO_3 \cdot VOSO_3 \cdot VO_2 \cdot 4\,H_2O$ . . . . . . 48 (V):   Hb/B2–550

$O_{12}S_3V_2$ . . . . $V_2(SO_4)_3$ . . . . 48 (V):   Hb/A2–380

                                                         Hb/B1–284

–. . . . . $V_2(SO_4)_3 \cdot 3\,H_2O$ . . . . 48 (V):   Hb/B1–285/6

–. . . . . $V_2(SO_4)_3 \cdot 4\,H_2O$ . . . . 48 (V):   Hb/B1–285

–. . . . . $V_2(SO_4)_3 \cdot nH_2O$ . . . . 48 (V):   Hb/B1–285

Hb = Main Series Volume; GD = Gmelin-Durrer;   *Ab = Appended Volume to Main Series;*
Eb = Supplement Volume to Main Series;   *Erg. W. = New Supplement Series Volume*

$O_{12}S_3V_2$ . . . . $V_2(SO_4)_3$ solutions:

                                     $V_2(SO_4)_3-H_2O$ . . . . 48 (V):   Hb/A2−335, 476

                                                                      Hb/B1−287/8

$O_{12}S_3ZnZr$ . . . . . $ZnO \cdot ZrO_2 \cdot (SO_3)_3 \cdot 6\,H_2O$ . . . . 42 (Zr):   Hb−442

$O_{12}S_4Th$ . . . . $Th(S_2O_6)_2 \cdot 4\,H_2O$ . . . . 44 (Th):   Hb−293

$O_{12}Se_2U^{6-}$ . . . . . . $[UO_4(SeO_4)_2]^{6-}$ . . . . 55 (U):   Hb−158

$O_{12}SnTa_4$ . . . . $SnTa_4O_{12}$ . . . . 50 (Ta):   Hb/B2−231, 232

$O_{12}Ta_2Ti_3Zn$ . . . . $ZnTa_2O_6 \cdot 3\,TiO_2$ . . . . 50 (Ta):   Hb/B2−226

$O_{12}Ta_4Th$ . . . . $ThTa_4O_{12}$ . . . . 50 (Ta):   Hb/B2−229/30

−. . . . . $ThTa_4O_{12}$ solid solutions:

                                    $ThTa_4O_{12}-CaTa_2O_6$ . . . . 50 (Ta):   Hb/B2−230

                                    $Th(Ta,Nb)_4O_{12}$ . . . . 50 (Ta):   Hb/B2−266

$O_{12}Ta_4U$ . . . . . . $U(TaO_3)_4$ . . . . 50 (Ta):   Hb/B1−51

$O_{12}Tb_7$ . . . . $Tb_7O_{12} = TbO_{1.714}$ . . . . 39 (SE):   Hb/C1−387, 394/5

$O_{12}ThV_4$ . . . . $ThO_2 \cdot 2\,V_2O_5 = Th(VO_3)_4$ . . . . . . 48 (V):   Hb/B2−565/6

$O_{12}TiV_9$ . . . . TiV_9O_{12} . . . . 48 (V):   Hb/B2−558

$O_{12}Ti_2V_8$ . . . . $Ti_2V_8O_{12}$ . . . . 48 (V):   Hb/B2−558

$O_{12}Ti_3V_7$ . . . . $Ti_3V_7O_{12}$ . . . . 48 (V):   Hb/B2−558

$O_{12}Ti_7$ . . . . Ti_7O_{12} . . . . 41 (Ti):   Hb−224/6, 473

$O_{12}Ti_9V$ . . . . Ti_9VO_{12} . . . . 48 (V):   Hb/B2−558

$O_{12}UW_3$ . . . . $UO_3 \cdot 3\,WO_3 \cdot 5\,H_2O$ . . . . 55 (U):   Hb−244

$O_{12}U_3$ . . . . $2\,UO_3 \cdot UO_6 \cdot 6\,H_2O$ . . . . 55 (U):   Hb−106

$O_{12}V_4{}^{4-}$ . . . . $[V_4O_{12}]^{4-}$ . . . . 48 (V):   Hb/A2−498

                                                                      Hb/B1−132/45, 151/2

$O_{12}V_6$ . . . . V_6O_{12} . . . . 48 (V):   Hb/B1−84

$O_{12}Zn_7$ . . . . $10\,ZnO_2 \cdot 4\,ZnO \cdot 5\,H_2O$ . . . . . . 32 (Zn):   Hb−146

$O_{13}P_2Pb_8$ . . . . . . $Pb_3(PO_4)_2 \cdot 5\,PbO = 8\,PbO \cdot P_2O_5$. . . . . 47 (Pb):   Hb/C3−869

$O_{13}P_2Rb_4Se_2$ . . . . $2\,Rb_2O \cdot P_2O_5 \cdot 2\,SeO_3 \cdot 3\,H_2O$ . . . . . . 24 (Rb):   Hb−234

$O_{13}P_4{}^{6-}$ . . . . $[P_4O_{13}]^{6-}$ . . . . 16 (P):   Hb/C−249/51

$O_{13}P_4Pb_3$ . . . . . . $Pb_3P_4O_{13}$ . . . . 47 (Pb):   Hb/C3−875

$O_{13}P_4Rh_2$ . . . . . . $Rh_2O_3 \cdot 2\,P_2O_5 \cdot 6\,H_2O$ . . . . . . 64 (Rh):   Hb−72

$O_{13}P_4S$ . . . . . . $P_4O_{10} \cdot SO_3$ . . . . 16 (P):   Hb/C−583

$O_{13}P_4Zn_3$ . . . . . . $Zn_3P_4O_{13}$ . . . . 32 (Zn):   Eb−996

$O_{13}PbW_4$ . . . . . . $PbO \cdot 4\,WO_3 \cdot 7\,H_2O$ . . . . 54 (W):   Hb−319/20

$O_{13}Pb_3Sb_4$ . . . . . $Pb_3Sb_4O_{13} = 3\,PbO \cdot 2\,Sb_2O_5$. . . . . . 47 (Pb):   Hb/C3−932

$O_{13}Pb_3Ta_4$ . . . . . $Pb_3Ta_4O_{13}$ . . . . 50 (Ta):   Hb/B2−232, 235, 237

$O_{13}Pb_3V_4$ . . . . . . $3\,PbO \cdot 2\,V_2O_5 \cdot 2\,H_2O$

                                       $= Pb_3V_4O_{13} \cdot 2\,H_2O$ . . . . 48 (V):   Hb/B2−570

$O_{13}Pb_4SSi_3$ . . . . . $3\,PbSiO_3 \cdot PbSO_4$ . . . . . . 47 (Pb):   Hb/C3−856

$O_{13}Pb_6SSi_2$ . . . . . $3\,PbO \cdot 2\,PbSiO_3 \cdot PbSO_4$ . . . . . . 47 (Pb):   Hb/C3−856

$O_{13}Pb_8V_2$ . . . . . . $8\,PbO \cdot V_2O_5 = 5\,PbO \cdot Pb_3(VO_4)_2$ . . . . 48 (V):   Hb/B2−568

$O_{13}Pr_6$ . . . . $Pr_6O_{13} \cdot nH_2O$. . . . . 39 (SE):   Hb/C2−106

$O_{13}Pu_2S_3$ . . . . . . $Pu_2O(SO_4)_3 \cdot 8\,H_2O$ . . . . . . Erg. W. 4−182/3

$O_{13}Rb_2W_4$ . . . . . . $Rb_2O \cdot 4\,WO_3 \cdot 8\,H_2O$

                                       $= Rb_6H_4[H_2(W_2O_7)_6] \cdot 21\,H_2O$ . . . . 54 (W):   Hb−282

$O_{13}STi_5$ . . . . $5\,TiO_2 \cdot SO_3 \cdot 5\,H_2O$ . . . . 41 (Ti):   Hb−349

$O_{13}SV_4$ . . . . $2\,V_2O_5 \cdot SO_3$ . . . . 48 (V):   Hb/B1−302

$O_{13}S_3Ti_2$ . . . . $2\,TiO_2 \cdot 3\,SO_3 \cdot 3\,H_2O$ . . . . 41 (Ti):   Hb−348

See the "Instructions for the user"
at the beginning of the volume

$O_{13}S_3Ti_2$	$Ti_2O(SO_4)_3 \cdot 5\,H_2O$	41 (Ti):	Hb−346
$O_{13}S_3V^{3-}$	$[VO(SO_4)_3]^{3-}$	48 (V):	Hb/B1−305
$O_{13}S_3Zr_2$	$2\,ZrO_2 \cdot 3\,SO_3 \cdot 5\,H_2O$		
	$= H_4[Zr_4(SO_4)_6(OH)_8] \cdot 4\,H_2O$.	42 (Zr):	Hb−344
$O_{13}Sb_6$	$Sb_6O_{13} = Sb_2O_3 \cdot 2\,Sb_2O_5$		
	$= 2\,Sb_2O_4 \cdot Sb_2O_5$	18 (Sb):	Hb/B2−347, 370
$O_{13}SrW_4$	$SrO \cdot 4\,WO_3 \cdot 8\,H_2O$		
	$= Sr_3H_4[H_2(W_2O_7)_6] \cdot 21\,H_2O$	54 (W):	Hb−297
$O_{13}Sr_3V_4$	$3\,SrO \cdot 2\,V_2O_5 \cdot 2\,H_2O$	48 (V):	Hb/B2−520
$O_{13}Ta_4Tl_6$	$Tl_6Ta_4O_{13} \cdot 4\,H_2O$.	50 (Ta):	Hb/B2−201
$O_{13}Tl_2W_4$	$Tl_2O \cdot 4\,WO_3 \cdot 3\,H_2O$	54 (W):	Hb−314
$O_{13}Tl_6V_4$	$3\,Tl_2O \cdot 2\,V_2O_5 = Tl_6V_4O_{13}$	48 (V):	Hb/B2−389, 547/8
$O_{13}UV_4$	$UO_3 \cdot 2\,V_2O_5 \cdot xH_2O$	55 (U):	Hb−241
$O_{13}U_4Zn$	$ZnO \cdot 4\,UO_3$	55 (U):	Hb−236
$O_{13}V_4{}^{6-}$	$[V_4O_{13}]^{6-}$	48 (V):	Hb/B1−144, 148, 152
			Hb/B2−389
$O_{13}V_6$	$V_6O_{13} = VO_{2.17}$.	48 (V):	Hb/B1−17, 61/3
$O_{13}V_7$	$V_7O_{13} = VO_{1.86}$.	48 (V):	Hb/B1−13, 30, 50/1
$O_{13}W_4Zn$	$ZnO \cdot 4\,WO_3 \cdot 8\,H_2O$		
	$= Zn_3H_4[H_2(W_2O_7)_6] \cdot 21\,H_2O$	54 (W):	Hb−306/7
$O_{14}P_2S_3$	$P_2O_5 \cdot 3\,SO_3$	16 (P):	Hb/C−583
$O_{14}P_2U_3$	$(UO_2)_3(PO_4)_2$	55 (U):	Hb−181
$O_{14}P_4Pb^{6-}$	$[Pb(P_2O_7)_2]^{6-}$	16 (P):	Hb/C−239
		47 (Pb):	Hb/B1−367
			Hb/C3−874
$O_{14}P_4Tl^{7-}$	$[Tl(P_2O_7)_2]^{7-}$	16 (P):	Hb/C−238
$O_{14}P_4V_2$	$V_2O_4 \cdot 2\,P_2O_5 \cdot 6\,H_2O$.	48 (V):	Hb/B1−361
$O_{14}P_4Zn^{6-}$	$[Zn(P_2O_7)_2]^{6-}$	16 (P):	Hb/C−238, 239
$O_{14}P_7$	P_7O_{14}	16 (P):	Hb/C−76
$O_{14}Pb_5Si_4Zn$	$Pb_5Zn(Si_2O_7)_2$	47 (Pb):	Hb/C4−1229
$O_{14}Pb_7SSi_2$	$4\,PbO \cdot 2\,PbSiO_3 \cdot PbSO_4$	47 (Pb):	Hb/C3−856/7
$O_{14}Pb_7Sb_4$	$Pb_7Sb_4O_{14} = 7\,PbO \cdot Sb_2O_3 \cdot Sb_2O_4$ . . .	47 (Pb):	Hb/C3−930
$O_{14}Pr_6$.	$Pr_6O_{14} \cdot nH_2O$.	39 (SE):	Hb/C2−106
$O_{14}S_2Zr_4$	$Zr(SO_4)_2 \cdot 3\,ZrO_2 = ZrOSO_4 \cdot ZrO_2$	42 (Zr):	Hb−350
$O_{14}S_3Ta_2$	$Ta_2O_2(SO_4)_3 = Ta_2O_5(SO_3)_3$		
	$= (TaO_2)_2S_3O_{10}$	50 (Ta):	Hb/B1−177
$O_{14}S_3V_2$	$V_2O_5 \cdot 3\,SO_3$	48 (V):	Hb/B1−302/3
−.	$V_2O_5 \cdot 3\,SO_3 \cdot 3\,H_2O$	48 (V):	Hb/B1−302
$O_{14}S_3V_4$	$2\,VO_2 \cdot 1.5\,SO_2 \cdot nH_2O$	48 (V):	Hb/B1−281
$O_{14}S_4Ti$	$Ti(S_2O_7)_2 = TiO_2 \cdot 4\,SO_3$	41 (Ti):	Hb−346
−.	$TiO_2 \cdot 4\,SO_3 \cdot 2\,H_2O$	41 (Ti):	Hb−346
$O_{14}Se_3Zr_4$.	$3\,ZrOSeO_3 \cdot ZrO_2 \cdot 18\,H_2O$		
	$= 4\,ZrO_2 \cdot 3\,SeO_2 \cdot 18\,H_2O$	42 (Zr):	Hb−354/5
$O_{14}Ta_2Ti_4Zn$	$ZnTa_2O_6 \cdot 4\,TiO_2$	50 (Ta):	Hb/B2−226
$O_{14}Ta_2W_3$	$3\,WO_3 \cdot Ta_2O_5$	50 (Ta):	Hb/B1−51
$O_{14}Ti_5Zn_4$	$4\,ZnO \cdot 5\,TiO_2$	41 (Ti):	Hb−462
$O_{14}V_5{}^{3-}$	$[V_5O_{14}]^{3-}$	48 (V):	Hb/B1−140/1
$O_{14}V_6$	V_6O_{14}	48 (V):	Hb/B1−84

Hb = Main Series Volume; GD = Gmelin-Durrer;   Ab = Appended Volume to Main Series;
Eb = Supplement Volume to Main Series;   Erg. W. = New Supplement Series Volume

$O_{14}V_6$ . . . . $V_6O_{14} \cdot 4\,H_2O = VO_{2.33} \cdot 0.65\,H_2O$
                 $= V_3O_5(OH)_4$ . . . . 48 (V):   Hb/B1−70

−. . . . . $V_6O_{14} \cdot 6\,H_2O = V_6O_{13}(OH)_2 \cdot 5\,H_2O$ . . . 48 (V):   Hb/B1−70

$O_{14}W_5$ . . . . W_5O_{14} . . . . 54 (W):   Hb−117

$O_{14.5}Se_{3.5}V_2$ . . . . . $2\,VO_2 \cdot 3.5\,SeO_3 \cdot H_2O$ . . . . 48 (V):   Hb/B1−314

$O_{15}P_2Ta_4$ . . . . . . $2\,Ta_2O_5 \cdot P_2O_5 \cdot 6\,H_2O$
                 $= (TaO_2)_2HPO_4 \cdot 2.5\,H_2O$ . . . . 50 (Ta):   Hb/B1−268/9

$O_{15}P_2V_4$ . . . . $2\,V_2O_5 \cdot P_2O_5$ . . . . 48 (V):   Hb/B1−362

$O_{15}P_4Pb_5$ . . . . . . $5\,PbO \cdot 2\,P_2O_5$ . . . . 47 (Pb):   Hb/C3−870

$O_{15}P_4Zn_5$ . . . . . . $10\,ZnO \cdot 4\,P_2O_5 \cdot 10\,H_2O$. . . . . . . 32 (Zn):   Hb−278

$O_{15}Pb_2S_{10}{}^{6-}$ . . . . . $[Pb_2(S_2O_3)_5]^{6-}$ . . . . 47 (Pb):   Hb/C2−598

$O_{15}Pb_5Ta_4$ . . . . . $Pb_5Ta_4O_{15}$ . . . . 50 (Ta):   Hb/B2−232/3, 237/8

$O_{15}Pr_6$. . . . . $Pr_6O_{15} \cdot nH_2O$. . . . . 39 (SE):   Hb/C2−106, 108

$O_{15}Rb_2S_3V_4$ . . . . . $Rb_2SO_3 \cdot 2\,VOSO_3 \cdot 2\,VO_2 \cdot 4\,H_2O$. . . . 48 (V):   Hb/B2−489

$O_{15}Rb_4W_4$ . . . . . . $2\,Rb_2O \cdot 4\,WO_3 \cdot O \cdot 3\,H_2O$. . . . . 54 (W):   Hb−283

$O_{15}Rh_4S_3$ . . . . . . $Rh_2(SO_4)_3 \cdot Rh_2O_3$ . . . . 64 (Rh):   Hb−67

$O_{15}SU_4$ . . . . $4\,UO_3 \cdot SO_3 \cdot 7\,H_2O$ . . . . 55 (U):   Hb−156

$O_{15}S_4$ . . . . S_4O_{15} . . . . 9 (S):   Hb/B1−367

$O_{15}S_4Sb_2$ . . . . . . $Sb_2O_3 \cdot 4\,SO_3 = Sb_2(SO_4)_3 \cdot SO_3$ . . . . . 18 (Sb):   Hb/B3−544

$O_{15}S_5Sb_2$ . . . . . . $Sb_2(SO_3)_5$. . . . . 18 (Sb):   Hb/B2−449

$O_{15}Se_3U^{6-}$ . . . . . . $[UO_3(SeO_4)_3]^{6-}$ . . . . 55 (U):   Hb−158

$O_{15}SnSr_3Ta_4$ . . . . . $3\,SrO \cdot SnO_2 \cdot 2\,Ta_2O_5$
                 $= Sr_{1.2}(Ta_{0.8}Sn_{0.2})_2O_6$. . . . . 50 (Ta):   Hb/B2−232

$O_{15}Sr_3Ta_4Zr$ . . . . . $3\,SrO \cdot ZrO_2 \cdot 2\,Ta_2O_5$
                 $= Sr_{1.2}(Zr_{0.2}Ta_{0.8})_2O_6$. . . . . 50 (Ta):   Hb/B2−228/9

$O_{15}Sr_5Ta_4$. . . . . . $Sr_5Ta_4O_{15}$. . . . . 50 (Ta):   Hb/B2−181, 183

−. . . . . $Sr_5Ta_4O_{15}$ solid solutions:
            $(Sr,Ba)_5Ta_4O_{15}$. . . . . 50 (Ta):   Hb/B2−193

$O_{15}V_8$ . . . . $V_8O_{15} = VO_{1.87}$. . . . . 48 (V):   Hb/B1−13, 17, 50/1

$O_{16}Os_4$ . . . . $(OsO_4)_4$ . . . . 66 (Os):   Hb−48

$O_{16}P_4Pu_3$ . . . . . . $Pu_3(PO_4)_4 \cdot nH_2O$. . . . . Erg. W. 4−253

$O_{16}P_4Si_3$ . . . . . . $2\,P_2O_5 \cdot 3\,SiO_2 = Si_3(PO_4)_4$ . . . . 16 (P):   Hb/C−639

$O_{16}P_4Th_3$ . . . . . . $Th_3(PO_4)_4$. . . . . 44 (Th):   Hb−316

−. . . . . $Th_3(PO_4)_4 \cdot xH_2O$. . . . . 44 (Th):   Hb−317/8

−. . . . . $Th_3(PO_4)_4$ systems:
            $Th_3(PO_4)_4 - H_3PO_4 - H_2O$ . . . . 44 (Th):   Hb−318

$O_{16}P_4U_3$. . . . . . . $U_3(PO_4)_4$ . . . . 55 (U):   Hb−181

$O_{16}P_4Zr_3$ . . . . . . $3\,ZrO_2 \cdot 2\,P_2O_5 \cdot nH_2O$ . . . . 42 (Zr):   Hb−393

$O_{16}P_6Pb$. . . . . . . $PbO \cdot 3\,P_2O_5$ . . . . 47 (Pb):   Hb/C3−877

$O_{16}P_6Pt$ . . . . $PtO \cdot 3\,P_2O_5$ . . . . 68 (Pt):   Hb/C1−137

$O_{16}P_8$ . . . . P_8O_{16} . . . . 16 (P):   Hb/C−76

$O_{16}Pr_6$. . . . . $Pr_6O_{16} \cdot nH_2O$. . . . . 39 (SE):   Hb/C2−106

$O_{16}Pr_9$. . . . . $Pr_9O_{16} = PrO_{1.778}$ . . . . 39 (SE):   Hb/C1−285

$O_{16}Pt_6Rb_6S_4$ . . . . . $Rb_6Pt_6(SO_4)_4 \cdot 17\,H_2O$ . . . . 68 (Pt):   Hb/C2−255

$O_{16}PuRb_4S_4$. . . . . . $Rb_4Pu(SO_4)_4$ . . . . Erg. W. 4−183/4

−. . . . . $Rb_4Pu(SO_4)_4 \cdot 2\,H_2O$ . . . . Erg. W. 4−183/4

$O_{16}PuS_4Tl_5$ . . . . . . $Tl_5Pu(SO_4)_4$. . . . . Erg. W. 4−176, 178

$O_{16}Rb_2Rh_2S_4$ . . . . . $Rb_2SO_4 \cdot Rh_2(SO_4)_3 \cdot 24\,H_2O$ . . . . . . 64 (Rh):   Hb−95

See the "Instructions for the user"
at the beginning of the volume

$O_{16}Rb_2S_4Tl_2$ . . . . . $Rb_2SO_4 \cdot Tl_2(SO_4)_3 = RbTl(SO_4)_2$ . . . . . 38 (Tl):   Hb/3−469

−. . . . . $Rb_2SO_4 \cdot Tl_2(SO_4)_3 \cdot 8\ H_2O$ . . . . 38 (Tl):   Hb/3−469

$O_{16}Rb_2V_6$ . . . . . . $Rb_2O \cdot 3\ V_2O_5 = Rb_2V_6O_{16}$ . . . . 48 (V):   Hb/B2−487

$O_{16}Rb_2W_5$ . . . . . . $Rb_2O \cdot 5\ WO_3$ . . . . 54 (W):   Hb−282/3

$O_{16}Rb_7Ta_5$ . . . . . . $Rb_7Ta_5O_{16} \cdot 11\ H_2O$ . . . . 50 (Ta):   Hb/B2−171

$O_{16}Rh_2S_4Tl_2$ . . . . . $Tl_2SO_4 \cdot Rh_2(SO_4)_3 \cdot 24\ H_2O$ . . . . 64 (Rh):   Hb−99

$O_{16}S_2Zr_5$ . . . . $5\ ZrO_2 \cdot 2\ SO_3 \cdot xH_2O$ . . . . 42 (Zr):   Hb−351/2

$O_{16}S_3V_2$ . . . . $V_2(O_2)_2(SO_4)_3$ . . . . 48 (V):   Hb/B2−393

$O_{16}S_4Sb_2Sr$ . . . . . $SrSO_4 \cdot Sb_2(SO_4)_3 \cdot 6\ H_2O$ . . . . . . 29 (Sr):   Hb−221

$O_{16}S_4Tl_4$ . . . . $Tl_2SO_4 \cdot Tl_2(SO_4)_3$ . . . . 38 (Tl):   Hb/3−363

$O_{16}S_4Tl_4Zr$ . . . . . $2\ Tl_2SO_4 \cdot Zr(SO_4)_2 \cdot 4\ H_2O$ . . . . . . 42 (Zr):   Hb−446

$O_{16}S_4Zn_2{}^{4-}$ . . . . . $[Zn_2(SO_4)_4]^{4-}$ . . . . 32 (Zn):   Eb−1007

$O_{16}ScTaTe_6$ . . . . . . $ScTa(Te_3O_8)_2$ . . . . 50 (Ta):   Hb/B2−319

$O_{16}Se_4Tl_4$ . . . . $Tl_2SeO_4 \cdot Tl_2(SeO_4)_3$ . . . . 38 (Tl):   Hb/3−375

$O_{16}Si_{10}$ . . . . $Si_{10}O_{16}$ . . . . 15 (Si):   Hb/B−591

$O_{16}Sn_5Ta_2Zn$ . . . . $ZnTa_2O_6 \cdot 5\ SnO_2$ . . . . 50 (Ta):   Hb/B2−232

$O_{16}SrV_6$ . . . . $SrO \cdot 3\ V_2O_5$ . . . . 48 (V):   Hb/B2−522

−. . . . . $SrO \cdot 3\ V_2O_5 \cdot 13.5\ H_2O$ . . . . 48 (V):   Hb/B2 522

$O_{16}Ta_5{}^{7-}$ . . . . . . $[Ta_5O_{16}]^{7-}$ . . . . 50 (Ta):   Hb/B1−59/60

$O_{16}Th_3V_4$ . . . . . . $3\ ThO_2 \cdot 2\ V_2O_5 = Th_3(VO_4)_4$ . . . . . . 48 (V):   Hb/B2−565

−. . . . . $Th_3(VO_4)_4$ solid solutions:

                                $Ce_{4(1-x)}Th_{3x}(VO_4)_4$ . . . . 48 (V):   Hb/B2−566

                                $Na_{1.33}Th_{3-0.33x}(VO_4)_4$ . . . . . . 48 (V):   Hb/B2−566

$O_{16}V_4Zr_3$ . . . . $3\ ZrO_2 \cdot 2\ V_2O_5 \cdot 9\ H_2O$

                                $= (ZrO)_3V_4O_{13} \cdot 9\ H_2O$ . . . . 48 (V):   Hb/B2−562/3

$O_{16}V_6Zn_{1+x}$ . . . . . $Zn_{1+y}(V_3O_8)_2$ . . . . 48 (V):   Hb/B2−534

$O_{17}P_2Sb_8$ . . . . . . $4\ Sb_2O_3 \cdot P_2O_5$ . . . . 18 (Sb):   Hb/B3−561

$O_{17}P_2Zr_6$ . . . . . . $6\ ZrO_2 \cdot P_2O_5$ . . . . 42 (Zr):   Hb−393

$O_{17}P_6Pb_2$ . . . . . . $2\ PbO \cdot 3\ P_2O_5$ . . . . 47 (Pb):   Hb/C3−877

$O_{17}P_6Sr_2$ . . . . . . $2\ SrO \cdot 3\ P_2O_5 \cdot xH_2O$ . . . . 29 (Sr):   Hb−215

$O_{17}Pb_2V_6$ . . . . . . $2\ PbO \cdot 3\ V_2O_5 = Pb_2V_6O_{17}$ . . . . . . 48 (V):   Hb/B2−570

$O_{17}Pb_{10}SSi_2$ . . . . . $10\ PbO \cdot 2\ SiO_2 \cdot SO_3$ . . . . 47 (Pb):   Hb/C3−856

$O_{17}Pb_{10}TiV_2$. . . . . . $10\ PbO \cdot V_2O_5 \cdot TiO_2$ . . . . 48 (V):   Hb/B2−574

$O_{17}Pb_{12}$ . . . . $Pb_{12}O_{17} = PbO_{1.41}$ . . . . 47 (Pb):   Hb/C1−139/44

$O_{17}Pr_6$ . . . . $Pr_6O_{17} \cdot nH_2O$. . . . . 39 (SE):   Hb/C2−106

$O_{17}Pt_8S$ . . . . $Pt_8O_{13}SO_4 \cdot 16\ H_2O$ . . . . 68 (Pt):   Hb/C1−111

$O_{17}S_2U_6$ . . . . $6\ UO_2 \cdot S_2O_5 \cdot 10\ H_2O$ . . . . 55 (U):   Hb−156

$O_{17}S_3Ti_4$ . . . . $4\ TiO_2 \cdot 3\ SO_3$ . . . . 41 (Ti):   Hb−349

$O_{17}S_3Zr_4$ . . . . $4\ ZrO_2 \cdot 3\ SO_3 = (ZrOSO_4)_3 \cdot ZrO_2$ . . . . 42 (Zr):   Hb−346/7

−. . . . . $4\ ZrO_2 \cdot 3\ SO_3 \cdot 5\ H_2O$

                                $= Zr_4(SO_4)_3(OH)_{10}$ . . . . 42 (Zr):   Hb−340, 348

−. . . . . $4\ ZrO_2 \cdot 3\ SO_3 \cdot 7\ H_2O$ . . . . 42 (Zr):   Hb−348

−. . . . . $4\ ZrO_2 \cdot 3\ SO_3 \cdot 8\ H_2O$ . . . . 42 (Zr):   Hb−348

−. . . . . $4\ ZrO_2 \cdot 3\ SO_3 \cdot 9\ H_2O$ . . . . 42 (Zr):   Hb−348

−. . . . . $4\ ZrO_2 \cdot 3\ SO_3 \cdot 15\ H_2O$ . . . . 42 (Zr):   Hb−347/8

$O_{17}S_4V_2$ . . . . $V_2O_5 \cdot 4\ SO_3$ . . . . 48 (V):   Hb/B1−303

−. . . . . $V_2O_5 \cdot 4\ SO_3 \cdot 4\ H_2O$ . . . . 48 (V):   Hb/B1−304

$O_{17}Si_7U$ . . . . $UO_3 \cdot 7\ SiO_2$ . . . . 55 (U):   Hb−179

Hb = Main Series Volume; GD = Gmelin-Durrer;   *Ab = Appended Volume to Main Series;*
Eb = Supplement Volume to Main Series;   *Erg. W. = New Supplement Series Volume*

$O_{17}Sr_2V_6$ . . . . 2 SrO · 3 V_2O_5 . . . . 48 (V):   Hb/B2−521
$O_{17}Sr_2W_5$ . . . . 4 SrO · 10 WO_3 · 26 H_2O . . . . 54 (W):   Hb−297
$O_{17}Ta_2W_4$ . . . . 4 WO_3 · Ta_2O_5 . . . . 50 (Ta):   Hb/B1−51
$O_{17}Ta_2Zr_6$ . . . . $Zr_6Ta_2O_{17}$ = Ta_2O_5 · 6 ZrO_2 . . . . 50 (Ta):   Hb/B2−49, 228
−. . . . . $Zr_6Ta_2O_{17}$ solid solutions:
                                   $Zr_6Ta_2O_{17}$−$ZrTiO_4$. . . . . 50 (Ta):   Hb/B2−229
$O_{17}Ta_6Tl_4$ . . . . $Tl_4Ta_6O_{17}$ · 6 H_2O. . . . . 50 (Ta):   Hb/B2−201
$O_{17}V_2W_4$ . . . . V_2O_5 · 4 WO_3 · 10.25 H_2O
                                   = 4 V_2O_5 · 16 WO_3 · 41 H_2O . . . . . 54 (W):   Hb−385/6
$O_{17}V_4Zn_7$ . . . . 7 ZnO · 2 V_2O_5 . . . . 48 (V):   Hb/B2−535
$O_{17}V_6{}^{4-}$ . . . . $[V_6O_{17}]^{4-}$ . . . . 48 (V):   Hb/B1−141, 153
$O_{17}V_6Zn_2$ . . . . 2 ZnO · 3 V_2O_5 · 15 H_2O
                                   = $Zn_2V_6O_{17}$ · 15 H_2O . . . . 48 (V):   Hb/B2−535/6
$O_{17}W_5Zn_2$ . . . . 4 ZnO · 10 WO_3 · 18 H_2O. . . . . 54 (W):   Hb−306
$O_{17}Zn_{10}$ . . . . $Zn_{10}O_{17}$ . . . . 32 (Zn):   Hb−145
$O_{18}P_6Pb_3$ . . . . $Pb_3[(PO_3)_3]_2$ · 3 H_2O . . . . 47 (Pb):   Hb/C3−876/7
$O_{18}P_6S_2$ . . . . 3 P_2O_4 · 2 SO_3 . . . . 16 (P):   Hb/C−583
$O_{18}P_6Sr_3$ . . . . $Sr_3(PO_3)_6$ . . . . 29 (Sr):   Hb−216
$O_{18}P_6Zn_3$ . . . . $Zn_3(PO_3)_6$. . . . . 32 (Zn):   Eb−997
−. . . . . $Zn_3(PO_3)_6$ · 9 H_2O . . . . 32 (Zn):   Hb−280
$O_{18}PbS_{12}{}^{10-}$ . . . . . $[Pb(S_2O_3)_6]^{10-}$ . . . . 47 (Pb):   Hb/C2−598
$O_{18}Pr_6$. . . . . Pr_6O_{18} · nH_2O. . . . . 39 (SE):   Hb/C2−106
$O_{18}Pr_{10}$ . . . . $Pr_{10}O_{18}$ = $PrO_{1.80}$ . . . . 39 (SE):   Hb/C1−285
$O_{18}STa_{10}$ . . . . 5 Ta_2O_3 · SO_3 · 8 H_2O . . . . 50 (Ta):   Hb/B1−177
$O_{18}S_4Ti_3$ . . . . 3 TiO_2 · 4 SO_3 · 15 H_2O. . . . . 41 (Ti):   Hb−350
$O_{18}S_4V_2Zn$ . . . . . . $Zn(VO)_2(SO_4)_4$ . . . . 48 (V):   Hb/B2−536
$O_{18}S_4Zr_3$ . . . . 3 ZrO_2 · 4 SO_3 · 15 H_2O . . . . 42 (Zr):   Hb−345
$O_{18}Sb_{12}W_5$ . . . . . . Sb_2O_3 · 5 WO_3 · 4 H_2O . . . . 54 (W):   Hb−207
$O_{18}Se_3U_2{}^{6-}$ . . . . . . $[UO_2(UO_4)(SeO_4)_3]^{6-}$ . . . . 55 (U):   Hb−158
$O_{18}Si_2Zr_7$ . . . . 7 ZrO_2 · 2 SiO_2 . . . . 42 (Zr):   Hb−389
$O_{19}P_4Tl_6$. . . . . $Tl_6P_4O_{19}$ · 12 H_2O. . . . . 38 (Tl):   Hb/3−430
$O_{19}P_6Si_2$ . . . . 3 P_2O_5 · 2 SiO_2 · H_2O . . . . 16 (P):   Hb/C−640
$O_{19}P_6V_2$ . . . . V_2O_4 · 3 P_2O_5 · 10 H_2O . . . . 48 (V):   Hb/B1−361
$O_{19}Pb_4Ta_6$ . . . . . . $Pb_4Ta_6O_{19}$ · 16.5 H_2O . . . . 50 (Ta):   Hb/B2−237
$O_{19}Pb_4U_5$ . . . . 4 PbO · 5 UO_3 · 4 H_2O . . . . 55 (U):   Hb−239
$O_{19}Pb_8SSi_4$ . . . . . . 3 PbO · 4 $PbSiO_3$ · $PbSO_4$ . . . . 47 (Pb):   Hb/C3−856
−. . . . . 8 PbO · 4 SiO_2 · SO_3 . . . . 47 (Pb):   Hb/C3−857
$O_{19}Pb_{12}$ . . . . $Pb_{12}O_{19}$ . . . . 47 (Pb):   Hb/C1−144/7
−. . . . . $Pb_{12}O_{19}$ systems:
                                   $Pb_{12}O_{19}$−PbO_2. . . . . 47 (Pb):   Hb/C1−55
                                   $Pb_{12}O_{19}$−Pb_3O_4 . . . . 47 (Pb):   Hb/C1−54/5
$O_{19}PrTa_7$ . . . . $PrTa_7O_{19}$ . . . . 50 (Ta):   Hb/B2−201
$O_{19}Pr_6$ . . . . Pr_6O_{19} · nH_2O. . . . . 39 (SE):   Hb/C2−106
$O_{19}Rb_8Ta_6$ . . . . . . $Rb_8Ta_6O_{19}$ · 14 H_2O . . . . 50 (Ta):   Hb/B2−171
$O_{19}S_2U_7$ . . . . 7 UO_2 · S_2O_5 · 8 H_2O . . . . 55 (U):   Hb−156
$O_{19}S_3Zr_5$ . . . . 5 ZrO_2 · 3 SO_3 · 13 H_2O . . . . 42 (Zr):   Hb−350
$O_{19}S_5Ti_2$ . . . . 2 TiO_2 · 5 SO_3 · 5 H_2O . . . . 41 (Ti):   Hb−347
$O_{19}Se_5V_2$ . . . . 2 VO_2 · 5 SeO_3 · 10 H_2O . . . . 48 (V):   Hb/B1−314

See the "Instructions for the user"
at the beginning of the volume

$O_{19}SmTa_7$ . . . . . . $SmTa_7O_{19}$ . . . . 50 (Ta): Hb/B2–201, 214

$O_{19}Sr_4Ta_6$ . . . . . . $Sr_4Ta_6O_{19} \cdot 16.5\ H_2O$ . . . . 50 (Ta): Hb/B2–182/3

$O_{19}Ta_6^{8-}$ . . . . $[Ta_6O_{19}]^{8-}$ . . . . 50 (Ta): Hb/B1–59/61

Hb/B2–167

$O_{20}P_2Ta_6$ . . . . $3\ Ta_2O_5 \cdot P_2O_5$ . . . . 50 (Ta): Hb/B1–268

$O_{20}P_5Ta_3$ . . . . $Ta_3(PO_4)_5$ . . . . 50 (Ta): Hb/B1–269

$O_{20}P_6Pb_5$ . . . . $Pb_5(P_3O_{10})_2$ . . . . 47 (Pb): Hb/C3–874/5

$O_{20}P_6Sr_5$ . . . . $Sr_5(P_3O_{10})_2$ . . . . 29 (Sr): Eb–286

$O_{20}P_6V_2$ . . . . $V_2O_5 \cdot 3\ P_2O_5$ . . . . 48 (V): Hb/B1–362

$O_{20}P_6Zn_5$ . . . . $Zn_5(P_3O_{10})_2$ . . . . 32 (Zn): Hb–279

$O_{20}Pr_6$ . . . . $Pr_6O_{20} \cdot nH_2O$ . . . . 39 (SE): Hb/C2–106

$O_{20}Pr_{11}$ . . . . $Pr_{11}O_{20} = PrO_{1.818}$ . . . . 39 (SE): Hb/C1–285

$O_{20}S_2Ti_7$ . . . . $7\ TiO_2 \cdot 2\ SO_3 \cdot xH_2O$ . . . . . . 41 (Ti): Hb–349

$O_{20}S_5Ta_2$ . . . . $Ta_2(SO_4)_5 = (TaO_2)_2S_5O_{16}$ . . . 50 (Ta): Hb/B1–176/7

$O_{20}S_5Tl_6$ . . . . $2\ Tl_2SO_4 \cdot Tl_2(SO_4)_3$ . . . . 38 (Tl): Hb/3–362

$O_{20}Tb_{11}$ . . . . $Tb_{11}O_{20} = TbO_{1.818}$ . . . . 39 (SE): Hb/C1–387, 395/6

$O_{20}U_7$ . . . . U_7O_{20} . . . . 55 (U): Hb–82, 94

$O_{20}W_6Zn_2$ . . . . $2\ ZnO \cdot 6\ WO_3 \cdot 10\ H_2O$ . . . . 54 (W): Hb–306

$O_{20}Zn_{12}$ . . . . $8\ ZnO_2 \cdot 4\ ZnO \cdot 7\ H_2O$ . . . . 32 (Zn): Hb–146

$O_{21}P_6Sb_4$ . . . . $Sb_4(P_2O_7)_3 \cdot 2\ H_2O$ . . . . 18 (Sb): Hb/B3–561

$O_{21}P_6V_4$ . . . . $V_4(P_2O_7)_3 \cdot 30\ H_2O$ . . . . 48 (V): Hb/B1–360

$O_{21}Pb_4Rb_2Si_8$ . . . . . $Rb_2Pb_4Si_8O_{21}$ . . . . 47 (Pb): Hb/C3–1145

$O_{21}Pb_4Si_8Tl_2$ . . . . . $Tl_2Pb_4Si_8O_{21}$ . . . . 47 (Pb): Hb/C4–1354

$O_{21}Pb_8Si_6Sr$ . . . . . $Pb_8Sr(Si_2O_7)_3$ . . . . 47 (Pb): Hb/C3–1199

$O_{21}Pb_8Si_6Zn$ . . . . . $Pb_8Zn(Si_2O_7)_3$ . . . . 47 (Pb): Hb/C4–1229

$O_{21}Pb_{8.5}Si_6Sr_{0.5}$ . . . $Pb_{8.5}Sr_{0.5}(Si_2O_7)_3$ . . . . 47 (Pb): Hb/C3–1199

$O_{21}Pr_6$ . . . . $Pr_6O_{21} \cdot nH_2O$ . . . . 39 (SE): Hb/C2–106

$O_{21}S_2Sb_{10}$ . . . . . . $5\ Sb_2O_3 \cdot 2\ SO_3 \cdot 7\ H_2O$ . . . . . . 18 (Sb): Hb/B3–543, 545

$O_{21}S_2U_8$ . . . . $8\ UO_2 \cdot S_2O_5 \cdot 21\ H_2O$ . . . . 55 (U): Hb–156

$O_{21}S_4Tl_2V_6$ . . . . . $Tl_2SO_3 \cdot 3\ VOSO_3 \cdot 3\ VO_2 \cdot 8\ H_2O$ . . . . . 48 (V): Hb/B2–550

$O_{21}Sb_2W_6$ . . . . $Sb_2O_3 \cdot 6\ WO_3 \cdot 8\ H_2O$ . . . . 54 (W): Hb–207

$O_{21}SrV_8$ . . . . $SrO \cdot 4\ V_2O_5 \cdot 11\ H_2O$ . . . . 48 (V): Hb/B2–522

$O_{21}U_2V_6$ . . . . $2\ UO_3 \cdot 3\ V_2O_5$ . . . . 48 (V): Hb/B1–33

$O_{21}W_6^{6-}$ . . . . $[W_6O_{21}]^{6-}$ . . . . 54 (W): Hb–144

$O_{22}Pr_6$ . . . . $Pr_6O_{22} \cdot nH_2O$ . . . . 39 (SE): Hb/C2–106

$O_{22}S_4U_5$ . . . . $4\ UOSO_4 \cdot UO_2 \cdot 8\ H_2O$ . . . . 55 (U): Hb–151

$O_{22}S_7Tl_8$ . . . . $3\ Tl_2S_2O_6 \cdot Tl_2SO_4$ . . . . 38 (Tl): Hb/3–368/9

$O_{23}P_4SrU_4$ . . . . . $SrO \cdot 4\ UO_3 \cdot 2\ P_2O_5 \cdot 21\ H_2O$

$= Sr(UO_2)_4H_2(PO_4)_4 \cdot 20\ H_2O$ . . . . . . 55 (U): Hb–232

$O_{23}Pr_6$ . . . . $Pr_6O_{23} \cdot nH_2O$ . . . . 39 (SE): Hb/C2–106

$O_{23}S_3Zr_7$ . . . . $7\ ZrO_2 \cdot 3\ SO_3 \cdot 14\ H_2O$ . . . . . . 42 (Zr): Hb–351

$O_{23}S_5Ti_4$ . . . . $4\ TiO_2 \cdot 5\ SO_3 \cdot 12.5\ H_2O$ . . . . . . 41 (Ti): Hb–350

$O_{23}S_5U_4$ . . . . $3\ UOSO_4 \cdot U(SO_4)_2 \cdot xH_2O$ . . . . . . 55 (U): Hb–151

$O_{23}Sr_3V_8$ . . . . $3\ SrO \cdot 4\ V_2O_5 \cdot 14\ H_2O$ . . . . 48 (V): Hb/B2–521

$O_{24}PbS_{16}^{14-}$ . . . . . $[Pb(S_2O_3)_8]^{14-}$ . . . . 47 (Pb): Hb/C2–598

$O_{24}Pb_3W_7$ . . . . $3\ PbO \cdot 7\ WO_3 \cdot 10\ H_2O$ . . . . 54 (W): Hb–319

$O_{24}Pr_6$ . . . . $Pr_6O_{24} \cdot nH_2O$ . . . . 39 (SE): Hb/C2–106

$O_{24}Rb_4V_{10}$ . . . . . $Rb_4[V_{10}O_{24}] \cdot xH_2O$

$= 2\ Rb_2O \cdot 3\ V_2O_4 \cdot 2\ V_2O_5 \cdot xH_2O$ . . . 48 (V): Hb/B2–486

Hb = Main Series Volume; GD = Gmelin-Durrer; Ab = Appended Volume to Main Series;
Eb = Supplement Volume to Main Series; Erg. W. = New Supplement Series Volume

$O_{24}S_5Sb_6$ . . . . 3 $Sb_2O_3 \cdot 5\ SO_3 \cdot H_2O$ . . . . 18 (Sb):   Hb/B3–544/5

$O_{24}S_6Tl_8$ . . . . . . 3 $Tl_2SO_4 \cdot Tl_2(SO_4)_3$ . . . . 38 (Tl):   Hb/3–362

–. . . . . 3 $Tl_2SO_4 \cdot Tl_2(SO_4)_3$ solid solutions:
                        $(NH_4, Tl)_3Tl(SO_4)_3$ . . . . 38 (Tl):   Hb/3–463

$O_{24}Sr_3W_7$ . . . . 3 $SrO \cdot 7\ WO_3 \cdot 16\ H_2O$ . . . . 54 (W):   Hb–297

$O_{24}V_{10}{}^{4-}$ . . . . $[V_{10}O_{24}]^{4-}$ . . . . 48 (V):   Hb/B1–161

$O_{24}Zn_{14}$ . . . . 10 $ZnO_2 \cdot 4\ ZnO \cdot 5\ H_2O$ . . . . 32 (Zn):   Hb–146

$O_{25}P_2Pb_{16}Si_2$ . . . . . 16 $PbO \cdot P_2O_5 \cdot 2\ SiO_2$ . . . . 47 (Pb):   Hb/C3–886

$O_{25}P_6Pb_{10}$ . . . . . . 3 $Pb_3(PO_4)_2 \cdot PbO = 10\ PbO \cdot 3\ P_2O_5$ . . . 47 (Pb):   Hb/C3–870

–. . . . . 3 $Pb_3(PO_4)_2 \cdot PbO$ systems:
                        3 $Pb_3(PO_4)_2 \cdot PbO–Pb_5F(PO_4)_3$ . . . . . 47 (Pb):   Hb/C3–881

$O_{25}P_6V_4$ . . . . 2 $V_2O_5 \cdot 3\ P_2O_5$ . . . . 48 (V):   Hb/B1–363

–. . . . . 2 $V_2O_5 \cdot 3\ P_2O_5 \cdot 9\ H_2O$. . . . . 48 (V):   Hb/B1–363

–. . . . . 2 $V_2O_5 \cdot 3\ P_2O_5 \cdot 13\ H_2O$
                        = $(VO)_2(HPO_4)_3 \cdot 5\ H_2O$
                        = $[VO_2H_2PO_4]_2H_3PO_4 \cdot 3\ H_2O$ . . . . 48 (V):   Hb/B1–362

$O_{25}P_6Zr_5$ . . . . 5 $ZrO_2 \cdot 3\ P_2O_5 \cdot 9\ H_2O$. . . . . 42 (Zr):   Hb–393

$O_{25}Pb_{10}Sb_6$ . . . . . . 3 $Pb_3(SbO_4)_2 \cdot PbO$. . . . . 47 (Pb):   Hb/C3–934

$O_{25}Rb_2S_8$ . . . . $Rb_2S_2O_7 \cdot 6\ SO_3 = Rb_2SO_4 \cdot 7\ SO_3$ . . . 24 (Rb):   Hb–214

$O_{25}Rb_2W_8$. . . . . . . $Rb_2O \cdot 8\ WO_3$. . . . . 54 (W):   Hb–283

$O_{25}Rb_4V_{10}$. . . . . . . $Rb_4[V_{10}O_{25}] \cdot xH_2O$
                        = 2 $Rb_2O \cdot 2\ V_2O_4 \cdot 3\ V_2O_5 \cdot xH_2O$ . . . 48 (V):   Hb/B2–486

$O_{25}S_8Tl_2$ . . . . $Tl_2S_2O_7 \cdot 6\ SO_3 = Tl_2SO_4 \cdot 7\ SO_3$ . . . . . 38 (Tl):   Hb/3–367

$O_{25}Sb_6U_5$ . . . . 5 $UO_2 \cdot 3\ Sb_2O_5 \cdot 15\ H_2O$ . . . . 55 (U):   Hb–185

$O_{25}Sr_{10}V_6$ . . . . 10 $SrO \cdot 3\ V_2O_5$. . . . . 48 (V):   Hb/B2–519

$O_{25}Ta_9V$. . . . . VTa_9O_{25} . . . . 50 (Ta):   Hb/B2–244, 245

$O_{25}V_{10}{}^{4-}$ . . . . $[V_{10}O_{25}]^{4-}$ . . . . 48 (V):   Hb/B1–161

$O_{26}Pb_8SSi_6$ . . . . . . $Pb_8(SiO_4)_2(Si_2O_7)_2SO_4$. . . . . 47 (Pb):   Hb/C3–856

$O_{26}V_{10}{}^{4-}$ . . . . $[V_{10}O_{26}]^{4-}$ . . . . 48 (V):   Hb/B1–161

$O_{26}V_{10}Zn$ . . . . $ZnO \cdot 5\ V_2O_5 = ZnV_{10}O_{26}$ . . . . 48 (V):   Hb/B2–536

$O_{26.19}P_{5.98}V_{5.12}$ . . . . 3.12 $VO_2 \cdot V_2O_5 \cdot 2.99\ P_2O_5 \cdot 21.35\ H_2O$ . . 48 (V):   Hb/B1–363

$O_{27}P_4Sr_2U_5$ . . . . . . 2 $SrO \cdot 5\ UO_3 \cdot 2\ P_2O_5 \cdot 24\ H_2O$ . . . . . 55 (U):   Hb–232

$O_{27}P_6Rh_8$ . . . . 4 $Rh_2O_3 \cdot 3\ P_2O_5 \cdot 32\ H_2O$ . . . . 64 (Rh):   Hb–72

$O_{27}P_6Tl_8$. . . . . $Tl_8P_6O_{27} \cdot 13\ H_2O$ . . . . 38 (Tl):   Hb/3–430

$O_{27}P_6U_4$. . . . . 4 $UO_3 \cdot 3\ P_2O_5$ . . . . 55 (U):   Hb–184

$O_{27}S_2Sb_{14}$ . . . . . . 7 $Sb_2O_3 \cdot 2\ SO_3$. . . . . 18 (Sb):   Hb/B3–545

–. . . . . 7 $Sb_2O_3 \cdot 2\ SO_3 \cdot nH_2O$. . . . . 18 (Sb):   Hb/B3–545

$O_{27}S_5Ti_6$ . . . . 6 $TiO_2 \cdot 5\ SO_3 \cdot 4\ H_2O$ . . . . 41 (Ti):   Hb–347

$O_{27}S_5Zr_6$ . . . . $(ZrOSO_4)_5 \cdot ZrO_2 = 5\ Zr(SO_4)_2 \cdot 7\ ZrO_2$ . . 42 (Zr):   Hb–346

–. . . . . $(ZrOSO_4)_5 \cdot ZrO_2 \cdot 12\ H_2O$. . . . . . . 42 (Zr):   Hb–346

$O_{27}S_7$ . . . . S_7O_{27} . . . . 9 (S):   Hb/B1–367

$O_{27}S_8Sb_2$ . . . . $Sb_2O_3 \cdot 8\ SO_3$ . . . . 18 (Sb):   Hb/B3–544

$O_{27}Sb_6W_4$ . . . . 3 $Sb_2O_5 \cdot 4\ WO_3 \cdot 11\ H_2O$ . . . . 54 (W):   Hb–380

$O_{27}V_{10}{}^{4-}$ . . . . $[V_{10}O_{27}]^{4-}$ . . . . 48 (V):   Hb/B1–152, 155

$O_{27}V_{10}Zn_2$. . . . . . . 2 $ZnO \cdot 5\ V_2O_5 = Zn_2V_{10}O_{27}$. . . . . . . 48 (V):   Hb/B2–536

$O_{28}Rb_6V_{10}$. . . . . . . 3 $Rb_2O \cdot 5\ V_2O_5 \cdot 8.5\ H_2O$
                        = $Rb_6V_{10}O_{28} \cdot 8.5\ H_2O$ . . . . 48 (V):   Hb/B2–487

See the "Instructions for the user"
at the beginning of the volume

$O_{28}Sr_3V_{10}$ . . . . 3 SrO · 5 V_2O_5 · 24 H_2O

                   = $Sr_3V_{10}O_{28}$ · 24 H_2O . . . . 48 (V):   Hb/B2–521

–. . . . . 3 SrO · 5 V_2O_5 · xH_2O . . . . 48 (V):   Hb/B2–520

$O_{28}V_{10}^{6-}$ . . . . $[V_{10}O_{28}]^{6-}$ . . . . 48 (V):   Hb/A2–498

                                                             Hb/B1–132/45, 153/7

$O_{28}V_{10}Zn_3$ . . . . . . 3 ZnO · 5 V_2O_5 · 8 H_2O

                   = $Zn_3V_{10}O_{28}$ · 8 H_2O . . . . 48 (V):   Hb/B2–536

–. . . . . 3 ZnO · 5 V_2O_5 · 16 H_2O . . . . 48 (V):   Hb/B2–536

$O_{29}S_5Zr_7$ . . . . 7 ZrO_2 · 5 SO_3 · 30 H_2O . . . . 42 (Zr):   Hb–348/9

$O_{29}S_7Th_4$ . . . . ($ThOSO_4$ · 2 H_2O) · (3 Th(SO_4)$_2$ · 2 H_2O). . 44 (Th):   Hb–292

$O_{30}P_8Zn_{10}$ . . . . 10 ZnO · 4 P_2O_5 · 10 H_2O. . . . . 32 (Zn):   Hb–278

$O_{30}P_8Zr_5$ . . . . 5 ZrO_2 · 4 P_2O_5 · 8 H_2O. . . . . 42 (Zr):   Hb–393

$O_{30}P_{10}Sr_5$ . . . . . . 5 $Sr_5(PO_3)_{10}$ · 13 H_2O . . . . 29 (Sr):   Hb–216

$O_{30}S_9Sb_2$ . . . . Sb_2O_3 · 9 SO_3. . . . . 18 (Sb):   Hb/B3–544

$O_{31}S_5Zr_8$ . . . . 8 ZrO_2 · 5 SO_3 · 19 H_2O

                   = ($Zr_4(SO_4)_2(OH)_{11})_2SO_4$ · 8 H_2O . . . 42 (Zr):   Hb–340, 350

$O_{31}Tl_{12}V_{10}$ . . . . . 6 Tl_2O · 5 V_2O_5 = $Tl_{12}V_{10}O_{31}$ . . . . . . 48 (V):   Hb/B2–548

$O_{31}W_7Zr_5$ . . . . 5 ZrO_2 · 7 WO_3 · 21 H_2O . . . . 54 (W):   Hb–315

$O_{32}Pb_5U_9$ . . . . 5 PbO · 9 UO_3 · 10 H_2O. . . . . 55 (U):   Hb–240

$O_{32}PtW_{10}$ . . . . PtO_2 · 10 WO_3 · xH_2O . . . . 68 (Pt):   Hb/C3–321

$O_{32}S_6Zr_7$ . . . . 7 ZrO_2 · 6 SO_3 · xH_2O . . . . 42 (Zr):   Hb–346

$O_{32}S_8Tl_{12}$ . . . . 5 Tl_2SO_4 · $Tl_2(SO_4)_3$ . . . . 38 (Tl):   Hb/3–363

$O_{32}Se_8Tl_{12}$ . . . . 5 Tl_2SeO_4 · $Tl_2(SeO_4)_3$ . . . . 38 (Tl):   Hb/3–375

$O_{32}SiW_{10}$ . . . . SiO_2 · 10 WO_3 · 7 H_2O

                   = $H_8[SiO(W_2O_7)_5]$ · 3 H_2O . . . . . . 54 (W):   Hb–345

$O_{32}Ta_{12}Zr$ . . . . . . ZrO_2 · 6 Ta_2O_5 . . . . 50 (Ta):   Hb/B2–349

$O_{32}TiW_{10}$ . . . . 4 H_2O · TiO_2 · 10 WO_3 · xH_2O . . . . . . 54 (W):   Hb–383

$O_{33}S_7Zr_6$ . . . . 6 ZrO_2 · 7 SO_3 · 19 H_2O . . . . 42 (Zr):   Hb–345

$O_{33}V_{12}^{6-}$ . . . . $[V_{12}O_{33}]^{6-}$ . . . . 48 (V):   Hb/B2–397

$O_{34}Sr_4W_{10}$ . . . . . 4 SrO · 10 WO_3 · 26 H_2O . . . . 54 (W):   Hb–297

$O_{34}V_{10}Yb_6$ . . . . . 3 Yb_2O_3 · 5 V_2O_5 . . . . 48 (V):   Hb/B2–556

$O_{34}W_{10}Zn_4$ . . . . . 4 ZnO · 10 WO_3 · 18 H_2O. . . . . 54 (W):   Hb–306

$O_{35}Ta_{12}U_2$ . . . . . . U_2O_5 · 6 Ta_2O_5 . . . . 50 (Ta):   Hb/B1–51

$O_{35}Ta_{12}V_2$ . . . . . . $V_2Ta_{12}O_{35}$ . . . . 50 (Ta):   Hb/B2–244

$O_{35}V_2W_{10}$ . . . . . . V_2O_5 · 10 WO_3 · 22 H_2O . . . . 54 (W):   Hb–385

$O_{36}PV_{12}^{7-}$ . . . . . . $[P(V_3O_9)_4]^{7-}$ . . . . 48 (V):   Hb/B1–362

                                                             Hb/B2–444

$O_{37}W_9Zr_5$ . . . . 5 ZrO_2 · 9 WO_3 · 33 H_2O . . . . 54 (W):   Hb–315

$O_{38}SiW_{12}$ . . . . SiO_2 · 12 WO_3 · 8 H_2O

                   = $H_8[Si(W_2O_7)_6]$ · 4 H_2O . . . . 54 (W):   Hb–343/4

–. . . . . SiO_2 · 12 WO_3 · 9 H_2O . . . . 54 (W):   Hb–343/4

–. . . . . SiO_2 · 12 WO_3 · 19 H_2O . . . . 54 (W):   Hb–343

–. . . . . SiO_2 · 12 WO_3 · 24 H_2O . . . . 54 (W):   Hb–344/5

–. . . . . SiO_2 · 12 WO_3 · 26 H_2O . . . . 54 (W):   Hb–343

–. . . . . SiO_2 · 12 WO_3 · 32 H_2O . . . . 54 (W):   Hb–340/3

$O_{38}TiW_{12}$ . . . . 4 H_2O · TiO_2 · 12 WO_3 · xH_2O . . . . . . 54 (W):   Hb–383

$O_{39}Rb_2SiW_{12}$ . . . . . Rb_2O · SiO_2 · 12 WO_3 · 6 H_2O

                   = $Rb_2H_6[Si(W_2O_7)_6]$ · 3 H_2O . . . . . . 54 (W):   Hb–349

Hb = Main Series Volume; GD = Gmelin-Durrer;   *Ab = Appended Volume to Main Series;*
Eb = Supplement Volume to Main Series;   *Erg. W. = New Supplement Series Volume*

$O_{39}SiTl_2W_{12}$ . . . . . $Tl_2O \cdot SiO_2 \cdot 12\ WO_3 \cdot 10\ H_2O$

$= Tl_2H_6[Si(W_2O_7)_6] \cdot 7\ H_2O$ . . . . 54 (W):   Hb–353

$O_{39}Sr_4V_{14}$ . . . . 4 $SrO \cdot 7\ V_2O_5 \cdot 30\ H_2O$ . . . . 48 (V):   Hb/B2–521

$O_{40}Pb_2SiW_{12}$ . . . . . 2 $PbO \cdot SiO_2 \cdot 12\ WO_3 \cdot 21\ H_2O$

$= Pb_2H_4[Si(W_2O_7)_6] \cdot 19\ H_2O$ . . . . . . 54 (W):   Hb–353

$O_{40}Rb_2S_{10}Ti_6$ . . . . . $Rb_2SO_4 \cdot 3\ Ti_2(SO_4)_3 \cdot 24\ H_2O$ . . . . . . 41 (Ti):   Hb–413

$O_{40}S_{10}Tl_{16}$ . . . . . . 7 $Tl_2SO_4 \cdot Tl_2(SO_4)_3 \cdot 6\ H_2O$ . . . . . . 38 (Tl):   Hb/3–362

$O_{40}SiSr_2W_{12}$ . . . . . 2 $SrO \cdot SiO_2 \cdot 12\ WO_3 \cdot 16\ H_2O$

$= Sr_2H_4[Si(W_2O_7)_6] \cdot 14\ H_2O$ . . . . . 54 (W):   Hb–351

–. . . . . 2 $SrO \cdot SiO_2 \cdot 12\ WO_3 \cdot 17\ H_2O$ . . . . . 54 (W):   Hb–351

–. . . . . 2 $SrO \cdot SiO_2 \cdot 12\ WO_3 \cdot 23\ H_2O$ . . . . . 54 (W):   Hb–350

–. . . . . 2 $SrO \cdot SiO_2 \cdot 12\ WO_3 \cdot 27\ H_2O$ . . . . . 54 (W):   Hb–350

$O_{40}SiThW_{12}$ . . . . . . $ThO_2 \cdot SiO_2 \cdot 12\ WO_3 \cdot 27\ H_2O$

$= ThH_4[Si(W_2O_7)_6] \cdot 25\ H_2O$ . . . . . 54 (W):   Hb–353

–. . . . . $ThO_2 \cdot SiO_2 \cdot 12\ WO_3 \cdot 30\ H_2O$. . . . . 54 (W):   Hb–353

$O_{40}SiW_{12}Zn_2$ . . . . . 2 $ZnO \cdot SiO_2 \cdot 12\ WO_3 \cdot 18\ H_2O$

$= Zn_2H_4[Si(W_2O_7)_6] \cdot 16\ H_2O$ . . . . . 54 (W):   Hb–352

–. . . . . 2 $ZnO \cdot SiO_2 \cdot 12\ WO_3 \cdot 27\ H_2O$ . . . . . 54 (W):   Hb–351/2

–. . . . . 2 $ZnO \cdot SiO_2 \cdot 12\ WO_3 \cdot 29\ H_2O$ . . . . . 54 (W):   Hb–351

$O_{41}P_2W_{12}$ . . . . $P_2O_5 \cdot 12\ WO_3 \cdot 42\ H_2O$ . . . . 54 (W):   Hb–374

$O_{41}Rb_{10}W_{12}$ . . . . . . 5 $Rb_2O \cdot 12\ WO_3 \cdot 18\ H_2O$ . . . . 54 (W):   Hb–282

$O_{41}SiUW_{12}$ . . . . . . 3 $UO_3 \cdot 3(SiO_2 \cdot 12\ WO_3) \cdot 88\ H_2O$ . . . . . 55 (U):   Hb–244

$O_{41}Sr_6V_{14}$ . . . . 6 $SrO \cdot 7\ V_2O_5$ . . . . 48 (V):   Hb/B2–521

$O_{41}Tl_{10}W_{12}$ . . . . . . 5 $Tl_2O \cdot 12\ WO_3 \cdot 9\ H_2O$ . . . . 54 (W):   Hb–314

$O_{41}W_{12}Zn_5$ . . . . . . 5 $ZnO \cdot 12\ WO_3 \cdot 35\ H_2O$. . . . . 54 (W):   Hb–306

$O_{42}P_2Tl_2W_{12}$ . . . . . $Tl_2O \cdot P_2O_5 \cdot 12\ WO_3 \cdot 4\ H_2O$. . . . . . . 54 (W):   Hb–375

$O_{42}Pb_4SiW_{12}$ . . . . . 4 $PbO \cdot SiO_2 \cdot 12\ WO_3 \cdot 20\ H_2O$

$= Pb_4[Si(W_2O_7)_6] \cdot 20\ H_2O$ . . . . . . 54 (W):   Hb–353

$O_{42}Rb_8SiW_{12}$ . . . . . 4 $Rb_2O \cdot SiO_2 \cdot 12\ WO_3$

$= Rb_8[Si(W_2O_7)_6]$ . . . . 54 (W):   Hb–349

$O_{43}P_2Pb_2W_{12}$ . . . . . 2 $PbO \cdot P_2O_5 \cdot 12\ WO_3 \cdot 6\ H_2O$. . . . . . 54 (W):   Hb–375

$O_{43}P_2Sr_2W_{12}$ . . . . . 2 $SrO \cdot P_2O_5 \cdot 12\ WO_3 \cdot 17\ H_2O$ . . . . . 54 (W):   Hb–375

$O_{43}P_2W_{12}Zn_2$ . . . . . 2 $ZnO \cdot P_2O_5 \cdot 12\ WO_3 \cdot 7\ H_2O$ . . . . . 54 (W):   Hb–375

$O_{44}Rb_{10}W_{12}$ . . . . . . 5 $Rb_2O \cdot 12\ WO_3 \cdot 3\ O \cdot 12\ H_2O$ . . . . . 54 (W):   Hb–283

$O_{44}S_{11}Th_2Tl_{14}$ . . . . 7 $Tl_2SO_4 \cdot 2\ Th(SO_4)_2$ . . . . 44 (Th):   Hb–352

$O_{44}S_{11}Tl_{10}$ . . . . . . 2 $Tl_2SO_4 \cdot 3\ Tl_2(SO_4)_3 \cdot 22\ H_2O$ . . . . . 38 (Tl):   Hb/3–362

$O_{44}S_{11}Tl_{14}Zr_2$ . . . . 7 $Tl_2SO_4 \cdot 2\ Zr(SO_4)_2$. . . . . 42 (Zr):   Hb–446

$O_{44}SiU_2W_{12}$ . . . . . . 6 $UO_3 \cdot 3(SiO_2 \cdot 12\ WO_3) \cdot 91\ H_2O$ . . . . . 55 (U):   Hb–244

$O_{47}S_{12}Tl_{10}$ . . . . . . 2 $Tl_2O \cdot 3\ Tl_2O_3 \cdot 12\ SO_3 \cdot 25\ H_2O$ . . . . . 38 (Tl):   Hb/3–362

$O_{50}P_2Ta_{18}$ . . . . . . 9 $Ta_2O_5 \cdot P_2O_5$ . . . . 50 (Ta):   Hb/B1–268

$O_{50}Ta_{17}V_3$ . . . . $V_3Ta_{17}O_{50}$. . . . . 50 (Ta):   Hb/B2–244, 245

$O_{51}Pr_4V_{18}$ . . . . $Pr_4(V_6O_{17})_3 \cdot 48\ H_2O$ . . . . 48 (V):   Hb/B2–555/6

$O_{51}STh_{24}$ . . . . 24 $ThO_2 \cdot SO_3 \cdot 12\ H_2O$ . . . . 44 (Th):   Hb–293

$O_{51}Sm_4V_{18}$ . . . . . . $Sm_4(V_6O_{17})_3 \cdot 48\ H_2O$ . . . . 48 (V):   Hb/B2–555/6

$O_{51}V_{18}Y_4$ . . . . $Y_4(V_6O_{17})_3 \cdot 48\ H_2O$ . . . . 48 (V):   Hb/B2–555/6

$O_{53}P_2W_{16}$ . . . . $P_2O_5 \cdot 16\ WO_3 \cdot 69\ H_2O$ . . . . 54 (W):   Hb–374

$O_{54}S_{10}Zr_{12}$ . . . . . . 5 $Zr(SO_4)_2 \cdot 7\ ZrO_2 = (ZrOSO_4)_5 \cdot ZrO_2$. . 42 (Zr):   Hb–346

$O_{55}P_2Sr_{10}V_{16}$ . . . . . 10 $SrO \cdot 8\ V_2O_5 \cdot P_2O_5 \cdot xH_2O$ . . . . 48 (V):   Hb/B2–523

See the "Instructions for the user"
at the beginning of the volume

8*

$O_{59}P_2W_{18}$ . . . . $P_2O_5 \cdot 18\ WO_3 \cdot 42\ H_2O$

          $= H_{10}[P_2(OH)_2(W_2O_7)_9] \cdot 36\ H_2O$ . . . . 54 (W):   Hb−365

$O_{59}V_2W_{18}$ . . . . $V_2O_5 \cdot 18\ WO_3 \cdot 36\ H_2O$ 54 (W):   Hb−385

$O_{60}P_2Rb_{10}V_{20}$ . . . . . 5 $Rb_2O \cdot 10\ V_2O_5 \cdot P_2O_5 \cdot 46\ H_2O$ . . . . . 48 (V):   Hb/B2−491/2

$O_{62}PtW_{20}$ . . . . $PtO_2 \cdot 20\ WO_3 \cdot xH_2O$ . . . . 68 (Pt):   Hb/C3−321

$O_{65}P_2W_{20}$ . . . . $P_2O_5 \cdot 20\ WO_3 \cdot xH_2O$ . . . . 54 (W):   Hb−364

$O_{65}P_{14}V_{12}$ . . . . 6 $V_2O_5 \cdot 7\ P_2O_5 \cdot 37\ H_2O$ . . . . 48 (V):   Hb/B1−363

$O_{67}P_2Pb_2W_{20}$ . . . . . 2 $PbO \cdot P_2O_5 \cdot 20\ WO_3 \cdot 6\ H_2O$ . . . . . . 54 (W):   Hb−370

$O_{68}P_2W_{21}$ . . . . $P_2O_5 \cdot 21\ WO_3 \cdot xH_2O$ . . . . 54 (W):   Hb−364

$O_{68}V_8W_{16}$ . . . . 4 $V_2O_5 \cdot 16\ WO_3 \cdot 41\ H_2O$

          $= V_2O_5 \cdot 4\ WO_3 \cdot 10.25\ H_2O$ . . . . . 54 (W):   Hb−385/6

$O_{71}P_2W_{22}$ . . . . $P_2O_5 \cdot 22\ WO_3 \cdot xH_2O$ . . . . 54 (W):   Hb−363

$O_{72}P_2Rb_{14}V_{24}$ . . . . . 7 $Rb_2O \cdot 12\ V_2O_5 \cdot P_2O_5 \cdot 32\ H_2O$ . . . . . 48 (V):   Hb/B2−491/2

$O_{72}P_{20}Zr_{11}$ . . . . . . 11 $ZrO_2 \cdot 10\ P_2O_5 \cdot 13\ H_2O$ . . . . 42 (Zr):   Hb−393

$O_{75}W_{22}Zn_9$ . . . . . . 9 $ZnO \cdot 22\ WO_3 \cdot 66\ H_2O$ . . . . 54 (W):   Hb−306

$O_{77}P_2W_{24}$ . . . . $P_2O_5 \cdot 24\ WO_3 \cdot 45\ H_2O$

          $= H_7[P(W_2O_7)_6] \cdot 19\ H_2O$ . . . . 54 (W):   Hb−363

−. . . . . $P_2O_5 \cdot 24\ WO_3 \cdot 51\ H_2O$ . . . . 54 (W):   Hb−363

−. . . . . $P_2O_5 \cdot 24\ WO_3 \cdot 63\ H_2O$ . . . . 54 (W):   Hb−359/63

$O_{78}V_{30}Yb_2$ . . . . . . $Yb_2O_3 \cdot 15\ V_2O_5$ . . . . 48 (V):   Hb/B2−556

$O_{79}Rb_6Si_2W_{24}$ . . . . . 3 $Rb_2O \cdot 2\ (SiO_2 \cdot 12\ WO_3) \cdot 23\ H_2O$

          $= Rb_3H_5[Si(W_2O_7)_6] \cdot 9\ H_2O$ . . . . . . 54 (W):   Hb−349

$O_{79}Si_2UW_{24}$ . . . . . . 3 $UO_3 \cdot 6(SiO_2 \cdot 12\ WO_3) \cdot 177\ H_2O$ . . . . 55 (U):   Hb−244

$O_{100}Ta_{34}U_6$ . . . . . . 3 $U_2O_5 \cdot 17\ Ta_2O_5$ . . . . 50 (Ta):   Hb/B1−51

$O_{105}P_2V_{40}$ . . . . 20 $V_2O_5 \cdot P_2O_5 \cdot 59\ H_2O$ . . . . 48 (V):   Hb/B1−363

$O_{116}Si_3ThW_{36}$ . . . . . $ThO_2 \cdot 3\ (SiO_2 \cdot 12\ WO_3) \cdot 49\ H_2O$

          $= ThH_{20}[Si(W_2O_7)_6]_3 \cdot 39\ H_2O$ . . . . . 54 (W):   Hb−353

$O_{116}Si_3UW_{36}$ . . . . . $UO_2 \cdot 3(SiO_2 \cdot 12\ WO_3) \cdot 90\ H_2O$ . . . . . . 54 (W):   Hb−354

$O_{118}Si_3U_2W_{36}$ . . . . . 2 $UO_2 \cdot 3(SiO_2 \cdot 12\ WO_3) \cdot 87\ H_2O$ . . . . . 54 (W):   Hb−354

$O_{120}Si_3U_2W_{36}$ . . . . . 2 $UO_3 \cdot 3(SiO_2 \cdot 12\ WO_3) \cdot 87\ H_2O$ . . . . 55 (U):   Hb−244

$O_{123}Si_3U_3W_{36}$ . . . . . 3 $UO_3 \cdot 3(SiO_2 \cdot 12\ WO_3) \cdot 88\ H_2O$ . . . . 55 (U):   Hb−244

$O_{129}Si_3U_5W_{36}$ . . . . . 5 $UO_3 \cdot 3(SiO_2 \cdot 12\ WO_3) \cdot 90\ H_2O$ . . . . 55 (U):   Hb−244

$O_{132}Si_3U_6W_{36}$ . . . . . 6 $UO_3 \cdot 3(SiO_2 \cdot 12\ WO_3) \cdot 91\ H_2O$ . . . . 55 (U):   Hb−244

$O_{177}Sr_{19}V_{20}W_{36}$ . . . . 19 $SrO \cdot 10\ V_2O_5 \cdot 36\ WO_3 \cdot 122\ H_2O$ . . . 54 (W):   Hb−389

$O_{237}Si_6U_3W_{72}$ . . . . . 3 $UO_3 \cdot 6(SiO_2 \cdot 12\ WO_3) \cdot 177\ H_2O$ . . . . 55 (U):   Hb−244

$O_{354}Si_9U_4W_{108}$ . . . . 4 $UO_3 \cdot 9\ (SiO_2 \cdot 12\ WO_3) \cdot 267\ H_2O$ . . . . 55 (U):   Hb−244

$O_{357}Si_9U_5W_{108}$ . . . . 5 $UO_3 \cdot 9\ (SiO_2 \cdot 12\ WO_3) \cdot 264\ H_2O$ . . . . 55 (U):   Hb−244

O_xPb . . . . PbO_x . . . . 47 (Pb):   Hb/C1−139/47

O_xPr . . . . PrO_x (x = 1.5 to 2). . . . . 39 (SE):   Hb/C1−279/90

−. . . . . PrO_x solid solutions:

         PrO_x-TbO_x . . . . 39 (SE):   Hb/C1−402/4

−. . . . . PrO_x systems:

         $PrO_x-La_2O_3$ . . . . 39 (SE):   Hb/C1−294/5

         $PrO_x-Nd_2O_3$ . . . . 39 (SE):   Hb/C1−294/5, 310/1

         PrO_x-O . . . . 39 (SE):   Hb/C1−275/8

         $PrO_x-Sm_2O_3$ . . . . 39 (SE):   Hb/C1−294/5

         $PrO_x-Y_2O_3$ . . . . 39 (SE):   Hb/C1−294/5

O_xRuZr_2 . . . . Zr_2RuO_x . . . . 63 (Ru):   Eb−259

Hb = Main Series Volume; GD = Gmelin-Durrer;   *Ab = Appended Volume to Main Series;*
Eb = Supplement Volume to Main Series;   *Erg. W. = New Supplement Series Volume*

O_xSm . . . . SmO_x (x = 0.4 to 0.6) . . . . 39 (SE):   Hb/C1–314
O_xTa . . . . TaO_x . . . . 50 (Ta):   Hb/B1–17/26, 30
O_xTb . . . . TbO_x (x = 1.5 to 2) . . . . 39 (SE):   Hb/C1–107, 387, 391/8
–. . . . . TbO_x solid solutions:
                         $TbO_x–CeO_2$ . . . . 39 (SE):   Hb/C1–401/2
                         $TbO_x–Gd_2O_3$ . . . . 39 (SE):   Hb/C1–404
                         $TbO_x–PrO_x$ . . . . 39 (SE):   Hb/C1–402/4
                         $TbO_x–Y_2O_3$ . . . . 39 (SE):   Hb/C1–400/1
–. . . . . TbO_x systems:
                         $TbO_x–Er_2O_3$ . . . . 39 (SE):   Hb/C1–421/2
                         $TbO_x–O$ . . . . 39 (SE):   Hb/C1–391/4
O_xV . . . . VO_x . . . . 48 (V):   Hb/B1–11/20
$O_{x+6}Ta_2Tl_2$ . . . . . . $Tl_2Ta_2O_{6+x}$ . . . . 50 (Ta):   Hb/B2–201
$O_{x+7}Sn_{x+2}Ta_2$ . . . . $Sn_{2+x}Ta_2O_{7+x}$ . . . . 50 (Ta):   Hb/B2–231
$O_{2x+3}S_xSb_2$ . . . . . . $Sb_2O_3 \cdot xSO_2$ . . . . 18 (Sb):   Hb/B3–541

Os . . . .   Osmium:
                         Abundance . . . . 68 (Pt):   Hb/A1–24/7
                         Atom . . . . 66 (Os):   Hb–3/7
                         Atomic nucleus . . . . 66 (Os):   Hb–1/3
                         Catalytic properties . . . . 66 (Os):   Hb–23/4
                                                              68 (Pt):   Hb/A3–429/30
                         Chemical reactions . . . . 66 (Os):   Hb–21/4
                         Colloidal forms . . . . 68 (Pt):   Hb/A3–427/30
                         Cosmochemistry . . . . 68 (Pt):   Hb/A1–15/22
                         Crystallographic properties . . . . . 66 (Os):   Hb–8/9
                         Detection . . . . 68 (Pt):   Hb/A4–432, 447/51,
                                                                          490/1
                         – in iron . . . . 59 (Fe):   Hb/FII1–41
                         Determination . . . . 68 (Pt):   Hb/A4–472/6, 490/1
                         – in special substances . . . . 68 (Pt):   Hb/A4–525/32
                         Economic deposits
                             see Pt . . . Platinum
                         Economics . . . . 68 (Pt):   Hb/A2–302/10
                         Electrical properties . . . . 66 (Os):   Hb–20
                         Electrochemical behavior . . . . . . 66 (Os):   Hb–20/1
                         Geochemical character . . . . 68 (Pt):   Hb/A1–23/8
                         Geochemistry . . . . 68 (Pt):   Hb/A1–22/49
                                                                          Hb/A2–275/6
                         History . . . . 68 (Pt):   Hb/A1–12/3
                         Magnetic properties . . . . 66 (Os):   Hb–19/20
                         Mechanical properties . . . . 66 (Os):   Hb–9/11
                         Mineralogy . . . . 68 (Pt):   Hb/A2–276/301
                         Occurrence, extraterrestrial . . . . . 68 (Pt):   Hb/A1–15/22
                         –, terrestrial . . . . 68 (Pt):   Hb/A1–22/49
                                                                          Hb/A2–275/6
                         – in meteorites . . . . 68 (Pt):   Hb/A1–16/22

See the "Instructions for the user"
at the beginning of the volume

Os.	Osmium:		
	Occurrence in the biosphere	68 (Pt):	Hb/A2−275
	− in the hydrosphere	68 (Pt):	Hb/A2−275
	− in the lithosphere	68 (Pt):	Hb/A1−28/49
	Optical properties.	66 (Os):	Hb−13/9
	Physiological action	68 (Pt):	Hb/A2−275/6
	Position in the periodic system	66 (Os):	Hb−8
	Preparation	68 (Pt):	Hb/A3−311/430
	− from minerals.	68 (Pt):	Hb/A3−312/50
	− from wastes and residues	68 (Pt):	Hb/A3−350/62
	− in pure state.	68 (Pt):	Hb/A3−375/6
	− of dense osmium	68 (Pt):	Hb/A3−383
	Purity testing	68 (Pt):	Hb/A4−520
	Separation, analytical	68 (Pt):	Hb/A4−493/5, 517/8
	Special forms	68 (Pt):	Hb/A3−401/4, 427/30
	Spectra	66 (Os):	Hb−13/9
	Thermal properties	66 (Os):	Hb−11/3
	Toxicity	68 (Pt):	Hb/A2−275/6
	Uses	68 (Pt):	Hb/A3−385/91
	Valence	66 (Os):	Hb−7/8
−.	Osmium ions:		
	Properties of the ion	66 (Os):	Hb−5/7
	Spectra	66 (Os):	Hb−14/5
−.	Osmium isotopes	66 (Os):	Hb−1/2
−.	Osmium systems:		
	Os−Fe	68 (Pt):	Hb/A6−725/6
	Os−Fe−S	68 (Pt):	Hb/A6−726
	Os−H	66 (Os):	Hb−25
	Os−P.	66 (Os):	Hb−62/3
	Os−S	66 (Os):	Hb−58
	Os−Ti	68 (Pt):	Hb/A6−721
Os^+	Os^+.	66 (Os):	Hb−5
Os^{2+}	Os^{2+}	66 (Os):	Hb−5
Os^{3+}	Os^{3+}	66 (Os):	Hb−5/6
Os^{4+}	Os^{4+}	66 (Os):	Hb−5/7
Os^{6+}	Os^{6+}	66 (Os):	Hb−6
Os^{8+}	Os^{8+}	66 (Os):	Hb−5
OsP_2	OsP_2	66 (Os):	Hb−63
OsS	OsS	66 (Os):	Hb−58
OsS_2	OsS_2	66 (Os):	Hb−59
OsS_3	OsS_3	66 (Os):	Hb−58
OsS_4	OsS_4	66 (Os):	Hb−59/60
$OsSe_2$	$OsSe_2$	66 (Os):	Hb−61
$OsTe_2$	$OsTe_2$	66 (Os):	Hb−61
$OsTi$.	$OsTi$	68 (Pt):	Hb/A6−721
Os_2S_3	Os_2S_3.	66 (Os):	Hb−58
Os_2Zr	$ZrOs_2$	68 (Pt):	Hb/A6−722

Hb = *Main Series Volume*; GD = *Gmelin-Durrer*;   Ab = *Appended Volume to Main Series*;
Eb = *Supplement Volume to Main Series*;   Erg. W. = *New Supplement Series Volume*

P . . . . Phosphorus:

Absorption of X- and gamma rays . .	16 (P):	Hb/B−191/8
Abundance	16 (P):	Hb/A−104
Atom.	16 (P):	Hb/B−161/208
Atomic nucleus	16 (P):	Hb/B−159/61
Atomic radius	16 (P):	Hb/B−203/4
Atomic refraction	16 (P):	Hb/B−207/8
Atomic scattering factors.	16 (P):	Hb/B−162/3
Atomic volume	16 (P):	Hb/B−204
Atomic weight	16 (P):	Hb/B−99/104
Behavior as solvent.	16 (P):	Hb/B−339
Bond properties.	16 (P):	Hb/B−199/203
Chemical reactions	16 (P):	Hb/B−258/339
– in electrical discharges	16 (P):	Hb/B−259
– on heating.	16 (P):	Hb/B−260/3
– on irradiation	16 (P):	Hb/B−258/9
– with acids	16 (P):	Hb/B−279/81
– with air	16 (P):	Hb/B−263/76
– with metal compounds.	16 (P):	Hb/B−310/6
– with metals	16 (P):	Hb/B−288/304
– with nonmetal compounds	16 (P):	Hb/B−276/81, 304/10
– with nonmetals	16 (P):	Hb/B−281/8
– with organic compounds.	16 (P):	Hb/B−329/31
– with oxygen.	16 (P):	Hb/B−263/76
– with solutions of metal compounds	16 (P):	Hb/B−316/29
– with water.	16 (P):	Hb/B−276/8
Chemical shift.	16 (P):	Hb/B−200/1
Colloidal forms	16 (P):	Hb/B−107
Constitution.	16 (P):	Hb/B−215/7
– of phosphorus vapor	16 (P):	Hb/B−233/4
Cosmochemistry	16 (P):	Hb/A−53/68
Crystal chemistry	16 (P):	Hb/A−69/103
Crystallographic properties	16 (P):	Hb/B−215/8, 227/32, 238/40, 247/8, 251/2
Detection	16 (P):	Hb/B−339/52
– in iron	59 (Fe):	Hb/FI−369/72
– in special substances.	16 (P):	Hb/B−352/9
Determination.	16 (P):	Hb/B−359/81
– in inorganic substances	16 (P):	Hb/B−396/420, 449/54
– in iron	16 (P):	Hb/B−399/400
	59 (Fe):	Hb/FI−372/411 Hb/FII2−314, 349
– in organic substances	16 (P):	Hb/B−396/7, 420/39
– in soils.	16 (P):	Hb/B−396/7, 439/49
Diadochy	16 (P):	Hb/A−85/101
Diffusion in iron.	59 (Fe):	Hb/A8−1797

See the "Instructions for the user"
at the beginning of the volume

P . . . . Phosphorus:
              Diffusion in silicon . . . . 16 (P):   Hb/C–635/6
              Discovery . . . . 16 (P):   Hb/A–6/14
              Economics. . . . . 16 (P):   Hb/A–240/2
                                             28 (Ca):   Hb/B1–202/4
              Electrical properties . . . . 16 (P):   Hb/B–238, 245,
                                                          250/1, 254/5
              Electrochemical behavior . . . . 16 (P):   Hb/B–256/8
              Electron affinity . . . . 16 (P):   Hb/B–205
              Electron configuration . . . . 16 (P):   Hb/B–161
              Electronegativity . . . . 16 (P):   Hb/B–201/3
              Formation . . . . 16 (P):   Hb/B–104/5
              g factor . . . . 16 (P):   Hb/B–161/2
              Geochemical character . . . . 16 (P):   Hb/A–104
              Geochemical cycle . . . . 16 (P):   Hb/A–237/40
              Geochemistry . . . . 16 (P):   Hb/A–69/240
              Handling. . . . . 16 (P):   Hb/B–84/5
              History. . . . . 16 (P):   Hb/A–1/52
              Hydrosols . . . . 16 (P):   Hb/B–107
               Increments . . . . 16 (P):   Hb/B–207/8
               Industrial hazards. . . . . 16 (P):   Hb/B–84/5
               Interaction with electrons . . . . . . 16 (P):   Hb/B–205/7
              – with ions. . . . . 16 (P):   Hb/B–207
              Ionization potentials . . . . 16 (P):   Hb/B–205/7
              Magnetic moment . . . . 16 (P):   Hb/B–161/2
              Magnetic properties . . . . 16 (P):   Hb/B–244/5, 250/1, 254
              Mass absorption coefficient . . . . . 16 (P):   Hb/B–192/7
              Mass photoabsorption coefficient. . . 16 (P):   Hb/B–197
              Mass scattering coefficient. . . . . . 16 (P):   Hb/B–197/8
              Mechanical properties . . . . 16 (P):   Hb/B–235, 239/43,
                                                           248/9, 252/3
              Mesonic atom. . . . . 16 (P):   Hb/B–188/90
              Mineralogy:
                  Anhydrous phosphates
                    with foreign anions . . . . 16 (P):   Hb/A–350/431
                  – without foreign anions. . . . . . 16 (P):   Hb/A–251/98
                  Apatite group. . . . . 16 (P):   Hb/A–350/82
                  Hydrogen phosphates . . . . 16 (P):   Hb/A–243/51
                  Hydrous phosphates with
                    foreign anions . . . . 16 (P):   Hb/A–431/509
                  – without foreign anions. . . . . . 16 (P):   Hb/A–298/350
                  Structures of
                    phosphorus minerals . . . . 16 (P):   Hb/A–101/3
                  Unclassified phosphates. . . . . . 16 (P):   Hb/A–509/10
              Modifications:
                  Black phosphorus:
                    Formation . . . . 16 (P):   Hb/B–105, 227/9
                      Historical data . . . . 16 (P):   Hb/A–52

P . . . . Phosphorus:
              Modifications:
                Black phosphorus:
                    Oxidation. . . . . 16 (P):   Hb/B−276
                    Phase transitions . . . . 16 (P):   Hb/B−224
                    Physical properties . . . . 16 (P):   Hb/B−98, 247/51

                  Brown phosphorus. . . . . 16 (P):   Hb/B−254/5
                  Gray phosphorus. . . . . 16 (P):   Hb/B−98, 247/51
                  Orange-red phosphorus . . . . . . 16 (P):   Hb/B−97, 216, 255
                  Purple phosphorus. . . . . 16 (P):   Hb/B−232, 255
                  Red phosphorus:
                    Constitution . . . . 16 (P):   Hb/B−216/7
                    Deposition from waste gases. . . 16 (P):   Hb/B−42
                    Determination . . . . 16 (P):   Hb/B−361/2
                    Formation . . . . 16 (P):   Hb/B−104/5
                    Historical data . . . . 16 (P):   Hb/A−46/52
                    Oxidation. . . . . 16 (P):   Hb/B−275/6
                    Phase transitions . . . . 16 (P):   Hb/B−224/7
                    Physical properties . . . . 16 (P):   Hb/B−251/6
                    Preparation . . . . 16 (P):   Hb/B−81/3, 86
                    Purification. . . . . 16 (P):   Hb/B−83
                    Purity testing. . . . . 16 (P):   Hb/B−83/4
                    Stabilization . . . . 16 (P):   Hb/B−83

                  Ruby-red phosphorus . . . . 16 (P):   Hb/B−232
                  Violet phosphorus:
                    Constitution . . . . 16 (P):   Hb/B−216/7
                    Oxidation. . . . . 16 (P):   Hb/B−275/6
                    Phase transitions . . . . 16 (P):   Hb/B−224/7
                    Physical properties . . . . 16 (P):   Hb/B−98, 251/6

                  White phosphorus:
                    Aftertreatment. . . . . 16 (P):   Hb/B−80/1
                    Constitution . . . . 16 (P):   Hb/B−215
                    Deposition from waste gases. . . 16 (P):   Hb/B−40/2
                    Determination . . . . 16 (P):   Hb/B−81, 382, 398
                    Formation . . . . 16 (P):   Hb/B−104
                    Luminescence . . . . 16 (P):   Hb/B−265/74
                    Oxidation. . . . . 16 (P):   Hb/B−263/75
                    Phase transitions . . . . 16 (P):   Hb/B−217/23
                    Physical properties . . . . 16 (P):   Hb/B−238/47
                    Preparation . . . . 16 (P):   Hb/B−40/3, 80/1,
                                                                      85/6, 105/7
                    Purification. . . . . 16 (P):   Hb/B−80, 106/7
                    Purity testing. . . . . 16 (P):   Hb/B−81
                  Yellow phosphorus . . . . 16 (P):   Hb/B−97, 240,
                                                                      245/6, 255

                Molecule. . . . . 16 (P):   Hb/B−208/15

See the "Instructions for the user"
at the beginning of the volume

P . . . . Phosphorus:

Occurrence, extraterrestrial . . . . . . 16 (P):   Hb/A−53/68
−, terrestrial . . . . 16 (P):   Hb/A−69/240
− in meteorites . . . . 16 (P):   Hb/A−55/68
− in minerals . . . . 16 (P):   Hb/A−69/85
− in soils . . . . 16 (P):   Hb/A−156/76
− in the atmosphere . . . . 16 (P):   Hb/A−214/5
− in the biosphere . . . . 16 (P):   Hb/A−215/37
− in the hydrosphere . . . . 16 (P):   Hb/A−182/214
− in the lithosphere . . . . 16 (P):   Hb/A−104/82
− in the magmatic cycle . . . . 16 (P):   Hb/A−104/23
− in the metamorphic cycle . . . . . 16 (P):   Hb/A−176/82
− in the sedimentary cycle . . . . . . 16 (P):   Hb/A−123/56
Optical properties . . . . 16 (P):   Hb/B−238, 245/7,
                                                                  251, 255/6

Organosols . . . . 16 (P):   Hb/B−107
Oxidation . . . . 16 (P):   Hb/B−263/76
Parachor . . . . 16 (P):   Hb/B−207
Phase diagram . . . . 16 (P):   Hb/B−215
Phase transitions . . . . 16 (P):   Hb/B−215/33
Polymorphism . . . . 16 (P):   Hb/B−215/33
Position in the electrochemical
    series . . . . 16 (P):   Hb/B−256
− in the periodic system . . . . 16 (P):   Hb/B−98
Preparation . . . . 16 (P):   Hb/B−105/7
−, historical data . . . . 16 (P):   Hb/A−23/37
−, industrial . . . . 16 (P):   Hb/B−1/86
−, −, oxidation of PH₃ . . . . 16 (P):   Hb/B−58/9
−, −, thermal methods . . . . 16 (P):   Hb/B−21/50
−, −, wet dressing . . . . 16 (P):   Hb/B−3/21
− of special forms . . . . 16 (P):   Hb/B−107
Purification . . . . 16 (P):   Hb/B−106/7
Scattering of X- and gamma rays . . . 16 (P):   Hb/B−191/8
Separation, analytical . . . . 16 (P):   Hb/B−382/96
Solubility . . . . 16 (P):   Hb/B−331/9
Spectra:
    Arc spectrum . . . . 16 (P):   Hb/B−163/7
    Band spectrum . . . . 16 (P):   Hb/B−210/1
    Radiation of the mesonic atom . . . 16 (P):   Hb/B−188/90
    Spark spectra . . . . 16 (P):   Hb/B−167/74
    X-ray spectrum . . . . 16 (P):   Hb/B−174/88

Technology
    see Preparation, industrial

Thermal properties . . . . 16 (P):   Hb/B−218/27, 232/8,
                                                                  240, 243/4,
                                                                  249, 253/4

Toxicity . . . . 16 (P):   Hb/B−87/91
Uses, historical data . . . . 16 (P):   Hb/A−37/46

Hb = Main Series Volume; GD = Gmelin-Durrer;   Ab = Appended Volume to Main Series;
Eb = Supplement Volume to Main Series;   Erg. W. = New Supplement Series Volume

P . . . . Phosphorus:
Valence . . . . 16 (P):   Hb/B−98
Vapor pressure . . . . 16 (P):   Hb/B−221, 226

−. . . . . Phosphorus ions
see P⁺ ... P⁺, etc.

−. . . . . Phosphorus isotopes:
Detection and determination. . . . . . 16 (P):   Hb/B−455/8
Enrichment, preparation,
and separation . . . . 16 (P):   Hb/B−108/59, 455/6
Nuclear properties . . . . 16 (P):   Hb/B−159/61
Toxicity . . . . 16 (P):   Hb/B−97
Uses . . . . 16 (P):   Hb/B−108, 149

−. . . . . Phosphorus solutions:
P−inorganic solvents. . . . . 16 (P):   Hb/B−332/3
P−organic solvents . . . . 16 (P):   Hb/B−333/9

−. . . . . Phosphorus systems:
P−Ag . . . . 61 (Ag):   Hb/B4−1/3
P−Ag−Cu . . . . 61 (Ag):   Hb/B4−484/5
P−As. . . . . 17 (As):   Hb−473
P−Au . . . . 62 (Au):   Hb/3−728
P−Bi . . . . 19 (Bi):   Eb−847
P−Br . . . . 16 (P):   Hb/C−490/1
P−C−Fe . . . . 59 (Fe):   Hb/A5−1095/6
P−C−Fe−Si . . . . 59 (Fe):   Hb/A8−1798
P−Co. . . . . 58 (Co):   Hb/A2−388
                                                    Eb/A−738
P−Cr. . . . . 52 (Cr):   Hb/B−437
P−Cr−Cu . . . . 60 (Cu):   Hb/B3−1228
P−Cr−Ni. . . . . 57 (Ni):   Hb/B1−222
                                                    Hb/B3−1215
P−Cu. . . . . 60 (Cu):   Hb/B2−904/6
P−Cu−Fe . . . . 60 (Cu):   Hb/B3−1315
P−Cu−Mn. . . . . 60 (Cu):   Hb/B3−1248
P−Cu−Ni . . . . 60 (Cu):   Hb/B3−1258/9
P−Cu−O. . . . . 60 (Cu):   Hb/B2−918
P−Cu−Si . . . . 60 (Cu):   Hb/B2−933
P−Cu−Sn . . . . 60 (Cu):   Hb/B3−1196
P−Cu−Zn . . . . 60 (Cu):   Hb/B3−1151
P−Fe. . . . . 59 (Fe):   Hb/A8−1781/4
P−Fe−O . . . . 59 (Fe):   Hb/A5−923/35
P−Fe−Si. . . . . 59 (Fe):   Hb/A8−1797/8
P−Ge . . . . 45 (Ge):   Eb−558
P−Hg . . . . 34 (Hg):   Hb/B4−1357
P−I. . . . . 16 (P):   Hb/C−518
P−Ir . . . . 68 (Pt):   Hb/A6−729/30
P−Mn−Ni . . . . 57 (Ni):   Hb/B1−311
P−Ni. . . . . 57 (Ni):   Hb/B3−946/8

See the "Instructions for the user"
at the beginning of the volume

P . . . . Phosphorus systems:
           P–Os. . . . . 66 (Os):   Hb–62/3
           P–Pd. . . . . 65 (Pd):   Hb/2–302/5
           P–Pt . . . . 68 (Pt):   Hb/C1–133/5
           P–Re. . . . . 70 (Re):   Hb–123
           P–Rh. . . . . 68 (Pt):   Hb/A5–547/9
           P–Ru. . . . . 68 (Pt):   Hb/A5–534/5
           P–S . . . . 16 (P):   Hb/C–567
           P–Se. . . . . 16 (P):   Hb/C–605/6
           P–Si . . . . 16 (P):   Hb/C–635/6
           P–Ti . . . . 41 (Ti):   Hb–378
           P–Tl . . . . 38 (Tl):   Hb/3–425/6
           P–V . . . . 48 (V):   Hb/B1–358
           P–Zr . . . . 42 (Zr):   Hb–389/90

P^+ . . . . P^+:
           Formation . . . . 16 (P):   Hb/B–208
                                           Hb/C–24
           Ionization potentials . . . . 16 (P):   Hb/B–205/6
           Properties of the ion . . . . 16 (P):   Hb/B–161/2, 204/5
           Spectrum . . . . 16 (P):   Hb/B–167/9

P^{2+} . . . . P^{2+} . . . . 16 (P):   Hb/B–161/2, 169/70,
                                                205/6

P^{3+} . . . . P^{3+}:
           Ionic susceptibility . . . . 16 (P):   Hb/B–208
           Ionization potentials . . . . 16 (P):   Hb/B–205/6
           Properties of the ion . . . . 16 (P):   Hb/B–161/2
           Spectrum . . . . 16 (P):   Hb/B–170/1

P^{4+} . . . . P^{4+} . . . . 16 (P):   Hb/B–161/2, 171/2,
                                                205/6

P^{5+} . . . . P^{5+}:
           Increments . . . . 16 (P):   Hb/B–208
           Ionic mobility . . . . 16 (P):   Hb/B–256
           Ionization potentials . . . . ۔ . . . 16 (P):   Hb/B–205/6
           Properties of the ion . . . . 16 (P):   Hb/B–162, 204/5
           Scattering factors . . . . 16 (P):   Hb/B–163
           Spectrum . . . . 16 (P):   Hb/B–172

P^{n+} . . . . P^{n+} (n = 6 to 12) . . . . 16 (P):   Hb/B–162, 173/4,
                                              205/6

P^- . . . . P^- . . . . 16 (P):   Hb/B–161, 208
                                           Hb/C–24
P^{2-} . . . . P^{2-} . . . . 16 (P):   Hb/B–161
P^{3-} . . . . P^{3-} . . . . 16 (P):   Hb/B–161, 163, 205, 208

PPd_3 . . . . Pd_3P . . . . 65 (Pd):   Hb/2–303/4, 306
PPd_5 . . . . Pd_5P . . . . 65 (Pd):   Hb/2–303, 305
PPt . . . . PtP . . . . 68 (Pt):   Hb/C1–135
PPt_2 . . . . Pt_2P . . . . 68 (Pt):   Hb/C1–135
PPu . . . . PuP . . . . Erg. W. 4–249/51

Hb = Main Series Volume; GD = Gmelin-Durrer;   *Ab = Appended Volume to Main Series;*
Eb = Supplement Volume to Main Series;   *Erg. W. = New Supplement Series Volume*

PRe	ReP	70 (Re):	Hb–123/4
PRe$_2$.	Re$_2$P	70 (Re):	Hb–123/4
PRh$_2$	Rh$_2$P	68 (Pt):	Hb/A5–547/9
PRu	RuP	63 (Ru):	Eb–335
		68 (Pt):	Hb/A5–534/5
PRuS	RuPS	63 (Ru):	Eb–335
PRuSe	RuPSe	63 (Ru):	Eb–335
PRu$_2$	Ru$_2$P	63 (Ru):	Eb–334
		68 (Pt):	Hb/A5–534/5
PS$_{0.25}$	PS$_{0.25}$ = P$_4$S	16 (P):	Hb/C–581
PS$_{0.5}$	PS$_{0.5}$ = P$_4$S$_2$	16 (P):	Hb/C–581
PS.	PS.	16 (P):	Hb/C–572/3
–.	(PS)$_n$	16 (P):	Hb/C–573
PS$^+$	PS$^+$	16 (P):	Hb/C–572/3
PSTa	TaPS	50 (Ta):	Hb/B1–271
–.	TaPS solid solutions:		
	TaPS–NbPS.	50 (Ta):	Hb/B1–271
	TaPS–NbPSe	49 (Nb):	Hb/B1–414
		50 (Ta):	Hb/B1–271
PS$_{1.5}$	PS$_{1.5}$ = P$_2$S$_3$	16 (P):	Hb/B–96/7
			Hb/C–581
PS$_{1.66}$	PS$_{1.66}$ = P$_3$S$_5$	16 (P):	Hb/C–581
PS$_2$	PS$_2$	16 (P):	Hb/C–582
PS$_3$	PS$_3$	16 (P):	Hb/C–582
PS$_4$	PS$_4$	16 (P):	Hb/C–582
PS$_4$Sb	SbPS$_4$.	18 (Sb):	Hb/B3–561/2
PS$_4$Tl$_3$	Tl$_3$PS$_4$.	38 (Tl):	Hb/3–432
PS$_5$	PS$_5$	16 (P):	Hb/C–582
PSe$_3$.	PSe$_3$	16 (P):	Hb/C–606
PSe$_5$.	PSe$_5$	16 (P):	Hb/C–607
PSi	PSi	16 (P):	Hb/C–636/7
PSi$_2$	PSi$_2$	16 (P):	Hb/C–637
PTa	TaP	50 (Ta):	Hb/B1–267
PTa$_2$.	Ta$_2$P	50 (Ta):	Hb/B1–266
PTa$_3$.	Ta$_3$P	50 (Ta):	Hb/B1–266
PTi$_2$	Ti$_2$P	41 (Ti):	Hb–378
PTl$_3$	Tl$_3$P	38 (Tl):	Hb/3–426
PU.	UP	55 (U):	Hb–179
PV	VP.	48 (V):	Hb/B1–358/9
PV$_2$	V$_2$P	48 (V):	Hb/B1–358
PV$_3$	V$_3$P	48 (V):	Hb/B1–358
PW	WP	54 (W):	Hb–206
PW$_2$	W$_2$P.	54 (W):	Hb–206
PZn	ZnP	32 (Zn):	Hb–275
PZr	ZrP	42 (Zr):	Hb–391
PZr$_3$	Zr$_3$P	42 (Zr):	Hb–391
P$_{1.33}$Zr.	ZrP$_{1.33}$	42 (Zr):	Hb–390
P$_{1.46}$Zr.	ZrP$_{1.46}$	42 (Zr):	Hb–389, 390

See the "Instructions for the user"
at the beginning of the volume

P_2	P_2	16 (P):	Hb/B−209/12
P_2^+	P_2^+	16 (P):	Hb/B−212
P_2^-	P_2^-	16 (P):	Hb/B−208
$P_2Pb_2S_6$	$Pb_2P_2S_6$	47 (Pb):	Hb/C3−884
$P_2Pb_2S_7$	$Pb_2P_2S_7$	47 (Pb):	Hb/C3−885
P_2Pb_3	Pb_3P_2	47 (Pb):	Hb/C3−861
$P_2Pb_3S_8$	$Pb_3(PS_4)_2$	47 (Pb):	Hb/C3−884
P_2Pd	PdP_2	65 (Pd):	Hb/2−303/4, 306
P_2Pd_5	Pd_5P_2	65 (Pd):	Hb/2−303/4, 306
P_2Pt	PtP_2	68 (Pt):	Hb/C1−133, 135/6
P_2Re	ReP_2	70 (Re):	Hb−123/4
P_2Rh	RhP_2	68 (Pt):	Hb/A5−547/9
P_2Ru	RuP_2	63 (Ru):	Eb−335
		68 (Pt):	Hb/A5−534/5
P_2S_3	$P_2S_3 = P_4S_6$	16 (P):	Hb/B−96/7
			Hb/C−581
P_2S_4	P_2S_4	16 (P):	Hb/C−582
P_2S_5	P_2S_5		
	see $P_4S_{10} \dots P_4S_{10}$		
$P_2S_6Zn_2$	$Zn_2P_2S_6$	32 (Zn):	Hb−281
$P_2S_6Zn_3$	$Zn_3(PS_3)_2$	32 (Zn):	Hb−281
$P_2S_7Zn_2$	$Zn_2P_2S_7$	32 (Zn):	Hb−281
$P_2S_8Zn_3$	$Zn_3(PS_4)_2$	32 (Zn):	Hb−281
P_2S_{11}	P_2S_{11}	16 (P):	Hb/C−582
P_2Se	P_2Se	16 (P):	Hb/C−606
P_2Se_3	P_2Se_3	16 (P):	Hb/C−606
P_2Se_5	P_2Se_5	16 (P):	Hb/C−607/8
P_2Sr_3	Sr_3P_2	29 (Sr):	Hb−211
			Eb−283
P_2Ta	TaP_2	50 (Ta):	Hb/B1−268
P_2Te_3	P_2Te_3	16 (P):	Hb/C−608/9
P_2V	VP_2	48 (V):	Hb/B1−359
P_2W	WP_2	54 (W):	Hb−206
P_2Zn	ZnP_2	32 (Zn):	Hb−275/6
			Eb−995
P_2Zn_3	Zn_3P_2	16 (P):	Hb/B−92
		32 (Zn):	Hb−275
			Eb−994/5
P_2Zr	ZrP_2	42 (Zr):	Hb−391/2
P_3	P_3	16 (P):	Hb/B−213
P_3^+	P_3^+	16 (P):	Hb/B−208
P_3^-	P_3^-	16 (P):	Hb/B−208
P_3Pb	PbP_3	47 (Pb):	Hb/C3−861
P_3Re	ReP_3	70 (Re):	Hb−123/4
P_3Rh	RhP_3	68 (Pt):	Hb/A5−547/9
P_3S_2Zn	ZnP_3S_2	32 (Zn):	Hb−281
P_3S_5	$P_3S_5 = PS_{1.66}$	16 (P):	Hb/C−581
P_3S_6	P_3S_6	16 (P):	Hb/C−582

Hb = *Main Series Volume;* GD = *Gmelin-Durrer;*   Ab = *Appended Volume to Main Series;*
Eb = *Supplement Volume to Main Series;*   Erg. W. = *New Supplement Series Volume*

P_4	P_4	16 (P):	Hb/B−213/5
P_4^+	P_4^+	16 (P):	Hb/B−208
P_4^-	P_4^-	16 (P):	Hb/B−208
P_4Rh_5	Rh_5P_4	68 (Pt):	Hb/A5−547/9
P_4S	$P_4S = PS_{0.25}$	16 (P):	Hb/C−581
P_4SSe_2	P_4SSe_2	16 (P):	Hb/C−608
P_4S_2	$P_4S_2 = PS_{0.5}$	16 (P):	Hb/C−581
P_4S_2Se	P_4S_2Se	16 (P):	Hb/C−608
P_4S_3	P_4S_3	16 (P):	Hb/B−96
			Hb/C−568/72
P_4S_5	P_4S_5	16 (P):	Hb/C−573/4
P_4S_6	$P_4S_6 = P_2S_3$	16 (P):	Hb/B−96/7
			Hb/C−581
P_4S_7	P_4S_7	16 (P):	Hb/C−574/6
P_4S_{10}	$P_4S_{10} = P_2S_5$	16 (P):	Hb/B−97
			Hb/C−576/81
−.	P_4S_{10} systems:		
	$P_2S_5−POCl_3−PSCl_3$	16 (P):	Hb/C−597/8
	$P_2S_5−POCl_3−PSCl_3−P_2O_5$	16 (P):	Hb/C−597/8
P_4S_{11}	P_4S_{11}	16 (P):	Hb/C−582
P_4Se	P_4Se	16 (P):	Hb/C−606
P_4Se_3	P_4Se_3	16 (P):	Hb/C−606/7
P_4Ta_5	Ta_5P_4	50 (Ta):	Hb/B1−266
P_4Th_3	Th_3P_4	44 (Th):	Hb−315
P_4U_3	U_3P_4	55 (U):	Hb−179
P_4W_3	W_3P_4	54 (W):	Hb−206
P_4Zn	ZnP_4	32 (Zn):	Hb−276
P_4Zn_3	Zn_3P_4	32 (Zn):	Hb−276
P_5Pb	PbP_5	47 (Pb):	Hb/C3−861
P_5Pt_3	Pt_3P_5	68 (Pt):	Hb/C1−135
P_5Rb_2	Rb_2P_5	24 (Rb):	Hb−233
P_5Sr	SrP_5	29 (Sr):	Hb−211
P_6	P_6	16 (P):	Hb/B−208
P_6Zn	ZnP_6	32 (Zn):	Hb−276
P_7Pt_{20}	$Pt_{20}P_7$	68 (Pt):	Hb/C1−133, 135
P_8	P_8	16 (P):	Hb/B−208
P_8^+	P_8^+	16 (P):	Hb/B−208
P_8S_7	P_8S_7	16 (P):	Hb/C−575
P_8S_{11}	$P_8S_{11} = P_4S_3 \cdot 2 P_2S_4$	16 (P):	Hb/C−581
$P_{14}PbZn$	$PbZnP_{14}$	47 (Pb):	Hb/C4−1229
P_xTh	ThP_x	44 (Th):	Hb−314
P_xTi	TiP_x	41 (Ti):	Hb−378/9
Pa	Protactinium:		
	Atom	51 (Pa):	Hb−43/7, 77, 94
		Erg. W. 8−1/34	
	Atomic nucleus	51 (Pa):	Hb−37/43, 47/51,
			74/87, 93/7

See the "Instructions for the user"
at the beginning of the volume

Pa . . . . Protactinium:

Atomic radius . . . . Erg. W. 8–4/5
Chemical reactions . . . . 51 (Pa):   Hb–53/61, 87/8, 91/3

Detection . . . . 51 (Pa):   Hb–62/4, 88, 97/8
Determination. . . . . 51 (Pa):   Hb–64/6, 88, 98/9
Electrochemical behavior . . . . . . 51 (Pa):   Hb–52/3, 87, 97
Electron configuration . . . . Erg. W. 8–3/23
Enrichment . . . . 51 (Pa):   Hb–13/28
Formation . . . . 51 (Pa):   Hb–11/2, 37, 71/2, 90, 93

History . . . . 51 (Pa):   Hb–1/10, 70/1, 89/90

Ionization potentials . . . . Erg. W. 8–23
Nuclear decay,
  alpha radioactivity . . . . 51 (Pa):   Hb–49
                                          Erg. W. 7a–85/96
–, beta radioactivity. . . . . 51 (Pa):   Hb–49/50, 75/6, 79/84, 94/6

–, decay scheme . . . . Erg. W. 7a–99
–, gamma radioactivity . . . . 51 (Pa):   Hb–50/1, 75/7, 84/7, 96/7
                                          Erg. W. 7a–80
Nuclear fission by neutrons . . . . . Erg. W. 7a–140
Occurrence . . . . 51 (Pa):   Hb–10/1, 71, 90
Oxidation states. . . . . Erg. W. 8–8/11
Position in the periodic system . . . . 51 (Pa):   Hb–46/7, 77, 94
Preparation . . . . 51 (Pa):   Hb–12/37, 72/4, 91/3

– in charged particle reactions . . . . . Erg. W. 7b–145
Purification . . . . 51 (Pa):   Hb–29/32
Spectra:
  Emission spectra . . . . 51 (Pa):   Hb–51
  Mössbauer spectra. . . . . Erg. W. 8–89/90
  X-ray spectra . . . . 51 (Pa):   Hb–51/2

Uses . . . . 51 (Pa):   Hb–36/7, 74, 93
Valence . . . . 51 (Pa):   Hb–46, 77, 94

–. . . . . Protactinium ions:
Chemical reactions . . . . 51 (Pa):   Hb–56/61, 87/8, 91/3

Electrochemical behavior . . . . . . 51 (Pa):   Hb–52/3, 87
Ionization potentials . . . . Erg. W. 8–23
Properties of the ion . . . . 51 (Pa):   Hb–44
                                          Erg. W. 8–5/28

–. . . . . Protactinium isotopes
                        see Pa ... Protactinium
Pa$^+$ . . . . Pa$^+$ . . . . Erg. W. 8–23

Hb = Main Series Volume; GD = Gmelin-Durrer;   Ab = Appended Volume to Main Series;
Eb = Supplement Volume to Main Series;   Erg. W. = New Supplement Series Volume

Pa^{2+}	Pa^{2+}	Erg. W. 8–11/23
Pa^{3+}	Pa^{3+}	Erg. W. 8–5, 11/25
Pa^{4+}	Pa^{4+}	Erg. W. 8–5, 11/23
Pa^{5+}	Pa^{5+}	51 (Pa): Hb–44, 52/3
		Erg. W. 8–5, 23
Pa^{6+}	Pa^{6+}	Erg. W. 8–23
Pa^{n+}	Pa^{n+} (n = 7 to 10)	Erg. W. 8–23

Pb . . . . Lead:

Absorption of light	47 (Pb):	Hb/B1–300/7
– of X-rays	47 (Pb):	Hb/B1–94/9
Abundance	47 (Pb):	Hb/A2a–1/2, 278/80
Anodic dissolution	47 (Pb):	Hb/B2–219/26, 227
Anodic protective layers	47 (Pb):	Hb/B2–197/219
Atom	47 (Pb):	Hb/B1–55/102
Atomic field	47 (Pb):	Hb/B1–55/8
Atomic nucleus	47 (Pb):	Hb/B1–47/55
Atomic radius	47 (Pb):	Hb/B1–101/2
Atomic refraction	47 (Pb):	Hb/B1–138
Atomic scattering factors	47 (Pb):	Hb/B1–160, 207/8
Atomic susceptibility	47 (Pb):	Hb/B1–102
Atomic volume	47 (Pb):	Hb/B1–101/2
Atomic weight	47 (Pb):	Hb/A1–49/50
		Hb/B1–3/9
Behavior as anode	47 (Pb):	Hb/B2–150/227
– as cathode	47 (Pb):	Hb/B2–133/50
– as electrode	47 (Pb):	Hb/B2–108/33
Boiling point	47 (Pb):	Hb/B1–239
Chemical constant	47 (Pb):	Hb/B1–251
Chemical reactions	47 (Pb):	Hb/B1–307/54
– with acids	47 (Pb):	Hb/B1–335/53
– with air	47 (Pb):	Hb/B1–308/14
– with metal compounds	47 (Pb):	Hb/B1–325/35
– with metals	47 (Pb):	Hb/B1–321
– with nonmetal compounds	47 (Pb):	Hb/B1–314/8, 321/5
– with nonmetals	47 (Pb):	Hb/B1–308/14, 318/21
– with organic substances	47 (Pb):	Hb/B1–350/4
– with water	47 (Pb):	Hb/B1–314/8
Compressibility	47 (Pb):	Hb/B1–222/3
Condensation	47 (Pb):	Hb/B1–237
Conduction mechanism	47 (Pb):	Hb/B1–260/2
Constitution of vapor	47 (Pb):	Hb/B1–237
Corrosion, electrochemical	47 (Pb):	Hb/B2–415
Cosmochemistry	47 (Pb):	Hb/A2a–1/56
Creep properties	47 (Pb):	Hb/B1–230/2
Critical constants	47 (Pb):	Hb/B1–239
Crystal chemistry	47 (Pb):	Hb/A2a–63/244
Crystallization	47 (Pb):	Hb/B1–142/58

See the "Instructions for the user"
at the beginning of the volume

Pb . . . . Lead:

Crystallographic properties	47 (Pb):	Hb/B1−140/212
Debye temperature	47 (Pb):	Hb/B1−246/8
Deformation	47 (Pb):	Hb/B1−174/85
Density	47 (Pb):	Hb/B1−214/6
Deposition equivalent	47 (Pb):	Hb/B2−229
Detection	47 (Pb):	Hb/B1−377/89
– in iron	59 (Fe):	Hb/FII1−54/6
– in special substances	47 (Pb):	Hb/B1−390/9
Determination	47 (Pb):	Hb/B1−399/400
– in iron	59 (Fe):	Hb/FII1−56/7
		Hb/FII2−300, 350
– in special substances	47 (Pb):	Hb/B1−400/77
Diffusion in Bi	47 (Pb):	Hb/C3−942/3
– in Bi−Pb alloys	47 (Pb):	Hb/C3−942/3
– in Cd−Pb alloys	47 (Pb):	Hb/C4−1232
– in Hg	47 (Pb):	Hb/C4−1261
– in Hg−Pb alloys	47 (Pb):	Hb/C4−1261
– in In	47 (Pb):	Hb/C4−1307
– in In−Pb alloys	47 (Pb):	Hb/C4−1307
– in Pb−Sb alloys	47 (Pb):	Hb/C3−905
– in Pb−Sn alloys	47 (Pb):	Hb/C4−1482/6
– in Pb−Tl alloys	47 (Pb):	Hb/C4−1327/9
– in Pb−Zn alloys	47 (Pb):	Hb/C4−1213
– in Sn	47 (Pb):	Hb/C4−1482/6
Disorder	47 (Pb):	Hb/B1−164/74
Economic deposits:		
Africa	47 (Pb):	Hb/A4−1/29
Arctic	47 (Pb):	Hb/A4−101/2
Asia	47 (Pb):	Hb/A4−62/89
Australia	47 (Pb):	Hb/A4−89/100
Central America and Mexico	47 (Pb):	Hb/A4−159/65
Europe	47 (Pb):	Hb/A3−132/89
North America	47 (Pb):	Hb/A4−102/58
Oceania	47 (Pb):	Hb/A4−101
South America	47 (Pb):	Hb/A4−165/80
Soviet Union	47 (Pb):	Hb/A4−29/62
Economics	47 (Pb):	Hb/A3−115/32
Elasticity	47 (Pb):	Hb/B1−219/22
Electrical conductivity	47 (Pb):	Hb/B1−262/72
Electrical double layer	47 (Pb):	Hb/B2−119/33
Electrical properties	47 (Pb):	Hb/B1−251/99
Electrochemical behavior	47 (Pb):	Hb/B2−1/416
Electrokinetics	47 (Pb):	Hb/B2−67/8
Electrolytic deposition	47 (Pb):	Hb/B2−227/414
		Hb/C1−31/3
– from aqueous solutions	47 (Pb):	Hb/B2−229/397
		Hb/C1−31/3
– from aqueous suspensions	47 (Pb):	Hb/B2−397/400

Hb = Main Series Volume; GD = Gmelin-Durrer;   Ab = Appended Volume to Main Series;
Eb = Supplement Volume to Main Series;   Erg. W. = New Supplement Series Volume

Pb . . . . Lead:

Electrolytic deposition from melts . . .	47 (Pb):	Hb/B2−402/14
		Hb/C1−33
− from nonaqueous solutions	47 (Pb):	Hb/B2−400/1
− from solid lead compounds	47 (Pb):	Hb/B2−414
Electrolytic solution pressure	47 (Pb):	Hb/B2−11
Electron configuration	47 (Pb):	Hb/B1−58/9
Electronegativity	47 (Pb):	Hb/B1−100/1
Electron emission.	47 (Pb):	Hb/B1−296/9
Electropolishing.	47 (Pb):	Hb/B2−225/6
Energy band structure	47 (Pb):	Hb/B1−251/6
Equation of state	47 (Pb):	Hb/B1−212/4
Expansion coefficient.	47 (Pb):	Hb/B1−217/8
Galvanomagnetic effects.	47 (Pb):	Hb/B1−286/90
Geochemical character.	47 (Pb):	Hb/A2a−277/8
Geochemical cycle	47 (Pb):	Hb/A2a−57/62
Geochemistry	47 (Pb):	Hb/A2a−57/280
		Hb/A2b−1/276
		Hb/A2c−1/185
Hall effect	47 (Pb):	Hb/B1−286/9
Hardness	47 (Pb):	Hb/B1−228/30
Heat of fusion.	47 (Pb):	Hb/B1−241
− of sublimation.	47 (Pb):	Hb/B1−241
− of vaporization	47 (Pb):	Hb/B1−241
History:		
Elemental lead	47 (Pb):	Hb/A1−6/50
Lead compounds.	47 (Pb):	Hb/A1−99/218
Lead ores, dressing and recovery. .	47 (Pb):	Hb/A1−50/75
Lead recovery, ore smelting. . . .	47 (Pb):	Hb/A1−75/87
Names and symbols	47 (Pb):	Hb/A1−1/6
Preparation in pure state	47 (Pb):	Hb/A1−87/98
Interaction with electrons	47 (Pb):	Hb/B1−99/100
Interfacial tension.	47 (Pb):	Hb/B1−234/7
Internal friction	47 (Pb):	Hb/B1−224/8
Ionic mobility	47 (Pb):	Hb/B2−11/3
Ionization	47 (Pb):	Hb/B1−99
Lattice defects.	47 (Pb):	Hb/B1−164/74
Lattice energy.	47 (Pb):	Hb/B1−160/2
Lattice structure.	47 (Pb):	Hb/B1−158/63
Lattice vibrations	47 (Pb):	Hb/B1−196/200
Lead storage battery	47 (Pb):	Hb/B2−415/6
Magnetic properties	47 (Pb):	Hb/B1−251/99
Mass absorption coefficient	47 (Pb):	Hb/B1−94/7
Mass energy transfer coefficient. . . .	47 (Pb):	Hb/B1−98
Mechanical properties	47 (Pb):	Hb/B1−212/37
Melting point	47 (Pb):	Hb/B1−239/40
Mineral deposits		
see Economic deposits		
Mineralogy	47 (Pb):	Hb/A3−1/114

See the "Instructions for the user"
at the beginning of the volume

Pb . . . . Lead:

Modifications	47 (Pb):	Hb/B1–140/2
Molecule.	47 (Pb):	Hb/B1–138/9
Names and symbols	47 (Pb):	Hb/A1–1/6
Nuclear magnetic resonance.	47 (Pb):	Hb/B1–257/8
Nuclear moments.	47 (Pb):	Hb/B1–48/9
Occurrence, extraterrestrial	47 (Pb):	Hb/A2a–1/56
–, terrestrial	47 (Pb):	Hb/A2a–57/280
		Hb/A2b–1/276
		Hb/A2c–1/185
– in meteorites	47 (Pb):	Hb/A2a–2/25
– in minerals	47 (Pb):	Hb/A2a–63/244
– in pegmatites and pneumatolytes . .	47 (Pb):	Hb/A2b–169/91
– in recent volcanism.	47 (Pb):	Hb/A2b–270/6
– in the atmosphere	47 (Pb):	Hb/A2c–171/85
– in the hydrosphere	47 (Pb):	Hb/A2c–143/70
– in the hydrothermal phase	47 (Pb):	Hb/A2b–192/269
– in the lithosphere	47 (Pb):	Hb/A2b–1/276
– in the magmatic cycle	47 (Pb):	Hb/A2b–1/276
– in the metamorphic cycle	47 (Pb):	Hb/A2c–75/142
– in the orthomagmatic phase	47 (Pb):	Hb/A2b–1/168
– in the sedimentary cycle	47 (Pb):	Hb/A2c–1/74
– on the moon	47 (Pb):	Hb/A2a–25/56
Optical properties.	47 (Pb):	Hb/B1–300/7
Overvoltage of hydrogen.	47 (Pb):	Hb/B2–69/102
– of oxygen	47 (Pb):	Hb/B2–102/8
Passivation by anodic		
protective layers	47 (Pb):	Hb/B2–197/219
Photoabsorption coefficient	47 (Pb):	Hb/B1–98/9
Photoelectric effect	47 (Pb):	Hb/B1–297/9
Photoelectron emission	47 (Pb):	Hb/B1–297/8
Plasma vibrations.	47 (Pb):	Hb/B1–256/7
Plasticity.	47 (Pb):	Hb/B1–230/2
Polymorphism	47 (Pb):	Hb/B1–140/2
Position in the periodic system	47 (Pb):	Hb/B1–1
Potentials vs. acid solutions	47 (Pb):	Hb/B2–28/35
– vs. bases.	47 (Pb):	Hb/B2–35/6
– vs. foreign ionic salt solutions	47 (Pb):	Hb/B2–37/45
– vs. lead-free gels	47 (Pb):	Hb/B2–47
– vs. lead-free melts	47 (Pb):	Hb/B2–62/3
– vs. lead salt solutions.	47 (Pb):	Hb/B2–13/28
– vs. melts containing lead salts	47 (Pb):	Hb/B2–48/62
– vs. nonaqueous solutions		
and fluids	47 (Pb):	Hb/B2–47
– vs. solid lead salts	47 (Pb):	Hb/B2–63/4
– vs. water.	47 (Pb):	Hb/B2–45/7
Preparation, industrial	47 (Pb):	Hb/C1–1/46
–, –, dry processes	47 (Pb):	Hb/C1–1/30
–, –, electrolytic processes	47 (Pb):	Hb/C1–31/3

Hb = Main Series Volume; GD = Gmelin-Durrer;   *Ab = Appended Volume to Main Series;*
Eb = Supplement Volume to Main Series;   *Erg. W. = New Supplement Series Volume*

Pb . . . . Lead:

Preparation, industrial,
  historical data . . . . 47 (Pb):   Hb/A1 – 50/98
–, –, recovery from residues . . . . . . 47 (Pb):   Hb/C1 – 33/5
–, –, refining. . . . . 47 (Pb):   Hb/C1 – 35/46
–, –, wet processes . . . . 47 (Pb):   Hb/C1 – 30/1
– in pure state. . . . . 47 (Pb):   Hb/A1 – 87/98
Raffination. . . . . 47 (Pb):   Hb/C1 – 35/46
Recrystallization . . . . 47 (Pb):   Hb/B1 – 186/96
Redox potentials
  in aqueous solutions. . . . . 47 (Pb):   Hb/B2 – 47/8
– in melts . . . . 47 (Pb):   Hb/B2 – 63
Redox reactions. . . . . 47 (Pb):   Hb/B1 – 375/7
Reflectivity . . . . 47 (Pb):   Hb/B1 – 300/7
Refractive index. . . . . 47 (Pb):   Hb/B1 – 300/7
Scattering of electrons . . . . 47 (Pb):   Hb/B1 – 100
Self-diffusion . . . . 47 (Pb):   Hb/B1 – 208/12
Separation, analytical . . . . 47 (Pb):   Hb/B1 – 400
Solidification point . . . . 47 (Pb):   Hb/B1 – 240
Sound velocity . . . . 47 (Pb):   Hb/B1 – 223/4
Spectra:
  Arc spectrum . . . . 47 (Pb):   Hb/B1 – 65/80
  Band spectrum . . . . 47 (Pb):   Hb/B1 – 139
  Spark spectra. . . . . 47 (Pb):   Hb/B1 – 109/15
  Terms. . . . . 47 (Pb):   Hb/B1 – 58/65
  X-ray spectrum. . . . . 47 (Pb):   Hb/B1 – 81/99

Standard potential . . . . 47 (Pb):   Hb/B2 – 1/11
Strain release . . . . 47 (Pb):   Hb/B1 – 185/96
Structure of the melt . . . . 47 (Pb):   Hb/B1 – 200/8
Superconductivity . . . . 47 (Pb):   Hb/B1 – 272/86
Surface structure . . . . 47 (Pb):   Hb/B1 – 163
Surface tension . . . . 47 (Pb):   Hb/B1 – 234/7
Susceptibility . . . . 47 (Pb):   Hb/B1 – 258/60
Thermal conductivity. . . . . 47 (Pb):   Hb/B1 – 248/51
Thermal diffusivity . . . . 47 (Pb):   Hb/B1 – 251
Thermal expansion . . . . 47 (Pb):   Hb/B1 – 217/8
Thermal properties . . . . 47 (Pb):   Hb/B1 – 212/51
Thermal radiation. . . . . 47 (Pb):   Hb/B1 – 307
Thermodynamic functions . . . . . . 47 (Pb):   Hb/B1 – 242/6
Thermoelectric effects . . . . 47 (Pb):   Hb/B1 – 290/6
Thermomagnetic effects . . . . . . 47 (Pb):   Hb/B1 – 296
Toxicity . . . . 47 (Pb):   Hb/B1 – 482/97
Ultrasound attenuation. . . . . . . 47 (Pb):   Hb/B1 – 224/8
Vaporization. . . . . 47 (Pb):   Hb/B1 – 237
Vapor pressure . . . . 47 (Pb):   Hb/B1 – 237/8
Viscosity of the melt . . . . 47 (Pb):   Hb/B1 – 232/4
Wave functions . . . . 47 (Pb):   Hb/B1 – 55/8
Wetting . . . . 47 (Pb):   Hb/B1 – 236/7
Zero charge potential. . . . . 47 (Pb):   Hb/B2 – 64/7

See the "Instructions for the user"
at the beginning of the volume

Pb . . . . Lead ions
                        see Pb⁺ ... Pb⁺, etc.

–. . . . . Lead isotopes:
                        Atomic nucleus . . . . 47 (Pb):   Hb/B1–47/55
                        Detection and determination. . . . . . 47 (Pb):   Hb/B1–477/82
                        Enrichment and separation
                            of stable Pb isotopes. . . . . 47 (Pb):   Hb/B1–9/12
                        –, preparation, and separation
                            of unstable Pb isotopes . . . . . . 47 (Pb):   Hb/B1–12/47
                        Formation of natural
                            lead isotopes . . . . 47 (Pb):   Hb/A2a–246/8
                        Isotope geochemistry . . . . 47 (Pb):   Hb/A2a–245/77
                        Isotopic abundance. . . . . 47 (Pb):   Hb/B1–1/2
                        Isotopic ratios in lunar rocks . . . . . 47 (Pb):   Hb/A2a–42/51
                        – in meteorites . . . . 47 (Pb):   Hb/A2a–8/25
                        – in minerals . . . . 47 (Pb):   Hb/A2a–136/52, 241/4
                        – in rocks . . . . 47 (Pb):   Hb/A2b–33/4, 116/26
                                                                             Hb/A2c–66/9

                        Occurrence in fumarolic products
                            and vapor condensates . . . . . . 47 (Pb):   Hb/A2b–272/6
                        – in mineral waters . . . . 47 (Pb):   Hb/A2c–169/70
                        – in the atmosphere . . . . 47 (Pb):   Hb/A2c–171/85

–. . . . . Lead systems:
                        Pb–Ag. . . . . 61 (Ag):   Hb/C–292/6
                        Pb–Ag–Ag₂S–PbS. . . . . 61 (Ag):   Hb/B4–263/5
                        Pb–Ag–Al. . . . . 35 (Al):   Hb/A6–1087/9
                                                               61 (Ag):   Hb/C–310/1
                        Pb–Ag–As . . . . 61 (Ag):   Hb/B4–273
                        Pb–Ag–Cu–S. . . . . 61 (Ag):   Hb/B4–470
                        Pb–Ag–Se . . . . 61 (Ag):   Hb/B4–267/8
                        Pb–Ag–Te . . . . 61 (Ag):   Hb/B4–269
                        Pb–Ag–Zn . . . . 61 (Ag):   Hb/C–308/10
                        Pb–Al . . . . 35 (Al):   Hb/A5–858/9
                                                               47 (Pb):   Hb/C4–1278/82
                        Pb–Al–Bi . . . . 47 (Pb):   Hb/C4–1287
                        Pb–Al–Bi–Sb. . . . . 47 (Pb):   Hb/C4–1287/8
                        Pb–Al–Ce . . . . 47 (Pb):   Hb/C4–1365
                        Pb–Al–Er . . . . 47 (Pb):   Hb/C4–1371
                        Pb–Al–Ga. . . . . 47 (Pb):   Hb/C4–1305
                        Pb–Al–Mg . . . . 35 (Al):   Hb/A5–863
                                                               47 (Pb):   Hb/C4–1288/90
                        Pb–Al–Sb. . . . . 35 (Al):   Hb/A5–861/3
                                                               47 (Pb):   Hb/C4–1286/7
                        Pb–Al–Sn. . . . . 35 (Al):   Hb/A5–863/4
                                                               47 (Pb):   Hb/C4–1529
                        Pb–Al–Zn. . . . . 47 (Pb):   Hb/C4–1290/1
                        Pb–As. . . . . 47 (Pb):   Hb/C3–887/9
                        Pb–As–S . . . . 47 (Pb):   Hb/C3–901/3

Hb = Main Series Volume; GD = Gmelin-Durrer;   *Ab = Appended Volume to Main Series;*
Eb = Supplement Volume to Main Series;   *Erg. W. = New Supplement Series Volume*

Pb . . . . Lead systems:

Pb–As–Sb	47 (Pb):	Hb/C3–941/2
Pb–As–Se	47 (Pb):	Hb/C3–904/5
Pb–As–Zn	47 (Pb):	Hb/C4–1223
Pb–Au	62 (Au):	Hb/3–848/9
Pb–Au–Zn	62 (Au):	Hb/3–798
Pb–B	47 (Pb):	Hb/C2–685
Pb–Ba	47 (Pb):	Hb/C3–1200/2
Pb–Bi	47 (Pb):	Hb/C3–943/4
Pb–Bi–Cd	47 (Pb):	Hb/C4–1246/8
Pb–Bi–Cd–Sn	47 (Pb):	Hb/C4–1525/6
Pb–Bi–Cd–Sn–Zn	47 (Pb):	Hb/C4–1528/9
Pb–Bi–Cd–Zn	47 (Pb):	Hb/C4–1251
Pb–Bi–Cl	47 (Pb):	Hb/C3–968
Pb–Bi–Cu	60 (Cu):	Hb/B3–1206
Pb–Bi–F	47 (Pb):	Hb/C3–968
Pb–Bi–Ga	47 (Pb):	Hb/C4–1304
Pb–Bi–Ga–Sn	47 (Pb):	Hb/C4–1529
Pb–Bi–O	47 (Pb):	Hb/C3–996/7
Pb–Bi–S	47 (Pb):	Hb/C3–969/70
Pb–Bi–Sb	47 (Pb):	Hb/C3–977/8
Pb–Bi–Se	47 (Pb):	Hb/C3–971/2
Pb–Bi–Sn	47 (Pb):	Hb/C4–1518/9
Pb–Bi–Te	47 (Pb):	Hb/C3–973/5
Pb–Bi–Th	47 (Pb):	Hb/C4–1470
Pb–Bi–Tl	47 (Pb):	Hb/C4–1347
Pb–Bi–Zn	47 (Pb):	Hb/C4–1224/5
Pb–Br–Sn	47 (Pb):	Hb/C4–1584
Pb–C–Ni	57 (Ni):	Hb/B1–152
Pb–Ca	47 (Pb):	Hb/C3–1179/80
Pb–Cd	47 (Pb):	Hb/C4–1232/3
Pb–Cd–Ga	47 (Pb):	Hb/C4–1304
Pb–Cd–Hg	47 (Pb):	Hb/C4–1275
Pb–Cd–Mg	27 (Mg):	Hb/A4–655/6
	47 (Pb):	Hb/C4–1248
Pb–Cd–Ni	57 (Ni):	Hb/B1–153
Pb–Cd–Sb	47 (Pb):	Hb/C4–1245/6
Pb–Cd–Sn	47 (Pb):	Hb/C4–1524/5
Pb–Cd–Sn–Zn	47 (Pb):	Hb/C4–1528/9
Pb–Cd–Tl	47 (Pb):	Hb/C4–1347
Pb–Cd–Zn	47 (Pb):	Hb/C4–1248/9
Pb–Ce	47 (Pb):	Hb/C4–1364/5
Pb–Ce–Zn	47 (Pb):	Hb/C4–1363, 1365
Pb–Cl–S–Sn	47 (Pb):	Hb/C4–1586
Pb–Cl–Sn	47 (Pb):	Hb/C4–1583/4
Pb–Co	58 (Co):	Hb/A1–196
		Eb/A–415
Pb–Cr	52 (Cr):	Hb/A2–727
Pb–Cs–Zn	47 (Pb):	Hb/C4–1225

See the "Instructions for the user"
at the beginning of the volume

Pb . . . . Lead systems:
Pb–Cu . . . . 60 (Cu):   Hb/B3–1197
Pb–Cu–Cu₂S–PbS . . . . 60 (Cu):   Hb/B3–1201
Pb–Cu–O . . . . 60 (Cu):   Hb/B3–1197/8
Pb–Cu–S . . . . 60 (Cu):   Hb/B3–1200/2
Pb–Cu–Sb . . . . 60 (Cu):   Hb/B3–1204
Pb–Cu–Te . . . . 60 (Cu):   Hb/B3–1202
Pb–Dy . . . . 47 (Pb):   Hb/C4–1369/70
Pb–Dy–Zn . . . . 47 (Pb):   Hb/C4–1363, 1370
Pb–Er . . . . 47 (Pb):   Hb/C4–1370/1
Pb–Er–Zn . . . . 47 (Pb):   Hb/C4–1371
Pb–Eu . . . . 47 (Pb):   Hb/C4–1367/8
Pb–Eu–Zn . . . . 47 (Pb):   Hb/C4–1363, 1368
Pb–Ga . . . . 36 (Ga):   Hb–63
                                        47 (Pb):   Hb/C4–1303/4
Pb–Ga–Ge . . . . 47 (Pb):   Hb/C4–1473
Pb–Ga–Hg . . . . 47 (Pb):   Hb/C4–1305
Pb–Ga–Sb . . . . 47 (Pb):   Hb/C4–1305
Pb–Ga–Se . . . . 47 (Pb):   Hb/C4–1306
Pb–Ga–Sn–Zn . . . . 47 (Pb):   Hb/C4–1529
Pb–Ga–Zn . . . . 47 (Pb):   Hb/C4–1305
Pb–Gd . . . . 47 (Pb):   Hb/C4–1368/9
Pb–Gd–Zn . . . . 47 (Pb):   Hb/C4–1363, 1369
Pb–Ge . . . . 47 (Pb):   Hb/C4–1472/3
Pb–Ge–Mg . . . . 47 (Pb):   Hb/C4–1473
Pb–Ge–Ni . . . . 57 (Ni):   Hb/B1–153
Pb–Ge–Ni–Sn . . . . 57 (Ni):   Hb/B1–153
Pb–H . . . . 47 (Pb):   Hb/C1–47/8
Pb–Hg . . . . 47 (Pb):   Hb/C4–1261/2
Pb–Hg–Na . . . . 47 (Pb):   Hb/C4–1274/5
Pb–Ho . . . . 47 (Pb):   Hb/C4–1370
Pb–Ho–Zn . . . . 47 (Pb):   Hb/C4–1363, 1370
Pb–In . . . . 37 (In):   Hb–60/1
                                        47 (Pb):   Hb/C4–1307/9
Pb–In–Sb . . . . 47 (Pb):   Hb/C4–1324
Pb–In–Sn . . . . 47 (Pb):   Hb/C4–1529/30
Pb–In–Tl . . . . 47 (Pb):   Hb/C4–1347
Pb–K . . . . 47 (Pb):   Hb/C3–1052/5
Pb–K–Na . . . . 47 (Pb):   Hb/C3–1055/6
Pb–La . . . . 47 (Pb):   Hb/C4–1361/3
Pb–La–Zn . . . . 47 (Pb):   Hb/C4–1363
Pb–Li . . . . 47 (Pb):   Hb/C3–979/80
Pb–Lu . . . . 47 (Pb):   Hb/C4–1372/3
Pb–Lu–Zn . . . . 47 (Pb):   Hb/C4–1363, 1373
Pb–Mg . . . . 27 (Mg):   Hb/A4–637/40
                                        47 (Pb):   Hb/C3–1161/2
Pb–Mg–Na . . . . 27 (Mg):   Hb/A4–654
Pb–Mg–Sb . . . . 27 (Mg):   Hb/A4–653

Hb = Main Series Volume; GD = Gmelin-Durrer;   *Ab = Appended Volume to Main Series;*
Eb = Supplement Volume to Main Series;   *Erg. W. = New Supplement Series Volume*

Pb . . . . Lead systems:

Pb–Mg–Sn	27 (Mg):	Hb/A4–656/7
	47 (Pb):	Hb/C4–1522
Pb–Mg–Zn	47 (Pb):	Hb/C4–1225
Pb–Na	47 (Pb):	Hb/C3–993/7
Pb–Na–Zn	47 (Pb):	Hb/C4–1225
Pb–Nb	49 (Nb):	Hb/B2–241/2
Pb–Nd	47 (Pb):	Hb/C4–1366/7
Pb–Nd–Zn	47 (Pb):	Hb/C4–1363, 1367
Pb–Ni	57 (Ni):	Hb/B1–148/9
Pb–Ni–S	57 (Ni):	Hb/B3–1195
Pb–Ni–Sb	57 (Ni):	Hb/B1–152
Pb–O	47 (Pb):	Hb/C1–51/2
Pb–O–S	9 (S):	Hb/B1–183
	47 (Pb):	Hb/C2–533/4
Pb–O–Sb	47 (Pb):	Hb/C3–926/9
Pb–O–Ta	50 (Ta):	Hb/B2–232
Pb–O–Te	47 (Pb):	Hb/C2–678/9
Pb–PbBr$_2$	47 (Pb):	Hb/C2–348
Pb–PbCl$_2$	47 (Pb):	Hb/C1–282/3
Pb–PbI$_2$–Tl–TlI	47 (Pb):	Hb/C4–1358
Pb–PbO	47 (Pb):	Hb/C1–52/4
Pb–PbS–Sn–SnS	47 (Pb):	Hb/C4–1584
Pb–PbSe–PbTe	47 (Pb):	Hb/C2–681
Pb–Pd	68 (Pt):	Hb/A5–588/93
Pb–Po	47 (Pb):	Hb/C2–685
Pb–Pr	47 (Pb):	Hb/C4–1365/6
Pb–Pr–Zn	47 (Pb):	Hb/C4–1363, 1366
Pb–Pt	68 (Pt):	Hb/A6–763/7
Pb–Pt–Tl	68 (Pt):	Hb/A6–767
Pb–S	47 (Pb):	Hb/C2–413/4
Pb–S–Sb	47 (Pb):	Hb/C3–935/6
Pb–S–Se	47 (Pb):	Hb/C2–639/40
Pb–S–Si	47 (Pb):	Hb/C3–855
Pb–S–Sn	47 (Pb):	Hb/C4–1584
Pb–Sb	47 (Pb):	Hb/C3–905/7
Pb–Sb–Se	47 (Pb):	Hb/C3–940
Pb–Sb–Si	47 (Pb):	Hb/C3–941
Pb–Sb–Sn	47 (Pb):	Hb/C4–1515
Pb–Sb–Te	47 (Pb):	Hb/C3–940/1
Pb–Sb–Zn	47 (Pb):	Hb/C4–1223/4
Pb–Sc	47 (Pb):	Hb/C4–1360
Pb–Se	47 (Pb):	Hb/C2–608/9
Pb–Se–Sn–Te	47 (Pb):	Hb/C4–1594/5
Pb–Si	47 (Pb):	Hb/C3–807/8
Pb–Sm	47 (Pb):	Hb/C4–1367
Pb–Sm–Zn	47 (Pb):	Hb/C4–1363, 1367
Pb–Sn	47 (Pb):	Hb/C4–1487/9

See the "Instructions for the user"
at the beginning of the volume

Pb . . . . Lead systems:

Pb–Sn–Te	47 (Pb):	Hb/C4–1588
Pb–Sn–Tl	47 (Pb):	Hb/C4–1530
Pb–Sn–Zn	47 (Pb):	Hb/C4–1522/4
Pb–Sr	47 (Pb):	Hb/C3–1193/4
Pb–Sr–Zn	47 (Pb):	Hb/C4–1225
Pb–Tb	47 (Pb):	Hb/C4–1369
Pb–Te	47 (Pb):	Hb/C2–640/3
Pb–Th	47 (Pb):	Hb/C4–1468/70
Pb–Ti	47 (Pb):	Hb/C4–1377/8
Pb–Ti–Zr	47 (Pb):	Hb/C4–1423
Pb–Tl	47 (Pb):	Hb/C4–1329/32
Pb–Tm	47 (Pb):	Hb/C4–1371
Pb–V	48 (V):	Hb/B2–687
Pb–W	54 (W):	Hb–104
Pb–Y	47 (Pb):	Hb/C4–1360/1
Pb–Y–Zn	47 (Pb):	Hb/C4–1361, 1363
Pb–Yb	47 (Pb):	Hb/C4–1371/2
Pb–Zn	47 (Pb):	Hb/C4–1213/4
Pb–Zr	47 (Pb):	Hb/C4–1420/2

Pb^+ . . . . Pb^+:

Ionic radius	47 (Pb):	Hb/B1–136
Ionization	47 (Pb):	Hb/B1–115
Scattering of electrons	47 (Pb):	Hb/B1–116
Spectra:		
Crystal spectra	21 (Na):	Eb/6–337
Spark spectrum	47 (Pb):	Hb/B1–109/12
Standard potential	47 (Pb):	Hb/B2–4
Terms	47 (Pb):	Hb/B1–103/5

Pb^{2+} . . . . Pb^{2+}:

Chemical reactions	47 (Pb):	Hb/B1–354/75
– with inorganic compounds.	47 (Pb):	Hb/B1–361/70
– with organic compounds.	47 (Pb):	Hb/B1–370
Diffusion in sodium halides	21 (Na):	Eb/6–293
Electrochemical		
deposition equivalent	47 (Pb):	Hb/B2–229
Electrolysis	47 (Pb):	Hb/B2–227/414
Exchange reactions.	47 (Pb):	Hb/B1–370/5
Faraday effect.	47 (Pb):	Hb/B1–138
Hydration	47 (Pb):	Hb/B1–354/8
Hydrolysis	47 (Pb):	Hb/B1–359/61
Ionic activity.	47 (Pb):	Hb/B1–358/9
Ionic mobility	47 (Pb):	Hb/B2–11/3
Ionic radius	47 (Pb):	Hb/B1–136/7
Ionic refraction	47 (Pb):	Hb/B1–138
Ionic susceptibility	47 (Pb):	Hb/B1–137/8
Ionic volume	47 (Pb):	Hb/B1–136/7
Ionization	47 (Pb):	Hb/B1–115

Hb = Main Series Volume; GD = Gmelin-Durrer;   *Ab = Appended Volume to Main Series;*
Eb = Supplement Volume to Main Series;   *Erg. W. = New Supplement Series Volume*

Pb^{2+}	Pb^{2+}:		
	Redox reactions.	47 (Pb):	Hb/B1−375/7
	Solvation	47 (Pb):	Hb/B1−354/8
	Sorption	47 (Pb):	Hb/B1−371/5
	Spectra:		
	Absorption spectra.	47 (Pb):	Hb/B1−121/6, 132/6
	Crystal spectra	21 (Na):	Eb/6−333/4, 337
	Luminescence spectra	21 (Na):	Eb/6−347
		47 (Pb):	Hb/B1−116/21, 126/30
	Spark spectrum	47 (Pb):	Hb/B1−112/4
	Standard potential	47 (Pb):	Hb/B2−4/5, 10
	Terms	47 (Pb):	Hb/B1−105/7
−.	$[Pb(H_2O)_n]^{2+}$ (n = 1 to 5) . . .	47 (Pb):	Hb/C4−1609/10
Pb^{3+}	Pb^{3+}:		
	Ionization	47 (Pb):	Hb/B1−115
	Paramagnetic resonance absorption .	47 (Pb):	Hb/B1−130/2
	Spectra	47 (Pb):	Hb/B1−114
	Terms	47 (Pb):	Hb/B1−107/8
Pb^{4+}	Pb^{4+}:		
	Chemical reactions	47 (Pb):	Hb/B1−354
	Ionic radius	47 (Pb):	Hb/B1−136/7
	Ionic refraction	47 (Pb):	Hb/B1−138
	Ionic susceptibility	47 (Pb):	Hb/B1−138
	Ionization	47 (Pb):	Hb/B1−115
	Spectra	47 (Pb):	Hb/B1−115, 136
	Standard potential	47 (Pb):	Hb/B2−10
	Terms	47 (Pb):	Hb/B1−108/9
Pb^{5+}	Pb^{5+}	47 (Pb):	Hb/B1−115
Pb^{n+}	Pb^{n+} (n = 6 to 81)	47 (Pb):	Hb/B1−115/6
Pb^-	Pb^-	47 (Pb):	Hb/B1−136
Pb^{4-}	Pb^{4-}	47 (Pb):	Hb/B1−137
PbPd	PdPb	68 (Pt):	Hb/A5−589/93
$PbPd_2$	Pd_2Pb	68 (Pt):	Hb/A5−589/93
$PbPd_3$	Pd_3Pb	68 (Pt):	Hb/A5−589/93
PbPo	PbPo	47 (Pb):	Hb/C2−685
PbPr.	PrPb	47 (Pb):	Hb/C4−1366
$PbPr_2$	Pr_2Pb	47 (Pb):	Hb/C4−1366
$PbPr_2S_4$	$PbPr_2S_4$	47 (Pb):	Hb/C4−1374
$PbPr_2Se_4$	$PbPr_2Se_4$	47 (Pb):	Hb/C4−1375
PbPt.	PtPb	68 (Pt):	Hb/A6−764, 766/7
$PbPt_3$	Pt_3Pb	68 (Pt):	Hb/A6−764, 766
PbRb	RbPb	47 (Pb):	Hb/C3−1137
PbS	PbS:		
	Chemical reactions	47 (Pb):	Hb/C2−486/529
	− on heating.	47 (Pb):	Hb/C2−486
	− with acids	47 (Pb):	Hb/C2−516/9
	− with alkaline solutions	47 (Pb):	Hb/C2−519

See the "Instructions for the user"
at the beginning of the volume

PbS . . . . PbS:

    Chemical reactions

      with metal compounds . . . . 47 (Pb):   Hb/C2–511/6, 520/7

    – with metal compounds . . . . 47 (Pb):   Hb/C2–511/6, 520/7

    – with metals . . . . 47 (Pb):   Hb/C2–509/11, 520

    – with nonmetal compounds . . . . 47 (Pb):   Hb/C2–506/9, 520

    – with nonmetals . . . . 47 (Pb):   Hb/C2–503/6, 519/20

    – with organic compounds . . . . . . 47 (Pb):   Hb/C2–527/9

    – with oxygen . . . . 47 (Pb):   Hb/C2–486/503

    Colloidal forms . . . . 47 (Pb):   Hb/C2–440/3

    Crystallographic properties . . . . . 47 (Pb):   Hb/C2–444/51

    Electrical properties . . . . 47 (Pb):   Hb/C2–456/76

    Electrochemical behavior . . . . . . 47 (Pb):   Hb/B2–403, 413

                                            Hb/C2–481/5

    Formation . . . . 47 (Pb):   Hb/C2–414/23

    History . . . . 47 (Pb):   Hb/A1–131/9

    Hydrosols . . . . 47 (Pb):   Hb/C2–440/1

    Magnetic properties . . . . 47 (Pb):   Hb/C2–458, 472, 475

    Mechanical properties . . . . 47 (Pb):   Hb/C2–451/3

    Molecule . . . . 47 (Pb):   Hb/C2–443/4

    Optical properties . . . . 47 (Pb):   Hb/C2–476/80

    Organosols . . . . 47 (Pb):   Hb/C2–441

    Photoconductivity . . . . 47 (Pb):   Hb/C2–464/71

    Photosensitive PbS films . . . . 47 (Pb):   Hb/C2–438/40

    Photovoltaic effect . . . . 47 (Pb):   Hb/C2–471/2

    Preparation . . . . 47 (Pb):   Hb/C2–414/23

    – in pure state . . . . 47 (Pb):   Hb/C2–423/6

    – of special forms . . . . 47 (Pb):   Hb/C2–426/32

    Rectifier effect . . . . 47 (Pb):   Hb/C2–463/4

    Semiconductor properties . . . . . . 47 (Pb):   Hb/C2–426/8, 434,

                                            438/40, 456/80

    Single crystals . . . . 47 (Pb):   Hb/C2–432/4

    Sorption of oxygen . . . . 47 (Pb):   Hb/C2–498/9

    Spectra . . . . 47 (Pb):   Hb/C2–443/4

    Suspensions . . . . 47 (Pb):   Hb/C2–442/3

    Thermal properties . . . . 47 (Pb):   Hb/C2–451/6

    Thin layers . . . . 47 (Pb):   Hb/C2–435/8, 449/50

    Toxicity . . . . 47 (Pb):   Hb/B1–494

–. . . . . PbS melts:

    $PbS-Ag_2S-Cu_2S$ . . . . 61 (Ag):   Hb/B4–471

–. . . . . PbS solid solutions:

    PbS–AgBr . . . . 61 (Ag):   Hb/B4–265

    $PbS-Ce_2S_3$ . . . . 47 (Pb):   Hb/C4–1374

    $PbS-Dy_2S_3$ . . . . 47 (Pb):   Hb/C4–1374

    $PbS-Er_2S_3$ . . . . 47 (Pb):   Hb/C4–1374

    $PbS-Gd_2S_3$ . . . . 47 (Pb):   Hb/C4–1374

    $PbS-Ho_2S_3$ . . . . 47 (Pb):   Hb/C4–1374

    $PbS-La_2S_3$ . . . . 47 (Pb):   Hb/C4–1374

    $PbS-Lu_2S_3$ . . . . 47 (Pb):   Hb/C4–1374

Hb = Main Series Volume; GD = Gmelin-Durrer;   *Ab = Appended Volume to Main Series;*
Eb = Supplement Volume to Main Series;   *Erg. W. = New Supplement Series Volume*

PbS . . . .   PbS solid solutions:
                           PbS−Nd$_2$S$_3$ . . . . 47 (Pb):   Hb/C4−1374
                           [(Ni,Pb)S]$_n$ . . . . 57 (Ni):   Hb/B3−1196
                           PbS−PbSe . . . . 47 (Pb):   Hb/C2−639/40
                           PbS−PbTe−SnTe . . . . 47 (Pb):   Hb/C4−1594
                           PbS−Pr$_2$S$_3$ . . . . 47 (Pb):   Hb/C4−1374
                           PbS−Sm$_2$S$_3$ . . . . 47 (Pb):   Hb/C4−1374
                           PbS−SnS . . . . 47 (Pb):   Hb/C4−1585
                           PbS−Tb$_2$S$_3$ . . . . 47 (Pb):   Hb/C4−1374
                           PbS−Tm$_2$S$_3$ . . . . 47 (Pb):   Hb/C4−1374
                           PbS−Y$_2$S$_3$ . . . . 47 (Pb):   Hb/C4−1374
                           PbS−Yb$_2$S$_3$ . . . . 47 (Pb):   Hb/C4−1374

−. . . . .   PbS solutions:
                           PbS−H$_2$O . . . . 47 (Pb):   Hb/C2−529/30
                           PbS−H$_2$S−H$_2$O . . . . 47 (Pb):   Hb/C2−530/1

−. . . . .   PbS systems:
                           PbS−Ag−Ag$_2$S−Pb . . . . 61 (Ag):   Hb/B4−263/5
                           PbS−AgBiS$_2$ . . . . 61 (Ag):   Hb/B4−266
                           PbS−AgSbS$_2$ . . . . 61 (Ag):   Hb/B4−265/6
                           PbS−Ag$_2$S . . . . 61 (Ag):   Hb/B4−264
                           PbS−Ag$_2$S−Bi$_2$S$_3$ . . . . 61 (Ag):   Hb/B4−266
                           PbS−Ag$_2$S−Cu$_2$S . . . . 61 (Ag):   Hb/B4−470/1
                           PbS−Al$_2$S$_3$ . . . . 47 (Pb):   Hb/C4−1294
                           PbS−As$_2$S$_3$ . . . . 47 (Pb):   Hb/C3−901/3
                           PbS−B$_2$S$_3$ . . . . 47 (Pb):   Hb/C2−696
                           PbS−Bi$_2$S$_3$ . . . . 47 (Pb):   Hb/C3−969/70
                           PbS−Cu−Cu$_2$S−Pb . . . . 60 (Cu):   Hb/B3−1201
                           PbS−Cu$_2$S−Cu$_3$As−Cu$_3$Sb . . . . . . 60 (Cu):   Hb/B3−1206
                           PbS−Cu$_2$S−FeS . . . . 60 (Cu):   Hb/B3−1329
                           PbS−Cu$_2$S−FeS−ZnS . . . . 60 (Cu):   Hb/B3−1329
                           PbS−Cu$_2$S−ZnS . . . . 60 (Cu):   Hb/B3−1208
                           PbS−FeS . . . . 59 (Fe):   Hb/B5−1132
                           PbS−FeS−FeSiO$_3$−PbSiO$_3$ . . . . . . 59 (Fe):   Hb/B5−1137
                           PbS−GeS$_2$ . . . . 47 (Pb):   Hb/C4−1477
                           PbS−GeSe . . . . 47 (Pb):   Hb/C4−1478
                           PbS−KCl−PbCl$_2$ . . . . 47 (Pb):   Hb/C3−1096
                           PbS−NaCl−PbCl$_2$ . . . . 47 (Pb):   Hb/C3−1032
                           PbS−Pb−Sn−SnS . . . . 47 (Pb):   Hb/C4−1584
                           PbS−PbCl$_2$ . . . . 47 (Pb):   Hb/C2−605
                           PbS−PbCl$_2$−SnCl$_2$−SnS . . . . . . 47 (Pb):   Hb/C4−1586
                           PbS−PbO . . . . 47 (Pb):   Hb/C2−534
                           PbS−PbSe−PbTe . . . . 47 (Pb):   Hb/C2−684
                           PbS−PbTe . . . . 47 (Pb):   Hb/C2−679/80
                           PbS−Sb$_2$S$_3$ . . . . 47 (Pb):   Hb/C3−935/6
                           PbS−SnS . . . . 47 (Pb):   Hb/C4−1585
                           PbS−SnSe . . . . 47 (Pb):   Hb/C4−1588
                           PbS−Sn$_2$S$_3$ . . . . 47 (Pb):   Hb/C4−1585
                           PbS−Tl$_2$S . . . . 47 (Pb):   Hb/C4−1352/3
                           PbS−ZnS . . . . 47 (Pb):   Hb/C4−1227

See the "Instructions for the user"
at the beginning of the volume

PbS$_2$ . . . . PbS$_2$ . . . . 47 (Pb): Hb/C2−532
PbS$_2$$^{2-}$ . . . . [PbS$_2$]$^{2-}$ . . . . 47 (Pb): Hb/A2b−226
PbS$_2$Sn . . . . . . PbSnS$_2$ . . . . 47 (Pb): Hb/C4−1585/6
PbS$_3$ . . . . PbS$_3$ . . . . 47 (Pb): Hb/C2−532
PbS$_3$Ta . . . . . . PbTaS$_3$ . . . . 50 (Ta): Hb/B2−318
PbS$_3$Ti . . . . . . PbTiS$_3$ . . . . 47 (Pb): Hb/C4−1391
PbS$_3$Zr . . . . . . PbZrS$_3$ . . . . 47 (Pb): Hb/C4−1431
PbS$_4$Sb$_2$ . . . . . . Pb(SbS$_2$)$_2$ = PbS · Sb$_2$S$_3$ . . . . . 47 (Pb): Hb/C3−939
PbS$_4$Sm$_2$ . . . . . . PbSm$_2$S$_4$ . . . . 47 (Pb): Hb/C4−1374
PbS$_4$Tm$_2$ . . . . . . PbTm$_2$S$_4$ . . . . 47 (Pb): Hb/C4−1374
PbS$_4$Y$_2$ . . . . . . PbY$_2$S$_4$ . . . . 47 (Pb): Hb/C2−515
PbS$_4$Yb$_2$ . . . . . . PbYb$_2$S$_4$ . . . . 47 (Pb): Hb/C4−1374
PbS$_5$ . . . . PbS$_5$ . . . . 47 (Pb): Hb/C2−533
PbS$_5$Sn$_4$ . . . . . . PbSn$_4$S$_5$ . . . . 47 (Pb): Hb/C4−1586
PbSb$_2$Se$_4$ . . . . . . Pb(SbSe$_2$)$_2$ = PbSe · Sb$_2$Se$_3$ . . . . . . 47 (Pb): Hb/C3−940

PbSe . . . . PbSe:
                    Chemical reactions . . . . 47 (Pb): Hb/C2−630/2
                    Crystallographic properties . . . . . . 47 (Pb): Hb/C2−614/7
                    Electrical properties . . . . 47 (Pb): Hb/C2−619/29
                    Formation and preparation . . . . . 47 (Pb): Hb/C2−610/3
                    Magnetic properties . . . . 47 (Pb): Hb/C2−622, 627,
                                                                        628/9
                    Mechanical properties . . . . 47 (Pb): Hb/C2−617/9
                    Molecule. . . . . 47 (Pb): Hb/C2−613/4
                    Optical properties. . . . . 47 (Pb): Hb/C2−629/30
                    Special forms . . . . 47 (Pb): Hb/C2−612/3
                    Spectra . . . . 47 (Pb): Hb/C2−614
                    Thermal properties . . . . 47 (Pb): Hb/C2−617/9
−. . . . . PbSe solid solutions:
                    PbSe−BiSe . . . . 47 (Pb): Hb/C3−972
                    PbSe−Bi$_2$Se$_3$ . . . . 47 (Pb): Hb/C3−972
                    PbSe−CdSe . . . . 47 (Pb): Hb/C4−1256
                    PbSe−Ce$_2$Se$_3$ . . . . 47 (Pb): Hb/C4−1374
                    PbSe−Dy$_2$Se$_3$ . . . . 47 (Pb): Hb/C4−1374/5
                    PbSe−Er$_2$Se$_3$ . . . . 47 (Pb): Hb/C4−1374/5
                    PbSe−Gd$_2$Se$_3$ . . . . 47 (Pb): Hb/C4−1374/5
                    PbSe−Ho$_2$Se$_3$ . . . . 47 (Pb): Hb/C4−1374/5
                    PbSe−La$_2$Se$_3$ . . . . 47 (Pb): Hb/C4−1374
                    PbSe−Lu$_2$Se$_3$ . . . . 47 (Pb): Hb/C4−1374/5
                    PbSe−Nd$_2$Se$_3$ . . . . 47 (Pb): Hb/C4−1374
                    PbSe−PbS . . . . 47 (Pb): Hb/C2−639/40
                    PbSe−Pr$_2$Se$_3$ . . . . 47 (Pb): Hb/C4−1374
                    PbSe−Sm$_2$Se$_3$. . . . . 47 (Pb): Hb/C4−1374
                    PbSe−SnTe . . . . 47 (Pb): Hb/C4−1595
                    PbSe−Tb$_2$Se$_3$ . . . . 47 (Pb): Hb/C4−1374/5
                    PbSe−Tm$_2$Se$_3$. . . . . 47 (Pb): Hb/C4−1374/5
                    PbSe−Yb$_2$Se$_3$ . . . . 47 (Pb): Hb/C4−1374/5

PbSe . . . . PbSe systems:
                    PbSe–AgBiSe$_2$ . . . . 61 (Ag):   Hb/B4–268/9
                    PbSe–AgSbSe$_2$ . . . . 61 (Ag):   Hb/B4–268
                    PbSe–AgSbTe$_2$ . . . . 61 (Ag):   Hb/B4–273
                    PbSe–Ag$_2$Se . . . . 61 (Ag):   Hb/B4–268
                    PbSe–Bi$_2$Se$_3$ . . . . 47 (Pb):   Hb/C3–971/2
                    PbSe–CdSe . . . . 47 (Pb):   Hb/C4–1255/6
                    PbSe–GeS . . . . 47 (Pb):   Hb/C4–1479
                    PbSe–GeSe . . . . 47 (Pb):   Hb/C4–1477/8
                    PbSe–GeTe . . . . 47 (Pb):   Hb/C4–1480
                    PbSe–InSe . . . . 47 (Pb):   Hb/C4–1326/7
                    PbSe–Pb–PbTe . . . . 47 (Pb):   Hb/C2–681
                    PbSe–PbO . . . . 47 (Pb):   Hb/C2–632
                    PbSe–PbS–PbTe . . . . 47 (Pb):   Hb/C2–684
                    PbSe–Sb$_2$Se$_3$ . . . . 47 (Pb):   Hb/C3–940
                    PbSe–SnS . . . . 47 (Pb):   Hb/C4–1587/8
                    PbSe–SnSe . . . . 47 (Pb):   Hb/C4–1586/7
PbSe$_2$ . . . . PbSe$_2$ . . . . 47 (Pb):   Hb/C2–632
PbSe$_4$Sm$_2$ . . . . . . PbSm$_2$Se$_4$ . . . . 47 (Pb):   Hb/C4–1375
PbSe$_4$Tm$_2$ . . . . . . PbTm$_2$Se$_4$ . . . . 47 (Pb):   Hb/C4–1375
PbSe$_4$Y$_2$ . . . . . . PbY$_2$Se$_4$ . . . . 47 (Pb):   Hb/C4–1375
PbSe$_4$Yb$_2$ . . . . . . PbYb$_2$Se$_4$ . . . . 47 (Pb):   Hb/C4–1375
PbSr . . . . SrPb . . . . 47 (Pb):   Hb/C3–1194
PbSr$_2$ . . . . Sr$_2$Pb . . . . 47 (Pb):   Hb/C3–1193/4
PbTe . . . . PbTe:
                    Chemical reactions . . . . 47 (Pb):   Hb/C2–677/8
                    Crystallographic properties . . . . . 47 (Pb):   Hb/C2–651/4
                    Electrical properties . . . . 47 (Pb):   Hb/C2–658/74
                    Electrochemical behavior . . . . . . 47 (Pb):   Hb/C2–677
                    Formation and preparation . . . . . 47 (Pb):   Hb/C2–648/51
                    Magnetic properties . . . . 47 (Pb):   Hb/C2–663, 669/70,
                                                                                                         674
                    Mechanical properties . . . . 47 (Pb):   Hb/C2–654/5
                    Molecule . . . . 47 (Pb):   Hb/C2–651
                    Optical properties . . . . 47 (Pb):   Hb/C2–675/7
                    Spectra . . . . 47 (Pb):   Hb/C2–651
                    Thermal properties . . . . 47 (Pb):   Hb/C2–654/8
–. . . . . PbTe solid solutions:
                    PbTe–AgSbTe$_2$–GeTe . . . . 61 (Ag):   Hb/B4–273
                    PbTe–AgSbTe$_2$–SnTe . . . . 61 (Ag):   Hb/B4–272
                    PbTe–BiTe . . . . 47 (Pb):   Hb/C3–475
                    PbTe–Bi$_2$Te$_3$ . . . . 47 (Pb):   Hb/C3–975
                    PbTe–CdTe . . . . 47 (Pb):   Hb/C4–1256
                    PbTe–EuTe . . . . 47 (Pb):   Hb/C4–1375
                    PbTe–PbS–SnTe . . . . 47 (Pb):   Hb/C4–1594
                    PbTe–SnSe . . . . 47 (Pb):   Hb/C4–1595
                    PbTe–SnSe–SnTe . . . . 47 (Pb):   Hb/C4–1595
                    PbTe–SnTe . . . . 47 (Pb):   Hb/C4–1589/94

See the "Instructions for the user"
at the beginning of the volume

PbTe . . . . PbTe systems:
                     PbTe–AgBiSe$_2$ . . . . 61 (Ag):   Hb/B4–273
                     PbTe–AgBiTe$_2$ . . . . 61 (Ag):   Hb/B4–270/2
                     PbTe–AgSbSe$_2$ . . . . 61 (Ag):   Hb/B4–273
                     PbTe–AgSbTe$_2$ . . . . 61 (Ag):   Hb/B4–269/70
                     PbTe–Ag$_2$Te–Sb$_2$Te$_3$ . . . . 61 (Ag):   Hb/B4–269/70
                     PbTe–Bi$_2$Te$_3$ . . . . 47 (Pb):   Hb/C3–973/5
                     PbTe–CdTe . . . . 47 (Pb):   Hb/C4–1256
                     PbTe–GeS. . . . . 47 (Pb):   Hb/C4–1480
                     PbTe–GeSe . . . . 47 (Pb):   Hb/C4–1480
                     PbTe–GeTe . . . . 47 (Pb):   Hb/C4–1479/80
                     PbTe–InTe . . . . 47 (Pb):   Hb/C4–1327
                     PbTe–Pb–PbSe . . . . 47 (Pb):   Hb/C2–681
                     PbTe–PbS . . . . 47 (Pb):   Hb/C2–679/80
                     PbTe–PbS–PbSe . . . . 47 (Pb):   Hb/C2–684
                     PbTe–SnSe–SnTe . . . . 47 (Pb):   Hb/C4–1594/5
                     PbTe–Tl$_2$Te . . . . 47 (Pb):   Hb/C4–1353/4

PbTe$_3$Tl$_4$ . . . . Tl$_4$PbTe$_3$ . . . . 47 (Pb):   Hb/C4–1354
PbTh . . . . ThPb . . . . 47 (Pb):   Hb/C4–1468, 1470
PbTi$_2$ . . . . Ti$_2$Pb . . . . 47 (Pb):   Hb/C4–1377
PbTi$_4$ . . . . Ti$_4$Pb . . . . 47 (Pb):   Hb/C4–1378
–. . . . . Ti$_4$Pb solid solutions:
                     (Zr,Ti)$_4$Pb . . . . 47 (Pb):   Hb/C4–1423

PbTl$_2$ . . . . PbTl$_2$ . . . . 47 (Pb):   Hb/C4–1330, 1347
PbV$_3$ . . . . V$_3$Pb . . . . 48 (V):   Hb/B2–687
PbYb . . . . YbPb . . . . 47 (Pb):   Hb/C4–1372
PbYb$_2$. . . . . Yb$_2$Pb. . . . . 47 (Pb):   Hb/C4–1372
PbZr$_4$ . . . . Zr$_4$Pb solid solutions:
                     (Zr,Ti)$_4$Pb . . . . 47 (Pb):   Hb/C4–1423

PbZr$_5$ . . . . Zr$_5$Pb . . . . 47 (Pb):   Hb/C4–1420/1

Pb$_2$ . . . . Pb$_2$ . . . . 47 (Pb):   Hb/B1–138/9
Pb$_2$$^+$ . . . . Pb$_2$$^+$ . . . . 47 (Pb):   Hb/B1–237
Pb$_2$Pd . . . . PdPb$_2$ . . . . 68 (Pt):   Hb/A5–589/93
Pb$_2$Pd$_3$ . . . . Pd$_3$Pb$_2$ . . . . 68 (Pt):   Hb/A5–589/93
Pb$_2$Pt . . . . PtPb$_2$ . . . . 68 (Pt):   Hb/A6–764, 767
Pb$_2$Pt$_3$S$_6$ . . . . 2 PbS · 2 PtS · PtS$_2$ = Pb$_2$Pt$_2$[PtS$_6$] . . . . 68 (Pt):   Hb/C3–315
Pb$_2$Rh . . . . RhPb$_2$. . . . . 68 (Pt):   Hb/A5–557
Pb$_2$S . . . . Pb$_2$S . . . . 47 (Pb):   Hb/C2–414
Pb$_2$S$_3$ . . . . Pb$_2$S$_3$ . . . . 47 (Pb):   Hb/C2–414
Pb$_2$S$_4$Si . . . . . . Pb$_2$SiS$_4$. . . . . 47 (Pb):   Hb/C3–855/6
Pb$_2$S$_5$Sb$_2$ . . . . . . Pb$_2$Sb$_2$S$_5$ = 2 PbS · Sb$_2$S$_3$ . . . . 47 (Pb):   Hb/C3–938
Pb$_2$S$_{11}$Sb$_6$ . . . . . Pb$_2$Sb$_6$S$_{11}$ = 2 PbS · 3 Sb$_2$S$_3$ . . . . 47 (Pb):   Hb/C3–939
Pb$_2$Te$_3$ . . . . Pb$_2$Te$_3$ · 4 H$_2$O . . . . 47 (Pb):   Hb/C2–678
Pb$_2$Th . . . . ThPb$_2$ . . . . 47 (Pb):   Hb/C4–1468
Pb$_2$Y . . . . YPb$_2$ . . . . 47 (Pb):   Hb/C4–1360
Pb$_3$Pr . . . . PrPb$_3$ . . . . 47 (Pb):   Hb/C4–1360, 1366
Pb$_3$Pr$_5$. . . . . Pr$_5$Pb$_3$. . . . . 47 (Pb):   Hb/C4–1359, 1365/6

Hb = Main Series Volume; GD = Gmelin-Durrer;   *Ab = Appended Volume to Main Series;*
Eb = Supplement Volume to Main Series;   *Erg. W. = New Supplement Series Volume*

$Pb_3S_6Sb_2$ . . . . . . $Pb_3(SbS_3)_2 = 3\,PbS \cdot Sb_2S_3$ . . . . 47 (Pb):   Hb/C3–937

$Pb_3S_7Si_2$ . . . . . . $Pb_3Si_2S_7$ . . . . 47 (Pb):   Hb/C3–855/6

$Pb_3S_8Sb_2$ . . . . . . $Pb_3(SbS_4)_2 = 3\,PbS \cdot Sb_2S_5$ . . . . 47 (Pb):   Hb/C3–939

$Pb_3S_9Sb_4$ . . . . . . $Pb_3Sb_4S_9 = 3\,PbS \cdot 2\,Sb_2S_3$ . . . . . 47 (Pb):   Hb/C3–938

$Pb_3S_{14}Sb_2Sn_4$ . . . . $Pb_3Sn_4Sb_2S_{14}$ . . . . 47 (Pb):   Hb/C4–1595

Pb_3Sc . . . . $ScPb_3$ . . . . 47 (Pb):   Hb/C4–1360

Pb_3Sc_5 . . . . Sc_5Pb_3 . . . . 47 (Pb):   Hb/C4–1359/60

Pb_3Se . . . . Pb_3Se . . . . 47 (Pb):   Hb/C2–632

Pb_3Sm . . . . $SmPb_3$ . . . . 47 (Pb):   Hb/C4–1360, 1367

Pb_3Sm_5 . . . . Sm_5Pb_3 . . . . 47 (Pb):   Hb/C4–1359, 1367

Pb_3Sr . . . . $SrPb_3$ . . . . 47 (Pb):   Hb/C3–1194

Pb_3Tb . . . . $TbPb_3$ . . . . 47 (Pb):   Hb/C4–1360, 1369

Pb_3Tb_5 . . . . Tb_5Pb_3 . . . . 47 (Pb):   Hb/C4–1359, 1369

Pb_3Te . . . . Pb_3Te . . . . 47 (Pb):   Hb/C2–678

Pb_3Th . . . . $ThPb_3$ . . . . 47 (Pb):   Hb/C4–1468, 1470

Pb_3Ti_5 . . . . Ti_5Pb_3 . . . . 47 (Pb):   Hb/C4–1377/8

–. . . . . Ti_5Pb_3 solid solutions:

                        $(Zr,Ti)_5Pb_3$ . . . . 47 (Pb):   Hb/C4–1423

Pb_3Tm . . . . $TmPb_3$ . . . . 47 (Pb):   Hb/C4–1360, 1371

Pb_3Tm_5 . . . . Tm_5Pb_3 . . . . 47 (Pb):   Hb/C4–1359, 1371

Pb_3Y . . . . YPb_3 . . . . 47 (Pb):   Hb/C4–1360

Pb_3Y_5 . . . . Y_5Pb_3 . . . . 47 (Pb):   Hb/C4–1359, 1360/1

Pb_3Yb . . . . $YbPb_3$ . . . . 47 (Pb):   Hb/C4–1360, 1372

Pb_3Yb_5 . . . . Yb_5Pb_3 . . . . 47 (Pb):   Hb/C4–1359, 1372

Pb_3Zr_5 . . . . Zr_5Pb_3 . . . . 47 (Pb):   Hb/C4–1420/1

–. . . . . Zr_5Pb_3 solid solutions:

                        $(Zr,Ti)_5Pb_3$ . . . . 47 (Pb):   Hb/C4–1423

Pb_4Pt . . . . $PtPb_4$ . . . . 68 (Pt):   Hb/A6–764, 767

Pb_4S . . . . Pb_4S . . . . 47 (Pb):   Hb/C2–414

$Pb_4S_7Sb_2$ . . . . . . $Pb_4Sb_2S_7 = 4\,PbS \cdot Sb_2S_3$ . . . . . . 47 (Pb):   Hb/C3–937

Pb_4Th . . . . $ThPb_4$ . . . . 47 (Pb):   Hb/C4–1468

Pb_4Y_5 . . . . Y_5Pb_4 . . . . 47 (Pb):   Hb/C4–1360

$Pb_5S_8Sb_2$ . . . . . $Pb_5(SbS_4)_2 = 5\,PbS \cdot Sb_2S_3$ . . . . . . 47 (Pb):   Hb/C3–937

$Pb_5S_{11}Sb_4$ . . . . . $Pb_5Sb_4S_{11} = 5\,PbS \cdot 2\,Sb_2S_3$ . . . . 47 (Pb):   Hb/C3–937/8

$Pb_5S_{14}Sb_3Sn_3$ . . . . $Pb_5Sn_3Sb_3S_{14}$ . . . . 47 (Pb):   Hb/C4–1595

$Pb_5S_{14}Sb_6$ . . . . . . $Pb_5Sb_6S_{14} = 5\,PbS \cdot 3\,Sb_2S_3$ . . . . . 47 (Pb):   Hb/C3–938

$Pb_5S_{17}Sb_8$ . . . . . . $Pb_5Sb_8S_{17} = 5\,PbS \cdot 4\,Sb_2S_3$ . . . . . 47 (Pb):   Hb/C3–938

$Pb_6S_9Sb_2$ . . . . . . $Pb_6Sb_2S_9 = 6\,PbS \cdot Sb_2S_3$ . . . . 47 (Pb):   Hb/C3–937

$Pb_7S_{22}Sb_{10}$ . . . . . . $Pb_7Sb_{10}S_{22} = 7\,PbS \cdot 5\,Sb_2S_3$. . . . . 47 (Pb):   Hb/C3–938

$Pb_7S_{25}Sb_{12}$ . . . . . . $Pb_7Sb_{12}S_{25} = 7\,PbS \cdot 6\,Sb_2S_3$ . . . . . 47 (Pb):   Hb/C3–939

$Pb_9S_{21}Sb_8$ . . . . . . $Pb_9Sb_8S_{21} = 9\,PbS \cdot 4\,Sb_2S_3$ . . . . . 47 (Pb):   Hb/C3–938

$Pb_9S_{30}Sb_{14}$ . . . . . . $Pb_9Sb_{14}S_{30} = 9\,PbS \cdot 7\,Sb_2S_3$ . . . . . 47 (Pb):   Hb/C3–938

Pd . . . . Palladium:

                        Absorption of light . . . . 65 (Pd):   Hb/1–50

                        – of X-rays. . . . . 65 (Pd):   Hb/1–62/3

                        Abundance . . . . 68 (Pt):   Hb/A1–24/7

                        Atom. . . . . 65 (Pd):   Hb/1–7/13

*See the "Instructions for the user"
at the beginning of the volume*

Pd . . . . Palladium:

Atomic nucleus . . . . 65 (Pd):   Hb/1–1/7
Atomic radius . . . . 65 (Pd):   Hb/1–8/9
Atomic volume . . . . 65 (Pd):   Hb/1–11/2
Atomic weight . . . . 65 (Pd):   Hb/1–12/3
Behavior as anode . . . . 65 (Pd):   Hb/1–95/9, 105/8
– as cathode . . . . 65 (Pd):   Hb/1–94/5, 99/105
– at alternating current
    electrolysis . . . . 65 (Pd):   Hb/1–99
Catalytic properties . . . . 68 (Pt):   Hb/A3–429/30
Cells . . . . 65 (Pd):   Hb/1–89/94
Chemical constant . . . . 65 (Pd):   Hb/1–48
Chemical reactions . . . . 65 (Pd):   Hb/1–109/14
Colloidal forms . . . . 68 (Pt):   Hb/A3–425/7, 429/30
Color . . . . 65 (Pd):   Hb/1–49
Compressibility . . . . 65 (Pd):   Hb/1–35/6
Contact electricity . . . . 65 (Pd):   Hb/1–77
Cosmochemistry . . . . 68 (Pt):   Hb/A1–15/22
Critical constants . . . . 65 (Pd):   Hb/1–39
Crystallographic properties . . . . . . 65 (Pd):   Hb/1–14/25
Debye temperature . . . . 65 (Pd):   Hb/1–47
Deformation . . . . 65 (Pd):   Hb/1–22/3
Density . . . . 65 (Pd):   Hb/1–25/6
Detection . . . . 68 (Pt):   Hb/A4–432, 439/46,
                                                          489/90
– in iron . . . . 59 (Fe):   Hb/FII1–41
Determination . . . . 68 (Pt):   Hb/A4–467/71,
                                                          489/90
– in special substances . . . . 68 (Pt):   Hb/A4–525/32
Diffusion . . . . 65 (Pd):   Hb/1–37/9
– of hydrogen . . . . 65 (Pd):   Hb/1–116/33
Discharge phenomena . . . . 65 (Pd):   Hb/1–81/2
Economic deposits
    see Pt ... Platinum

Economics . . . . 68 (Pt):   Hb/A2–302/10
Elasticity . . . . 65 (Pd):   Hb/1–31/3
Electrical conductivity . . . . 65 (Pd):   Hb/1–69/77
Electrical properties . . . . 65 (Pd):   Hb/1–69/86
Electrochemical behavior . . . . 65 (Pd):   Hb/1–87/109
Electrode sputtering . . . . 65 (Pd):   Hb/1–82/3
Electrolytic deposition . . . . 68 (Pt):   Hb/A3–390
Electron configuration . . . . 65 (Pd):   Hb/1–7/8
Electron emission . . . . 65 (Pd):   Hb/1–83/5
Equation of state . . . . 65 (Pd):   Hb/1–39
Ettingshausen effect . . . . 65 (Pd):   Hb/1–67/8
Faraday effect . . . . 65 (Pd):   Hb/1–51
Geochemical character . . . . 68 (Pt):   Hb/A1–23/8
Geochemistry . . . . 68 (Pt):   Hb/A1–22/49
                                                          Hb/A2–275/6

Hb = Main Series Volume; GD = Gmelin-Durrer;   *Ab = Appended Volume to Main Series;*
Eb = Supplement Volume to Main Series;   *Erg. W. = New Supplement Series Volume*

Pd . . . . Palladium:

Hall effect	65 (Pd):	Hb/1−67
Hardness	65 (Pd):	Hb/1−29/31
Heat of transformation	65 (Pd):	Hb/1−42/4
History	68 (Pt):	Hb/A1−9/11
Interfacial tension	65 (Pd):	Hb/1−37
Internal friction	65 (Pd):	Hb/1−37
Internal pressure	65 (Pd):	Hb/1−36
Ion emission	65 (Pd):	Hb/1−85/6
Ionization potentials	65 (Pd):	Hb/1−86
Lattice structure	65 (Pd):	Hb/1−15/20
Leduc effect	65 (Pd):	Hb/1−68/9
Magnetic moment	65 (Pd):	Hb/1−9/11
Magnetic properties	65 (Pd):	Hb/1−63/9
Mechanical properties	65 (Pd):	Hb/1−25/39
Mineralogy	68 (Pt):	Hb/A2−276/301
Nernst effect	65 (Pd):	Hb/1−68
Occurrence, extraterrestrial	68 (Pt):	Hb/A1−15/22
− terrestrial	68 (Pt):	Hb/A1−22/49
		Hb/A2−275/6
− in the biosphere	68 (Pt):	Hb/A2−275
− in the hydrosphere	68 (Pt):	Hb/A2−275
− in the lithosphere	68 (Pt):	Hb/A1−28/49
Optical properties	65 (Pd):	Hb/1−49/63
Overvoltage phenomena	65 (Pd):	Hb/1−99/109
Passage of elementary particles	65 (Pd):	Hb/1−85/6
Physiological action	68 (Pt):	Hb/A2−275/6
Polymorphism	65 (Pd):	Hb/1−14/5
Position in the electrochemical		
series	65 (Pd):	Hb/1−87
− in the periodic system	65 (Pd):	Hb/1−14
Potentials	65 (Pd):	Hb/1−87/9
Preparation	68 (Pt):	Hb/A3−311/430
− from minerals	68 (Pt):	Hb/A3−312/50
− from wastes and residues	68 (Pt):	Hb/A3−350/62
− in pure state	68 (Pt):	Hb/A3−374/5
− of dense palladium	68 (Pt):	Hb/A3−384/5
Purity testing	68 (Pt):	Hb/A4−519/20
Recrystallization	65 (Pd):	Hb/1−24/5
Reflection	65 (Pd):	Hb/1−49/50
Refraction	65 (Pd):	Hb/1−50
Scattering of X-rays	65 (Pd):	Hb/1−62/3
Separation, analytical	68 (Pt):	Hb/A4−497/500,
		517/8
Sorption	65 (Pd):	Hb/1−39
− of hydrogen	65 (Pd):	Hb/2−133/256
− of nitrogen	65 (Pd):	Hb/2−268
− of oxygen	65 (Pd):	Hb/2−257/9
− of rare gases	65 (Pd):	Hb/2−115/6

See the "Instructions for the user"
at the beginning of the volume

Pd . . . . Palladium:
              Special forms . . . . 68 (Pt):   Hb/A3−399/404,
                                                      425/7, 429/30
              Specific heat . . . . 65 (Pd):   Hb/1−44/6
              Spectra:
                  Arc spectrum . . . . 65 (Pd):   Hb/1−51/8
                  Spark spectra . . . . 65 (Pd):   Hb/1−51/8
                  Terms . . . . 65 (Pd):   Hb/1−54/6
                  X-ray spectrum . . . . 65 (Pd):   Hb/1−58/62
              Standard potential . . . . 65 (Pd):   Hb/1−87
              Strength . . . . 65 (Pd):   Hb/1−33/5
              Structure . . . . 65 (Pd):   Hb/1−23/4
              − of thin layers . . . . 65 (Pd):   Hb/1−20/2
              Surface tension . . . . 65 (Pd):   Hb/1−37
              Susceptibility . . . . 65 (Pd):   Hb/1−64/7
              Thermal conductivity . . . . 65 (Pd):   Hb/1−48/9
              Thermal expansion . . . . 65 (Pd):   Hb/1−26/9
              Thermal properties . . . . 65 (Pd):   Hb/1−39/49
              Thermal radiation . . . . 65 (Pd):   Hb/1−51
              Thermodynamic functions . . . . . . 65 (Pd):   Hb/1−47/8
              Thermoelectric effects . . . . 65 (Pd):   Hb/1−78/81
              Toxicity . . . . 68 (Pt):   Hb/A2−275/6
              Transformation points . . . . 65 (Pd):   Hb/1−40/2
              Triboelectricity . . . . 65 (Pd):   Hb/1−77
              Uses . . . . 68 (Pt):   Hb/A3−385/91
              Valence . . . . 65 (Pd):   Hb/1−13/4
              Vapor pressure . . . . 65 (Pd):   Hb/1−39/40
              Vibration frequency . . . . 65 (Pd):   Hb/1−46/7

−. . . . . Palladium ions:
              Diffusion . . . . 21 (Na):   Eb/6−293
              Ionic refraction . . . . 65 (Pd):   Hb/1−50
              Ionization potentials . . . . 65 (Pd):   Hb/1−86
              Properties of the ion . . . . 65 (Pd):   Hb/1−9/10, 12
              Spectra . . . . 65 (Pd):   Hb/1−56

−. . . . . Palladium isotopes . . . . 65 (Pd):   Hb/1−1/2
−. . . . . Palladium systems:
              Pd−Ag . . . . 68 (Pt):   Hb/A5−653
              Pd−Ag−Au . . . . 68 (Pt):   Hb/A5−710
              Pd−Ag−Cu . . . . 68 (Pt):   Hb/A5−686/8
              Pd−Ag−H . . . . 68 (Pt):   Hb/A5−666/78
              Pd−Au . . . . 68 (Pt):   Hb/A5−691
              Pd−Au−Cu . . . . 68 (Pt):   Hb/A5−709/10
              Pd−Au−H . . . . 68 (Pt):   Hb/A5−701/9
              Pd−Au−Ni . . . . 68 (Pt):   Hb/A5−709
              Pd−Au−Pt . . . . 68 (Pt):   Hb/A6−840/1
              Pd−Be . . . . 68 (Pt):   Hb/A5−579/80
              Pd−Bi . . . . 68 (Pt):   Hb/A5−575/8
              Pd−Cd . . . . 68 (Pt):   Hb/A5−583

Hb = Main Series Volume; GD = Gmelin-Durrer;   *Ab = Appended Volume to Main Series;*
Eb = Supplement Volume to Main Series;   *Erg. W. = New Supplement Series Volume*

Pd . . . . Palladium systems:

	Pd–Co	68 (Pt):	Hb/A5–613/5
	Pd–Cr	68 (Pt):	Hb/A5–594/6
	Pd–Cu	68 (Pt):	Hb/A5–624/31
	Pd–Fe	68 (Pt):	Hb/A5–617/9
	Pd–Fe–S	68 (Pt):	Hb/A5–624
	Pd–H	65 (Pd):	Hb/2–133/216
	Pd–D	65 (Pd):	Hb/2–209/16
	Pd–H–Pt	68 (Pt):	Hb/A6–839/40
	Pd–In	68 (Pt):	Hb/A5–584/5
	Pd–Mn	68 (Pt):	Hb/A5–598/607
	Pd–Ni	68 (Pt):	Hb/A5–607/8
	Pd–O–S	9 (S):	Hb/B1–185
	Pd–P	65 (Pd):	Hb/2–302/5
	Pd–Pb	68 (Pt):	Hb/A5–588/93
	Pd–PdS	65 (Pd):	Hb/2–289/91
	Pd–Pt	68 (Pt):	Hb/A6–836
	Pd–Ru	68 (Pt):	Hb/A5–715/6
	Pd–S	65 (Pd):	Hb/2–289
	Pd–Sb	68 (Pt):	Hb/A5–570/5
	Pd–Si	65 (Pd):	Hb/2–302
	Pd–Sn	68 (Pt):	Hb/A5–586/8
	Pd–Ti	68 (Pt):	Hb/A5–585
Pd^+	Pd^+	65 (Pd):	Hb/1–9, 56, 86
Pd^{2+}	Pd^{2+}	21 (Na):	Eb/6–293
		65 (Pd):	Hb/1–9/10, 12, 50
Pd^{3+}	Pd^{3+}	65 (Pd):	Hb/1–9
Pd^{4+}	Pd^{4+}	65 (Pd):	Hb/1–9/10
PdS	PdS	65 (Pd):	Hb/2–292/4
–.	PdS systems:		
	PdS–Pd	65 (Pd):	Hb/2–289/91
PdS_2	PdS_2	65 (Pd):	Hb/2–294/5
PdSb	PdSb	68 (Pt):	Hb/A5–574
$PdSb_2$.	$PdSb_2$.	68 (Pt):	Hb/A5–574/5
PdSe	PdSe	65 (Pd):	Hb/2–296
$PdSe_2$	$PdSe_2$.	65 (Pd):	Hb/2–296
PdSi.	PdSi	65 (Pd):	Hb/2–302
PdSn	PdSn	68 (Pt):	Hb/A5–587/8
$PdSn_2$.	$PdSn_2$.	68 (Pt):	Hb/A5–587/8
$PdSn_4$.	$PdSn_4$.	68 (Pt):	Hb/A5–587/8
PdTe	PdTe	65 (Pd):	Hb/2–297
$PdTe_2$	$PdTe_2$.	65 (Pd):	Hb/2–297/8
PdTi.	PdTi.	68 (Pt):	Hb/A5–585
$PdTi_2$	$PdTi_2$	68 (Pt):	Hb/A5–585
Pd_2S	Pd_2S	65 (Pd):	Hb/2–292
Pd_2Si	Pd_2Si	65 (Pd):	Hb/2–302
Pd_2Sn.	Pd_2Sn.	68 (Pt):	Hb/A5–587/8
Pd_2Tl	Pd_2Tl	68 (Pt):	Hb/A5–585

See the "Instructions for the user"
at the beginning of the volume

Pd_3Pu	$PuPd_3$	Erg. W. 4–209	
Pd_3S	Pd_3S	: 65 (Pd):	Hb/2–292
Pd_3S_4	$Pd_3S_4 = 2\ PdS \cdot PdS_2$	65 (Pd):	Hb/2–294
Pd_3Sb	Pd_3Sb	68 (Pt):	Hb/A5–573/4
Pd_3Sn	Pd_3Sn	68 (Pt):	Hb/A5–587/8
Pd_3Sn_2	Pd_3Sn_2	68 (Pt):	Hb/A5–587/8
Pd_3Ti	Pd_3Ti	68 (Pt):	Hb/A5–585
Pd_3Ti_2	Pd_3Ti_2	68 (Pt):	Hb/A5–585
Pd_3Zr	Pd_3Zr	68 (Pt):	Hb/A5–585
Pd_4S	Pd_4S	65 (Pd):	Hb/2–292
Pd_4Se	Pd_4Se	65 (Pd):	Hb/2–296
Pd_5Sb_3	Pd_5Sb_3	68 (Pt):	Hb/A5–574
Pd_5Zn_{21}	Pd_5Zn_{21}	68 (Pt):	Hb/A5–582

Pm	Promethium:		
	Abundance	39 (SE):	Hb/A1–34
	Crystallographic properties	39 (SE):	Hb/B3–69, 77/8
	Diffusion.	47 (Pb):	Hb/C4–1364
	Electrical properties	39 (SE):	Hb/B3–236, 282/3
	History	39 (SE):	Hb/A1–18
	Magnetic properties	39 (SE):	Hb/B3–294
	Mechanical properties	39 (SE):	Hb/B3–77/8, 208/9
	Occurrence	39 (SE):	Hb/A1–34/42
	Position in the periodic system	39 (SE):	Hb/A1–2/4
	Thermal properties	39 (SE):	Hb/B3– 84/5, 103/12, 118/20, 139/40, 187, 198/9
–.	Promethium ions	39 (SE):	Hb/B3–284, 294
–.	Promethium isotopes	Erg. W. 8–207	
–.	Promethium systems:		
	Pm–N–O	39 (SE):	Hb/C2–180
Pm^{3+}	Pm^{3+}	39 (SE):	Hb/B3–284, 294

Po	Polonium:		
	Atom	12 (Po):	Hb–32/7, 60/1, 145, 174
	Atomic nucleus	12 (Po):	Hb–29/32, 141/5, 153/4, 158/64, 176/7, 185
	Chemical reactions	12 (Po):	Hb–93/9, 147/8
	Crystallographic properties	12 (Po):	Hb–61/2
	Detection	12 (Po):	Hb–124/6
	Determination	12 (Po):	Hb–126/30
	Electrical properties	12 (Po):	Hb–69, 146/7, 152, 156, 183
	Electrochemical behavior	12 (Po):	Hb–69/93, 147
	Electrochemical deposition	12 (Po):	Hb–72/93, 147
	Enrichment	12 (Po):	Hb–7/18, 140/1

Hb = Main Series Volume; GD = Gmelin-Durrer;   *Ab = Appended Volume to Main Series;*
Eb = Supplement Volume to Main Series;   *Erg. W. = New Supplement Series Volume*

Po . . . . Polonium:

Formation . . . . 12 (Po):   Hb–6, 29, 139, 149,
                                                     153, 157/8,
                                                     175, 184

History . . . . 12 (Po):   Hb–1/5, 138, 148,
                                                     152/3, 157,
                                                     175, 184

Mechanical properties . . . . 12 (Po):   Hb–62/4
Molecule . . . . 12 (Po):   Hb–60
Nuclear decay,
   alpha radiation . . . . 12 (Po):   Hb–38/56, 143/5,
                                                     150/2, 154/6,
                                                     164/73, 177/87
                                                 Erg. W. 7a–85/95
–, beta radiation . . . . 12 (Po):   Hb–56, 183
–, decay scheme . . . . 12 (Po):   Hb–139
–, gamma radiation . . . . 12 (Po):   Hb–56, 183
Occurrence . . . . 12 (Po):   Hb–5/6, 138/40,
                                                     149, 153, 157,
                                                     175, 184

Optical properties . . . . 12 (Po):   Hb–66/8
Position in the periodic system . . . . 12 (Po):   Hb–37/8
Potentials . . . . 12 (Po):   Hb–69/80
Preparation . . . . 12 (Po):   Hb–6/29, 140/1,
                                                     149, 153

– in pure state . . . . 12 (Po):   Hb–19/26
– of special forms . . . . 12 (Po):   Hb–27/8
Separation, analytical . . . . 12 (Po):   Hb–130
Spectra . . . . 12 (Po):   Hb–67/8
Thermal properties . . . . 12 (Po):   Hb–64/6, 146
Toxicity . . . . 12 (Po):   Hb–123/4
Uses . . . . 12 (Po):   Hb–28/9
                                                 Erg. W. 8–207, 224
Valence . . . . 12 (Po):   Hb–34/7

–. . . . . Polonium ions:
Chemical reactions . . . . 12 (Po):   Hb–99/123
Colloidchemical behavior . . . . 12 (Po):   Hb–106/9
Electrochemical behavior . . . . 12 (Po):   Hb–72/93
Properties of the ion . . . . 12 (Po):   Hb–33/4
Sorption . . . . 12 (Po):   Hb–108/9

–. . . . . Polonium isotopes
                          see Po ... Polonium

–. . . . . Polonium systems:
Po–Bi . . . . 19 (Bi):   Eb–828/9
Po–Hg . . . . 34 (Hg):   Hb/B3–1180
Po–Ni . . . . 57 (Ni):   Hb/B2–764
Po–Pb . . . . 47 (Pb):   Hb/C2–685

See the "Instructions for the user"
at the beginning of the volume

Po^{n+} . . . .   Po^{n+} (n = 1 to 6)
                                see Po ... Polonium ions

PoS . . . .   PoS . . . . 12 (Po):   Hb−135
PoSe . . . .   PoSe . . . . 12 (Po):   Hb−135
PoTe . . . .   PoTe . . . . 12 (Po):   Hb−135
Po$_2$ . . . .   Po$_2$ . . . . 12 (Po):   Hb−60
Po$_2$S$_3$ . . . .   Po$_2$S$_3$ . . . . 12 (Po):   Hb−135

Pr . . . .   Praseodymium:
                                Abundance . . . . 39 (SE):   Hb/A1−34/5
                                Band model . . . . 39 (SE):   Hb/B3−211/5
                                Boiling point . . . . 39 (SE):   Hb/B3−105
                                Characteristic electron energy
                                  losses . . . . 39 (SE):   Hb/B3−225/6
                                Charge carrier concentration. . . . . 39 (SE):   Hb/B3−228
                                Cosmochemistry . . . . 39 (SE):   Hb/A1−23/7
                                Crystal chemistry . . . . 39 (SE):   Hb/A1−28/32
                                Crystal field parameters . . . . 39 (SE):   Hb/B3−219
                                Crystallographic properties . . . . . . 39 (SE):   Hb/B3−60/1, 66/9,
                                                                                    76/8
                                Crystal structure . . . . 39 (SE):   Hb/B3−66/9
                                Debye temperature . . . . 39 (SE):   Hb/B3−185/6
                                Density . . . . 39 (SE):   Hb/B3−85/7
                                Diffusion. . . . . 39 (SE):   Hb/B3−77/8
                                Disorder . . . . 39 (SE):   Hb/B3−76
                                Economic deposits . . . . 39 (SE):   Hb/A1−110/22
                                Elasticity. . . . . 39 (SE):   Hb/B3−92/8
                                Electrical properties . . . . 39 (SE):   Hb/B3−209/83
                                Electrical resistance. . . . . 39 (SE):   Hb/B3−232/4, 258
                                Electron density. . . . . 39 (SE):   Hb/B3−216
                                Electron emission . . . . 39 (SE):   Hb/B3−278/83
                                Enthalpy of fusion . . . . 39 (SE):   Hb/B3−120
                                − of sublimation. . . . . 39 (SE):   Hb/B3−110/1
                                − of transformation . . . . 39 (SE):   Hb/B3−118
                                − of vaporization . . . . 39 (SE):   Hb/B3−110/1
                                Equation of state . . . . 39 (SE):   Hb/B3−83/4
                                Geochemistry . . . . 39 (SE):   Hb/A1−28/54
                                Grüneisen constant. . . . . 39 (SE):   Hb/B3−84/5
                                Hall effect . . . . 39 (SE):   Hb/B3−265
                                Hardness . . . . 39 (SE):   Hb/B3−103
                                History. . . . . 39 (SE):   Hb/A1−4/22
                                Hyperfine interactions . . . . 39 (SE):   Hb/B3−219/20
                                Isomer shift . . . . 39 (SE):   Hb/B3−216
                                Magnetic properties . . . . 39 (SE):   Hb/B3−289/92
                                Mechanical properties . . . . 39 (SE):   Hb/B3−85/103, 193/5,
                                                                                    207/9
                                Melting point . . . . 39 (SE):   Hb/B3−118/9
                                Mineralogy . . . . 39 (SE):   Hb/A1−54/110

Pr . . . . Praseodymium:
             Modifications . . . . 39 (SE):   Hb/B3−60/1
             Molar volume . . . . 39 (SE):   Hb/B3−85/7
             Occurrence, extraterrestrial . . . . . 39 (SE):   Hb/A1−22/7
             −, terrestrial . . . . 39 (SE):   Hb/A1−27/54
             − in the biosphere . . . . 39 (SE):   Hb/A1−53/4
             − in the lithosphere . . . . 39 (SE):   Hb/A1−27/54
             Optical properties . . . . 39 (SE):   Hb/B3−338/9, 343
             Polymorphism . . . . 39 (SE):   Hb/B3−60/1
             Position in the periodic system . . . . 39 (SE):   Hb/A1−2/4
             Spectral emissivity . . . . 39 (SE):   Hb/B3−343
             Superconductivity . . . . 39 (SE):   Hb/B3−254
             Surface tension . . . . 39 (SE):   Hb/B3−208/9
             Susceptibility . . . . 39 (SE):   Hb/B3−289/92
             Thermal conductivity . . . . 39 (SE):   Hb/B3−198/9
             Thermal diffusivity . . . . 39 (SE):   Hb/B3−206
             Thermal expansion . . . . 39 (SE):   Hb/B3−88/92
             Thermal properties . . . . 39 (SE):   Hb/B3−83/92, 103/20,
                                                      129/34, 185,
                                                      198/9, 206
             Thermodynamic functions . . . . . . 39 (SE):   Hb/B3−129/34
             Thermoelectric power . . . . 39 (SE):   Hb/B3−274
             Ultrasound absorption . . . . 39 (SE):   Hb/B3−193
             Valence . . . . 39 (SE):   Hb/A1−4
             Vapor pressure . . . . 39 (SE):   Hb/B3−103/5
             Viscosity . . . . 39 (SE):   Hb/B3−207/8
−. . . . . Praseodymium ions
             see Pr^{3+} ... Pr^{3+}
−. . . . . Praseodymium systems:
             Pr−Ag . . . . 61 (Ag):   Hb/C−217/8
             Pr−Al . . . . 35 (Al):   Hb/A5−825/6
             Pr−Gd−N . . . . 39 (SE):   Hb/C2−175
             Pr−H . . . . 39 (SE):   Hb/C1−2/3, 50/1
             Pr−Mg . . . . 27 (Mg):   Hb/A4−599/601
             Pr−Ni . . . . 57 (Ni):   Hb/B1−73/4
             Pr−Pb . . . . 47 (Pb):   Hb/C4−1365/6
             Pr−Pb−Zn . . . . 47 (Pb):   Hb/C4−1363, 1366
             Pr−Sn . . . . 46 (Sn):   Hb/D−307
             Pr−V . . . . 48 (V):   Hb/B2−600/1
Pr^{3+} . . . . Pr^{3+}:
             Magnetic moment . . . . 39 (SE):   Hb/B3−284
             Spectra . . . . 49 (Nb):   Hb/B3−133, 135
                                         50 (Ta):   Hb/B2−203
$PrRu_2$ . . . . Ru_2Pr . . . . 63 (Ru):   Eb−183
$PrSn_3$ . . . . $PrSn_3$ . . . . 46 (Sn):   Hb/D−307/10
−. . . . . $PrSn_3$ solid solutions:
             $La_{1-x}Pr_xSn_3$ . . . . 46 (Sn):   Hb/D−310

See the "Instructions for the user"
at the beginning of the volume

Pr_2S_3 . . . . Pr_2S_3 solid solutions:
               $Pr_2S_3-Ag_2S$ . . . . 61 (Ag): Hb/B4-232
               Pr_2S_3-PbS. . . . . 47 (Pb): Hb/C4-1374

Pr_2Se_3 . . . . Pr_2Se_3 solid solutions:
               $Pr_2Se_3-Ag_2Se$. . . . . 61 (Ag): Hb/B4-234/5
               Pr_2Se_3-PbSe . . . . 47 (Pb): Hb/C4-1374

Pr_2Sn . . . . Pr_2Sn . . . . 46 (Sn): Hb/D-307
Pr_2Sn_3. . . . . Pr_2Sn_3 . . . . 46 (Sn): Hb/D-307
Pr_3Sn . . . . Pr_3Sn . . . . 46 (Sn): Hb/D-307/10
Pr_5Sn_3 . . . . Pr_5Sn_3 . . . . 46 (Sn): Hb/D-307/10
Pr_5Sn_4 . . . . Pr_5Sn_4 . . . . 46 (Sn): Hb/D-307/10

Pt . . . . Platinum:
               Abrasion. . . . . 68 (Pt): Hb/B1-45
               Absorption of light . . . . 68 (Pt): Hb/B2-77/8
               - of X- and gamma rays . . . . 68 (Pt): Hb/B2-101/3
               Abundance . . . . 68 (Pt): Hb/A1-24/7
               Aerosols. . . . . 68 (Pt): Hb/A3-405
               Anodic polarization. . . . . 68 (Pt): Hb/B3-205/8
               Atom. . . . . 68 (Pt): Hb/B1-7/12
               Atomic nucleus . . . . 68 (Pt): Hb/B1-1/7
                                                 Hb/B2-180
               Atomic number . . . . 68 (Pt): Hb/B1-2
               Atomic radius . . . . 68 (Pt): Hb/B1-8
                Atomic volume . . . . 68 (Pt): Hb/B1-10/1
               Atomic weight . . . . 68 (Pt): Hb/B1-11/2
               Becquerel effect. . . . . 68 (Pt): Hb/B4-272/9
               Behavior as anode . . . . 68 (Pt): Hb/B4-298/308
                                                  Hb/C1-34/5
               - as cathode. . . . . 68 (Pt): Hb/B4-295/8
                                                  Hb/C1-9
               - at alternating current electrolysis . . 68 (Pt): Hb/B4-308/14
               Benedicks effect. . . . . 68 (Pt): Hb/B2-153
                Boiling point . . . . 68 (Pt): Hb/B1-57
                Catalytic properties. . . . . 68 (Pt): Hb/A3-429/30
                                                  Hb/B4-338
               Cathodic polarization. . . . . 68 (Pt): Hb/B3-228/9
                Cells . . . . 68 (Pt): Hb/B4-267/70
               Chemical constant . . . . 68 (Pt): Hb/B1-70
               Chemical reactions . . . . 68 (Pt): Hb/B4-319/38
               - on heating. . . . . 68 (Pt): Hb/B4-319/20
               - with air . . . . 68 (Pt): Hb/B4-319
               - with elements. . . . . 68 (Pt): Hb/B4-320/3
               - with metal compounds. . . . . . . 68 (Pt): Hb/B4-331/8
               - with nonmetal compounds . . . . 68 (Pt): Hb/B4-323/31
               - with organic compounds. . . . . . 68 (Pt): Hb/B4-329, 330/1
               Colloidal forms . . . . 68 (Pt): Hb/A3-405/23,
                                                  429/30

Hb = Main Series Volume; GD = Gmelin-Durrer;   Ab = Appended Volume to Main Series;
Eb = Supplement Volume to Main Series;   Erg. W. = New Supplement Series Volume

Pt . . . . Platinum:

Color.	68 (Pt):	Hb/B2–73
Compressibility	68 (Pt):	Hb/B1–45
Contact electricity.	68 (Pt):	Hb/B2–138/40
Corbino effect.	68 (Pt):	Hb/B2–108
Cosmochemistry	68 (Pt):	Hb/A1–15/22
Critical constants	68 (Pt):	Hb/B1–55
Crystallographic properties	68 (Pt):	Hb/B1–12/26
Debye temperature	68 (Pt):	Hb/B1–68/9
Deformation.	68 (Pt):	Hb/B1–21/2
Density	68 (Pt):	Hb/B1–26/9
Deposition potential	68 (Pt):	Hb/B4–285/6
Detection	68 (Pt):	Hb/A4–432, 455/62, 491/2
– in iron	59 (Fe):	Hb/FII1–41
Determination.	68 (Pt):	Hb/A4–480/8, 491/2
– in special substances	68 (Pt):	Hb/A4–525/32
Diffusion.	68 (Pt):	Hb/B1–47/55
Discharge phenomena	68 (Pt):	Hb/B2–154/7
Economic deposits of platinum and platinum metals:		
Africa	68 (Pt):	Hb/A2–190/263
Asia	68 (Pt):	Hb/A1–143/4 Hb/A2–145/6
Australia and Oceania	68 (Pt):	Hb/A2–263/74
Central America	68 (Pt):	Hb/A2–179
Europe	68 (Pt):	Hb/A1–50/3
North America	68 (Pt):	Hb/A2–146/79
South America	68 (Pt):	Hb/A2–179/90
Soviet Union	68 (Pt):	Hb/A1–53/143
Economics.	68 (Pt):	Hb/A2–302/10
Elasticity.	68 (Pt):	Hb/B1–38/41
Electrical conductivity	68 (Pt):	Hb/B2–109/38
– of thin layers	68 (Pt):	Hb/B2–132/7
Electrical properties	68 (Pt):	Hb/B2–109/79
Electrochemical behavior	68 (Pt):	Hb/B3–181/262 Hb/B4–263/318
Electrode sputtering	68 (Pt):	Hb/B1–20/1 Hb/B2–157/9
Electrolytic deposition	68 (Pt):	Hb/A3–387/9 Hb/B4–279/95
Electrolytic solution pressure	68 (Pt):	Hb/B4–266
Electron configuration	68 (Pt):	Hb/B1–7/8
Electron emission.	68 (Pt):	Hb/B2–159/72
Enthalpy.	68 (Pt):	Hb/B1–69
Entropy	68 (Pt):	Hb/B1–69/70
Equation of state	68 (Pt):	Hb/B1–55
Faraday effect.	68 (Pt):	Hb/B2–78
Free energy	68 (Pt):	Hb/B1–69

*See the "Instructions for the user"
at the beginning of the volume*

Pt . . . . Platinum:

Geochemical character	68 (Pt):	Hb/A1−23/8
Geochemistry	68 (Pt):	Hb/A1−22/49
		Hb/A2−275/6
Hall effect	68 (Pt):	Hb/B2−107/8
Hardness	68 (Pt):	Hb/B1−34/8
Heat content.	68 (Pt):	Hb/B1−69
Heat of fusion	68 (Pt):	Hb/B1−60
− of sublimation.	68 (Pt):	Hb/B1−59/60
− of vaporization	68 (Pt):	Hb/B1−59
History.	68 (Pt):	Hb/A1−1/9
Hydrosols	68 (Pt):	Hb/A3−406/21
Interaction with electrons	68 (Pt):	Hb/B2−84, 172/6
− with positive ions	68 (Pt):	Hb/B2−176/9
− with positrons.	68 (Pt):	Hb/B2−176
Interfacial tension.	68 (Pt):	Hb/B1−46/7
Internal energy	68 (Pt):	Hb/B1−69
Internal pressure	68 (Pt):	Hb/B1−46
Ionic refraction	68 (Pt):	Hb/B2−77
Ionization potentials	68 (Pt):	Hb/B2−179
Lattice structure.	68 (Pt):	Hb/B1−14/7
Lattice vibrations	68 (Pt):	Hb/B1−67/8
Lorenz number	68 (Pt):	Hb/B2−137/8
Magnetic moment	68 (Pt):	Hb/B1−9/10
Magnetic properties	68 (Pt):	Hb/B2−103/7
Mechanical properties	68 (Pt):	Hb/B1−26/55
Melting point	68 (Pt):	Hb/B1−57/9
Mineralogy	68 (Pt):	Hb/A2−276/301
Nuclear properties	68 (Pt):	Hb/B1−1/7
		Hb/B2−180
Occurrence, extraterrestrial	68 (Pt):	Hb/A1−15/22
−, terrestrial	68 (Pt):	Hb/A1−22/49
		Hb/A2−275/6
− in meteorites	68 (Pt):	Hb/A1−16/22
− in the biosphere.	68 (Pt):	Hb/A2−275
− in the hydrosphere	68 (Pt):	Hb/A2−275
− in the lithosphere	68 (Pt):	Hb/A1−28/49
Optical properties.	68 (Pt):	Hb/B2−73/103
Organosols	68 (Pt):	Hb/A3−421/3
Orientation of thin layers.	68 (Pt):	Hb/B1−17/20
Overvoltage phenomena, theoretical .	68 (Pt):	Hb/B3−232/51, 262
− during anodic deposition		
of halogen	68 (Pt):	Hb/B3−252/62
− − of oxygen	68 (Pt):	Hb/B3−209/28
− during anodic solution		
of hydrogen	68 (Pt):	Hb/B3−205/8
− during cathodic deposition		
of hydrogen	68 (Pt):	Hb/B3−181/204
− during cathodic solution of oxygen .	68 (Pt):	Hb/B3−228/9

Hb = Main Series Volume; GD = Gmelin-Durrer;   Ab = Appended Volume to Main Series;
Eb = Supplement Volume to Main Series;   Erg. W. = New Supplement Series Volume

Pt . . . . Platinum:
        Overvoltage phenomena
          during oxygen electrode-
          hydrogen electrode transition. . . . 68 (Pt):   Hb/B3–229/31
        Peltier effect. . . . . 68 (Pt):   Hb/B2–153/4
        Photoelectric electron emission . . . . 68 (Pt):   Hb/B2–159/67
        Physiological action . . . . 68 (Pt):   Hb/A2–275/6
        Plasticity. . . . . 68 (Pt):   Hb/B1–41/2
        Platinum oxygen electrode. . . . . . . 68 (Pt):   Hb/B3–229/31
        Platinum resistance thermometer . . . 68 (Pt):   Hb/B2–115/24
        Polarization . . . . 68 (Pt):   Hb/B4–314/8
        Polymorphism . . . . 68 (Pt):   Hb/B1–12/3
        Position in the electrochemical series. 68 (Pt):   Hb/B2–140
                                              Hb/B4–265

        – in the periodic system . . . . 68 (Pt):   Hb/B1–12
        Potentials . . . . 68 (Pt):   Hb/B4–266/7
        Preparation . . . . 68 (Pt):   Hb/A3–311/430
        – from minerals. . . . . 68 (Pt):   Hb/A3–312/50
        – from wastes and residues . . . . . . 68 (Pt):   Hb/A3–350/62
        – in pure state. . . . . 68 (Pt):   Hb/A3–367/71
        – of dense platinum . . . . 68 (Pt):   Hb/A3–378/83
        Purity testing . . . . 68 (Pt):   Hb/A3–365/7
                                              Hb/A4–521/4

        Recrystallization . . . . 68 (Pt):   Hb/B1–23/6
        Reflection of light. . . . . 68 (Pt):   Hb/B2–73/5
        – of X-rays. . . . . 68 (Pt):   Hb/B2–100/1
        Refraction of light. . . . . 68 (Pt):   Hb/B2–75/7
        – of X-rays. . . . . 68 (Pt):   Hb/B2–100/1
        Scattering of X- and gamma rays . . . 68 (Pt):   Hb/B2–101/3
        Separation, analytical . . . . 68 (Pt):   Hb/A4–504/18
        Sorption of hydrogen. . . . . 68 (Pt):   Hb/C1–5/25
        – of nitrogen . . . . 68 (Pt):   Hb/C1–54/6
        – of oxygen . . . . 68 (Pt):   Hb/C1–26/41
        – of rare gases . . . . 68 (Pt):   Hb/C1–1/4
        Special forms . . . . 68 (Pt):   Hb/A3–391/8, 401/23,
                                              429/30

        Specific heat . . . . 68 (Pt):   Hb/B1–60/7
        Spectra:
          Arc spectrum . . . . 68 (Pt):   Hb/B2–84/92
          Spark spectra. . . . . 68 (Pt):   Hb/B2–92/3
          X-ray spectrum . . . . 68 (Pt):   Hb/B2–93/100

        Standard potential . . . . 68 (Pt):   Hb/B4–263/5
        Static electricity. . . . . 68 (Pt):   Hb/B2–138
        Strength . . . . 68 (Pt):   Hb/B1–42/4
        Structure . . . . 68 (Pt):   Hb/B1–22/3
        – of thin layers . . . . 68 (Pt):   Hb/B1–17/20
        Superconductivity . . . . 68 (Pt):   Hb/B2–114/5
        Surface tension . . . . 68 (Pt):   Hb/B1–46/7
        Susceptibility . . . . 68 (Pt):   Hb/B2–103/7

See the "Instructions for the user"
at the beginning of the volume

Pt . . . . Platinum:
    Thermal conductivity . . . . 68 (Pt):   Hb/B1−70/2
    Thermal electron emission . . . . . . 68 (Pt):   Hb/B2−167/9
    Thermal expansion . . . . 68 (Pt):   Hb/B1−29/34
    Thermal properties . . . . 68 (Pt):   Hb/B1−55/72
    Thermal radiation . . . . 68 (Pt):   Hb/B2−78/84
    Thermocouples . . . . 68 (Pt):   Hb/B2−141/51
    Thermoelectric power . . . . 68 (Pt):   Hb/B2−140/53
    Thomson effect . . . . 68 (Pt):   Hb/B2−154
    Toxicity . . . . 68 (Pt):   Hb/A2−275/6
    Uses . . . . 68 (Pt):   Hb/A3−385/91
                                                Hb/B2−115/24
                                              Hb/B4−270/1
    Valence . . . . 68 (Pt):   Hb/B1−12
    Vapor pressure . . . . 68 (Pt):   Hb/B1−55/6
    Vibration frequency . . . . 68 (Pt):   Hb/B1−67/8
    Viscosity . . . . 68 (Pt):   Hb/B1−47
    Wiedemann-Franz number . . . . . . 68 (Pt):   Hb/B2−137/8
    Work function of electrons . . . . . . 68 (Pt):   Hb/B2−170/2
    X-ray photoelectric effect . . . . . . 68 (Pt):   Hb/B2−166/7

−. . . . . Platinum ions
    see Pt⁺ ... Pt⁺, etc.

−. . . . . Platinum isotopes . . . . 68 (Pt):   Hb/B1−1/2
                                              Hb/B2−179, 180

−. . . . . Platinum systems:
    Pt−Ag . . . . 68 (Pt):   Hb/A6−810/2
    Pt−Al . . . . 35 (Al):   Hb/A6−1107/8
    Pt−As . . . . 68 (Pt):   Hb/C1−137/8
    Pt−Au . . . . 68 (Pt):   Hb/A6−818/9
    Pt−Au−Ge . . . . 68 (Pt):   Hb/A6−825/6
    Pt−Au−Pd . . . . 68 (Pt):   Hb/A6−840/1
    Pt−Au−Rh . . . . 68 (Pt):   Hb/A6−836
    Pt−C . . . . 68 (Pt):   Hb/C1−115
    Pt−Cd . . . . 68 (Pt):   Hb/A6−751/2
    Pt−Co . . . . 68 (Pt):   Hb/A6−785/6
    Pt−Cr . . . . 68 (Pt):   Hb/A6−769
    Pt−Cu . . . . 68 (Pt):   Hb/A6−799
    Pt−Cu−Fe . . . . 68 (Pt):   Hb/A6−809
    Pt−Cu−Ni . . . . 68 (Pt):   Hb/A6−809
    Pt−Fe . . . . 68 (Pt):   Hb/A6−791/3
    Pt−Fe−S . . . . 68 (Pt):   Hb/A6−798
    Pt−H . . . . 68 (Pt):   Hb/C1−5/25
    Pt−H−Pd . . . . 68 (Pt):   Hb/A6−839/40
    Pt−Hg . . . . 68 (Pt):   Hb/A6−752/3
    Pt−Ir . . . . 68 (Pt):   Hb/A6−842/3
    Pt−Mn . . . . 68 (Pt):   Hb/A6−778
    Pt−Mo . . . . 68 (Pt):   Hb/A6−773/4
    Pt−N . . . . 68 (Pt):   Hb/C1−54/6

Hb = Main Series Volume; GD = Gmelin-Durrer;   *Ab = Appended Volume to Main Series;*
Eb = Supplement Volume to Main Series;   *Erg. W. = New Supplement Series Volume*

Pt . . . . Platinum systems:
                          Pt−Nb . . . . 68 (Pt):   Hb/A6−768
                          Pt−Ni . . . . 68 (Pt):   Hb/A6−780/1
                          Pt−O . . . . 68 (Pt):   Hb/C1−26/41
                          Pt−P . . . . 68 (Pt):   Hb/C1−133/5
                          Pt−Pb . . . . 68 (Pt):   Hb/A6−763/7
                          Pt−Pb−Tl . . . . 68 (Pt):   Hb/A6−767
                          Pt−Pd . . . . 68 (Pt):   Hb/A6−836
                          Pt−Rh . . . . 68 (Pt):   Hb/A6−829/30
                          Pt−Ru . . . . 68 (Pt):   Hb/A6−826
                          Pt−S . . . . 68 (Pt):   Hb/C1−103/4
                          Pt−Sb . . . . 68 (Pt):   Hb/A6−743/4
                          Pt−Si . . . . 68 (Pt):   Hb/C1−128/31
                          Pt−Sn . . . . 68 (Pt):   Hb/A6−758/63
                          Pt−Tl . . . . 68 (Pt):   Hb/A6−755/7
                          Pt−W . . . . 68 (Pt):   Hb/A6−774/5

Pt^+ . . . . Pt^+ . . . . 68 (Pt):   Hb/B1−8
                                                                       Hb/B2−179
Pt^{2+} . . . . Pt^{2+} . . . . 68 (Pt):   Hb/B1−7/9
                                                                       Hb/B2−92/3, 107, 179
Pt^{3+} . . . . Pt^{3+} . . . . 68 (Pt):   Hb/B1−7/8
Pt^{4+} . . . . Pt^{4+} . . . . 68 (Pt):   Hb/B1−7/8, 9/10
                                                                       Hb/B2−77
Pt^{5+} . . . . Pt^{5+} . . . . 68 (Pt):   Hb/B1−7/8
Pt^{6+} . . . . Pt^{6+} . . . . 68 (Pt):   Hb/B1−7/8

PtS . . . . PtS . . . . 68 (Pt):   Hb/C1−103/5
PtS_2 . . . . PtS_2 . . . . 68 (Pt):   Hb/C1−106/8
PtSb . . . . PtSb . . . . 68 (Pt):   Hb/A6−743/4, 746
$PtSb_2$ . . . . $PtSb_2$ . . . . 68 (Pt):   Hb/A6−743/4, 746
PtSe . . . . PtSe . . . . 68 (Pt):   Hb/C1−112
$PtSe_2$ . . . . $PtSe_2$ . . . . 68 (Pt):   Hb/C1−112/3
$PtSe_3$ . . . . $PtSe_3$ . . . . 68 (Pt):   Hb/C1−113
PtSi . . . . PtSi . . . . 68 (Pt):   Hb/C1−129, 132
$PtSi_{12}$ . . . . $PtSi_{12}$ . . . . 68 (Pt):   Hb/C1−131
PtSn . . . . PtSn . . . . 68 (Pt):   Hb/A6−758/60, 762
$PtSn_2$ . . . . $PtSn_2$ . . . . 68 (Pt):   Hb/A6−758/60, 763
$PtSn_3$ . . . . $PtSn_3$ . . . . 68 (Pt):   Hb/A6−759
$PtSn_4$ . . . . $PtSn_4$ . . . . 68 (Pt):   Hb/A6−758/60, 763
PtTe . . . . PtTe . . . . 68 (Pt):   Hb/C1−113
$PtTe_2$ . . . . $PtTe_2$ . . . . 68 (Pt):   Hb/C1−114
PtTi . . . . PtTi . . . . 68 (Pt):   Hb/A6−757
$PtTi_2$ . . . . $PtTi_2$ . . . . 68 (Pt):   Hb/A6−757
PtTl . . . . PtTl . . . . 68 (Pt):   Hb/A6−755/7
PtZn . . . . PtZn . . . . 68 (Pt):   Hb/A6−750/1
$PtZn_2$ . . . . $PtZn_2$ . . . . 68 (Pt):   Hb/A6−750/1
Pt_2S_3 . . . . $Pt_2S_3 = Pt_3[PtS_6]$ . . . . 68 (Pt):   Hb/C1−106
Pt_2Si . . . . Pt_2Si . . . . 68 (Pt):   Hb/C1−129, 132
Pt_2Sn_3 . . . . Pt_2Sn_3 . . . . 68 (Pt):   Hb/A6−758/60, 762/3
                                                                       Hb/C3−310

See the "Instructions for the user"
at the beginning of the volume

Pt_2Te	Pt_2Te	68 (Pt):	Hb/C1–113
Pt_2Zn_3	Pt_2Zn_3	68 (Pt):	Hb/A6–750/1
Pt_3S_6Sn	$2\ PtS \cdot PtS_2 \cdot SnS_2 = Pt_3[SnS_6]$	68 (Pt):	Hb/C3–313
$Pt_3S_6Tl_4$	$2\ Tl_2S \cdot 2\ PtS \cdot PtS_2 = Tl_4Pt_2[PtS_6]$	68 (Pt):	Hb/C3–307
$Pt_3S_6Zn_2$	$2\ ZnS \cdot 2\ PtS \cdot PtS_2 = Zn_2Pt_2[PtS_6]$	68 (Pt):	Hb/C3–296
Pt_3Si_2	Pt_3Si_2	68 (Pt):	Hb/C1–131
Pt_3Sn	Pt_3Sn	68 (Pt):	Hb/A6–758/62
Pt_3Sn_8	Pt_3Sn_8	68 (Pt):	Hb/A6–759, 763
Pt_3Ti	Pt_3Ti	68 (Pt):	Hb/A6–757
Pt_3Ti_2	Pt_3Ti_2	68 (Pt):	Hb/A6–757
Pt_3Zr	Pt_3Zr	68 (Pt):	Hb/A6–758
Pt_4S_6	$Pt_3[PtS_6] = Pt_2S_3$	68 (Pt):	Hb/C1–106
Pt_4S_6Sn	$4\ PtS \cdot SnS_2 = Pt_4[SnS_6]$	68 (Pt):	Hb/C3–312
Pt_4Sb	Pt_4Sb	68 (Pt):	Hb/A6–743/4
Pt_4Se_6Sn	$4\ PtSe \cdot SnSe_2 = Pt_4[SnSe_6]$	68 (Pt):	Hb/C3–313
Pt_4Si_3	Pt_4Si_3	68 (Pt):	Hb/C1–131
Pt_4Sn_3	Pt_4Sn_3	68 (Pt):	Hb/C3–310
Pt_5S_6	Pt_5S_6	68 (Pt):	Hb/C1–105/6
Pt_5Sb_2	Pt_5Sb_2	68 (Pt):	Hb/A6–743/4
Pt_5Si_2	Pt_5Si_2	68 (Pt):	Hb/C1–129, 131
Pt_5Zn_2	Pt_5Zn_2	68 (Pt):	Hb/A6–750
Pt_5Zn_{21}	Pt_5Zn_{21}	68 (Pt):	Hb/A6–751

Pu	Plutonium:	
	Atom.	Erg. W. 8–1/34
	Atomic mass	Erg. W. 7a–123/7, 128, 133
	Atomic nucleus	Erg. W. 7a–24/30, 61/170
	Atomic radius	Erg. W. 8–4/5
	Biological behavior	Erg. W. 8–403/15
	Criticality	Erg. W. 8–361/84
	Decomposition and solution methods	Erg. W. 8–124/6
	Detection in soils and fallout.	Erg. W. 7b–353/8
	Determination:	
	Complexometric methods.	Erg. W. 8–139/41
	Electrochemical methods	Erg. W. 8–132/9
	General	Erg. W. 8–119/30
	Gravimetric methods	Erg. W. 8–130/1
	Isotope analysis	Erg. W. 8–158/61
	Isotope dilution analysis.	Erg. W. 8–162/3
	Optical methods	Erg. W. 8–141/7
	Radiometric methods	Erg. W. 8–151/8
	Redox methods	Erg. W. 8–132/5
	Separation methods	Erg. W. 8–126/30
	X-ray optical methods	Erg. W. 8–147/51
	Other methods	Erg. W. 8–163/4
	Determination in biological materials	Erg. W. 8–179/85
	Diffusion.	Erg. W. 8–266/70

Hb = Main Series Volume; GD = Gmelin-Durrer;   *Ab = Appended Volume to Main Series;*
Eb = Supplement Volume to Main Series;   *Erg. W. = New Supplement Series Volume*

Pu . . . . Plutonium:

Electron configuration . . . . Erg. W. 8–3/23
Fabrication processes . . . . Erg. W. 8–234/50
Fire prevention . . . . Erg. W. 8–331/2
Formation in nature . . . . Erg. W. 7b–1/2
Handling techniques . . . . Erg. W. 8–123, 314, 349/61
Health physics . . . . Erg. W. 8–320/49
History . . . . Erg. W. 7a–1/11
– (Ekaosmium) . . . . 66 (Os):   Hb–98/100
In-pile behavior of nuclear fuels . . . . Erg. W. 8–259/87
Ionization potentials . . . . Erg. W. 8–23
Isotope separation . . . . Erg. W. 7b–359/70
Nuclear binding energies . . . . . . Erg. W. 7a–128
Nuclear decay,
    alpha radioactivity . . . . Erg. W. 7a–82/96
–, decay energies . . . . Erg. W. 7a–95, 123/6
–, decay properties . . . . Erg. W. 7a–24/30
–, decay scheme . . . . Erg. W. 7a–98/103, 107/12
–, gamma radioactivity . . . . Erg. W. 7a–78/81
Nuclear fission,
    fission by charged particles . . . . . Erg. W. 7a–151/63
–, – by neutrons . . . . Erg. W. 7a–138/50
–, spontaneous fission . . . . Erg. W. 7a–134/8, 140
–, theoretical . . . . Erg. W. 7a–163/70
Nuclear models . . . . Erg. W. 7a–61/82
Nuclear moments . . . . Erg. W. 8–37
Nuclear properties . . . . Erg. W. 7a–24/30, 61/170
Nuclear radiation,
    effect on gases . . . . Erg. W. 8–309/12
–, – on organic liquids . . . . Erg. W. 8–308/9
–, – on solids . . . . Erg. W. 8–292/300
–, – on solutions . . . . Erg. W. 8–302/7
–, – on the chemical properties . . . . Erg. W. 8–312/4
Occurrence, extraterrestrial . . . . . . Erg. W. 7a–14/7
–, terrestrial . . . . Erg. W. 7a–12/4
Oxidation states . . . . Erg. W. 8–8/11
Performance of heat sources . . . . . Erg. W. 8–288/90
Production, aqueous processing
    of irradiated fuels . . . . Erg. W. 7b–177/240
–, – –, chemistry of
    reprocessing solutions . . . . Erg. W. 7b–193/6
–, – –, general . . . . Erg. W. 7b–177/83
–, – –, precipitation . . . . Erg. W. 7b–196/7
–, – –, preparation of feed solutions . . Erg. W. 7b–184/93
–, – –, solvent extraction . . . . Erg. W. 7b–197/235
–, – –, sorption and ion exchange . . . Erg. W. 7b–235/40
–, nonaqueous processing
    of irradiated fuels . . . . Erg. W. 7b–240/65
–, recovery from scrap and wastes . . Erg. W. 7b–279/88

See the "Instructions for the user"
at the beginning of the volume

Pu . . . . Plutonium:
            Production in charged particle
                reactions, cross sections. . . . . . . Erg. W. 7b–104/16, 121/76
            – –, experiments . . . . Erg. W. 7b–37/104, 116/21
            – –, general features . . . . Erg. W. 7b–28/37
            – in nuclear explosions. . . . . Erg. W. 7b–21/8
            – in nuclear reactors . . . . Erg. W. 7b–3/18
            – of metallic plutonium. . . . . Erg. W. 8–234/42
            Radiotoxicity . . . . Erg. W. 8–320/5
            Recoil reactions . . . . Erg. W. 8–315/9
            Recovery of plutonium-238 . . . . . Erg. W. 7b–305/13
            – of plutonium-242 . . . . Erg. W. 7b–313/5
            Safety measures . . . . Erg. W. 8–385/93
            Sampling . . . . Erg. W. 8–123
            Separation from nuclear reaction
                products . . . . Erg. W. 7b–336/42, 347
            –, analytical . . . . Erg. W. 8–126/30
            Specifications . . . . Erg. W. 8–122
            Spectra:
                Absorption spectra. . . . . Erg. W. 8–51/7, 63/4, 67/76
                Emission spectra. . . . . Erg. W. 8–36/41
                 Luminescence spectra . . . . Erg. W. 8–57, 68
                Mössbauer spectra . . . . Erg. W. 8–89/90, 97/8
                X-ray spectra . . . . Erg. W. 8–80/8
            Standards . . . . Erg. W. 8–120/1
            Storage and control of wastes . . . . . Erg. W. 8–393/402
            – of samples. . . . . Erg. W. 8–123
            Therapy after contamination. . . . . Erg. W. 8–415/24
            Transition reactions . . . . Erg. W. 8–305/6
            Treatment and storage of wastes . . . Erg. W. 7b–265/79
            Uses of fissile plutonium isotopes. . . Erg. W. 8–189/201
            – of other plutonium isotopes . . . . . Erg. W. 8–201/13, 250
–. . . . . Plutonium fuel elements:
            In-pile behavior . . . . Erg. W. 8–259/83
            Production. . . . . Erg. W. 8–239/42
–. . . . . Plutonium ions:
            Ionization potentials . . . . Erg. W. 8–23
            Properties of the ion . . . . Erg. W. 8–5/28
            Spectra:
                Absorption spectra. . . . . Erg. W. 8–51/3, 55/7, 63/4,
                                                   67/72, 74/6
                Emission spectra. . . . . Erg. W. 8–40/1
                 Luminescence spectra . . . . Erg. W. 8–57, 68
–. . . . . Plutonium isotopes
            see Pu ... Plutonium
–. . . . . Plutonium systems:
            Pu–C. . . . . Erg. W. 4–201

Pu . . . . Plutonium systems:
           Pu–C–Fe . . . . Erg. W. 4–209
           Pu–C–N. . . . . Erg. W. 4–225/8
           Pu–C–O. . . . . Erg. W. 4–224/5
           Pu–C–Si . . . . Erg. W. 4–209
           Pu–C–Th . . . . Erg. W. 4–210/1
           Pu–C–U. . . . . Erg. W. 4–211/3
           Pu–H . . . . Erg. W. 4–3/5
           Pu–Nb–Si. . . . . Erg. W. 4–247/8
           Pu–O . . . . Erg. W. 4–10/2
           Pu–O–H_2O . . . . Erg. W. 4–10/2
           Pu–O–U. . . . . Erg. W. 4–59/60
           Pu–S. . . . . Erg. W. 4–159/61
           Pu–Si . . . . Erg. W. 4–245/6
Pu^+ . . . . Pu^+ . . . . Erg. W. 8–23, 40/1
Pu^{2+} . . . . Pu^{2+} . . . . Erg. W. 8–11/23
Pu^{3+} . . . . Pu^{3+} . . . . Erg. W. 8–5, 11/27, 51/3, 55/7
Pu^{4+} . . . . Pu^{4+} . . . . Erg. W. 8–5, 11/23,
                                          63/4, 67/8
Pu^{5+} . . . . Pu^{5+} . . . . Erg. W. 8–5, 11/23
Pu^{6+} . . . . Pu^{6+} . . . . Erg. W. 8–5, 11/23
Pu^{7+} . . . . Pu^{7+} . . . . Erg. W. 8–11/25
Pu^{n+} . . . . Pu^{n+} (n = 8 to 10). . . . . Erg. W. 8–23

$PuRh_3$ . . . . $PuRh_3$. . . . . Erg. W. 4–209
$PuRu_2$ . . . . $PuRu_2$. . . . . Erg. W. 4–209
PuS . . . . PuS . . . . Erg. W. 4–159/66
–. . . . . PuS solid solutions:
           (Pu,U)S . . . . Erg. W. 4–168/71
PuS_2 . . . . PuS_2 . . . . Erg. W. 4–159/66
–. . . . . PuS_2 solid solutions:
           $(Pu,U)S_2$ . . . . Erg. W. 4–170
PuS_{2-x} . . . . PuS_{2-x} . . . . Erg. W. 4–159/66
PuSe . . . . PuSe . . . . Erg. W. 4–189/90
$PuSe_{2-x}$ . . . . . . $PuSe_{2-x}$ . . . . Erg. W. 4–189/90
PuSi. . . . . PuSi . . . . Erg. W. 4–245/8
–. . . . . PuSi systems:
           PuSi–PuC . . . . Erg. W. 4–248
$PuSi_2$ . . . . $PuSi_2$ . . . . Erg. W. 4–245/8
PuTe . . . . PuTe . . . . Erg. W. 4–192
$PuTe_2$ . . . . $PuTe_2$. . . . . Erg. W. 4–192/3
$PuTe_{2-x}$ . . . . . . $PuTe_{2-x}$ . . . . Erg. W. 4–192/3
Pu_2S_3 . . . . Pu_2S_3. . . . . Erg. W. 4–159/66
–. . . . . Pu_2S_3 solid solutions:
           $(Pu,U)_2S_3$ . . . . Erg. W. 4–169/70
Pu_2S_5Th . . . . $ThPu_2S_5$ . . . . Erg. W. 4–170
Pu_2S_5U . . . . UPu_2S_5 . . . . Erg. W. 4–170

See the "Instructions for the user"
at the beginning of the volume

Pu₂Se₃ . . . . Pu₂Se₃ . . . . Erg. W. 4–189/90
Pu₂Te₃ . . . . Pu₂Te₃ . . . . Erg. W. 4–192/3
Pu₃S₄ . . . . Pu₃S₄ . . . . Erg. W. 4–159/66
Pu₃S₅ . . . . Pu₃S₅ . . . . Erg. W. 4–170
Pu₃Se₄ . . . . Pu₃Se₄ . . . . Erg. W. 4–189/90
Pu₃Si₂ . . . . Pu₃Si₂ . . . . Erg. W. 4–245/8
Pu₃Si₅ . . . . Pu₃Si₅ . . . . Erg. W. 4–245/8
Pu₅S₇ . . . . Pu₅S₇ . . . . Erg. W. 4–159/66
Pu₅Si₃ . . . . Pu₅Si₃ . . . . Erg. W. 4–245/8

Ra . . . . Radium:
                        Abundance . . . . 31 (Ra):   Hb–4
                        Atom. . . . . 31 (Ra):   Hb–3, 50/1
                        Atomic nucleus . . . . 31 (Ra):   Hb–2/3, 51/4, 68/9,
                                                                         74, 78/9
                        Chemical reactions . . . . 31 (Ra):   Hb–56, 69, 75/6,
                                                                         79/80
                        Detection . . . . 31 (Ra):   Hb–56, 69, 76, 80
                        Determination. . . . . 31 (Ra):   Hb–56/7, 70, 76, 80
                        Economic deposits . . . . 31 (Ra):   Hb–4/7, 65
                        Economics. . . . . 31 (Ra):   Hb–49/50, 67/8
                        Electrochemical behavior . . . . . . 31 (Ra):   Hb–56, 75, 79
                        Formation . . . . 31 (Ra):   Hb–50, 68, 72, 77
                        History. . . . . 31 (Ra):   Hb–3, 64, 72, 77
                        Mineralogy . . . . 31 (Ra):   Hb–7/26
                        Nuclear decay,
                           alpha radiation . . . . 31 (Ra):   Hb–51/4, 74, 78
                                                             Erg. W. 7a–85/95
                        –, beta radiation. . . . . 31 (Ra):   Hb–53, 68/9, 79
                        –, decay scheme . . . . 31 (Ra):   Hb–2
                        –, gamma radiation. . . . . 31 (Ra):   Hb–53/4, 68/9, 79
                                                             Erg. W. 7a–80
                        Occurrence . . . . 31 (Ra):   Hb–3/27, 64/5
                        Ore dressing . . . . 31 (Ra):   Hb–28/33, 66/7
                        Physical properties . . . . 31 (Ra):   Hb–50/6, 68/9,
                                                                         74, 78
                        Preparation . . . . 31 (Ra):   Hb–50, 72/3, 77/8
                        –, industrial . . . . 31 (Ra):   Hb–27/48, 65/7
                        – in pure state. . . . . 31 (Ra):   Hb–40/7
                        Standard preparations . . . . 31 (Ra):   Hb–47/8
                        Storage . . . . 31 (Ra):   Hb–47
                        Uses . . . . 31 (Ra):   Hb–63, 71, 76
                                                             Erg. W. 8–224, 230/1
                        Valence . . . . 31 (Ra):   Hb–51
–. . . . . Radium ions . . . . 31 (Ra):   Hb–56, 75, 79
–. . . . . Radium isotopes
                        see Ra ... Radium
Ra²⁺ . . . . Ra²⁺ . . . . 31 (Ra):   Hb–56, 75, 79

Hb = Main Series Volume; GD = Gmelin-Durrer;   *Ab = Appended Volume to Main Series;*
Eb = Supplement Volume to Main Series;   *Erg. W. = New Supplement Series Volume*

RaA . . . . Radium A = ^{218}Po
                see Po ... Polonium

RaB . . . . Radium B = ^{214}Pb
                see Pb ... Lead isotopes

RaC . . . . Radium C = ^{214}Bi
                see Bi ... Bismuth isotopes

RaC'. . . . . Radium C' = ^{214}Po
                see Po ... Polonium

RaC'' . . . . Radium C'' = ^{210}Tl
                see Tl ... Thallium isotopes

RaD . . . . Radium D = ^{210}Pb
                see Pb ... Lead isotopes

RaE . . . . Radium E = ^{210}Bi
                see Bi ... Bismuth isotopes

RaE'' . . . . Radium E'' = ^{206}Tl
                see Tl ... Thallium isotopes

RaEm . . . . Radium emanation = ^{222}Rn
                see Rn ... Radon

RaF . . . . Radium F = ^{210}Po
                see Po ... Polonium

RaG . . . . Radium G = ^{206}Pb
                see Pb ... Lead isotopes

Rb. . . . . Rubidium:
        Absorption of light . . . . 24 (Rb):   Hb–54
        – of X-rays. . . . . 24 (Rb):   Hb–71
        Abundance . . . . 24 (Rb):   Hb–2/3
        Affinity. . . . . 24 (Rb):   Hb–42
        Atom. . . . . 24 (Rb):   Hb–34/42
        Atomic nucleus . . . . 24 (Rb):   Hb–27/34
        Chemical reactions . . . . 24 (Rb):   Hb–88/9
        Cosmochemistry . . . . 24 (Rb):   Hb–1
        Crystallographic properties . . . . . . 24 (Rb):   Hb–44
        Detection . . . . 24 (Rb):   Hb–90/6
        – in iron . . . . 59 (Fe):   Hb/FII1–4
        – in special substances. . . . . 24 (Rb):   Hb–102/3
        Determination. . . . . 24 (Rb):   Hb–96/100
        – in special substances . . . . 24 (Rb):   Hb–102/3
        Economics. . . . . 24 (Rb):   Hb–23
        Electrical properties . . . . 24 (Rb):   Hb–73/85
        Electrochemical behavior . . . . 24 (Rb):   Hb–85/8
        Formation . . . . 24 (Rb):   Hb–23/6
        Geochemistry . . . . 24 (Rb):   Hb–2/13
        History. . . . . 24 (Rb):   Hb–1
        Ionization . . . . 24 (Rb):   Hb–83/5

See the "Instructions for the user"
at the beginning of the volume

Rb . . . . Rubidium:
         Magnetic properties . . . . 24 (Rb):   Hb–71/3
         Mechanical properties . . . . 24 (Rb):   Hb–45/7
         Mineralogy . . . . 24 (Rb):   Hb–13
         Molecule. . . . . 24 (Rb):   Hb–42/3
         Occurrence, extraterrestrial . . . . . . 24 (Rb):   Hb–1
         –, terrestrial . . . . 24 (Rb):   Hb–2/13
         – in the biosphere. . . . . 24 (Rb):   Hb–10/3
         – in the hydrosphere . . . . 24 (Rb):   Hb–8/10
         – in the lithosphere . . . . 24 (Rb):   Hb–2/8
         Optical properties. . . . . 24 (Rb):   Hb–53/71
         Position in the periodic system . . . . 24 (Rb):   Hb–42
         Preparation . . . . 24 (Rb):   Hb–23/6
         –, industrial . . . . 24 (Rb):   Hb–13/20
         – of special forms. . . . . 24 (Rb):   Hb–26
         Purification . . . . 24 (Rb):   Hb–20/3, 26
         Radioactivity . . . . 24 (Rb):   Hb–3, 30/4
         Scattering of X-rays . . . . 24 (Rb):   Hb–71
         Separation, analytical . . . . 24 (Rb):   Hb–100/2
         Spectra:
            Arc spectrum . . . . 24 (Rb):   Hb–55/65
            Band spectrum . . . . 24 (Rb):   Hb–68/9
            Spark spectra. . . . . 24 (Rb):   Hb–65/7
            X-ray spectrum . . . . 24 (Rb):   Hb–69/71
         Storage . . . . 24 (Rb):   Hb–26
         Thermal properties . . . . 24 (Rb):   Hb–47/53
         Valence . . . . 24 (Rb):   Hb–42

–. . . . . Rubidium ions
         see Rb^+ .... Rb^+, etc.

–. . . . . Rubidium isotopes . . . . 24 (Rb):   Hb–27/8
–. . . . . Rubidium systems:
         Rb–Al . . . . 35 (Al):   Hb/A4–637
         Rb–B–Ta . . . . 50 (Ta):   Hb/B2–320
         Rb–Cs . . . . 25 (Cs):   Hb/2–258
         Rb–Cu . . . . 60 (Cu):   Hb/B3–1107
         Rb–Hg . . . . 34 (Hg):   Hb/A2–1010/2
         Rb–K . . . . 24 (Rb):   Hb–245/6
         Rb–Na . . . . 24 (Rb):   Hb–242/3
         Rb–S . . . . 24 (Rb):   Hb–205
         Rb–V . . . . 48 (V):   Hb/B2–578

Rb^+ . . . . Rb^+:
         Chemical reactions . . . . 24 (Rb):   Hb–89/102
         Diffusion in sodium halides . . . . . . 21 (Na):   Eb/6–291
         Electrochemical behavior . . . . 24 (Rb):   Hb–86/8
         Interaction with gases . . . . 24 (Rb):   Hb–83/4
         Ionic susceptibility . . . . 24 (Rb):   Hb–72/3
         Ionization potentials . . . . 24 (Rb):   Hb–84/5

Hb = Main Series Volume; GD = Gmelin-Durrer;   *Ab = Appended Volume to Main Series;*
Eb = Supplement Volume to Main Series;   *Erg. W. = New Supplement Series Volume*

Rb^+ . . . . Rb^+:

                     Properties of the ion . . . . 24 (Rb):   Hb−37/9, 41

                     Spectra . . . . 24 (Rb):   Hb−65/7

Rb^{n+} . . . . Rb^{n+} (n = 2 to 8) . . . . 24 (Rb):   Hb−39, 67, 84/5

RbSn . . . . RbSn . . . . 46 (Sn):   Hb/D−78

Rb_2 . . . . Rb_2 . . . . 24 (Rb):   Hb−42/3

Rb_2S . . . . Rb_2S . . . . 24 (Rb):   Hb−204

−. . . . . $Rb_2S \cdot 2 H_2O$ . . . . 24 (Rb):   Hb−205

−. . . . . $Rb_2S \cdot 4 H_2O$ . . . . 24 (Rb):   Hb−205

Rb_2S_2 . . . . Rb_2S_2 . . . . 24 (Rb):   Hb−206

−. . . . . $Rb_2S_2 \cdot H_2O$ . . . . 24 (Rb):   Hb−206

−. . . . . $Rb_2S_2 \cdot 2 H_2O$ . . . . 24 (Rb):   Hb−206

−. . . . . $Rb_2S_2 \cdot 3 H_2O$ . . . . 24 (Rb):   Hb−206

Rb_2S_3 . . . . Rb_2S_3 . . . . 24 (Rb):   Hb−206

−. . . . . $Rb_2S_3 \cdot H_2O$ . . . . 24 (Rb):   Hb−206

Rb_2S_4 . . . . Rb_2S_4 . . . . 24 (Rb):   Hb−206

−. . . . . $Rb_2S_4 \cdot 2 H_2O$ . . . . 24 (Rb):   Hb−206

Rb_2S_5 . . . . Rb_2S_5 . . . . 24 (Rb):   Hb−206

Rb_2S_6 . . . . Rb_2S_6 . . . . 24 (Rb):   Hb−205

Rb_2Se . . . . Rb_2Se . . . . 24 (Rb):   Hb−216

Rb_2Te . . . . Rb_2Te . . . . 24 (Rb):   Hb−219

RdAc . . . . Radioactinium = ^{227}Th
                     see Th ... Thorium

RdTh . . . . Radiothorium = ^{228}Th
                     see Th ... Thorium

Re . . . . Rhenium:

                     Atom. . . . . 70 (Re):   Hb−44/6

                     Atomic nucleus . . . . 70 (Re):   Hb−40/4

                     Chemical reactions . . . . 70 (Re):   Hb−59/61

                     Colloidal forms . . . . 70 (Re):   Hb−36

                     Cosmochemistry . . . . 70 (Re):   Hb−11/2

                     Crystallographic properties . . . . . . 70 (Re):   Hb−47/8

                     Detection . . . . 70 (Re):   Hb−62/9

                     Determination. . . . . 70 (Re):   Hb−69/75, 79/81

                     Economics. . . . . 70 (Re):   Hb−30

                     Electrical properties . . . . 70 (Re):   Hb−57/8

                     Electrochemical behavior . . . . 70 (Re):   Hb−59

                     Formation . . . . 70 (Re):   Hb−31/2

                     Geochemical character. . . . . 70 (Re):   Hb−13/4

                     Geochemistry . . . . 70 (Re):   Hb−12/25

                     History. . . . . 70 (Re):   Hb−1/10

                     Magnetic properties . . . . 70 (Re):   Hb−56/7

                     Mechanical properties . . . . 70 (Re):   Hb−48/9

                     Occurrence, extraterrestrial . . . . . . 70 (Re):   Hb−11/2

                     −, terrestrial . . . . 70 (Re):   Hb−12/25

See the "Instructions for the user"
at the beginning of the volume

Re . . . . Rhenium:
      Occurrence in industrial products . . . 70 (Re): Hb−24
      – in the hydrosphere . . . . 70 (Re): Hb−25
      – in the lithosphere . . . . 70 (Re): Hb−14/24
      Optical properties. . . . . 70 (Re): Hb−50/6
      Physiological action . . . . 70 (Re): Hb−61
      Position in the periodic system . . . . 70 (Re): Hb−47
      Preparation . . . . 70 (Re): Hb−32/3
      –, industrial . . . . 70 (Re): Hb−25/30
      – of special forms. . . . . 70 (Re): Hb−34/6
      Purity testing . . . . 70 (Re): Hb−33/4
      Separation, analytical . . . . 70 (Re): Hb−75/9
      Spectra . . . . 70 (Re): Hb−51/6
      Thermal properties . . . . 70 (Re): Hb−49/50
      Uses . . . . 70 (Re): Hb−36/40, 83
      Valence . . . . 70 (Re): Hb−46/7
−. . . . . Rhenium ions . . . . 70 (Re): Hb−45/6, 57
−. . . . . Rhenium isotopes . . . . 70 (Re): Hb−40/1
−. . . . . Rhenium systems:
      Re−As . . . . 70 (Re): Hb−124
      Re−Fe . . . . 70 (Re): Hb−85/6
      Re−Nb. . . . . 49 (Nb): Hb/A−161
      Re−P. . . . . 70 (Re): Hb−123
      Re−Rh . . . . 70 (Re): Hb−86/7
      Re−S. . . . . 70 (Re): Hb−117
      Re−W . . . . 70 (Re): Hb−84/5
Re^{n+} . . . . Re^{n+} (n = 1 to 7) . . . . 70 (Re): Hb−45/6, 57
ReS_2 . . . . ReS_2 . . . . 70 (Re): Hb−117/8
ReS_4Tl . . . . $TlReS_4$ . . . . 70 (Re): Hb−149
$ReSe_2$ . . . . $ReSe_2$ . . . . 70 (Re): Hb−120
Re_2S_7 . . . . Re_2S_7 . . . . 70 (Re): Hb−118/9
Re_2Se_7 . . . . Re_2Se_7 . . . . 70 (Re): Hb−120
Re_3W_2. . . . . W_2Re_3 . . . . 70 (Re): Hb−84/5

Rf . . . . Rutherfordium (Element 104). . . . . . Erg. W. 7a−4
      see also this Volume, Appendix

Rh . . . . Rhodium:
      Abundance . . . . 68 (Pt): Hb/A1−24/7
      Atom. . . . . 64 (Rh): Hb−9/14
      Atomic nucleus . . . . 64 (Rh): Hb−1/9
      Catalytic properties. . . . . 68 (Pt): Hb/A3−429/30
      Chemical reactions . . . . 64 (Rh): Hb−42/4
      Colloidal forms . . . . 68 (Pt): Hb/A3−424, 429/30
      Cosmochemistry . . . . 68 (Pt): Hb/A1−15/22
      Crystallographic properties . . . . . . 64 (Rh): Hb−14/6
      Detection . . . . 68 (Pt): Hb/A4−436/8, 488/9
      – in iron . . . . 59 (Fe): Hb/FII1−41

Hb = Main Series Volume; GD = Gmelin-Durrer; *Ab = Appended Volume to Main Series;*
Eb = Supplement Volume to Main Series; *Erg. W. = New Supplement Series Volume*

Rh . . . . Rhodium:
    Determination. . . . . 68 (Pt):   Hb/A4−464/7, 488/9
    − in special substances . . . . 68 (Pt):   Hb/A4−525/32
    Economic deposits
      see Pt ... Platinum

    Economics. . . . . 68 (Pt):   Hb/A2−302/10
    Electrical properties . . . . 64 (Rh):   Hb−35/40
    Electrochemical behavior . . . . . . 64 (Rh):   Hb−40/2
    Electrolytic deposition . . . . 68 (Pt):   Hb/A3−389/90
    Geochemical character . . . . 68 (Pt):   Hb/A1−23/8
    Geochemistry . . . . 68 (Pt):   Hb/A1−22/49
                                                   Hb/A2−275/6
    History . . . . 68 (Pt):   Hb/A1−11/2
    Magnetic properties . . . . 64 (Rh):   Hb−33/4
    Mechanical properties . . . . 64 (Rh):   Hb−16/9
    Mineralogy . . . . 68 (Pt):   Hb/A2−276/301
    Occurrence, extraterrestrial . . . . . 68 (Pt):   Hb/A1−15/22
    −, terrestrial . . . . 68 (Pt):   Hb/A1−22/49
                                                 Hb/A2−275/6
    − in meteorites . . . . 68 (Pt):   Hb/A1−16/22
    − in the biosphere. . . . . 68 (Pt):   Hb/A2−275
    − in the hydrosphere . . . . 68 (Pt):   Hb/A2−275
    − in the lithosphere . . . . 68 (Pt):   Hb/A1−28/49
    Optical properties. . . . . 64 (Rh):   Hb−22/33
    Physiological action . . . . 68 (Pt):   Hb/A2−275/6
    Position in the periodic system . . . . 64 (Rh):   Hb−14
    Preparation . . . . 68 (Pt):   Hb/A3−311/430
    − from minerals . . . . 68 (Pt):   Hb/A3−312/50
    − from wastes and residues . . . . . 68 (Pt):   Hb/A3−350/62
    − in pure state. . . . . 68 (Pt):   Hb/A3−372/4
    − of dense rhodium . . . . 68 (Pt):   Hb/A3−383/4
    Purity testing . . . . 68 (Pt):   Hb/A3−365/7
                                                 Hb/A4−519
    Separation, analytical . . . . 68 (Pt):   Hb/A4−495/6, 517/8
    Special forms . . . . 68 (Pt):   Hb/A3−399, 401/4,
                                                    424, 429/30
    Spectra . . . . 64 (Rh):   Hb−23/33
    Thermal properties . . . . 64 (Rh):   Hb−19/22
    Toxicity . . . . 68 (Pt):   Hb/A2−275/6
    Uses . . . . 68 (Pt):   Hb/A3−385/91
    Valence . . . . 64 (Rh):   Hb−13/4
−. . . . . Rhodium ions:
    Electrochemical behavior . . . . . . 64 (Rh):   Hb−41/2
    Ionic susceptibility . . . . 64 (Rh):   Hb−34
    Properties of the ion . . . . 64 (Rh):   Hb−9/12
    Spectra . . . . 64 (Rh):   Hb−26/7
−. . . . . Rhodium isotopes . . . . 64 (Rh):   Hb−1/2
                                                   Erg. W. 7b−355/6

See the "Instructions for the user"
at the beginning of the volume

Rh . . . . Rhodium systems:

Rh–Ag . . . . 68 (Pt):   Hb/A5–566
Rh–Au . . . . 68 (Pt):   Hb/A5–566/8
Rh–Au–Pt . . . . 68 (Pt):   Hb/A6–836
Rh–Be . . . . 68 (Pt):   Hb/A5–552
Rh–Bi . . . . 68 (Pt):   Hb/A5–550/2
Rh–Cd . . . . 68 (Pt):   Hb/A5–554
Rh–Cu . . . . 68 (Pt):   Hb/A5–563/5
Rh–Fe . . . . 68 (Pt):   Hb/A5–560/2
Rh–Fe–S . . . . 68 (Pt):   Hb/A5–563
Rh–P . . . . 68 (Pt):   Hb/A5–547/9
Rh–Pt . . . . 68 (Pt):   Hb/A6–829/30
Rh–Re . . . . 70 (Re):   Hb–86/7
Rh–S . . . . 64 (Rh):   Hb–62/4
Rh–Sn . . . . 68 (Pt):   Hb/A5–555/6

Rh^+ . . . . Rh^+ . . . . 64 (Rh):   Hb–10/1, 26/7
Rh^{2+} . . . . Rh^{2+} . . . . 64 (Rh):   Hb–9, 11, 26
Rh^{3+} . . . . Rh^{3+} . . . . 64 (Rh):   Hb–9/12, 34, 41/2
– . . . . $[Rh(H_2O)_6]^{3+}$ . . . . 64 (Rh):   Hb–58
Rh^{4+} . . . . Rh^{4+} . . . . 64 (Rh):   Hb–9, 11
Rh^{6+} . . . . Rh^{6+} . . . . 64 (Rh):   Hb–9

RhS . . . . RhS . . . . 64 (Rh):   Hb–64
RhS_2 . . . . RhS_2 . . . . 64 (Rh):   Hb–66
$RhSn$ . . . . $RhSn$ . . . . 68 (Pt):   Hb/A5–555/6
$RhSn_2$ . . . . $RhSn_2$ . . . . 68 (Pt):   Hb/A5–555/6
$RhSn_3$ . . . . $RhSn_3$ . . . . 68 (Pt):   Hb/A5–555
$RhSn_4$ . . . . $RhSn_4$ . . . . 68 (Pt):   Hb/A5–555/6
$RhZn_2$ . . . . $RhZn_2$ . . . . 68 (Pt):   Hb/A5–553/4
Rh_2S_3 . . . . Rh_2S_3 . . . . 64 (Rh):   Hb–65
Rh_2S_5 . . . . Rh_2S_5 . . . . 64 (Rh):   Hb–66
Rh_2Se_5 . . . . Rh_2Se_5 . . . . 64 (Rh):   Hb–69
Rh_2Sn . . . . Rh_2Sn . . . . 68 (Pt):   Hb/A5–555/6
Rh_2Te_5 . . . . Rh_2Te_5 . . . . 64 (Rh):   Hb–69
Rh_2Ti . . . . Rh_2Ti . . . . 68 (Pt):   Hb/A5–555
Rh_3S_4 . . . . Rh_3S_4 . . . . 64 (Rh):   Hb–64
Rh_3Sn_2 . . . . Rh_3Sn_2 . . . . 68 (Pt):   Hb/A5–555/6
Rh_3Ti . . . . Rh_3Ti . . . . 68 (Pt):   Hb/A5–555
Rh_5Zn_{21} . . . . Rh_5Zn_{21} . . . . 68 (Pt):   Hb/A5–554
Rh_9S_8 . . . . Rh_9S_8 . . . . 64 (Rh):   Hb–64

Rn . . . . Radon:

Affinity . . . . 1 (EG):   Hb–223/4
Atom . . . . 1 (EG):   Hb–207/9
Atomic nucleus . . . . 1 (EG):   Hb–199/207
Chemical reactions . . . . 1 (EG):   Hb–223/4, 229/31
Cosmochemistry . . . . 1 (EG):   Hb–6/7
Detection . . . . 1 (EG):   Hb–241/8
Determination . . . . 1 (EG):   Hb–241/8

Rn . . . . Radon:
    Diffusion . . . . 1 (EG): Hb–211/2
    Electrical properties . . . . 1 (EG): Hb–222
    Formation . . . . 1 (EG): Hb–193
    Geochemistry . . . . 1 (EG): Hb–7/35
    History . . . . 1 (EG): Hb–3/6
    Ionization . . . . 1 (EG): Hb–222
    Mechanical properties . . . . 1 (EG): Hb–209/21
    Molecule . . . . 1 (EG): Hb–209
    Nuclear decay, alpha radiation . . . . Erg. W. 7a–87/95
    Occurrence, extraterrestrial . . . . . 1 (EG): Hb–6/7
    –, terrestrial . . . . 1 (EG): Hb–7/35
    – in the atmosphere . . . . 1 (EG): Hb–7/17
    – in the hydrosphere . . . . 1 (EG): Hb–17/28
    – in the lithosphere . . . . 1 (EG): Hb–28/35
    Optical properties . . . . 1 (EG): Hb–221/2
    Position in the periodic system . . . . 1 (EG): Hb–209, 223
    Preparation . . . . 1 (EG): Hb–193/8
    Solubility . . . . 1 (EG): Hb–212/7
    Sorption . . . . 1 (EG): Hb–217/21
    – on palladium . . . . 65 (Pd): Hb/2–115
    – on platinum . . . . 68 (Pt): Hb/C1–1/3
    Spectra . . . . 1 (EG): Hb–221/2
    Thermal properties . . . . 1 (EG): Hb–210/1
    Uses . . . . 1 (EG): Hb–251
    Valence . . . . 1 (EG): Hb–223/4

– . . . . Rn · xH₂O . . . . Erg. W. 1–160
– . . . . Radon compounds . . . . Erg. W. 1–129/30, 160
– . . . . Radon isotopes:
    Analysis . . . . 1 (EG): Hb–245/7
    Preparation . . . . 1 (EG): Hb–198

– . . . . Radon solid solutions:
    Rn–CO₂ . . . . 14 (C): Hb/C1–507
    Rn–H₂S . . . . 9 (S): Hb/B1–19, 107
    Rn–SO₂ . . . . 9 (S): Hb/B1–210
    Rn · 6 H₂O–SO₂ · 6 H₂O . . . . 9 (S): Hb/B2–1120/1

– . . . . Radon systems:
    Rn–HCl . . . . 6 (Cl): Eb/B1–1

Rn⁺ . . . . Rn⁺ . . . . 1 (EG): Hb–222
Rn₂ . . . . Rn₂ . . . . 1 (EG): Hb–209

Ru . . . . Ruthenium:
    Absorption of X-rays . . . . 63 (Ru): Hb–20
                                                               Eb–31/2
    Abundance . . . . 68 (Pt): Hb/A1–24/7
    Atom . . . . 63 (Ru): Hb–3/7
                                                               Eb–10/4

See the "Instructions for the user"
at the beginning of the volume

Ru . . . . Ruthenium:

Atomic nucleus . . . . 63 (Ru):   Hb−1/3
                                                  Eb−3/10
Atomic refraction . . . . 63 (Ru):   Eb−32
Atomic scattering factors. . . . . . . 63 (Ru):   Eb−11
Atomic susceptibility . . . . 63 (Ru):   Eb−32
Atomic weight . . . . 63 (Ru):   Hb−6/7
                                                  Eb−2
Chemical reactions . . . . 63 (Ru):   Hb−22/4
                                                  Eb−69/73
Colloidal forms . . . . 68 (Pt):   Hb/A3−423/4
Cosmochemistry . . . . 68 (Pt):   Hb/A1−15/22
Crystallographic properties . . . . . 63 (Ru):   Hb−7/9
                                                  Eb−32/5
Detection . . . . 63 (Ru):   Eb−91/6
                                    68 (Pt):   Hb/A4−432/6, 488
– in iron . . . . 59 (Fe):   Hb/FII1−41
– in special substances . . . . 63 (Ru):   Eb−96
Determination. . . . . 63 (Ru):   Eb−96/102
                                    68 (Pt):   Hb/A4−463/4, 488
– in special substances . . . . 63 (Ru):   Eb−104/7
                                    68 (Pt):   Hb/A4−525/32
Economic deposits
    see Pt ... Platinum
Economics. . . . . 68 (Pt):   Hb/A2−306/10
Electrical properties . . . . 63 (Ru):   Hb−21
                                                  Eb−45/51
Electrochemical behavior . . . . . . 63 (Ru):   Hb−22
                                                  Eb−51/69
Electrolytic deposition . . . . 63 (Ru):   Eb−65/9
                                    68 (Pt):   Hb/A3−389
Electron configuration . . . . 63 (Ru):   Hb−3/4
                                                  Eb−1, 10
Geochemical character. . . . . 68 (Pt):   Hb/A1−23/8
Geochemistry . . . . 68 (Pt):   Hb/A1−22/49
                                                  Hb/A2−275/6
History. . . . . 68 (Pt):   Hb/A1−13/4
Ionization . . . . 63 (Ru):   Hb−21
                                                  Eb−11/2
Magnetic properties . . . . 63 (Ru):   Hb−20/1
                                                  Eb−45/51
Mass absorption coefficient . . . . . 63 (Ru):   Eb−31/2
Mechanical properties . . . . 63 (Ru):   Hb−9/10
                                                  Eb−35/45
Mineralogy . . . . 68 (Pt):   Hb/A2−276/301
Nuclear properties . . . . 63 (Ru):   Hb−1/3
                                                  Eb−3/10
Occurrence, extraterrestrial . . . . . 68 (Pt):   Hb/A1−15/22

Hb = Main Series Volume; GD = Gmelin-Durrer;   *Ab = Appended Volume to Main Series;*
Eb = Supplement Volume to Main Series;   *Erg. W. = New Supplement Series Volume*

Ru . . . . Ruthenium:

       Occurrence, terrestrial . . . . 68 (Pt):   Hb/A1–22/49
                                                    Hb/A2–275/6
       – in meteorites . . . . 68 (Pt):   Hb/A1–16/22
       – in the biosphere. . . . . 68 (Pt):   Hb/A2–275
       – in the hydrosphere . . . . 68 (Pt):   Hb/A2–275
       – in the lithosphere . . . . 68 (Pt):   Hb/A1–28/49
       Optical properties. . . . . 63 (Ru):   Hb–12/20
                                                    Eb–14/32, 51
       Physiological action . . . . 68 (Pt):   Hb/A2–275/6
       Position in the periodic system . . . . 63 (Ru):   Hb–7
                                                    Eb–1
       Preparation . . . . 68 (Pt):   Hb/A3–311/430
       – from minerals. . . . . 68 (Pt):   Hb/A3–312/50
       – from wastes and residues . . . . . 68 (Pt):   Hb/A3–350/62
       – in pure state. . . . . 68 (Pt):   Hb/A3–372
       – of dense ruthenium. . . . . 68 (Pt):   Hb/A3–383
       Redox reactions. . . . . 63 (Ru):   Eb–74/91
       Scattering of X-rays . . . . 63 (Ru):   Hb–20
                                                    Eb–31
       Separation, analytical . . . . 63 (Ru):   Eb–102/4
                                        68 (Pt):   Hb/A4–493/5, 517/8
       Sorption of hydrogen. . . . . 63 (Ru):   Eb–231/2
       Special forms . . . . 68 (Pt):   Hb/A3–398/9, 401/4,
                                                    423/4

       Spectra:
          Arc spectrum . . . . 63 (Ru):   Hb–12/5
                                                    Eb–14/9
          Band spectrum . . . . 63 (Ru):   Hb–15/6
          Crystal spectra . . . . 63 (Ru):   Eb–21/6
          Spark spectra. . . . . 63 (Ru):   Hb–12/5
                                                    Eb–19/21
          X-ray spectrum . . . . 63 (Ru):   Hb–16/20
                                                    Eb–26/32
       Thermal properties . . . . 63 (Ru):   Hb–10/2
                                                    Eb–35/45
       Toxicity . . . . 68 (Pt):   Hb/A2–275/6
       Uses . . . . 63 (Ru):   Eb–172/6
                                        68 (Pt):   Hb/A3–385/91
       Valence . . . . 63 (Ru):   Hb–7
–. . . . . Ruthenium ions
       see Ru$^+$ ... Ru$^+$, etc.
–. . . . . Ruthenium isotopes:
       Abundance . . . . 63 (Ru):   Hb–1/2
                                                    Eb–2/3
       Atomic nucleus . . . . 63 (Ru):   Hb–1/3
                                                    Eb–3/10
       Deposition from radioactive wastes . . 63 (Ru):   Eb–69

See the "Instructions for the user"
at the beginning of the volume

Ru . . . . Ruthenium isotopes:

Detection . . . . 63 (Ru):   Eb–107/16

Erg. W. 7b–355/6

Determination. . . . . 63 (Ru):   Eb–107/16

Enrichment and separation
of stable Ru isotopes. . . . . 63 (Ru):   Eb–117

Handling. . . . . 63 (Ru):   Eb–118

Preparation, separation, and
enrichment of unstable
Ru isotopes from fallout . . . . . 63 (Ru):   Eb–166/72

–, –, – – from nuclear fuels . . . . . . 63 (Ru):   Eb–136/42

–, –, – – from other elements . . . . . 63 (Ru):   Eb–118/36

–, –, – – from radioactive wastes . . . 63 (Ru):   Eb–69, 142/64

–, –, – – from reactor material . . . . . 63 (Ru):   Eb–164/6

–. . . . . Ruthenium systems:

Ru–Ag. . . . . 63 (Ru):   Eb–228/30

Ru–Al . . . . 63 (Ru):   Eb–178/80

Ru–Au. . . . . 63 (Ru):   Eb–230

Ru–B . . . . 63 (Ru):   Eb–318/9

Ru–Be . . . . 63 (Ru):   Eb–177/8

68 (Pt):   Hb/A5–536

Ru–Bi . . . . 63 (Ru):   Eb–337

Ru–C . . . . 63 (Ru):   Eb–321/2

Ru–Ce . . . . 63 (Ru):   Eb–181/3

Ru–Co . . . . 63 (Ru):   Eb–223/4

Ru–Co–U . . . . 63 (Ru):   Eb–224

Ru–Cr . . . . 63 (Ru):   Eb–202/4

Ru–Cu. . . . . 63 (Ru):   Eb–228

Ru–Fe . . . . 63 (Ru):   Eb–224/7

68 (Pt):   Hb/A5–543

Ru–Fe–S . . . . 68 (Pt):   Hb/A5–544/5

Ru–Ga. . . . . 63 (Ru):   Eb–180

Ru–Ge. . . . . 63 (Ru):   Eb–193

Ru–Hf . . . . 63 (Ru):   Eb–191

Ru–In . . . . 63 (Ru):   Eb–180

Ru–La–Th. . . . . 63 (Ru):   Eb–192/3

Ru–Mn . . . . 63 (Ru):   Eb–219/20

Ru–Mo . . . . 63 (Ru):   Eb–206/8

Ru–N . . . . 63 (Ru):   Eb–260

Ru–Nb. . . . . 63 (Ru):   Eb–195/6

Ru–Ni . . . . 63 (Ru):   Eb–220/2

68 (Pt):   Hb/A5–541

Ru–O . . . . 63 (Ru):   Eb–232/3

Ru–O–H_2O . . . . 63 (Ru):   Eb–232/3

Ru–P. . . . . 68 (Pt):   Hb/A5–534/5

Ru–Pd . . . . 68 (Pt):   Hb/A5–715/6

Ru–Pt . . . . 68 (Pt):   Hb/A6–826

Ru–$RuCl_3$ . . . . 63 (Ru):   Eb–275

Hb = Main Series Volume; GD = Gmelin-Durrer;   Ab = Appended Volume to Main Series;
Eb = Supplement Volume to Main Series;   Erg. W. = New Supplement Series Volume

Ru . . . . Ruthenium systems:
       Ru–S . . . . 63 (Ru):   Hb–71
                       Eb–313
       Ru–Si . . . . 63 (Ru):   Eb–331/2
       Ru–Sn . . . . 63 (Ru):   Eb–193/4
                    68 (Pt):   Hb/A5–538/9
       Ru–Ta . . . . 63 (Ru):   Eb–198/9
       Ru–Th . . . . 63 (Ru):   Eb–191/2
       Ru–Ti . . . . 63 (Ru):   Eb–187/9
                    68 (Pt):   Hb/A5–538
       Ru–U . . . . 63 (Ru):   Eb–213/6
       Ru–V . . . . 63 (Ru):   Eb–194/5
       Ru–W . . . . 63 (Ru):   Eb–210/3
       Ru–Zr . . . . 63 (Ru):   Eb–189/91
Ru^+ . . . . Ru^+:
       Ionic susceptibility . . . . 63 (Ru):   Eb–32
       Ionization energy . . . . 63 (Ru):   Eb–12
       Properties of the ion . . . . 63 (Ru):   Hb–3/4
       Spectra . . . . 63 (Ru):   Hb–13/5
                       Eb–19/21
       Thermodynamic functions . . . . . . 63 (Ru):   Eb–44
Ru^{2+} . . . . Ru^{2+}:
       Chemical reactions . . . . 63 (Ru):   Eb–74/7
       Electrochemical behavior . . . . 63 (Ru):   Eb–65/9
       Ionic refraction . . . . 63 (Ru):   Eb–32
       Ionic susceptibility . . . . 63 (Ru):   Eb–32
       Ionization energy . . . . 63 (Ru):   Eb–12
       Magnetic moment . . . . 63 (Ru):   Eb–22
       Properties of the ion . . . . 63 (Ru):   Hb–3/4
       Spectra . . . . 63 (Ru):   Hb–14
                       Eb–21/2
       Thermodynamic functions . . . . . . 63 (Ru):   Eb–44
– . . . . $[Ru(H_2O)_6]^{2+}$ . . . . 63 (Ru):   Eb–247/8
Ru^{3+} . . . . Ru^{3+}:
       Chemical reactions . . . . 63 (Ru):   Eb–77/81
       Electrochemical behavior . . . . 63 (Ru):   Eb–65/9
       Ionic refraction . . . . 63 (Ru):   Eb–32
       Ionic susceptibility . . . . 63 (Ru):   Hb–21
                       Eb–32
       Ionization energy . . . . 63 (Ru):   Eb–12
       Paramagnetic resonance . . . . 63 (Ru):   Eb–23/4
       Properties of the ion . . . . 63 (Ru):   Hb–3/5
                       Eb–13/4, 21/3
       Spectra . . . . 63 (Ru):   Eb–21, 22/4
– . . . . $[Ru(H_2O)_6]^{3+}$ . . . . 63 (Ru):   Eb–285
Ru^{4+} . . . . Ru^{4+}:
       Chemical reactions . . . . 63 (Ru):   Eb–81/6
       Electrochemical behavior . . . . 63 (Ru):   Eb–65/9

See the "Instructions for the user"
at the beginning of the volume

Ru^{4+}	Ru^{4+}:		
	Ionic refraction	63 (Ru):	Eb–32
	Ionic susceptibility	63 (Ru):	Eb–32
	Ionization energy	63 (Ru):	Eb–12
	Paramagnetic resonance absorption .	63 (Ru):	Eb–25
	Properties of the ion	63 (Ru):	Hb–3/4
			Eb–13/4, 21, 24/5
	Spectra	63 (Ru):	Eb–21, 25
–.	$[Ru(H_2O)_6]^{4+}$	63 (Ru):	Eb–248
Ru^{5+}	Ru^{5+}:		
	Chemical reactions	63 (Ru):	Eb–86
	Ionic susceptibility	63 (Ru):	Eb–32
	Ionization energy	63 (Ru):	Eb–12
	Magnetic moment	63 (Ru):	Eb–25
	Properties of the ion	63 (Ru):	Hb–3
	Spectra	63 (Ru):	Eb–21, 25
Ru^{6+}	Ru^{6+}:		
	Chemical reactions	63 (Ru):	Eb–86/7
	Ionic susceptibility	63 (Ru):	Eb–32
	Ionization energy	63 (Ru):	Eb–12
	Properties of the ion	63 (Ru):	Hb–3
	Spectra	63 (Ru):	Eb–21, 25/6
Ru^{7+}	Ru^{7+}:		
	Chemical reactions	63 (Ru):	Eb–87/8
	Ionic susceptibility	63 (Ru):	Eb–32
	Ionization energy	63 (Ru):	Eb–12
	Properties of the ion	63 (Ru):	Hb–3
	Spectra	63 (Ru):	Eb–26
Ru^{8+}	Ru^{8+}:		
	Chemical reactions	63 (Ru):	Eb–88/91
	Ionic refraction	63 (Ru):	Eb–32
	Ionization energy	63 (Ru):	Eb–12
	Properties of the ion	63 (Ru):	Hb–3
	Spectra	63 (Ru):	Eb–26
Ru^-	Ru^-	63 (Ru):	Eb–22, 32
RuS	RuS	63 (Ru):	Eb–313
RuSSb	RuSbS	63 (Ru):	Eb–337
RuS_2	RuS_2	63 (Ru):	Hb–71/3
			Eb–313/4
RuS_3	RuS_3	63 (Ru):	Hb–73
RuS_6	RuS_6	63 (Ru):	Hb–73
RuSb	RuSb	63 (Ru):	Eb–336
RuSbSe	RuSbSe	63 (Ru):	Eb–337
RuSbTe	RuSbTe	63 (Ru):	Eb–337
$RuSb_2$	$RuSb_2$	63 (Ru):	Eb–336/7
RuSc	RuSc	63 (Ru):	Eb–181
RuSeTe	RuSeTe	63 (Ru):	Eb–318

Hb = Main Series Volume; GD = Gmelin-Durrer;   Ab = Appended Volume to Main Series;
Eb = Supplement Volume to Main Series;   Erg. W. = New Supplement Series Volume

$RuSe_2$	$RuSe_2$	63 (Ru):	Hb−79
			Eb−317
RuSi.	RuSi	63 (Ru):	Hb−85
			Eb−333
$RuSi_{1.8}$	$RuSi_{1.8}$	63 (Ru):	Eb−332
$RuSi_2$	$RuSi_2$	63 (Ru):	Eb−332
$RuSn_2$	$RuSn_2$	63 (Ru):	Eb−193/4
		68 (Pt):	Hb/A5−539
$RuSn_2^+$	Sn_2Ru^+	63 (Ru):	Eb−562
RuTa	RuTa	63 (Ru):	Eb−199
RuTe	RuTe	63 (Ru):	Hb−79
$RuTe_2$	$RuTe_2$	63 (Ru):	Hb−79
			Eb−317/8
RuTh	RuTh	63 (Ru):	Eb−191/2
RuTi.	RuTi	63 (Ru):	Eb−187/9
		68 (Pt):	Hb/A5−538
RuTm	RuTm	63 (Ru):	Eb−186/7
RuU	RuU	63 (Ru):	Eb−216
RuU_2	RuU_2	63 (Ru):	Eb−216
RuV	RuV	63 (Ru):	Eb−194
RuY	RuY	63 (Ru):	Eb−181
RuZr	RuZr	63 (Ru):	Eb−189/91
Ru_2S_3	Ru_2S_3	63 (Ru):	Eb−313
−.	$Ru_2S_3 \cdot 2 H_2O$.	63 (Ru):	Eb−313
Ru_2S_6Tl	$TlRu_2S_6$	63 (Ru):	Eb−337/8
Ru_2Sc	Ru_2Sc	63 (Ru):	Eb−181
Ru_2Se_3	$Ru_2Se_3 \cdot 2 H_2O$	63 (Ru):	Eb−317
Ru_2Si	Ru_2Si	63 (Ru):	Eb−332
Ru_2Si_3	Ru_2Si_3	63 (Ru):	Eb−334
Ru_2Sm	Ru_2Sm	63 (Ru):	Eb−184
Ru_2Sn_3	Ru_2Sn_3	63 (Ru):	Eb−193/4
Ru_2Tb.	Ru_2Tb.	63 (Ru):	Eb−186
Ru_2Th.	Ru_2Th.	63 (Ru):	Eb−191/2
Ru_2Th_3	Ru_2Th_3	63 (Ru):	Eb−191/2
Ru_2Ti	Ru_2Ti	68 (Pt):	Hb/A5−538
Ru_2Tm	Ru_2Tm	63 (Ru):	Eb−186/7
Ru_2V	Ru_2V	63 (Ru):	Eb−194
Ru_2Y	Ru_2Y	63 (Ru):	Eb−181
Ru_2Yb.	Ru_2Yb	63 (Ru):	Eb−187
Ru_2Zr	Ru_2Zr	63 (Ru):	Eb−189/91
		68 (Pt):	Hb/A5−538
Ru_3Si_2.	Ru_3Si_2	63 (Ru):	Eb−332
Ru_3Sn_7	Ru_3Sn_7	63 (Ru):	Eb−193/4
		68 (Pt):	Hb/A5−538/9
Ru_3Th_7	Ru_3Th_7	63 (Ru):	Eb−191/2
Ru_3Ti	Ru_3Ti	68 (Pt):	Hb/A5−538
Ru_3U	Ru_3U	63 (Ru):	Eb−216
Ru_3U_2.	Ru_3U_2	63 (Ru):	Eb−216
Ru_4Si_3.	Ru_4Si_3	63 (Ru):	Eb−332/3

See the "Instructions for the user"
at the beginning of the volume

Ru_4U_3 . . . . Ru_4U_3 . . . . 63 (Ru):   Eb−216
Ru_5U_3 . . . . Ru_5U_3 . . . . 63 (Ru):   Eb−216

$S_{0.95}V$ . . . . $VS_{0.95}$ . . . . 48 (V):   Hb/B1−268, 270/1
S . . . . Sulfur:
                         Absorption of light . . . . 9 (S):   Hb/A3−643/4
                         − of X-rays . . . . 9 (S):   Hb/A3−674/6
                         Abundance . . . . 9 (S):   Hb/A2−81
                         Adhesion . . . . 9 (S):   Hb/A3−612
                         Aerosols . . . . 9 (S):   Hb/A2−486
                         Atom . . . . 9 (S):   Hb/A3−546/60
                         Atomic nucleus . . . . 9 (S):   Hb/A3−537/46
                         Atomic radius . . . . 9 (S):   Hb/A3−548/53
                         Atomic refraction . . . . 9 (S):   Hb/A3−642/3
                         Atomic scattering factors . . . . 9 (S):   Hb/A3−600/1
                         Atomic volume . . . . 9 (S):   Hb/A3−553/6
                         Atomic weight . . . . 9 (S):   Hb/A3−556/9
                         Becquerel effect . . . . 9 (S):   Hb/A3−687
                         Behavior as anode . . . . 9 (S):   Hb/A3−689
                         − as cathode . . . . 9 (S):   Hb/A3−688/9
                         − as solvent . . . . 9 (S):   Hb/A3−762
                         − at the dropping
                           mercury electrode . . . . 9 (S):   Hb/A3−688
                         Boiling point . . . . 9 (S):   Hb/A3−623/5
                         Bond energy of the molecule . . . . . 9 (S):   Hb/A3−569
                         Cells . . . . 9 (S):   Hb/A3−686
                         Chemical constant . . . . 9 (S):   Hb/A3−637
                         Chemical reactions . . . . 9 (S):   Hb/A3−689/727
                         − on irradiation . . . . 9 (S):   Hb/A3−689
                         − with acids . . . . 9 (S):   Hb/A3−693/4
                         − with air . . . . 9 (S):   Hb/A3−689/90
                         − with alkaline solutions . . . . 9 (S):   Hb/A3−692
                         − with metal compounds . . . . 9 (S):   Hb/A3−705/26
                         − with metals . . . . 9 (S):   Hb/A3−696/9
                         − with nonmetal compounds . . . . . 9 (S):   Hb/A3−699/705
                         − with nonmetals . . . . 9 (S):   Hb/A3−694/5
                         − with organic compounds . . . . . . 9 (S):   Hb/A3−723, 726/7
                         − with oxygen . . . . 9 (S):   Hb/B1−155/66
                         − with water . . . . 9 (S):   Hb/A3−690/2
                         Cleavage . . . . 9 (S):   Hb/A3−591
                         Coagels . . . . 9 (S):   Hb/A2−501/2
                         Colloidal forms . . . . 9 (S):   Hb/A2−485/503
                         Color . . . . 9 (S):   Hb/A3−638/40
                         Commercial products . . . . 9 (S):   Hb/A2−283
                         Compressibility . . . . 9 (S):   Hb/A3−609
                         Cosmochemistry . . . . 9 (S):   Hb/A2−61/8
                         Critical constants . . . . 9 (S):   Hb/A3−622
                         Cryoscopic constant . . . . 9 (S):   Hb/A3−626
                         Crystal chemistry . . . . 9 (S):   Hb/A2−68/81

Hb = Main Series Volume; GD = Gmelin-Durrer;   *Ab = Appended Volume to Main Series;*
Eb = Supplement Volume to Main Series;   *Erg. W. = New Supplement Series Volume*

S . . . . Sulfur:
    Crystal form . . . . 9 (S):   Hb/A3–586/9
    Crystallization . . . . 9 (S):   Hb/A3–589/91
    Crystallographic properties . . . . . . 9 (S):   Hb/A3–571/601
    Debye temperature . . . . 9 (S):   Hb/A3–635
    Deformation . . . . 9 (S):   Hb/A3–601
    Density . . . . 9 (S):   Hb/A3–602/5
    Detection in iron . . . . 59 (Fe):   Hb/FI–341/2
    Determination in iron . . . . 59 (Fe):   Hb/FI–343/68
                                  Hb/FII2–315/6, 349/50
    Dielectric constant . . . . 9 (S):   Hb/A3–679/80
    Dielectric relaxation . . . . 9 (S):   Hb/A3–683
    Diffraction . . . . 9 (S):   Hb/A3–644
    Diffusion . . . . 9 (S):   Hb/A3–616/7
    Dissociation of the molecule . . . . . . 9 (S):   Hb/A3–560/7, 569/71
    Dressing of raw materials . . . . 9 (S):   Hb/A2–204/11
    Ebullioscopic constant . . . . 9 (S):   Hb/A3–625
    Economic deposits:
        Africa . . . . 9 (S):   Hb/A2–183/4
        America . . . . 9 (S):   Hb/A2–184/90
        Asia . . . . 9 (S):   Hb/A2–178/82
        Australia . . . . 9 (S):   Hb/A2–183
        Europe . . . . 9 (S):   Hb/A2–156/74
        Soviet Union . . . . 9 (S):   Hb/A2–174/8
    Economics . . . . 9 (S):   Hb/A2–199/203
    Elasticity . . . . 9 (S):   Hb/A3–605/7
    Electrical conductivity . . . . 9 (S):   Hb/A3–680/3
    Electrical moment of the molecule . . 9 (S):   Hb/A3–569
    Electrical properties . . . . 9 (S):   Hb/A3–679/85
    Electrochemical behavior . . . . 9 (S):   Hb/A3–685/9
    Electrokinetic phenomena . . . . 9 (S):   Hb/A3–687/8
    Electrolytic deposition . . . . 9 (S):   Hb/A3–688
    Electron configuration . . . . 9 (S):   Hb/A3–546/7
    Electronegativity . . . . 9 (S):   Hb/A3–559/60
    Electron emission . . . . 9 (S):   Hb/A3–683/4
    Enthalpy . . . . 9 (S):   Hb/A3–635/7
    Entropy . . . . 9 (S):   Hb/A3–635/7
    Equation of state . . . . 9 (S):   Hb/A3–622
    Expansion coefficient . . . . 9 (S):   Hb/A3–620/2
    Formation . . . . 9 (S):   Hb/A3–511, 517/23
                                  Hb/B1–278/83
    Free energy . . . . 9 (S):   Hb/A3–635/7
    Geochemical character . . . . 9 (S):   Hb/A2–81/2
    Geochemical cycle . . . . 9 (S):   Hb/A2–82/3
    Geochemistry . . . . 9 (S):   Hb/A2–68/155
    Hardness . . . . 9 (S):   Hb/A3–605
    Heat content . . . . 9 (S):   Hb/A3–635/6
    Heat of fusion . . . . 9 (S):   Hb/A3–628/9
    – of sublimation . . . . 9 (S):   Hb/A3–627/8

See the "Instructions for the user"
at the beginning of the volume

S . . . . Sulfur:

Heat of transformation	9 (S):	Hb/A3–629/31
– of vaporization	9 (S):	Hb/A3–627
History:		
Antiquity	9 (S):	Hb/A1–1/7
Middle ages	9 (S):	Hb/A1–8/21
Modern times	9 (S):	Hb/A1–22/60
Hydrosols	9 (S):	Hb/A2–486/501
Internal pressure	9 (S):	Hb/A3–610
Ionization	9 (S):	Hb/A3–684/5
Isomorphism	9 (S):	Hb/A3–591
Lattice structure.	9 (S):	Hb/A3–592/5
Luminescence.	9 (S):	Hb/A3–645
Magnetic moment of the atom	9 (S):	Hb/A3–547/8
of the molecule	9 (S):	Hb/A3–569
Magnetic properties	9 (S):	Hb/A3–676/9
Magnetooptical effects	9 (S):	Hb/A3–644
Mechanical properties	9 (S):	Hb/A3–602/20
Melting point	9 (S):	Hb/A3–625
Mineralogy	9 (S):	Hb/A2–190/9
Modifications	9 (S):	Hb/A3–517/23, 530/7, 571/86
Molar refraction.	9 (S):	Hb/A3–642/3
Molecule.	9 (S):	Hb/A3–560/71
Nuclear properties	9 (S):	Hb/A3–537/46
Occurrence, extraterrestrial	9 (S):	Hb/A2–61/8
–, terrestrial	9 (S):	Hb/A2–68/155
– in meteorites	9 (S):	Hb/A2–62/8
– in soils.	9 (S):	Hb/A2–149/55
– in the atmosphere	9 (S):	Hb/A2–127/34
– in the biosphere.	9 (S):	Hb/A2–135/49
– in the hydrosphere	9 (S):	Hb/A2–116/27
– in the lithosphere	9 (S):	Hb/A2–83/115
Optical properties.	9 (S):	Hb/A3–638/76
Organosols	9 (S):	Hb/A2–502/3
Phase transitions	9 (S):	Hb/A3–525/37
Phosphorescence.	9 (S):	Hb/A3–645
Photoelectric effect	9 (S):	Hb/A3–683/4
Photophoresis.	9 (S):	Hb/A3–645
Piezoelectricity	9 (S):	Hb/A3–683
Plasticity.	9 (S):	Hb/A3–605/7
Polarizability	9 (S):	Hb/A3–547
Polarization	9 (S):	Hb/A3–644
Polymorphism	9 (S):	Hb/A3–530/7, 571/86
Position in the		
electrochemical series	9 (S):	Hb/A3–686
– in the periodic system	9 (S):	Hb/A3–559
Potentials	9 (S):	Hb/A3–685/6
Preparation	9 (S):	Hb/A3–511/2

Hb = Main Series Volume; GD = Gmelin-Durrer;   Ab = Appended Volume to Main Series;
Eb = Supplement Volume to Main Series;   Erg. W. = New Supplement Series Volume

S . . . . Sulfur:

Preparation, industrial	9 (S):	Hb/A2 – 199/284
–, –, from coal.	9 (S):	Hb/A2 – 258/78
–, –, from gases	9 (S):	Hb/A2 – 258
–, –, from hydrogen sulfide	9 (S):	Hb/A2 – 241/58
–, –, from ores.	9 (S):	Hb/A2 – 211/23
–, –, from sulfur dioxide	9 (S):	Hb/A2 – 224/41
–, –, from other raw materials	9 (S):	Hb/A2 – 278/80
– of colloidal forms	9 (S):	Hb/A2 – 487/91
– of modifications.	9 (S):	Hb/A3 – 517/23
– of special forms.	9 (S):	Hb/A3 – 516/7
Purification	9 (S):	Hb/A2 – 280/3
		Hb/A3 – 512/5
Purity testing	9 (S):	Hb/A2 – 283/4
		Hb/A3 – 515/6
Reflection	9 (S):	Hb/A3 – 640
Refraction	9 (S):	Hb/A3 – 640/3
Scattering of X-rays	9 (S):	Hb/A3 – 674/6
Solidification point	9 (S):	Hb/A3 – 625
Solubility in inorganic solvents	9 (S):	Hb/A3 – 731/3
		Hb/B3 – 1748/52
– in organic solvents	9 (S):	Hb/A3 – 733/62
Sorption	9 (S):	Hb/A3 – 617/20
Special forms	9 (S):	Hb/A3 – 516/7
Specific heat	9 (S):	Hb/A3 – 631/5
Spectra:		
Arc spectrum	9 (S):	Hb/A3 – 645/9
Band spectrum	9 (S):	Hb/A3 – 654/63
Continuous spectrum	9 (S):	Hb/A3 – 663/4
Raman spectrum	9 (S):	Hb/A3 – 664/5
Spark spectra	9 (S):	Hb/A3 – 649/54
X-ray spectrum	9 (S):	Hb/A3 – 665/76
Strength	9 (S):	Hb/A3 – 607/8
Structure	9 (S):	Hb/A3 – 595/600
Surface tension	9 (S):	Hb/A3 – 610/2
Susceptibility	9 (S):	Hb/A3 – 676/9
Technology	9 (S):	Hb/A2 – 199/284
Texture	9 (S):	Hb/A3 – 601
Thermal conductivity	9 (S):	Hb/A3 – 637/8
Thermal expansion	9 (S):	Hb/A3 – 620/2
Thermal properties	9 (S):	Hb/A3 – 525/37,
		620/38
Toxicity	9 (S):	Hb/A2 – 503
Transition point	9 (S):	Hb/A3 – 626/7
Twinning	9 (S):	Hb/A3 – 591
Valence	9 (S):	Hb/A3 – 559
Vapor pressure	9 (S):	Hb/A3 – 525/9, 622/3
Vibrational frequency.	9 (S):	Hb/A3 – 635
Viscosity	9 (S):	Hb/A3 – 612/6

See the "Instructions for the user"
at the beginning of the volume

S . . . . Sulfur:
               Wettability. . . . . 9 (S):   Hb/A3−611/2
               X-ray photoelectric effect . . . . . . 9 (S):   Hb/A3−684

−. . . . . Sulfur ions
               see S^+ ... S^+, etc.

−. . . . . Sulfur isotopes:
               Abundance . . . . 9 (S):   Hb/A3−538
               Enrichment and separation . . . . . . 9 (S):   Hb/A3−523/5
               Formation . . . . 9 (S):   Hb/A3−541/2
               Nuclear properties . . . . 9 (S):   Hb/A3−537/46

−. . . . . Sulfur solutions:
               $S-AlBr_3$ . . . . 9 (S):   Hb/A3−728
               $S-AsCl_3$ . . . . 9 (S):   Hb/A3−729, 732
               $S-Br$ . . . . 9 (S):   Hb/A3−727, 731
               $S-CS_2$ . . . . 9 (S):   Hb/A3−729/31, 743/8
               $S-Cl$ . . . . 9 (S):   Hb/A3−729, 731
               $S-H_2S_2O_7$ . . . . 9 (S):   Hb/A3−728
               $S-HgBr_2$ . . . . 9 (S):   Hb/A3−728
               $S-HgCl_2$ . . . . 9 (S):   Hb/A3−728, 732
               $S-I$ . . . . 9 (S):   Hb/A3−727, 729, 731
               $S-NH_3$ . . . . 9 (S):   Hb/A3−732
               S−organic solvents . . . . 9 (S):   Hb/A3−728/30,
                                                     733/62
               $S-P$ . . . . 9 (S):   Hb/A3−727, 731
               $S-SO_3$ . . . . 9 (S):   Hb/A3−731
               $S-S_2Cl_2$ . . . . 9 (S):   Hb/A3−727, 729, 731/2
                                                     Hb/B3−1751/2
               $S-SbCl_3$ . . . . 9 (S):   Hb/A3−714, 727, 732
               $S-SeOCl_2$ . . . . 9 (S):   Hb/A3−732
                $S-SnCl_4$ . . . . 9 (S):   Hb/A3−729, 732

−. . . . . Sulfur systems:
               S−Ag . . . . 61 (Ag):   Hb/B3−2/8
               S−Ag−As . . . . 61 (Ag):   Hb/B4−53/6
               S−Ag−As−Cu. . . . . 61 (Ag):   Hb/B4−487
               S−Ag−Au . . . . 62 (Au):   Hb/3−766
               S−Ag−Bi . . . . 61 (Ag):   Hb/B4−127/8
               S−Ag−Cu . . . . 61 (Ag):   Hb/B4−458/63
               S−Ag−Cu−Fe. . . . . 61 (Ag):   Hb/B4−472
               S−Ag−Cu−Ni. . . . . 61 (Ag):   Hb/B4−471/2
               S−Ag−Cu−Pb. . . . . 61 (Ag):   Hb/B4−470
               S−Ag−Cu−Sb . . . . 61 (Ag):   Hb/B4−491
               S−Ag−Fe . . . . 61 (Ag):   Hb/B4−411/3
               S−Ag−O. . . . . 9 (S):   Hb/B1−185
                                          61 (Ag):   Hb/B3−61/2
               S−Ag−Sb . . . . 61 (Ag):   Hb/B4−90/2
               S−Ag−Tl . . . . 61 (Ag):   Hb/B4−225
               S−Al−C−Cu−Fe . . . . 60 (Cu):   Hb/B3−1327
               S−As. . . . . 17 (As):   Hb−415

S . . . . Sulfur systems:

S–As–Bi	19 (Bi):	Eb–858
S–As–Co–Cu.	60 (Cu):	Hb/B3–1269
S–As–Cu	60 (Cu):	Hb/B2–959/60
S–As–Cu–Fe	60 (Cu):	Hb/B3–1315
S–As–Cu–Ni	60 (Cu):	Hb/B3–1259
S–As–Cu–Sb	60 (Cu):	Hb/B2–973
S–As–Ni	57 (Ni):	Hb/B3–989/91
S–As–Pb	47 (Pb):	Hb/C3–901/3
S–Au–Cu	62 (Au):	Hb/3–764
S–Au–Fe	62 (Au):	Hb/3–764
S–Ba–O	9 (S):	Hb/B1–183
S–BaO–BaS–BaSO$_4$–SO$_2$	30 (Ba):	Eb–448
S–BaS	30 (Ba):	Hb–257
S–BaS–BaSO$_4$–SO$_2$	30 (Ba):	Eb–448
S–Bi	19 (Bi):	Hb–161
		Eb–725/9
S–Bi–Co	58 (Co):	Eb/A–756
S–Bi–Co–H	58 (Co):	Eb/A–756
S–Bi–Cu	60 (Cu):	Hb/B2–974/5
S–Bi–H	19 (Bi):	Hb–161
S–Bi–Ni	57 (Ni):	Hb/B1–14
		Hb/B3–996
S–Bi–O	9 (S):	Hb/B1–182
	19 (Bi):	Eb–745
S–Bi–Pb	47 (Pb):	Hb/C3–969/70
S–Br	9 (S):	Hb/B3–1864
S–Br–I	9 (S):	Hb/B3–1873
S–Ca–O	9 (S):	Hb/B1–182/3
	28 (Ca):	Hb/B3–657/60
S–CaO	28 (Ca):	Hb/B3–657
S–CaSO$_4$	28 (Ca):	Hb/B3–658
S–Cd	33 (Cd):	Eb–584
S–Cl	9 (S):	Hb/B3–1748/59
S–Cl–O–Ta	50 (Ta):	Hb/B1–178
S–Cl–Pb–Sn	47 (Pb):	Hb/C4–1586
S–Cl–Ta	50 (Ta):	Hb/B1–178
S–Co	58 (Co):	Hb/A2–318
		Eb/A–605/7
S–Co–Cu	60 (Cu):	Hb/B3–1266
S–Co–H–Sb	58 (Co):	Eb/A–755/6
S–Co–Ni	58 (Co):	Hb/A1–210
		Eb/A–881
S–Co–O	9 (S):	Hb/B1–183
	58 (Co):	Eb/A–626
S–Co–Sb	58 (Co):	Eb/A–754/5
S–CoS	58 (Co):	Eb/A–607
S–Cr	52 (Cr):	Hb/B–295/7
S–Cs	25 (Cs):	Hb/2–217

See the "Instructions for the user"
at the beginning of the volume

S . . . . Sulfur systems:

S–Cu	60 (Cu):	Hb/B1–432/4
S–Cu–Fe	60 (Cu):	Hb/B3–1282/4
S–Cu–Fe–Mn	60 (Cu):	Hb/B3–1330
S–Cu–Fe–O	60 (Cu):	Hb/B3–1295
S–Cu–Fe–Sb	60 (Cu):	Hb/B3–1315
S–Cu–Fe–Sn	60 (Cu):	Hb/B3–1327
S–Cu–Fe–Zn	60 (Cu):	Hb/B3–1325/6
S–Cu–Mn	60 (Cu):	Hb/B3–1246
S–Cu–Mn–O	60 (Cu):	Hb/B3–1246
S–Cu–Ni	60 (Cu):	Hb/B3–1255/6
S–Cu–O	9 (S):	Hb/B1–184/5
S–Cu–Pb	60 (Cu):	Hb/B3–1200/2
S–Cu–Sb	60 (Cu):	Hb/B2–971
S–Cu–Sn	60 (Cu):	Hb/B3–1195/6
S–Cu–Zn	60 (Cu):	Hb/B3–1148
S–Fe	59 (Fe):	Hb/A6–1168/70
S–Fe–FeO–Ni–NiO–SiO₂	59 (Fe):	Hb/B5–1160
S–Fe–H	59 (Fe):	Hb/A5–939/40
S–Fe–Ir	68 (Pt):	Hb/A6–738
S–Fe–Mn	59 (Fe):	Hb/A5–936/9
S–Fe–O	9 (S):	Hb/B1–183/4
S–Fe–Os	68 (Pt):	Hb/A6–726
S–Fe–Pd	68 (Pt):	Hb/A5–624
S–Fe–Pt	68 (Pt):	Hb/A6–798
S–Fe–Rh	68 (Pt):	Hb/A5–563
S–Fe–Ru	68 (Pt):	Hb/A5–544/5
S–FeS	59 (Fe):	Hb/A6–1170
S–H₂S	9 (S):	Hb/B1–107/8
S–I	9 (S):	Hb/B3–1870/1
S–I–Se	10 (Se):	Hb/B–193
S–In	37 (In):	Hb–96
S–Ir	67 (Ir):	Hb–74
S–Ir₂S₃	67 (Ir):	Hb–75
S–K	22 (K):	Hb/3–691/2
S–Li₂S	20 (Li):	Eb–473/4
S–Na	21 (Na):	Hb–484/5
S–Na–Ni	57 (Ni):	Hb/B3–1016
S–Na₂S	21 (Na):	Eb/3–1059/60
S–Nb	49 (Nb):	Hb/B1–257/8
S–Ni	57 (Ni):	Hb/B2–627/31
S–Ni–O	9 (S):	Hb/B1–183
	57 (Ni):	Hb/B2–676/7
S–Ni–Pb	57 (Ni):	Hb/B3–1195
S–Ni–Sb	57 (Ni):	Hb/B1–10
		Hb/B3–994/5
S–O–Pb	9 (S):	Hb/B1–183
	47 (Pb):	Hb/C2–533/4

S . . . . Sulfur systems:

S–O–Pd	9 (S):	Hb/B1–185
S–O–Sb	9 (S):	Hb/B1–182
S–O–Sr	9 (S):	Hb/B1–183
S–O–Ti	41 (Ti):	Hb–343
S–Os	66 (Os):	Hb–58
S–P	16 (P):	Hb/C–567
S–Pb	47 (Pb):	Hb/C2–413/4
S–Pb–Sb	47 (Pb):	Hb/C3–935/6
S–Pb–Se	47 (Pb):	Hb/C2–639/40
S–Pb–Si	47 (Pb):	Hb/C3–855
S–Pb–Sn	47 (Pb):	Hb/C4–1584
S–Pd	65 (Pd):	Hb/2–289
S–Pt	68 (Pt):	Hb/C1–103/4
S–Pu	Erg. W. 4–159/61	
S–Rb	24 (Rb):	Hb–205
S–Re	70 (Re):	Hb–117
S–Rh	64 (Rh):	Hb–62/4
S–Ru	63 (Ru):	Hb–71
		Eb–313
S–SO$_2$–SrO–SrS–SrSO$_4$	29 (Sr):	Eb–257
S–SO$_2$–SrS–SrSO$_4$	29 (Sr):	Eb–257
S–S$_2$Br$_2$	9 (S):	Hb/B3–1864
S–S$_2$Cl$_2$	9 (S):	Hb/B3–1748/52
S–Sb	18 (Sb):	Hb/B3–499/500
S–Se	10 (Se):	Hb/B–160/70
S–Se–Te	11 (Te):	Hb–363
S–Se$_2$Cl$_2$	10 (Se):	Hb/B–193
S–Ta	50 (Ta):	Hb/B1–169/70
S–Te	11 (Te):	Hb–350/1
S–Th	44 (Th):	Hb–277/8
S–Ti	41 (Ti):	Hb–337/9
S–Tl	38 (Tl):	Hb/3–339
S–V	48 (V):	Hb/B1–267/8
S–Zr	42 (Zr):	Hb–326

S$^+$ . . . . S$^+$:

Ionization potential	9 (S):	Hb/A3–684/5
Properties of the ion	9 (S):	Hb/A3–547/50, 554
Spectra	9 (S):	Hb/A3–649/50

S^{2+} . . . . S^{2+}:

Ionization potential	9 (S):	Hb/A3–684/5
Properties of the ion	9 (S):	Hb/A3–547/50, 554
Spectra	9 (S):	Hb/A3–650/1

S^{3+} . . . . S^{3+}:

Ionization potential	9 (S):	Hb/A3–684/5
Properties of the ion	9 (S):	Hb/A3–547/50, 554
Spectra	9 (S):	Hb/A3–651/2

See the "Instructions for the user"
at the beginning of the volume

S^{4+} . . . . S^{4+}:
Electrochemical behavior . . . . 9 (S): Hb/A3−688
Formation . . . . 9 (S): Hb/B1−310
Ionic refraction . . . . 9 (S): Hb/A3−642/3
Ionization potentials . . . . 9 (S): Hb/A3−684/5
Properties of the ion . . . . 9 (S): Hb/A3−547/50, 554/6
Scattering factors . . . . 9 (S): Hb/A3−601
Spectra . . . . 9 (S): Hb/A3−652

S^{5+} . . . . S^{5+}:
Ionization potential . . . . 9 (S): Hb/A3−684/5
Properties of the ion . . . . 9 (S): Hb/A3−547/50, 554
Spectra . . . . 9 (S): Hb/A3−652/3

S^{6+} . . . . S^{6+}:
Ionic refraction . . . . 9 (S): Hb/A3−642/3
Ionization potential . . . . 9 (S): Hb/A3−684/5
Properties of the ion . . . . 9 (S): Hb/A3−547/56
Scattering factors . . . . 9 (S): Hb/A3−600
Spectra . . . . 9 (S): Hb/A3−653

S^{n+} . . . . S^{n+} (n = 7 to 11) . . . . 9 (S): Hb/A3−547, 653/4, 684/5

S^- . . . . S^- . . . . 9 (S): Hb/A3−547/50, 554, 684/5
S^{2-} . . . . S^{2-}:
Chemical reactions . . . . 9 (S): Hb/B1−86/106
Electrochemical behavior . . . . 9 (S): Hb/A3−688
                                                 Hb/B1−83/6
Formation . . . . 9 (S): Hb/B1−81
Hydration . . . . 9 (S): Hb/B1−89
Ionic mobility . . . . 9 (S): Hb/B1−85
Ionic refraction . . . . 9 (S): Hb/A3−642/3
                                                 Hb/B1−37/8
Ionization potential . . . . 9 (S): Hb/A3−684/5
Properties of the ion . . . . 9 (S): Hb/A3−547/56
Scattering factors . . . . 9 (S): Hb/A3−600

SSb . . . . SbS . . . . 18 (Sb): Hb/B3−500
SSe . . . . SeS . . . . 10 (Se): Hb/B−172/3
SSe^{2-} . . . . $[SeS]^{2-}$ . . . . 10 (Se): Hb/B−175/6
SSeTa . . . . TaSeS . . . . 50 (Ta): Hb/B1−189
SSi . . . . SiS . . . . 15 (Si): Hb/B−743/6
SSiZn . . . . ZnSiS . . . . 32 (Zn): Hb−274
SSn . . . . SnS solid solutions:
SnS−PbS . . . . 47 (Pb): Hb/C4−1585

−. . . . . SnS systems:
SnS−FeS . . . . 59 (Fe): Hb/B5−1130
SnS−Pb−PbS−Sn . . . . 47 (Pb): Hb/C4−1584
SnS−PbCl$_2$−PbS−SnCl$_2$ . . . . 47 (Pb): Hb/C4−1586
SnS−PbS . . . . 47 (Pb): Hb/C4−1585
SnS−PbSe . . . . 47 (Pb): Hb/C4−1587/8

Hb = Main Series Volume; GD = Gmelin-Durrer; Ab = Appended Volume to Main Series;
Eb = Supplement Volume to Main Series; Erg. W. = New Supplement Series Volume

SSr	SrS	29 (Sr):	Hb−17/9, 21/2, 165/9
			Eb−146, 244/8
−.	SrS solutions:		
	SrS−H_2O	29 (Sr):	Hb−169
−.	SrS systems:		
	SrS−CaS	29 (Sr):	Eb−304
	SrS−MgS	29 (Sr):	Eb−303
	SrS−S−SO_2−SrO−$SrSO_4$	29 (Sr):	Eb−257
	SrS−S−SO_2−$SrSO_4$	29 (Sr):	Eb−257
	SrS−$SrCl_2$−SrO	29 (Sr):	Eb−249
SSr₂	Sr_2S	29 (Sr):	Hb−165
STaTe	TaTeS	50 (Ta):	Hb/B1−192
STa₂	Ta_2S	50 (Ta):	Hb/B1−170, 171/2
STa₆	Ta_6S	50 (Ta):	Hb/B1−170, 171/2
STe	TeS	11 (Te):	Hb−350, 353
STh	ThS	44 (Th):	Hb−277, 279
STi	TiS	41 (Ti):	Hb−337/9, 340
−.	TiS systems:		
	TiS−CoS−TiS_2	58 (Co):	Eb/A−870
	TiS−Fe−FeS−Fe_2Ti	59 (Fe):	Hb/A9−1874
	TiS−NiS−TiS_2	57 (Ni):	Hb/B3−1181/2
STi₂	Ti_2S	41 (Ti):	Hb−337/40
STl₂	Tl_2S	38 (Tl):	Hb/1−23
			Hb/3−339/45, 346/9
−.	Tl_2S systems:		
	Tl_2S−As_2S_3	38 (Tl):	Hb/3−435
	Tl_2S−PbS	47 (Pb):	Hb/C4−1352/3
SU	US	55 (U):	Hb−142
−.	US solid solutions:		
	(U,Pu)S	Erg. W. 4−168/71	
SV	VS	48 (V):	Hb/B1−271/3
−.	VS systems:		
	VS−Fe−FeS−V	59 (Fe):	Hb/A9−1924/5
SV₃	V_3S	48 (V):	Hb/B1−268, 269/70
SZn	ZnS:		
	Chemical reactions	32 (Zn):	Hb−201/6
			Eb−924/34
	Colloidal forms	32 (Zn):	Hb−206/7
			Eb−905
	Electrochemical behavior	32 (Zn):	Eb−924
	Formation	32 (Zn):	Hb−193/7, 198
			Eb−903/4, 914
	Lithopone	32 (Zn):	Hb−202/3, 217/8
			Eb−908/14,
			926/7, 929

See the "Instructions for the user"
at the beginning of the volume

SZn	ZnS:		
	Phosphors	32 (Zn):	Hb–202/3, 207/17
			Eb–904/5, 923,
			927, 929
	Physical properties	32 (Zn):	Hb–198/201,
			209/17
			Eb–914/24
	Pigments	32 (Zn):	Hb–217
			Eb–906/8
	Preparation	32 (Zn):	Hb–193/7
			Eb–903/5
	–, industrial	32 (Zn):	Hb–197
			Eb–905/14
	– of special forms	32 (Zn):	Hb–206/9, 217
			Eb–904/5, 906/14
	Sidot's blende	32 (Zn):	Hb–207/17
	Solubility	32 (Zn):	Hb–201
	Special forms	32 (Zn):	Hb–202/3, 206/18
			Eb–904/14, 923,
			926/7, 929
–	7 ZnS · H_2O	32 (Zn):	Hb–219
–	ZnS solid solutions:		
	ZnS–HgS	34 (Hg):	Hb/B4–1501/2
	$Zn_{1-x}Hg_xS_ySe_{1-y}$	34 (Hg):	Hb/B4–1502
	(Ni,Zn)S	57 (Ni):	Hb/B3–1136
–	ZnS systems:		
	ZnS–$AgAlS_2$	61 (Ag):	Hb/B4–185/6
	ZnS–$AgGaS_2$	61 (Ag):	Hb/B4–194
	ZnS–$AgInS_2$	61 (Ag):	Hb/B4–207
	ZnS–Ag_2S	61 (Ag):	Hb/B4–141
	ZnS–CdS	33 (Cd):	Eb–711/2
	ZnS–$CuGaS_2$	60 (Cu):	Hb/B3–1181
	ZnS–Cu_2S–FeS–PbS	60 (Cu):	Hb/B3–1329
	ZnS–Cu_2S–PbS	60 (Cu):	Hb/B3–1208
	ZnS–FeS	59 (Fe):	Hb/B5–1101/2
	ZnS–PbS	47 (Pb):	Hb/C4–1227
SZn_2	Zn_2S	32 (Zn):	Hb–193
SZr_4	Zr_4S	42 (Zr):	Hb–329
$S_{1.02}V$	$VS_{1.02}$	48 (V):	Hb/B1–268, 270/3
$S_{1.075}V$	$VS_{1.075}$	48 (V):	Hb/B1–273
$S_{1.1}V$	$VS_{1.1}$	48 (V):	Hb/B1–270/3
$S_{1.4}V$	$VS_{1.4}$	48 (V):	Hb/B1–273/4
$S_{1.67}V$	$VS_{1.67}$	48 (V):	Hb/B1–268, 273, 276
$S_{1.8}V$	$VS_{1.8}$	48 (V):	Hb/B1–268, 273, 276/7
S_2	S_2	9 (S):	Hb/A3–560/71, 654/63
	see also S ... Sulfur		
–	$^{32}S^{34}S$	9 (S):	Hb/A3–567/8

Hb = Main Series Volume; GD = Gmelin-Durrer; Ab = Appended Volume to Main Series;
Eb = Supplement Volume to Main Series; Erg. W. = New Supplement Series Volume

S_2^{2-} . . . . S_2^{2-} . . . . 9 (S):   Hb/B1−83

S_2Sb_2 . . . . Sb_2S_2 . . . . 18 (Sb):   Hb/B3−500

S_2Sb_2Se . . . . . . Sb_2S_2Se . . . . 18 (Sb):   Hb/B3−553

S_2Sb_3 . . . . Sb_3S_2 . . . . 18 (Sb):   Hb/B3−517

S_2Se . . . . SeS_2 . . . . 10 (Se):   Hb/B−173/4

S_2Si . . . . SiS_2 . . . . 15 (Si):   Hb/B−746/52

−. . . . . SiS_2 systems:

               SiS_2-Ag_2S . . . . 61 (Ag):   Hb/B3−387/8

S_2Sn . . . . SnS_2 . . . . 46 (Sn):   Hb/A−170/1

−. . . . . SnS_2 systems:

               SnS_2-Ag_2S . . . . 61 (Ag):   Hb/B4−247

S_2Ta . . . . TaS_2 . . . . 50 (Ta):   Hb/A2−517

                                               Hb/B1−169/70, 172/5

S_2Ta_{x+1} . . . . $Ta_{1+x}S_2$ . . . . 50 (Ta):   Hb/B1−170/1

S_2Te . . . . TeS_2 . . . . 11 (Te):   Hb−350, 353/4

S_2Th . . . . ThS_2 . . . . 44 (Th):   Hb−277, 280/2

S_2Ti . . . . TiS_2 . . . . 41 (Ti):   Hb−337/9, 341/3

−. . . . . TiS_2 systems:

               $TiS_2-CoS-TiS$ . . . . 58 (Co):   Eb/A−870

               $TiS_2-NiS-TiS$ . . . . 57 (Ni):   Hb/B3−1181/2

               TiS_2-TiO_2 . . . . 41 (Ti):   Hb−343

S_2Tl_2 . . . . $Tl_2S_2 = Tl_2S \cdot Tl_2S_3$ . . . . 38 (Tl):   Hb/3−345/6

S_2U . . . . US_2 . . . . 55 (U):   Hb−143

−. . . . . US_2 solid solutions:

               $(U,Pu)S_2$ . . . . Erg. W. 4−170

S_2V . . . . VS_2 . . . . 48 (V):   Hb/B1−268, 273, 277

S_2W . . . . WS_2 . . . . 54 (W):   Hb−184/5

S_2Zr . . . . ZrS_2 . . . . 42 (Zr):   Hb−326/7, 330/2

S_3^{2-} . . . . S_3^{2-} . . . . 9 (S):   Hb/B1−83

S_3Sb_2 . . . . Sb_2S_3:

               Chemical reactions . . . . 18 (Sb):   Hb/B3−517/24, 532/3

               Colloidal forms . . . . 18 (Sb):   Hb/B3−525/33

               Electrochemical behavior . . . . . . 18 (Sb):   Hb/B3−516/7, 532

               Formation . . . . 18 (Sb):   Hb/B3−503/8, 525/6

               History . . . . 18 (Sb):   Hb/A1−7/14, 32/3, 40/2

               Physical properties . . . . 18 (Sb):   Hb/B3−508/16, 531

               Preparation . . . . 18 (Sb):   Hb/B3−503/8, 525/7

               −, industrial . . . . 18 (Sb):   Hb/A3−311/2

                                                 Hb/B3−500/2

               Purity testing . . . . 18 (Sb):   Hb/B3−503

               Sols . . . . 18 (Sb):   Hb/B3−525/33

               Toxicity . . . . 18 (Sb):   Hb/B2−276

               Uses . . . . 18 (Sb):   Hb/A3−305

                                                 Hb/B3−502/3

        . . . . Sb_2S_3 systems:

               $Sb_2S_3-Ag_2S$ . . . . 61 (Ag):   Hb/B4−90/2

               $Sb_2S_3-Ag_2S-Cu_2S$ . . . . 61 (Ag):   Hb/B4−491

See the "Instructions for the user"
at the beginning of the volume

S_3Sb_2	Sb_2S_3 systems:		
	$Sb_2S_3-As_2S_3$	18 (Sb):	Hb/B3$-$564
	Sb_2S_3-BiS	19 (Bi):	Hb$-$202
	$Sb_2S_3-Bi_2S_3$	19 (Bi):	Eb$-$859/61
	Sb_2S_3-CaS	28 (Ca):	Hb/B3$-$1240/1
	Sb_2S_3-HgS	34 (Hg):	Hb/B4$-$1383
	$Sb_2S_3-Na_2S-H_2O$	21 (Na):	Eb/4$-$1720/1
	$Sb_2S_3-Na_2SO_4$	21 (Na):	Eb/5$-$2025
	$Sb_2S_3-PbO-Sb_2O_3$	47 (Pb):	Hb/C3$-$939
	Sb_2S_3-PbS	47 (Pb):	Hb/C3$-$935/6
	$Sb_2S_3-Sb_2O_3$	18 (Sb):	Hb/B3$-$539/40
$S_3Sb_2Se_2$	$Sb_2S_3Se_2$	18 (Sb):	Hb/B3$-$553
S_3Sc_2	Sc_2S_3 solid solutions:		
	$Sc_2S_3-Ag_2S$	61 (Ag):	Hb/B4$-$232
S_3Se	SeS_3	10 (Se):	Hb/B$-$174
S_3Se^{2-}	$[SeS_3]^{2-}$	10 (Se):	Hb/B$-$159
S_3Sm_2	Sm_2S_3 solid solutions:		
	$Sm_2S_3-Ag_2S$	61 (Ag):	Hb/B4$-$232
	Sm_2S_3-PbS	47 (Pb):	Hb/C4$-$1374
S_3Sn^{2-}	$[SnS_3]^{2-}$	46 (Sn):	Hb/B$-$393
S_3Sn_2	Sn_2S_3 systems:		
	Sn_2S_3-PbS	47 (Pb):	Hb/C4$-$1585
S_3SrTa	$SrTaS_3$	50 (Ta):	Hb/B2$-$317
S_3Ta	TaS_3	50 (Ta):	Hb/B1$-$175/6
$-$	TaS_3 systems:		
	$TaS_3-Ta-TaCl_5$	50 (Ta):	Hb/B1$-$178
S_3Tb_2	Tb_2S_3 solid solutions:		
	$Tb_2S_3-Ag_2S$	61 (Ag):	Hb/B4$-$232
	Tb_2S_3-PbS	47 (Pb):	Hb/C4$-$1374
S_3Te	TeS_3	11 (Te):	Hb$-$350, 354/5
S_3Th_2	Th_2S_3	44 (Th):	Hb$-$277, 279
S_3Ti	TiS_3	41 (Ti):	Hb$-$337/9, 343
S_3Ti_2	Ti_2S_3	41 (Ti):	Hb$-$337/9, 340/1
S_3Tl_2	Tl_2S_3	38 (Tl):	Hb/3$-$345
S_3Tm_2	Tm_2S_3 solid solutions:		
	$Tm_2S_3-Ag_2S$	61 (Ag):	Hb/B4$-$232
	Tm_2S_3-PbS	47 (Pb):	Hb/C4$-$1374
S_3U_2	U_2S_3	55 (U):	Hb$-$143
$-$	U_2S_3 solid solutions:		
	$(U,Pu)_2S_3$	Erg. W. 4$-$169/70	
S_3V^-	$[VS_3]^-$	48 (V):	Hb/B1$-$280
$-$	$[VS_3]_3^{3-}$	48 (V):	Hb/B1$-$280
$-$	$[VS_3(H_2O)]^-$	48 (V):	Hb/B1$-$281
S_3V_2	V_2S_3	48 (V):	Hb/A1$-$278, 294/7
			Hb/A2$-$380
			Hb/B1$-$267/8, 274/6

Hb = *Main Series Volume;* GD = *Gmelin-Durrer;* Ab = *Appended Volume to Main Series;*
Eb = *Supplement Volume to Main Series;* Erg. W. = *New Supplement Series Volume*

S_3W	WS_3	54 (W):	Hb−185/6
S_3Y_2	Y_2S_3 solid solutions:		
	Y_2S_3−Ag_2S	61 (Ag):	Hb/B4−232
	Y_2S_3−PbS	47 (Pb):	Hb/C4−1374
S_3Yb_2	Yb_2S_3 solid solutions:		
	Yb_2S_3−Ag_2S	61 (Ag):	Hb/B4−232
	Yb_2S_3−PbS	47 (Pb):	Hb/C4−1374
S_3Zr	ZrS_3	42 (Zr):	Hb−326/7, 332
S_3Zr_2	Zr_2S_3	42 (Zr):	Hb−329/30
S_3Zr_4	Zr_4S_3	42 (Zr):	Hb−329
S_4	S_4	9 (S):	Hb/A3−560/71
	see also S ... Sulfur		
S_4^{2-}	S_4^{2-}	9 (S):	Hb/B1−83
S_4Sb^{3-}	$[SbS_4]^{3-}$	18 (Sb):	Hb/B3−547
S_4Sb_2	Sb_2S_4	18 (Sb):	Hb/B3−500/3, 533
S_4Sb_2Sr	$Sr[Sb_2S_4]$	29 (Sr):	Eb−295
S_4Sn^{4-}	$[SnS_4]^{4-}$	46 (Sn):	Hb/B−342, 393
S_4Sr	SrS_4	29 (Sr):	Hb−169/70
−	$SrS_4 \cdot 2\,H_2O$	29 (Sr):	Hb−170
−	$SrS_4 \cdot 6\,H_2O$	29 (Sr):	Hb−170
S_4Ta^{3-}	$[TaS_4]^{3-}$	50 (Ta):	Hb/B2−318
S_4TaTi_2	Ti_2TaS_4	50 (Ta):	Hb/B2−318
S_4TaTl_3	Tl_3TaS_4	50 (Ta):	Hb/B2−318
S_4Ti_3	Ti_3S_4	41 (Ti):	Hb−337, 341
S_4Tl_3V	Tl_3VS_4	48 (V):	Hb/B2−549
S_4Tl_4	$Tl_2S \cdot Tl_2S_3 = Tl_2S_2$	38 (Tl):	Hb/3−345/6
S_4V	VS_4	48 (V):	Hb/B1−267/8, 278/80
S_4V^{3-}	$[VS_4]^{3-}$	48 (V):	Hb/B1−280
S_4V_3	V_3S_4	48 (V):	Hb/B1−273/4
S_4V_5	V_5S_4	48 (V):	Hb/B1−268, 270
S_5Sb_2	Sb_2S_5:		
	Chemical reactions	18 (Sb):	Hb/B3−536/9
	Colloidal forms	18 (Sb):	Hb/B3−539
	Formation	18 (Sb):	Hb/B3−534/6
	Physical properties	18 (Sb):	Hb/B3−536
	Preparation	18 (Sb):	Hb/B3−534/6
	−, industrial	18 (Sb):	Hb/B3−500/2
	Purity testing	18 (Sb):	Hb/B3−503
	Toxicity	18 (Sb):	Hb/B2−276
	Uses	18 (Sb):	Hb/A3−305
			Hb/B3−502/3
$S_5Sb_2Sr_2$	$Sr_2[Sb_2S_5]$	29 (Sr):	Eb−295
−	$Sr_2[Sb_2S_5] \cdot 15\,H_2O$	29 (Sr):	Hb−221
S_5Sr	SrS_5	29 (Sr):	Hb−169
			Eb−248
S_5Sr_3Te	Sr_3TeS_5	29 (Sr):	Hb−187

See the "Instructions for the user"
at the beginning of the volume

S_5Ti_3 . . . . Ti_3S_5 . . . . 41 (Ti):   Hb–337

S_5Ti_4 . . . . Ti_4S_5 . . . . 41 (Ti):   Hb–337

S_5Tl_2 . . . . Tl_2S_5 . . . . 38 (Tl):   Hb/3–346

S_5Tl_4 . . . . Tl_4S_5 . . . . 38 (Tl):   Hb/3–346

S_5V . . . . VS_5 . . . . 48 (V):   Hb/B1–268, 280

S_5V^- . . . . $[VS_5]_3^{3-}$ . . . . 48 (V):   Hb/B1–280

S_5V_2 . . . . V_2S_5 . . . . 48 (V):   Hb/A1–278, 282

    Hb/A2–380

    Hb/B1–268, 277/8

S_5V_4 . . . . V_4S_5 . . . . 48 (V):   Hb/B1–273/4

S_5Zn . . . . ZnS_5 . . . . 32 (Zn):   Hb–219

S_5Zr_3 . . . . Zr_3S_5 . . . . 42 (Zr):   Hb–330

S_6 . . . . S_6 . . . . 9 (S):   Hb/A3–560/71

           see also S ... Sulfur

$S_6Sb_2Sr_3$ . . . . . . $Sr_3Sb_2S_6 \cdot 10\ H_2O$ . . . . 29 (Sr):   Hb–221

$S_6Sb_2Zn_3$ . . . . . . $Zn_3(SbS_3)_2$ . . . . 32 (Zn):   Hb–291

$S_7Sb_2Sr_2$ . . . . . . $Sr_2Sb_2S_7 \cdot 11\ H_2O$ . . . . 29 (Sr):   Hb–221

S_7Th_3 . . . . Th_3S_7 . . . . 44 (Th):   Hb–277, 282

S_7Tl_6 . . . . $Tl_6S_7 = Tl_2S \cdot 2\ Tl_2S_3$ . . . . 38 (Tl):   Hb/3–346

S_7Tl_8 . . . . $Tl_8S_7 = 5\ Tl_2S \cdot 3\ Tl_2S_3$ . . . . 38 (Tl):   Hb/3–345/6

S_8 . . . . S_8 . . . . 9 (S):   Hb/A3–560/71

           see also S ... Sulfur

$S_8Sb_2Zn_3$ . . . . . . $Zn_3(SbS_4)_2$ . . . . 32 (Zn):   Hb–291

S_8V_5 . . . . V_5S_8 . . . . 48 (V):   Hb/B1–268, 276

$S_9V_2^{4-}$ . . . . $[V_2S_9]^{4-}$ . . . . 48 (V):   Hb/B1–280

$S_9V_3^{3-}$ . . . . $[VS_3]_3^{3-}$ . . . . 48 (V):   Hb/B1–280

$S_{12}Th_7$ . . . . Th_7S_{12} . . . . 44 (Th):   Hb–277, 279/80

$S_{14}Tl_{16}$ . . . . $5\ Tl_2S \cdot 3\ Tl_2S_3 = Tl_8S_7$ . . . . 38 (Tl):   Hb/3–345/6

$S_{15}V_3^{3-}$ . . . . $[VS_5]_3^{3-}$ . . . . 48 (V):   Hb/B1–280

S_xSr . . . . SrS_x . . . . 29 (Sr):   Eb–248/9

S_xV . . . . VS_x . . . . 48 (V):   Hb/B1–267/80

Sb . . . . Antimony:

    Absorption of light . . . . 18 (Sb):   Hb/B1–87/8

    Hb/B2–183/4

    – of X-rays. . . . . 18 (Sb):   Hb/B1–111/2

    Hb/B2–185

    Abundance . . . . 18 (Sb):   Hb/A1–55/60

    Aerosols. . . . . 18 (Sb):   Hb/A3–349

    Affinity. . . . . 18 (Sb):   Hb/B1–26

    Hb/B2–168

    Antimony electrode:

      Effects on applicability. . . . . . . 18 (Sb):   Hb/B2–226/47

      Field of employment. . . . . 18 (Sb):   Hb/B2–250/9

      General . . . . 18 (Sb):   Hb/B2–216/7

      Handling . . . . 18 (Sb):   Hb/B2–225/6

Hb = *Main Series Volume;* GD = *Gmelin-Durrer;*   Ab = *Appended Volume to Main Series;*
Eb = *Supplement Volume to Main Series;*   Erg. W. = *New Supplement Series Volume*

Sb . . . . Antimony:

    Antimony electrode:
        Manufacture and types . . . . 18 (Sb):   Hb/B2—217/24
        Potential . . . . 18 (Sb):   Hb/B2—241/7
        Theory . . . . 18 (Sb):   Hb/B2—247/50
    Atom. . . . . 18 (Sb):   Hb/B1—17/26
                                                       Hb/B2—166/8
    Atomic moments . . . . 18 (Sb):   Hb/B1—21
    Atomic nucleus . . . . 18 (Sb):   Hb/B1—1/17
                                                       Hb/B2—156/66
    Atomic number . . . . 18 (Sb):   Hb/B1—3
    Atomic radius . . . . 18 (Sb):   Hb/B1—18/9
                                                       Hb/B2—166
    Atomic refraction . . . . 18 (Sb):   Hb/B1—87
    Atomic volume . . . . 18 (Sb):   Hb/B1—21
                                                       Hb/B2—167
    Atomic weight . . . . 18 (Sb):   Hb/B1—22/5
                                                       Hb/B2—167
    Behavior as anode . . . . 18 (Sb):   Hb/B2—209/11
    – as cathode. . . . . 18 (Sb):   Hb/B2—207/9
    – as electrode . . . . 18 (Sb):   Hb/B2—211/3, 215/59
    – at alternating current
        electrolysis . . . . 18 (Sb):   Hb/B2—213/5
    Birefringence . . . . 18 (Sb):   Hb/B1—86/7
    Black antimony . . . . 18 (Sb):   Hb/A3—343/4
    Boiling point . . . . 18 (Sb):   Hb/B1—69
                                                       Hb/B2—180
    Bond energy . . . . 18 (Sb):   Hb/B1—27
                                                       Hb/B2—168
    Cells . . . . 18 (Sb):   Hb/B2—196/201
    Chemical constant . . . . 18 (Sb):   Hb/B1—83
    Chemical reactions . . . . 18 (Sb):   Hb/B2—260/74
    – on heating. . . . . 18 (Sb):   Hb/B2—260
    – with acids . . . . 18 (Sb):   Hb/B2—263/7
    – with air . . . . 18 (Sb):   Hb/B2—260
    – with elements . . . . 18 (Sb):   Hb/B2—261/3
    – with metal compounds. . . . . . . 18 (Sb):   Hb/B2—269/72
    – with nonmetal compounds . . . . 18 (Sb):   Hb/B2—267/9
    – with organic compounds. . . . . . 18 (Sb):   Hb/B2—272/4
    – with water. . . . . 18 (Sb):   Hb/B2—260/1
    Cleavage. . . . . 18 (Sb):   Hb/B1—37/8
                                                       Hb/B2—171
    Colloidal forms . . . . 18 (Sb):   Hb/A3—349/51
    Color. . . . . 18 (Sb):   Hb/B1—85
                                                       Hb/B2—183
    Compressibility . . . . 18 (Sb):   Hb/B1—62/4
                                                       Hb/B2—178
    Contact electricity. . . . . 18 (Sb):   Hb/B2—141

See the "Instructions for the user"
at the beginning of the volume

Sb . . . . Antimony:

Corbino effect	18 (Sb):	Hb/B1 – 118
Cosmochemistry	18 (Sb):	Hb/A1 – 49/50
Critical constants	18 (Sb):	Hb/B1 – 67
Cryoscopic constant	18 (Sb):	Hb/B1 – 71
Crystal chemistry	18 (Sb):	Hb/A1 – 51/3
Crystallization	18 (Sb):	Hb/B1 – 32/7
		Hb/B2 – 170/1
Crystallographic properties	18 (Sb):	Hb/B1 – 30/51
		Hb/B2 – 169/76
Debye temperature	18 (Sb):	Hb/B1 – 81/2
		Hb/B2 – 181
Deformation	18 (Sb):	Hb/B1 – 46/7
		Hb/B2 – 174/5
Density	18 (Sb):	Hb/B1 – 52/5
		Hb/B2 – 176
Detection	18 (Sb):	Hb/B2 – 277/307
– in special substances	18 (Sb):	Hb/B2 – 305/7
	59 (Fe):	Hb/FI – 424
Determination	18 (Sb):	Hb/B2 – 307/26
– in special substances	18 (Sb):	Hb/B2 – 329/33
	59 (Fe):	Hb/FI – 423/6
		Hb/FII – 314/5, 349
Dielectric properties	18 (Sb):	Hb/B2 – 186
Diffusion	18 (Sb):	Hb/B1 – 66/7
		Hb/B2 – 179
	46 (Sn):	Hb/D – 1
	47 (Pb):	Hb/C3 – 905
Discharge phenomena	18 (Sb):	Hb/B2 – 149/50
Dispersion	18 (Sb):	Hb/B1 – 87
Dissociation of the molecule	18 (Sb):	Hb/B1 – 28/9
		Hb/B2 – 169
Dressing of ores	18 (Sb):	Hb/A3 – 308/11
Economic deposits:		
Africa	18 (Sb):	Hb/A1 – 144/8
Asia	18 (Sb):	Hb/A1 – 123/39
Australia and Oceania	18 (Sb):	Hb/A1 – 139/44
Central America	18 (Sb):	Hb/A1 – 161/3
Europe	18 (Sb):	Hb/A1 – 88/123
North America	18 (Sb):	Hb/A1 – 148/61
South America	18 (Sb):	Hb/A1 – 163/72
Soviet Union	18 (Sb):	Hb/A1 – 172/94
Economics	18 (Sb):	Hb/A1 – 194/7
		Hb/A3 – 305/7
Elasticity	18 (Sb):	Hb/B1 – 60/1
		Hb/B2 – 177
Electrical conductivity	18 (Sb):	Hb/B1 – 120/8
		Hb/B2 – 129/40, 187/9

Hb = Main Series Volume; GD = Gmelin-Durrer;   *Ab = Appended Volume to Main Series;*
Eb = Supplement Volume to Main Series;   *Erg. W. = New Supplement Series Volume*

Sb . . . . Antimony:

Electrical properties	18 (Sb):	Hb/B1–116/8, 120/8
		Hb/B2–129/56, 186/9
Electrical relaxation.	18 (Sb):	Hb/B2–141/2
Electrocapillarity	18 (Sb):	Hb/B2–201
Electrochemical behavior	18 (Sb):	Hb/B2–189/259
Electrode sputtering	18 (Sb):	Hb/B2–150/1
Electrolytic deposition	18 (Sb):	Hb/B2–201/7
Electrolytic solution pressure	18 (Sb):	Hb/B2–190
Electron configuration	18 (Sb):	Hb/B1–17/8
		Hb/B2–166
Electron emission.	18 (Sb):	Hb/B2–153/4
Enthalpy	18 (Sb):	Hb/B1–82/3
Entropy	18 (Sb):	Hb/B1–82/3
Equation of state	18 (Sb):	Hb/B1–67
		Hb/B2–179
Ettingshausen effect	18 (Sb):	Hb/B1–118
Explosive antimony.	18 (Sb):	Hb/A3–344/8
Formation	18 (Sb):	Hb/A3–339/42
Free energy	18 (Sb):	Hb/B1–82/3
Frictional electricity.	18 (Sb):	Hb/B2–141
Galvanomagnetic effects.	18 (Sb):	Hb/B1–118/9
Geochemical character.	18 (Sb):	Hb/A1–54
Geochemistry	18 (Sb):	Hb/A1–51/192
Hall effect	18 (Sb):	Hb/B1–116/8
		Hb/B2–186
Hardness	18 (Sb):	Hb/B1–58/60
		Hb/B2–176/7
Heat content.	18 (Sb):	Hb/B1–82/3
Heat of fusion	18 (Sb):	Hb/B1–72/3
		Hb/B2–180
– of sublimation.	18 (Sb):	Hb/B1–72
		Hb/B2–180
– of transformation	18 (Sb):	Hb/B1–73/5
– of vaporization	18 (Sb):	Hb/B1–71/2
History.	18 (Sb):	Hb/A1–1/48
Hydrosols	18 (Sb):	Hb/A3–349/50
Interaction with ion beams.	18 (Sb):	Hb/B2–155
Interfacial tension.	18 (Sb):	Hb/B1–64/5
		Hb/B2–178/9
Internal pressure	18 (Sb):	Hb/B1–64
Ionization potentials	18 (Sb):	Hb/B2–155/6
Isomorphism	18 (Sb):	Hb/B1–38
		Hb/B2–171
Lattice structure.	18 (Sb):	Hb/B1–38/42
		Hb/B2–171/2
Leduc effect	18 (Sb):	Hb/B1–119
		Hb/B2–186
Luminescence.	18 (Sb):	Hb/B1–89

See the "Instructions for the user"
at the beginning of the volume

13*

Sb . . . . Antimony:

Machinability	18 (Sb):	Hb/B1–66
Magnetic moment	18 (Sb):	Hb/B1–21
Magnetic properties	18 (Sb):	Hb/B1–112/9
		Hb/B2–186
Magnetostriction	18 (Sb):	Hb/B1–119
Mechanical properties	18 (Sb):	Hb/B1–52/67
		Hb/B2–176/9
Melting point	18 (Sb):	Hb/B1–69/71
		Hb/B2–180
Mineralogy	18 (Sb):	Hb/A1–197/226
		Hb/A2–227/302
Molar polarization	18 (Sb):	Hb/B2–186
Molecule.	18 (Sb):	Hb/B1–26/9
		Hb/B2–168/9
Nernst effect	18 (Sb):	Hb/B1–118/9
Nuclear properties	18 (Sb):	Hb/B1–1/17
		Hb/B2–156/66
Occurrence, extraterrestrial	18 (Sb):	Hb/A1–49/50
–, terrestrial	18 (Sb):	Hb/A1–51/192
– in meteorites	18 (Sb):	Hb/A1–50
– in the biosphere.	18 (Sb):	Hb/A1–59/60
– in the hydrosphere	18 (Sb):	Hb/A1–58/9
– in the lithosphere	18 (Sb):	Hb/A1–56/8, 81/8
Optical properties.	18 (Sb):	Hb/B1–85/112
		Hb/B2–183/5
Organosols	18 (Sb):	Hb/A3–350/1
Orientation of thin layers.	18 (Sb):	Hb/B1–42/6
		Hb/B2–172/3
Overvoltage phenomena.	18 (Sb):	Hb/B2–211/3
Passage of alpha particles	18 (Sb):	Hb/B2–154/5
– of electrons	18 (Sb):	Hb/B2–153/4
Photoelectric effect	18 (Sb):	Hb/B2–152/3
Plasticity.	18 (Sb):	Hb/B1–61
Polymorphism	18 (Sb):	Hb/B1–30/2
		Hb/B2–169/70
Position in the electrochemical series	18 (Sb):	Hb/B2–190
– in the periodic system	18 (Sb):	Hb/B1–26
		Hb/B2–168
Potentials	18 (Sb):	Hb/B2–189/95
Preparation	18 (Sb):	Hb/A3–342/3
–, industrial	18 (Sb):	Hb/A3–303/39
–, –, dry methods	18 (Sb):	Hb/A3–312/30
–, –, wet methods.	18 (Sb):	Hb/A3–330/4
– of special forms.	18 (Sb):	Hb/A3–343/51
Purification	18 (Sb):	Hb/A3–334/8
Purity testing	18 (Sb):	Hb/A3–339
Recrystallization	18 (Sb):	Hb/B1–51

Hb = Main Series Volume; GD = Gmelin-Durrer;   Ab = Appended Volume to Main Series;
Eb = Supplement Volume to Main Series;   Erg. W. = New Supplement Series Volume

Sb . . . . Antimony:

Rectifying effect.	18 (Sb):	Hb/B2 – 142
Reflection	18 (Sb):	Hb/B1 – 85/6
		Hb/B2 – 183
Refraction	18 (Sb):	Hb/B1 – 86/7
		Hb/B2 – 183
Rotational constant.	18 (Sb):	Hb/B1 – 28
Scattering of light.	18 (Sb):	Hb/B1 – 87
– of X-rays.	18 (Sb):	Hb/B1 – 111/2
		Hb/B2 – 185
Secondary electron emission	18 (Sb):	Hb/B2 – 154, 155
Semiconductor properties	18 (Sb):	Hb/B2 – 137/9
Separation, analytical	18 (Sb):	Hb/B2 – 326/9
Shrinkage	18 (Sb):	Hb/B1 – 58
		Hb/B2 – 176
Sorption	18 (Sb):	Hb/B1 – 67
		Hb/B2 – 179
Special forms	18 (Sb):	Hb/B2 – 45/6, 173/4
Specific heat	18 (Sb):	Hb/B1 – 75/81
		Hb/B2 – 180/1
Spectra:		
Arc spectrum	18 (Sb):	Hb/B1 – 89/96
		Hb/B2 – 184
Band spectrum	18 (Sb):	Hb/B1 – 104/7
		Hb/B2 – 184/5
Continuous spectrum	18 (Sb):	Hb/B1 – 107
Spark spectra.	18 (Sb):	Hb/B1 – 96/102
		Hb/B2 – 184
X-ray spectrum	18 (Sb):	Hb/B1 – 107/11
		Hb/B2 – 185
Standard potential	18 (Sb):	Hb/B2 – 189/90
Strength	18 (Sb):	Hb/B1 – 61/2
Structure	18 (Sb):	Hb/B1 – 47/51
		Hb/B2 – 175/6
– of thin layers	18 (Sb):	Hb/B1 – 42/6
		Hb/B2 – 172/4
Superconductivity	18 (Sb):	Hb/B1 – 125
		Hb/B2 – 188
Surface tension	18 (Sb):	Hb/B1 – 64/5
		Hb/B2 – 178/9
Susceptibility	18 (Sb):	Hb/B1 – 112/6
		Hb/B2 – 186
Thermal conductivity.	18 (Sb):	Hb/B1 – 83/5
		Hb/B2 – 181/3
Thermal expansion	18 (Sb):	Hb/B1 – 55/8
		Hb/B2 – 176
Thermal properties	18 (Sb):	Hb/B1 – 55/8, 67/85
		Hb/B2 – 176, 179/83
Thermal radiation.	18 (Sb):	Hb/B1 – 88

See the "Instructions for the user"
at the beginning of the volume

Sb . . . . Antimony:
        Thermoelectric effects . . . . 18 (Sb):   Hb/B2–142/9
        Thermomagnetic effects . . . . 18 (Sb):   Hb/B1–119
                                                    Hb/B2–186
        Toxicity . . . . 18 (Sb):   Hb/B2–274/7
        Transformation point . . . . 18 (Sb):   Hb/B1–71
        Uses . . . . 18 (Sb):   Hb/A1–43/6
                                                    Hb/A3–304/5
        Valence . . . . 18 (Sb):   Hb/B1–25/6
                                                    Hb/B2–167/8
        Vapor pressure . . . . 18 (Sb):   Hb/B1–68/9
                                                    Hb/B2–179
        Vibrational frequency . . . . 18 (Sb):   Hb/B1–81/2
                                                    Hb/B2–181
        Viscosity . . . . 18 (Sb):   Hb/B1–65/6
        Wiedemann-Franz number . . . . . . 18 (Sb):   Hb/B2–139/40
        Yellow antimony . . . . 18 (Sb):   Hb/A3–344

–. . . . . Antimony ions
        see Sb^+ ... Sb^+, etc.

–. . . . . Antimony isotopes . . . . 18 (Sb):   Hb/B1–1/3, 27
                                                    Hb/B2–156/8
                                      Erg. W. 8–223/4

–. . . . . Antimony systems:
        Sb–Ag . . . . 61 (Ag):   Hb/B4–69/73
        Sb–Ag–Bi . . . . 61 (Ag):   Hb/B4–135
        Sb–Ag–Bi–Zn . . . . 61 (Ag):   Hb/B4–144
        Sb–Ag–Cd . . . . 61 (Ag):   Hb/B4–156/7
        Sb–Ag–Cu–S . . . . 61 (Ag):   Hb/B4–491
        Sb–Ag–S . . . . 61 (Ag):   Hb/B4–90/2
        Sb–Ag–Se . . . . 61 (Ag):   Hb/B4–101/2
        Sb–Ag–Sn . . . . 61 (Ag):   Hb/B4–252
        Sb–Ag–Zn . . . . 61 (Ag):   Hb/B4–142/3
        Sb–Al . . . . 35 (Al):   Hb/A4–629/32
        Sb–Al–Bi–Pb . . . . 47 (Pb):   Hb/C4–1287/8
        Sb–Al–Cu . . . . 60 (Cu):   Hb/B3–1177
        Sb–Al–Cu–Sn . . . . 60 (Cu):   Hb/B3–1196
        Sb–Al–Mg . . . . 35 (Al):   Hb/A4–665/6
        Sb–Al–Nb . . . . 49 (Nb):   Hb/B2–45
        Sb–Al–Pb . . . . 35 (Al):   Hb/A5–861/3
                                                    47 (Pb):   Hb/C4–1286/7
        Sb–Al–Sn . . . . 35 (Al):   Hb/A5–854
                                                    46 (Sn):   Hb/D–218
        Sb–As . . . . 18 (Sb):   Hb/B3–562/3
        Sb–As–Cu–S . . . . 60 (Cu):   Hb/B2–973
        Sb–As–Ni . . . . 57 (Ni):   Hb/B1–10
        Sb–As–Pb . . . . 47 (Pb):   Hb/C3–941/2
        Sb–Au . . . . 62 (Au):   Hb/3–770/1
        Sb–B–Ni . . . . 57 (Ni):   Hb/B3–996

Hb = Main Series Volume; GD = Gmelin-Durrer;   Ab = Appended Volume to Main Series;
Eb = Supplement Volume to Main Series;   Erg. W. = New Supplement Series Volume

Sb . . . . Antimony systems:

Sb–Ba	30 (Ba):	Eb–272	
Sb–Bi	19 (Bi):	Hb–198/202	
		Eb–601/2	
Sb–Bi–Cd	33 (Cd):	Hb–167	
Sb–Bi–Pb	47 (Pb):	Hb/C3–977/8	
Sb–Bi–Se	19 (Bi):	Eb–861	
Sb–Bi–Se–Te	19 (Bi):	Eb–865/6	
Sb–Bi–Sn	46 (Sn):	Hb/D–58/9	
Sb–Bi–Te	19 (Bi):	Eb–861/5	
Sb–Bi–Zn	32 (Zn):	Hb–293	
		Eb–617	
Sb–Ca	28 (Ca):	Hb/A2–479	
Sb–Cd	33 (Cd):	Hb–161/4	
		Eb–353/5	
Sb–Cd–Pb	47 (Pb):	Hb/C4–1245/6	
Sb–Cd–Sn	46 (Sn):	Hb/D–177/8	
Sb–Cd–Zn	33 (Cd):	Hb–213	
Sb–Co	58 (Co):	Hb/A1–185/7	
		Eb/A–376/87	
Sb–Co–H–S	58 (Co):	Eb/A–755/6	
Sb–Co–S	58 (Co):	Eb/A–754/5	
Sb–Cr	52 (Cr):	Hb/A2–710/2	
Sb–Cu–Fe	60 (Cu):	Hb/B3–1315	
Sb–Cu–Fe–S	60 (Cu):	Hb/B3–1315	
Sb–Cu–Ge	60 (Cu):	Hb/B3–1193	
Sb–Cu–Mg	27 (Mg):	Hb/A4–716/8	
	60 (Cu):	Hb/B3–1128	
Sb–Cu–Mn–Ni	60 (Cu):	Hb/B3–1249	
Sb–Cu–Ni	60 (Cu):	Hb/B3–1259/60	
Sb–Cu–Pb	60 (Cu):	Hb/B3–1204	
Sb–Cu–S	60 (Cu):	Hb/B2–971	
Sb–Cu–Sn	60 (Cu):	Hb/B3–1196	
Sb–Fe	59 (Fe):	Hb/A8–1805/6	
Sb–Ga–Ge–Sn	46 (Sn):	Hb/D–397	
Sb–Ga–In–Sn	46 (Sn):	Hb/D–277	
Sb–Ga–Pb	47 (Pb):	Hb/C4–1305	
Sb–Ge	45 (Ge):	Eb–475/7	
Sb–I	18 (Sb):	Hb/B2–486	
Sb–In–Pb	47 (Pb):	Hb/C4–1324	
Sb–In–Sn	46 (Sn):	Hb/D–264/5	
Sb–K	22 (K):	Hb/5–1035/6	
Sb–Mg	27 (Mg):	Hb/A3–406/7	
Sb–Mg–Pb	27 (Mg):	Hb/A4–653	
Sb–Mn–Ni	57 (Ni):	Hb/B1–311	
Sb–NH_3–Na	21 (Na):	Eb/4–1709/10	
Sb–Na	21 (Na):	Hb–965/6	
		Eb/4–1707/8	
Sb–Nb	49 (Nb):	Hb/B1–417/8	

See the "Instructions for the user"
at the beginning of the volume

Sb . . . . Antimony systems:
Sb–Ni	57 (Ni):	Hb/B1–2/10
Sb–Ni–Pb	57 (Ni):	Hb/B1–152
Sb–Ni–S	57 (Ni):	Hb/B1–10
		Hb/B3–994/5
Sb–O–Pb	47 (Pb):	Hb/C3–926/9
Sb–O–S	9 (S):	Hb/B1–182
Sb–Pb	47 (Pb):	Hb/C3–905/7
Sb–Pb–S	47 (Pb):	Hb/C3–935/6
Sb–Pb–Se	47 (Pb):	Hb/C3–940
Sb–Pb–Si	47 (Pb):	Hb/C3–941
Sb–Pb–Sn	47 (Pb):	Hb/C4–1515
Sb–Pb–Te	47 (Pb):	Hb/C3–940/1
Sb–Pb–Zn	47 (Pb):	Hb/C4–1223/4
Sb–Pd	68 (Pt):	Hb/A5–570/5
Sb–Pt	68 (Pt):	Hb/A6–743/4
Sb–S	18 (Sb):	Hb/B3–499/500
Sb–Se	18 (Sb):	Hb/B3–550/2
Sb–Si	18 (Sb):	Hb/B3–560
Sb–Sn	46 (Sn):	Hb/D–1/4
Sb–Sn–Zn	46 (Sn):	Hb/D–153
Sb–Te	18 (Sb):	Hb/B3–554/5
Sb–Tl	38 (Tl):	Hb/2–187/91
Sb–V	48 (V):	Hb/B2–575/7
Sb–Zn	32 (Zn):	Hb–286/90
		Eb–606/8

Sb^+ . . . . Sb^+ . . . . 18 (Sb): Hb/B1–20/2, 96/8, 103/4
                                                                                Hb/B2–155, 167

Sb^{2+} . . . . Sb^{2+} . . . . 18 (Sb): Hb/B1–20/2, 98/100, 103/4
                                                                                Hb/B2–155

Sb^{3+} . . . . Sb^{3+}:
Chemical reactions	18 (Sb):	Hb/B2–279/303
Electrochemical behavior	18 (Sb):	Hb/B2–201/7
Ionization potential	18 (Sb):	Hb/B2–155/6
Properties of the ion	18 (Sb):	Hb/B1–20, 21/2
		Hb/B2–167
Spectra	18 (Sb):	Hb/B1–100/1

Sb^{4+} . . . . Sb^{4+} . . . . 18 (Sb): Hb/B1–20/2, 101/2
                                                                                Hb/B2–155/6, 167

Sb^{5+} . . . . Sb^{5+}:
Chemical reactions	18 (Sb):	Hb/B2–279/303
Ionization potential	18 (Sb):	Hb/B2–155/6
Molar polarization	18 (Sb):	Hb/B2–186
Properties of the ion	18 (Sb):	Hb/B1–20, 21/2
		Hb/B2–167
Spectra	18 (Sb):	Hb/B1–102
Susceptibility	18 (Sb):	Hb/B1–116

Hb = Main Series Volume; GD = Gmelin-Durrer;   *Ab = Appended Volume to Main Series;*
Eb = Supplement Volume to Main Series;   *Erg. W. = New Supplement Series Volume*

Sb^{6+}	Sb^{6+}	18 (Sb):	Hb/B2−155
Sb^{7+}	Sb^{7+}	18 (Sb):	Hb/B2−155
Sb^{8+}	Sb^{8+}	18 (Sb):	Hb/B2−155
Sb^{22+}	Sb^{22+}	18 (Sb):	Hb/B2−184
Sb^{-}	Sb^{-}	18 (Sb):	Hb/B1−20, 22
			Hb/B2−167
Sb^{2-}	Sb^{2-}	18 (Sb):	Hb/B1−22
Sb^{3-}	Sb^{3-}	18 (Sb):	Hb/B1−116
			Hb/B2−167, 186
SbSe	SbSe	18 (Sb):	Hb/B3−550
SbSn	SnSb	46 (Sn):	Hb/D−4/5
$SbTa_3$	Ta_3Sb	50 (Ta):	Hb/B1−273
$SbTl_3$	Tl_3Sb	38 (Tl):	Hb/2−189
SbV	VSb	48 (V):	Hb/B2−575/7
SbV_3	V_3Sb	48 (V):	Hb/B2−575/7
SbZn	ZnSb	32 (Zn):	Hb−286/8, 290
			Eb−615
Sb_2	Sb_2	18 (Sb):	Hb/B1−26/9
			Hb/B2−168/9
		46 (Sn):	Hb/D−22
Sb_2Se_3	Sb_2Se_3	18 (Sb):	Hb/B3−552/3
−.	Sb_2Se_3 solid solutions:		
	$(Bi_{1-x}Sb_x)_2(Te_{1-y}Se_y)_3$	19 (Bi):	Eb−865
−.	Sb_2Se_3 systems:		
	Sb_2Se_3−Ag_2Se	61 (Ag):	Hb/B4−101/2
	Sb_2Se_3−PbSe	47 (Pb):	Hb/C3−940
Sb_2Se_5	Sb_2Se_5	18 (Sb):	Hb/B3−553
$Sb_2Se_8Zn_3$	$Zn_3(SbSe_4)_2$	32 (Zn):	Hb−291
Sb_2Sn_3	Sb_2Sn_3	46 (Sn):	Hb/D−2/3
Sb_2Ta	$TaSb_2$	50 (Ta):	Hb/B1−272, 273
Sb_2Te_3	Sb_2Te_3	18 (Sb):	Hb/B3−555
−.	Sb_2Te_3 solid solutions:		
	$(Bi_{1-x}Sb_x)_2(Te_{1-y}Se_y)_3$	19 (Bi):	Eb−865
	$(Bi_{1-x}Sb_x)_2Te_3$	19 (Bi):	Eb−862
−.	Sb_2Te_3 systems:		
	Sb_2Te_3−Ag_2Te	61 (Ag):	Hb/B4−105/8
	Sb_2Te_3−Ag_2Te−PbTe	61 (Ag):	Hb/B4−269/70
Sb_2Tl_7	Tl_7Sb_2	38 (Tl):	Hb/2−187/91
Sb_2V	VSb_2	48 (V):	Hb/B2−575/7
Sb_2Zn	$ZnSb_2$	32 (Zn):	Hb−287
Sb_2Zn_3	Zn_3Sb_2	32 (Zn):	Hb−286/8, 290
			Eb−615
Sb_3	Sb_3	18 (Sb):	Hb/B1−26
Sb_3Se_4	Sb_3Se_4	18 (Sb):	Hb/B3−550, 553
Sb_3Zn_4	Zn_4Sb_3	32 (Zn):	Eb−615
Sb_4	Sb_4	18 (Sb):	Hb/B1−26
			Hb/B2−168/9
		46 (Sn):	Hb/D−22

See the "Instructions for the user"
at the beginning of the volume

| Sb₄Se₅ | . . . . | Sb₄Se₅ | . . . . | 18 (Sb): | Hb/B3−550, 553 |

Sb_4Se_5 . . . . Sb_4Se_5 . . . . 18 (Sb): Hb/B3−550, 553
Sb_4Ta_5 . . . . Ta_5Sb_4 . . . . 50 (Ta): Hb/B1−273
Sb_4U_3 . . . . U_3Sb_4. . . . . 55 (U): Hb−185
Sb_8 . . . . Sb_8 . . . . 18 (Sb): Hb/B1−27

Sc . . . . Scandium:
                Abundance . . . . 39 (SE): Hb/A1−34/5
                                                 Hb/A2−17/8, 88
                Cosmochemistry . . . . 39 (SE): Hb/A1−23/7
                                                 Hb/A2−3/46
                Crystal chemistry . . . . 39 (SE): Hb/A1−28/32
                                                 Hb/A2−47/88
                Crystallographic properties . . . . . 39 (SE): Hb/B3−1/4
                Diadochy . . . . 39 (SE): Hb/A2−48/55
                Economic deposits . . . . 39 (SE): Hb/A1−110/22
                                                 Hb/A2−164/78
                Economics. . . . . 39 (SE): Hb/A2−180/1
                Electrical properties . . . . 39 (SE): Hb/B3−18/27
                Geochemical character. . . . . 39 (SE): Hb/A2−88
                Geochemical cycle . . . . 39 (SE): Hb/A2−149/51
                Geochemistry . . . . 39 (SE): Hb/A1−28/54
                                                 Hb/A2−47/151
                History. . . . . 39 (SE): Hb/A1−4/22
                                                 Hb/A2−1/2
                Magnetic properties . . . . 39 (SE): Hb/B3−18/27
                Mechanical properties . . . . 39 (SE): Hb/B3−5/10
                Mineralogy . . . . 39 (SE): Hb/A1−54/110
                                                 Hb/A2−152/9
                Occurrence, extraterrestrial . . . . . 39 (SE): Hb/A1−22/7
                                                 Hb/A2−3/46
                −, terrestrial . . . . 39 (SE): Hb/A1−27/54
                                               Hb/A2−47/151
                − in meteorites and tektites . . . . . 39 (SE): Hb/A1−25/7
                                               Hb/A2−18/34
                − in minerals . . . . 39 (SE): Hb/A1−50/1
                                             Hb/A2−55/88
                − in the atmosphere . . . . 39 (SE): Hb/A2−149
                − in the biosphere. . . . . 39 (SE): Hb/A1−53/4
                − in the hydrosphere . . . . 39 (SE): Hb/A2−147/9
                − in the lithosphere . . . . 39 (SE): Hb/A1−27/54
                                             Hb/A2−88/147
                − in the magmatic cycle . . . . 39 (SE): Hb/A1−28/32
                                           Hb/A2−88/120
                − in the metamorphic cycle . . . . . 39 (SE): Hb/A2−133/47
                − in the sedimentary cycle . . . . . . 39 (SE): Hb/A1−32/3
                                           Hb/A2−121/33
                − on the moon . . . . 39 (SE): Hb/A2−34/46
                Position in the periodic system . . . . 39 (SE): Hb/A1−2/4
                Preparation . . . . 39 (SE): Hb/A2−160/4

Hb = Main Series Volume; GD = Gmelin-Durrer;   Ab = Appended Volume to Main Series;
Eb = Supplement Volume to Main Series;   Erg. W. = New Supplement Series Volume

Sc . . . . Scandium:
                        Thermal properties . . . . 39 (SE):   Hb/B3–4/18
                        Uses . . . . 39 (SE):   Hb/A2–178/80
                        Valence . . . . 39 (SE):   Hb/A1–4
–. . . . . Scandium ions. . . . . 39 (SE):   Hb/A1–3
                                                                           Hb/B3–15/6
–. . . . . Scandium systems:
                        Sc–Ag . . . . 61 (Ag):   Hb/C–210/1
                        Sc–H. . . . . 39 (SE):   Hb/C1–2/3, 18/9
                        Sc–H–Y . . . . 39 (SE):   Hb/C1–31/3
                        Sc–Np–O . . . .   Erg. W. 4–45/6
                        Sc–O . . . . 39 (SE):   Hb/C1–180
                        Sc–Pb . . . . 47 (Pb):   Hb/C4–1360
                        Sc–Ta . . . . 50 (Ta):   Hb/B2–18/9
                        Sc–V. . . . . 48 (V):   Hb/B2–592/3
Sc^+ . . . . Sc^+ . . . . 39 (SE):   Hb/B3–16
Sc^{2+} . . . . Sc^{2+} . . . . 39 (SE):   Hb/B3–16
Sc^{3+} . . . . Sc^{3+} . . . . 39 (SE):   Hb/A1–3
                                                                           Hb/B3–16
–. . . . . $[Sc(H_2O)_6]^{3+}$ . . . . 39 (SE):   Hb/C2–87
$ScSn_3$ . . . . $ScSn_3$ . . . . 46 (Sn):   Hb/D–293
Sc_2Se_3 . . . . Sc_2Se_3 systems:
                        Sc_2Se_3–Ag_2Se . . . . 61 (Ag):   Hb/B4–234/5
Sc_5Sn_3 . . . . Sc_5Sn_3 . . . . 46 (Sn):   Hb/D–293

Se . . . . Selenium:
                        Absorption of light . . . . 10 (Se):   Hb/A1–206/10
                        – of X-rays. . . . . 10 (Se):   Hb/A1–210, 231/2
                        Abundance . . . . 10 (Se):   Hb/A1–8/9
                        Aerosols. . . . . 10 (Se):   Hb/A1–111/6
                        Atom. . . . . 10 (Se):   Hb/A1–140/50
                        Atomic moments . . . . 10 (Se):   Hb/A1–145
                        Atomic nucleus . . . . 10 (Se):   Hb/A1–131/40
                        Atomic number . . . . 10 (Se):   Hb/A1–132
                        Atomic radius . . . . 10 (Se):   Hb/A1–142/5
                        Atomic refraction . . . . 10 (Se):   Hb/A1–206
                        Atomic volume . . . . 10 (Se):   Hb/A1–145/6
                        Atomic weight . . . . 10 (Se):   Hb/A1–146/9
                        Becquerel effect. . . . . 10 (Se):   Hb/A1–238/42
                        Behavior as electrode . . . . 10 (Se):   Hb/A1–237/8
                        Boiling point . . . . 10 (Se):   Hb/A1–188/9
                        Cells . . . . 10 (Se):   Hb/A1–235/6
                        Chemical constant . . . . 10 (Se):   Hb/A1–197
                        Chemical reactions . . . . 10 (Se):   Hb/A1–242/59
                        – with elements. . . . . 10 (Se):   Hb/A1–242/6
                        – with inorganic compounds. . . . . . 10 (Se):   Hb/A1–246/55
                        – with organic compounds. . . . . . . 10 (Se):   Hb/A1–255/9

See the "Instructions for the user"
at the beginning of the volume

Se . . . . Selenium:
    Coagula . . . . 10 (Se): Hb/A1–129/30
    Colloidal forms . . . . 10 (Se): Hb/A1–118/30
    Color. . . . . 10 (Se): Hb/A1–199/200
    Compressibility . . . . 10 (Se): Hb/A1–185
    Contact electricity. . . . . 10 (Se): Hb/A2–295/6
    Cosmochemistry . . . . 10 (Se): Hb/A1–4/5
    Critical constants . . . . 10 (Se): Hb/A1–186
    Crystal form. . . . . 10 (Se): Hb/A1–160/3
    Crystallographic properties . . . . . 10 (Se): Hb/A1–155/75
    Debye temperature . . . . 10 (Se): Hb/A1–196
    Deformation. . . . . 10 (Se): Hb/A1–174
    Density . . . . 10 (Se): Hb/A1–175/80
    Detection . . . . 10 (Se): Hb/A1–261/9
    – in iron . . . . 59 (Fe): Hb/FII1–27/8
    – in special substances . . . . 10 (Se): Hb/A1–269/71
    Determination. . . . . 10 (Se): Hb/A1–271/83
    – in iron . . . . 59 (Fe): Hb/FII1–28/30
    – in special substances . . . . 10 (Se): Hb/A1–284/92
    Dielectric constant . . . . 10 (Se): Hb/A2–293/4
    Diffusion. . . . . 10 (Se): Hb/A1–186
    Discharge phenomena . . . . 10 (Se): Hb/A2–298/9
    Dissociation of the molecule. . . . . 10 (Se): Hb/A1–150/2, 154/5
    Economic deposits:
      America. . . . . 10 (Se): Hb/A1–43/8
      Asia . . . . 10 (Se): Hb/A1–35/42
      Australia . . . . 10 (Se): Hb/A1–42
      Europe . . . . 10 (Se): Hb/A1–28/37
    Elasticity. . . . . 10 (Se): Hb/A1–183
    Electrical conductivity . . . . 10 (Se): Hb/A2–301/414
    Electrical properties . . . . 10 (Se): Hb/A2–293/414
    Electrochemical behavior . . . . . . 10 (Se): Hb/A1–235/42
    Electron configuration . . . . 10 (Se): Hb/A1–140/1
    Electron emission. . . . . 10 (Se): Hb/A2–299/300
    Entropy . . . . 10 (Se): Hb/A1–197
    Equation of state . . . . 10 (Se): Hb/A1–186
    Faraday effect. . . . . 10 (Se): Hb/A1–210
    Formation . . . . 10 (Se): Hb/A1–76/7
    Free energy . . . . 10 (Se): Hb/A1–197
    Frictional electricity. . . . . 10 (Se): Hb/A2–295
    Geochemical character. . . . . 10 (Se): Hb/A1–5/8
    Geochemistry . . . . 10 (Se): Hb/A1–5/28
    Grain structure . . . . 10 (Se): Hb/A1–174/5
    Hall effect . . . . 10 (Se): Hb/A1–234
    Hardness . . . . 10 (Se): Hb/A1–182
    Heat content. . . . . 10 (Se): Hb/A1–197
    Heat of fusion . . . . 10 (Se): Hb/A1–192
    – of sublimation. . . . . 10 (Se): Hb/A1–191

Hb = Main Series Volume; GD = Gmelin-Durrer; *Ab = Appended Volume to Main Series;*
Eb = Supplement Volume to Main Series;   *Erg. W. = New Supplement Series Volume*

Se . . . . Selenium:

Heat of transition . . . . 10 (Se):   Hb/A1 – 192/3
– of vaporization . . . . 10 (Se):   Hb/A1 – 190/1
History . . . . 10 (Se):   Hb/A1 – 1/3
Hydrosols . . . . 10 (Se):   Hb/A1 – 116/27
Ionization . . . . 10 (Se):   Hb/A2 – 300/1
Isomorphism . . . . 10 (Se):   Hb/A1 – 166/7
Lattice structure . . . . 10 (Se):   Hb/A1 – 167/72
Magnetic properties . . . . 10 (Se):   Hb/A1 – 232/4
Mechanical properties . . . . 10 (Se):   Hb/A1 – 175/86
Melting point . . . . 10 (Se):   Hb/A1 – 189/90
Mineralogy . . . . 10 (Se):   Hb/A1 – 48/75
Molecule . . . . 10 (Se):   Hb/A1 – 150/5
Nuclear properties . . . . 10 (Se):   Hb/A1 – 131/40
Occurrence, extraterrestrial . . . . . 10 (Se):   Hb/A1 – 4/5
–, terrestrial . . . . 10 (Se):   Hb/A1 – 5/28
– in industrial products . . . . 10 (Se):   Hb/A1 – 25/8
– in soils . . . . 10 (Se):   Hb/A1 – 19/22
– in the biosphere . . . . 10 (Se):   Hb/A1 – 22/5
– in the hydrosphere . . . . 10 (Se):   Hb/A1 – 25
– in the lithosphere . . . . 10 (Se):   Hb/A1 – 10/9
Optical properties . . . . 10 (Se):   Hb/A1 – 199/232
Passage of radiation . . . . 10 (Se):   Hb/A2 – 299/300
Photoelectric effect . . . . 10 (Se):   Hb/A2 – 301/414
Piezoelectricity . . . . 10 (Se):   Hb/A2 – 294
Plasticity . . . . 10 (Se):   Hb/A1 – 183/4
Polymorphism . . . . 10 (Se):   Hb/A1 – 155/60
Position in the electrochemical
    series . . . . 10 (Se):   Hb/A1 – 235
– in the periodic system . . . . 10 (Se):   Hb/A1 – 150
Preparation . . . . 10 (Se):   Hb/A1 – 95/9
–, industrial . . . . 10 (Se):   Hb/A1 – 78/95
– in pure state . . . . 10 (Se):   Hb/A1 – 95/9
– of modifications . . . . 10 (Se):   Hb/A1 – 100/7
– of special forms . . . . 10 (Se):   Hb/A1 – 99/100
Radioactivity . . . . 10 (Se):   Hb/A1 – 140
Reflection . . . . 10 (Se):   Hb/A1 – 200/2
Refraction . . . . 10 (Se):   Hb/A1 – 202/6
Scattering of X-rays . . . . 10 (Se):   Hb/A1 – 231/2
Selenium photocell:
    Action of gamma rays
        and corpuscular radiation . . . . . 10 (Se):   Hb/A3 – 562/3
    – of infrared light . . . . 10 (Se):   Hb/A3 – 557/8
    – of spectrally
        decomposed visible light . . . . . 10 (Se):   Hb/A3 – 543/57
    – of ultraviolet light . . . . 10 (Se):   Hb/A3 – 558/9
    – of visible light . . . . 10 (Se):   Hb/A3 – 529/43
    – of X-rays . . . . 10 (Se):   Hb/A3 – 559/62
    Aging . . . . 10 (Se):   Hb/A3 – 578/80

See the "Instructions for the user"
at the beginning of the volume

Se . . . . Selenium:

Selenium photocell:

Application	10 (Se):	Hb/A3–590
Creep effect.	10 (Se):	Hb/A3–571
Electrical properties	10 (Se):	Hb/A3–525/63
Fatigue	10 (Se):	Hb/A3–571/7
General remarks	10 (Se):	Hb/A3–524/5
Inertia.	10 (Se):	Hb/A3–564/70
Instability	10 (Se):	Hb/A3–564/80
Optical properties	10 (Se):	Hb/A3–525/63
Recovery	10 (Se):	Hb/A3–577/8
Service	10 (Se):	Hb/A3–590/1
Technology.	10 (Se):	Hb/A3–580/90
Theory	10 (Se):	Hb/A3–591/8

Selenium photoresistive cell:

Action of light	10 (Se):	Hb/A2–315/30
– of X- and gamma rays	10 (Se):	Hb/A2–330/4
– of other factors.	10 (Se):	Hb/A2–334/59
Aging	10 (Se):	Hb/A2–393/4
Fabrication	10 (Se):	Hb/A2–395/414
Fatigue	10 (Se):	Hb/A2–387/92
General remarks	10 (Se):	Hb/A2–304/15
Industrial types.	10 (Se):	Hb/A2–396/400
Inertia.	10 (Se):	Hb/A2–362/77
Lifetime	10 (Se):	Hb/A2–393/4
Mechanism of dark conductivity . .	10 (Se):	Hb/A2–394/5
Negative photoeffect.	10 (Se):	Hb/A2–359/61
Polarization.	10 (Se):	Hb/A2–361/2
Recovery	10 (Se):	Hb/A2–377/87

Selenium rectifier:

Aging	10 (Se):	Hb/A3–437/8
Capacity.	10 (Se):	Hb/A3–426/9
Current-voltage characteristics . . .	10 (Se):	Hb/A3–417/9
Dependence on various factors . . .	10 (Se):	Hb/A3–439
Electrical properties	10 (Se):	Hb/A3–416/39
Fabrication	10 (Se):	Hb/A3–480
General remarks	10 (Se):	Hb/A3–415/6
Instability	10 (Se):	Hb/A3–435/8
Lifetime	10 (Se):	Hb/A3–438/9
Load characteristics	10 (Se):	Hb/A3–430/3
Magnitude of rectifier action	10 (Se):	Hb/A3–433/5
Rectifier circuits	10 (Se):	Hb/A3–419/20
Resistance	10 (Se):	Hb/A3–420/6
Service	10 (Se):	Hb/A3–481/2
Special types	10 (Se):	Hb/A3–478/80
Technology.	10 (Se):	Hb/A3–440/80
Theory	10 (Se):	Hb/A3–482/523
Utilization.	10 (Se):	Hb/A3–480/1

Hb = Main Series Volume; GD = Gmelin-Durrer;   Ab = Appended Volume to Main Series;
Eb = Supplement Volume to Main Series;   Erg. W. = New Supplement Series Volume

Se . . . . Selenium:
           Separation, analytical . . . . 10 (Se):   Hb/A1–283/4
           Sols . . . . 10 (Se):   Hb/A1–111/29
           Solubility . . . . 10 (Se):   Hb/A1–257/9
           Specific heat . . . . 10 (Se):   Hb/A1–193/6
           Spectra:
              Arc spectrum . . . . 10 (Se):   Hb/A1–210/5
              Band spectrum . . . . 10 (Se):   Hb/A1–219/27
              Continuous spectrum . . . . 10 (Se):   Hb/A1–227
              Spark spectra . . . . 10 (Se):   Hb/A1–215/9
              X-ray spectrum . . . . 10 (Se):   Hb/A1–228/32

           Standard potential . . . . 10 (Se):   Hb/A1–235
           Strength . . . . 10 (Se):   Hb/A1–184/5
           Structure . . . . 10 (Se):   Hb/A1–172/4
           Surface tension . . . . 10 (Se):   Hb/A1–185
           Susceptibility . . . . 10 (Se):   Hb/A1–232/4
           Thermal conductivity . . . . 10 (Se):   Hb/A1–197/9
           Thermal expansion . . . . 10 (Se):   Hb/A1–180/2
           Thermal properties . . . . 10 (Se):   Hb/A1–186/99
           Thermal radiation . . . . 10 (Se):   Hb/A1–210
           Thermoelectric effects . . . . 10 (Se):   Hb/A2–296/8
           Toxicity . . . . 10 (Se):   Hb/A1–259/60
           Transition point . . . . 10 (Se):   Hb/A1–190
           Triboluminescence . . . . 10 (Se):   Hb/A1–210
           Valence . . . . 10 (Se):   Hb/A1–149/50
           Vapor pressure . . . . 10 (Se):   Hb/A1–187/8
           Vibration frequencies . . . . 10 (Se):   Hb/A1–152/3, 196
           Viscosity . . . . 10 (Se):   Hb/A1–186
           Uses . . . . 10 (Se):   Hb/A1–108/10

–. . . . . Selenium ions
           see Se$^+$ ... Se$^+$, etc.

–. . . . . Selenium isotopes . . . . 10 (Se):   Hb/A1–131/2
–. . . . . Selenium systems:
           Se–Ag . . . . 61 (Ag):   Hb/B3–152/4
           Se–Ag–As . . . . 61 (Ag):   Hb/B4–66/7
           Se–Ag–Cu . . . . 61 (Ag):   Hb/B4–474/5
           Se–Ag–Pb . . . . 61 (Ag):   Hb/B4–267/8
           Se–Ag–Sb . . . . 61 (Ag):   Hb/B4–101/2
           Se–Al . . . . 35 (Al):   Hb/B1–279
           Se–As–Bi . . . . 19 (Bi):   Eb–858
           Se–As–Hg . . . . 34 (Hg):   Hb/B4–1378
           Se–As–Pb . . . . 47 (Pb):   Hb/C3–904/5
           Se–Bi . . . . 19 (Bi):   Hb–173/4
                                                  Eb–755/60
           Se–Bi–H . . . . 19 (Bi):   Hb–174
           Se–Bi–Pb . . . . 47 (Pb):   Hb/C3–971/2
            Se–Bi–Sb . . . . 19 (Bi):   Eb–861
           Se–Bi–Sb–Te . . . . 19 (Bi):   Eb–865/6

See the "Instructions for the user"
at the beginning of the volume

Se . . . .   Selenium systems:
                          Se–Co . . . . 58 (Co):   Eb/A–658/9
                          Se–Cr . . . . 52 (Cr):   Hb/B–337/8
                          Se–Cr–Te . . . . 52 (Cr):   Hb/B–343
                          Se–Cu . . . . 60 (Cu):   Hb/B1–600
                          Se–Cu–Fe . . . . 60 (Cu):   Hb/B3–1297
                          Se–Fe . . . . 59 (Fe):   Hb/A6–1171/2
                          Se–Ga–Pb . . . . 47 (Pb):   Hb/C4–1306
                          Se–Hg . . . . 34 (Hg):   Hb/B3–1046/7
                          Se–HgBr₂ . . . . 34 (Hg):   Hb/B2–757
                          Se–I . . . . 10 (Se):   Hb/B–157/8
                          Se–I–S . . . . 10 (Se):   Hb/B–193
                          Se–Na . . . . 21 (Na):   Hb–634
                          Se–Nb . . . . 49 (Nb):   Hb/B1–279/81
                          Se–Ni . . . . 57 (Ni):   Hb/B2–740/3
                          Se–P . . . . 16 (P):   Hb/C–605/6
                          Se–Pb . . . . 47 (Pb):   Hb/C2–608/9
                          Se–Pb–S . . . . 47 (Pb):   Hb/C2–639/40
                          Se–Pb–Sb . . . . 47 (Pb):   Hb/C3–940
                          Se–Pb–Sn–Te . . . . 47 (Pb):   Hb/C4–1594/5
                          Se–S . . . . 10 (Se):   Hb/B–160/70
                          Se–S–Te . . . . 11 (Te):   Hb–363
                          Se–S₂Cl₂ . . . . 10 (Se):   Hb/B–193
                          Se–Sb . . . . 18 (Sb):   Hb/B3–550/2
                          Se–Ta . . . . 50 (Ta):   Hb/B1–179/80
                          Se–Te . . . . 11 (Te):   Hb–361/2
                          Se–Th . . . . 44 (Th):   Hb–294
                          Se–Ti . . . . 41 (Ti):   Hb–352
                          Se–Tl . . . . 38 (Tl):   Hb/3–370/1
                          Se–V . . . . 48 (V):   Hb/B1–308/9
                          Se–Zn . . . . 32 (Zn):   Eb–967
Se⁺ . . . .   Se⁺ . . . . 10 (Se):   Hb/A1–144, 215/6
                                                                               Hb/A2–300/1
Se²⁺ . . . .   Se²⁺ . . . . 10 (Se):   Hb/A1–144, 216/7
                                                                               Hb/A2–300/1
Se³⁺ . . . .   Se³⁺ . . . . 10 (Se):   Hb/A1–144, 217/8
                                                                               Hb/A2–300/1
Se⁴⁺ . . . .   Se⁴⁺ . . . . 10 (Se):   Hb/A1–145/6, 218,
                                                                                    234
                                                                               Hb/A2–300/1
Se⁵⁺ . . . .   Se⁵⁺ . . . . 10 (Se):   Hb/A1–218/9
                                                                               Hb/A2–300/1
Se⁶⁺ . . . .   Se⁶⁺:
                          Increments . . . . 10 (Se):   Hb/A1–206, 234
                          Ionization potential . . . . 10 (Se):   Hb/A2–300/1
                          Properties of the ion . . . . 10 (Se):   Hb/A1–145/6
                          Spectra . . . . 10 (Se):   Hb/A1–219
Se⁷⁺ . . . .   Se⁷⁺ . . . . 10 (Se):   Hb/A2–300/1

Hb = Main Series Volume; GD = Gmelin-Durrer;   Ab = Appended Volume to Main Series;
Eb = Supplement Volume to Main Series;   Erg. W. = New Supplement Series Volume

Se^-	Se^-	10 (Se):	Hb/A1–144
Se^{2-}	Se^{2-}:		
	Chemical reactions	10 (Se):	Hb/B–17/9
	Electrochemical behavior	10 (Se):	Hb/A1–236
	Ionic refraction	10 (Se):	Hb/A1–206
	Properties of the ion	10 (Se):	Hb/A1–143/4, 146
			Hb/B–7
SeSi	SiSe	15 (Si):	Hb/B–754/5
SeSn	SnSe solid solutions:		
	SnSe–PbTe	47 (Pb):	Hb/C4–1595
	SnSe–PbTe–SnTe	47 (Pb):	Hb/C4–1595
–.	SnSe systems:		
	$SnSe–Ag_2Se–SnSe_2$	61 (Ag):	Hb/B4–249/50
	SnSe–PbS	47 (Pb):	Hb/C4–1588
	SnSe–PbSe	47 (Pb):	Hb/C4–1586/7
	SnSe–PbTe–SnTe	47 (Pb):	Hb/C4–1594/5
SeSr	SrSe	29 (Sr):	Hb–185/6
			Eb–260/1
–.	SrSe systems:		
	$SrSe–SrCl_2–SrO$	29 (Sr):	Eb–261
SeTa	TaSe	50 (Ta):	Hb/B1–181/7
SeTaTe	TaTeSe	50 (Ta):	Hb/B1–192
SeTe	TeSe	11 (Te):	Hb–361
SeTh	ThSe	44 (Th):	Hb–294
SeTi	TiSe	41 (Ti):	Hb–352/3
SeTl	$TlSe = Tl_2Se \cdot Tl_2Se_3$	38 (Tl):	Hb/3–372
$SeTl_2$	Tl_2Se	38 (Tl):	Hb/3–371/2
–.	Tl_2Se solid solutions:		
	$(Tl,Ag,Cu)_2Se$	61 (Ag):	Hb/B4–477
SeV	VSe	48 (V):	Hb/B1–308, 309/11
–.	VSe solid solutions:		
	(V,Cr)Se	48 (V):	Hb/B1–310
–.	VSe systems:		
	VSe–V	48 (V):	Hb/B1–308
SeZn	ZnSe	32 (Zn):	Hb–246
			Eb–967/8
–.	ZnSe solid solutions:		
	$Zn_{1-x}Hg_xS_ySe_{1-y}$	34 (Hg):	Hb/B4–1502
	ZnSe–HgSe	34 (Hg):	Hb/B4–1502
Se_2	Se_2	10 (Se):	Hb/A1–150/5
Se_2Si	$SiSe_2$	15 (Si):	Hb/B–755/6
–.	$SiSe_2$ systems:		
	$SiSe_2–Ag_2Se$	61 (Ag):	Hb/B3–388
Se_2Sn	$SnSe_2$ systems:		
	$SnSe_2–Ag_2Se$	61 (Ag):	Hb/B4–249
	$SnSe_2–Ag_2Se–SnSe$	61 (Ag):	Hb/B4–249/50

See the "Instructions for the user"
at the beginning of the volume

Se$_2$Ta . . . . TaSe$_2$ . . . . 50 (Ta):   Hb/B1–181/7

Se$_2$Ta$_{x+1}$ . . . . . . Ta$_{1+x}$Se$_2$ . . . . 50 (Ta):   Hb/B1–181/7

Se$_2$Th . . . . ThSe$_2$ . . . . 44 (Th):   Hb–294

Se$_2$Ti . . . . TiSe$_2$ . . . . 41 (Ti):   Hb–352

Se$_2$Ti$_x$ . . . . Ti$_x$Se$_2$ . . . . 41 (Ti):   Hb–353

Se$_2$U . . . . USe$_2$ . . . . 55 (U):   Hb–157

Se$_2$V . . . . VSe$_2$ . . . . 48 (V):   Hb/B1–309, 312/3

Se$_2$W . . . . WSe$_2$ . . . . 54 (W):   Hb–186

Se$_2$Zr . . . . ZrSe$_2$ . . . . 42 (Zr):   Hb–354

Se$_3$ . . . . Se$_3$ . . . . 10 (Se):   Hb/A1–150

Se$_3$Sm$_2$ . . . . Sm$_2$Se$_3$ solid solutions:

                   Sm$_2$Se$_3$–Ag$_2$Se . . . . 61 (Ag):   Hb/B4–234/5

                   Sm$_2$Se$_3$–PbSe . . . . 47 (Pb):   Hb/C4–1374

Se$_3$Ta . . . . TaSe$_3$ . . . . 50 (Ta):   Hb/B1–179/80, 187/9

Se$_3$Ta$_2$ . . . . Ta$_2$Se$_3$ . . . . 50 (Ta):   Hb/B1–182, 187

Se$_3$Tb$_2$ . . . . Tb$_2$Se$_3$ solid solutions:

                   Tb$_2$Se$_3$–Ag$_2$Se . . . . 61 (Ag):   Hb/B4–234/5

                   Tb$_2$Se$_3$–PbSe . . . . 47 (Pb):   Hb/C4–1374/5

Se$_3$Th$_2$ . . . . Th$_2$Se$_3$ . . . . 44 (Th):   Hb–294

Se$_3$Tl$_2$ . . . . Tl$_2$Se$_3$ . . . . 38 (Tl):   Hb/3–372/3

Se$_3$Tm$_2$ . . . . Tm$_2$Se$_3$ solid solutions:

                   Tm$_2$Se$_3$–Ag$_2$Se . . . . 61 (Ag):   Hb/B4–234/5

                   Tm$_2$Se$_3$–PbSe . . . . 47 (Pb):   Hb/C4–1374/5

Se$_3$U$_2$ . . . . U$_2$Se$_3$ . . . . 55 (U):   Hb–157

Se$_3$V$_2$ . . . . V$_2$Se$_3$ . . . . 48 (V):   Hb/B1–308, 311/2

Se$_3$W . . . . WSe$_3$ . . . . 54 (W):   Hb–186

Se$_3$Y$_2$ . . . . Y$_2$Se$_3$ solid solutions:

                   Y$_2$Se$_3$–Ag$_2$Se . . . . 61 (Ag):   Hb/B4–234/5

Se$_3$Yb$_2$ . . . . Yb$_2$Se$_3$ solid solutions:

                   Yb$_2$Se$_3$–Ag$_2$Se . . . . 61 (Ag):   Hb/B4–234/5

                   Yb$_2$Se$_3$–PbSe . . . . 47 (Pb):   Hb/C4–1374/5

Se$_4$ . . . . Se$_4$ . . . . 10 (Se):   Hb/A1–151

Se$_4$TaTl$_3$ . . . . . . Tl$_3$TaSe$_4$ . . . . 50 (Ta):   Hb/B2–319

Se$_4$Ti$_2$V . . . . Ti$_2$VSe$_4$ . . . . 48 (V):   Hb/B2–561

Se$_4$Tl$_3$V . . . . Tl$_3$VSe$_4$ . . . . 48 (V):   Hb/B2–551

Se$_4$Tl$_4$ . . . . Tl$_2$Se · Tl$_2$Se$_3$ = TlSe . . . . 38 (Tl):   Hb/3–372

Se$_4$V$_3$ . . . . V$_3$Se$_4$ . . . . 48 (V):   Hb/B1–308, 311

Se$_4$V$_5$ . . . . V$_5$Se$_4$ . . . . 48 (V):   Hb/B1–308, 309

Se$_5$Tl$_2$ . . . . Tl$_2$Se$_5$ . . . . 38 (Tl):   Hb/3–373

Se$_6$ . . . . Se$_6$ . . . . 10 (Se):   Hb/A1–150/1, 155

Se$_8$ . . . . Se$_8$ . . . . 10 (Se):   Hb/A1–150/1

Se$_8$V$_5$ . . . . V$_5$Se$_8$ . . . . 48 (V):   Hb/B1–308, 312

Se$_{10}$ . . . . Se$_{10}$ . . . . 10 (Se):   Hb/A1–150

Se$_{12}$Th$_7$ . . . . . . Th$_7$Se$_{12}$ . . . . 44 (Th):   Hb–294

Se$_{32}$V$_{29}$ . . . . . . V$_{29}$Se$_{32}$ . . . . 48 (V):   Hb/B1–308, 311

Se$_x$Ti . . . . TiSe$_x$ . . . . 41 (Ti):   Hb–352/3

Se$_x$V . . . . VSe$_x$ . . . . 48 (V):   Hb/B1–308/13

Hb = Main Series Volume; GD = Gmelin-Durrer;   Ab = Appended Volume to Main Series;
Eb = Supplement Volume to Main Series;   Erg. W. = New Supplement Series Volume

Si . . . . Silicon:

Absorption of light	15 (Si):	Hb/B−78
− of X-rays.	15 (Si):	Hb/B−91
Atom.	15 (Si):	Hb/B−17/33
Atomic nucleus	15 (Si):	Hb/B−16/7
Atomic radius	15 (Si):	Hb/B−19/23
Atomic scattering factors.	15 (Si):	Hb/B−46/8
Atomic volume	15 (Si):	Hb/B−26/7
Atomic weight	15 (Si):	Hb/B−27/31
Behavior as electrode	15 (Si):	Hb/B−115/6
Boiling point	15 (Si):	Hb/B−67
Brittleness.	15 (Si):	Hb/B−52
Catalytic properties.	15 (Si):	Hb/B−130
Cells	15 (Si):	Hb/B−112/4
Chemical constant	15 (Si):	Hb/B−76
Chemical reactions	15 (Si):	Hb/B−116/30, 922
Cleavage.	15 (Si):	Hb/B−37
Colloidal forms	15 (Si):	Hb/B−9/10
Color.	15 (Si):	Hb/B−76
Compressibility	15 (Si):	Hb/B−57/8
Critical constants	15 (Si):	Hb/B−62/3
Crystal form.	15 (Si):	Hb/B−35/7
Crystallographic properties	15 (Si):	Hb/B−34/52
Debye temperature	15 (Si):	Hb/B−74/5
Deformation.	15 (Si):	Hb/B−51/2
Density	15 (Si):	Hb/B−53/4
Detection	15 (Si):	Hb/B−135/6
− in iron	59 (Fe):	Hb/FI−314
− in special substances.	15 (Si):	Hb/B−136/49
Determination in iron.	59 (Fe):	Hb/FI−314/40
		Hb/FII2−308/10,
		345/7, 369
− in special substances.	15 (Si):	Hb/B−149/226
Dielectric constant	15 (Si):	Hb/B−110
Diffusion.	15 (Si):	Hb/B−59/62
Economics.	15 (Si):	Hb/B−11
Elasticity.	15 (Si):	Hb/B−56
Electrical conductivity	15 (Si):	Hb/B−93/104, 922/3
Electrical properties	15 (Si):	Hb/B−93/111, 921/3
Electrochemical behavior	15 (Si):	Hb/B−111/6
Electrolytic deposition	15 (Si):	Hb/B−116
Electron affinity.	15 (Si):	Hb/B−110
Electron configuration	15 (Si):	Hb/B−17
Electronegativity	15 (Si):	Hb/B−32/3
Electron emission.	15 (Si):	Hb/B−108/10
Enthalpy.	15 (Si):	Hb/B−75
Entropy	15 (Si):	Hb/B−75/6
Equation of state	15 (Si):	Hb/B−62
Expansion coefficient.	15 (Si):	Hb/B−63/5

See the "Instructions for the user"
at the beginning of the volume

Si . . . . Silicon:

Formability	15 (Si):	Hb/B–56
Formation	15 (Si):	Hb/B–1/2
Free energy	15 (Si):	Hb/B–75
Galvanomagnetic effects	15 (Si):	Hb/B–104/5
Hall effect	15 (Si):	Hb/B–104/5
Hardness	15 (Si):	Hb/B–55/6
Heat of fusion	15 (Si):	Hb/B–69/70
– of sublimation	15 (Si):	Hb/B–68/9
– of vaporization	15 (Si):	Hb/B–68/9
Interfacial tension	15 (Si):	Hb/B–58/9
Internal pressure	15 (Si):	Hb/B–58
Ionization potentials	15 (Si):	Hb/B–110/1
Isomorphism	15 (Si):	Hb/B–38
Lattice structure	15 (Si):	Hb/B–38/42
Magnetic moment	15 (Si):	Hb/B–18/9
Magnetic properties	15 (Si):	Hb/B–92/3
Mechanical properties	15 (Si):	Hb/B–53/62
Melting point	15 (Si):	Hb/B–67/8
Modifications	15 (Si):	Hb/B–34/5
Molecule	15 (Si):	Hb/B–33/4
Nuclear properties	15 (Si):	Hb/B–16/7
Optical constants	15 (Si):	Hb/B–76/8
Optical properties	15 (Si):	Hb/B–76/91
Parachor	15 (Si):	Hb/B–59
Photoelectric conductivity	15 (Si):	Hb/B–103/4
Plasticity	15 (Si):	Hb/B–56
Polarizability	15 (Si):	Hb/B–17/8
Polymorphism	15 (Si):	Hb/B–34/5
Position in the electrochemical series	15 (Si):	Hb/B–111
– in the periodic system	15 (Si):	Hb/B–31
Potentials	15 (Si):	Hb/B–112
Preparation	15 (Si):	Hb/B–2/14
–, industrial	15 (Si):	Hb/B–10/4
– in the purest state	15 (Si):	Hb/B–7
– of special forms	15 (Si):	Hb/B–7/10
Purification	15 (Si):	Hb/B–6/7, 13/4
Purity testing	15 (Si):	Hb/B–15
Scattering of X-rays	15 (Si):	Hb/B–91
Semiconductor properties	15 (Si):	Hb/B–93/102, 921/3
Silicon transistors	15 (Si):	Hb/B–921/3
Sorption	15 (Si):	Hb/B–62
Specific heat	15 (Si):	Hb/B–70/3
Spectra:		
Arc spectrum	15 (Si):	Hb/B–79/82
Band spectrum	15 (Si):	Hb/B–91
Microwave spectrum	15 (Si):	Hb/B–78

Si . . . . Silicon:
              Spectra:
                  Spark spectra . . . . 15 (Si):   Hb/B–83/8
                  X-ray spectrum . . . . 15 (Si):   Hb/B–88/91

              Standard potential . . . . 15 (Si):   Hb/B–111
              Strength . . . . 15 (Si):   Hb/B–57
              Structure . . . . 15 (Si):   Hb/B–48/51
              – of thin layers . . . . 15 (Si):   Hb/B–42/6
              Susceptibility . . . . 15 (Si):   Hb/B–92/3
              Thermal conductivity . . . . 15 (Si):   Hb/B–76
              Thermal expansion . . . . 15 (Si):   Hb/B–63/5
              Thermal properties . . . . 15 (Si):   Hb/B–62/76
              Thermoelectric effects . . . . 15 (Si):   Hb/B–106/8
              Thermomagnetic effects . . . . 15 (Si):   Hb/B–105/6
              Toxicity . . . . 15 (Si):   Hb/B–130/5
              Twinning . . . . 15 (Si):   Hb/B–37
              Uses . . . . 15 (Si):   Hb/B–15
              Valence . . . . 15 (Si):   Hb/B–31
              Vapor pressure . . . . 15 (Si):   Hb/B–65/7
              Vibrational frequency . . . . 15 (Si):   Hb/B–73/4
              Wettability . . . . 15 (Si):   Hb/B–58

–. . . . . Silicon ions
              see Si$^+$ ... Si$^+$, etc.

–. . . . . Silicon isotopes . . . . 15 (Si):   Hb/B–16/7
–. . . . . Silicon systems:
              Si–Ag . . . . 61 (Ag):   Hb/B3–372/4
              Si–Ag–Fe . . . . 61 (Ag):   Hb/B4–436
              Si–Ag–Ni . . . . 61 (Ag):   Hb/B4–386
              Si–Al . . . . 35 (Al):   Hb/A4–536/40
              Si–Al–Be . . . . 35 (Al):   Hb/A4–644/5
              Si–Al–Ca . . . . 35 (Al):   Hb/A4–681
              Si–Al–Cr . . . . 35 (Al):   Hb/A5–880
                                       52 (Cr):   Hb/B–895
              Si–Al–Cu . . . . 60 (Cu):   Hb/B3–1176
              Si–Al–Cu–Fe . . . . 60 (Cu):   Hb/B3–1327
              Si–Al–Cu–Mg . . . . 60 (Cu):   Hb/B3–1178/9
              Si–Al–Cu–Mn . . . . 60 (Cu):   Hb/B3–1249
              Si–Al–Fe . . . . 35 (Al):   Hb/A8–1334/62
              Si–Al–Mg . . . . 35 (Al):   Hb/A4–666/78
              Si–Al–Mg–Zn . . . . 35 (Al):   Hb/A5–783
              Si–Al–Mn . . . . 35 (Al):   Hb/A6–928/9
              Si–Al–Na . . . . 35 (Al):   Hb/A4–636
              Si–Al–Nb . . . . 49 (Nb):   Hb/B2–43/4
              Si–Al–Ni . . . . 35 (Al):   Hb/A6–1000
              Si–Al–Sn . . . . 35 (Al):   Hb/A5–853
                                       46 (Sn):   Hb/D–217/8
              Si–Al–Ta . . . . 50 (Ta):   Hb/B2–16/7

See the "Instructions for the user"
at the beginning of the volume

Si . . . . Silicon systems:

Si–Al–Zn . . . . 35 (Al):   Hb/A5–760/1
Si–As . . . . 17 (As):   Hb–471/2
Si–Au . . . . 62 (Au):   Hb/3–727
Si–B–Cr. . . . . 52 (Cr):   Hb/B–436
Si–B–Nb . . . . 49 (Nb):   Hb/B1–402/3
Si–B–Ta. . . . . 50 (Ta):   Hb/B1–264
Si–B–V . . . . 48 (V):   Hb/B1–357
Si–Be . . . . 26 (Be):   Hb–161
Si–Be–Cu . . . . 60 (Cu):   Hb/B3–1121/2
Si–Bi. . . . . 19 (Bi):   Eb–843/4
Si–C–Cr. . . . . 52 (Cr):   Hb/B–436
Si–C–Fe. . . . . 59 (Fe):   Hb/A8–1761/79
Si–C–Fe–P . . . . 59 (Fe):   Hb/A8–1798
Si–C–Fe–V . . . . 59 (Fe):   Hb/A9–1936
Si–C–Ge–V . . . . 48 (V):   Hb/B2–684
Si–C–Nb . . . . 49 (Nb):   Hb/B1–403/4
Si–C–Nb–V . . . . 49 (Nb):   Hb/B2–271
Si–C–Pu . . . . Erg. W. 4–209
Si–C–Ta. . . . . 50 (Ta):   Hb/B1–264/5
Si–Ca . . . . 28 (Ca):   Hb/B3–1029/30
Si–Ca–N . . . . 28 (Ca):   Hb/B3–1099
Si–Cd . . . . 33 (Cd):   Eb–569
Si–Co . . . . 58 (Co):   Hb/A2–384/5
                                                     Eb/A–725/8
Si–Co–Cu. . . . . 60 (Cu):   Hb/B3–1269
Si–Co–Mn . . . . 58 (Co):   Eb/A–880
Si–Co–Ni . . . . 58 (Co):   Eb/A–470
Si–Cr . . . . 52 (Cr):   Hb/B–426/8
Si–Cr–N. . . . . 52 (Cr):   Hb/B–436
Si–Cr–Ni . . . . 57 (Ni):   Hb/B1–221/2
                                                     Hb/B3–1214/5
Si–Cu . . . . 60 (Cu):   Hb/B2–883/6
Si–Cu–Fe. . . . . 60 (Cu):   Hb/B3–1314
Si–Cu–H–Mg. . . . . 60 (Cu):   Hb/B3–1127
Si–Cu–Mg . . . . 27 (Mg):   Hb/A4–714/5
                                       60 (Cu):   Hb/B3–1126
Si–Cu–Mn . . . . 60 (Cu):   Hb/B3–1248
Si–Cu–Ni . . . . 60 (Cu):   Hb/B3–1258
Si–Cu–P . . . . 60 (Cu):   Hb/B2–933
Si–Cu–Th. . . . . 60 (Cu):   Hb/B3–1192
Si–Cu–Ti . . . . 60 (Cu):   Hb/B3–1190
Si–Cu–Zn. . . . . 60 (Cu):   Hb/B3–1150
Si–Cu–Zr . . . . 60 (Cu):   Hb/B3–1191
Si–Fe . . . . 59 (Fe):   Hb/A8–1740/50
Si–Fe–Mg. . . . . 59 (Fe):   Hb/A9–1821/2
Si–Fe–Ni–O . . . . 59 (Fe):   Hb/A5–943
Si–Fe–O . . . . 59 (Fe):   Hb/A5–876/81

Hb = Main Series Volume; GD = Gmelin-Durrer;   *Ab = Appended Volume to Main Series;*
Eb = Supplement Volume to Main Series;   *Erg. W. = New Supplement Series Volume*

Si . . . . Silicon systems:

Si–Fe–P. . . . . 59 (Fe):   Hb/A8–1797/8
Si–Fe–Ti . . . . 59 (Fe):   Hb/A9–1888/93
Si–Ga–O–V . . . . 48 (V):   Hb/B2–592
Si–Ga–V . . . . 48 (V):   Hb/B2–592
Si–Ge . . . . 45 (Ge):   Eb–557
Si–Ge–V . . . . 48 (V):   Hb/B2–683/4
Si–Hg . . . . 34 (Hg):   Hb/B4–1355/6
Si–Li. . . . . 20 (Li):   Eb–513
Si–Mg. . . . . 27 (Mg):   Hb/A3–374/8
Si–Mg–Zn . . . . 27 (Mg):   Hb/A4–520/1
Si–Mn–Ni. . . . . 57 (Ni):   Hb/B1–310/1
                                                   Hb/B3–1240/1
Si–Mn–O . . . . 59 (Fe):   Hb/A5–941/2
Si–Mo–Ni. . . . . 57 (Ni):   Hb/B1–250
                                                   Hb/B3–1222/3
Si–N–Nb . . . . 49 (Nb):   Hb/B1–402
Si–N–Ta . . . . 50 (Ta):   Hb/B1–263/4
Si–Nb . . . . 49 (Nb):   Hb/B1–389/95
Si–Nb–Pu. . . . . Erg. W. 4–247/8
Si–Nb–Ta . . . . 49 (Nb):   Hb/B2–104
Si–Nb–V . . . . 49 (Nb):   Hb/B2–268
Si–NbB$_2$–Nb$_5$Si$_3$ . . . . 49 (Nb):   Hb/B1–402/3
Si–Ni . . . . 57 (Ni):   Hb/B3–914/9
Si–Ni–U. . . . . 57 (Ni):   Hb/B1–277
Si–Ni–V. . . . . 57 (Ni):   Hb/B1–159
Si–Ni–W . . . . 57 (Ni):   Hb/B1–269
Si–O–Ta . . . . 50 (Ta):   Hb/B1–263
Si–P . . . . 16 (P):   Hb/C–635/6
Si–Pb . . . . 47 (Pb):   Hb/C3–807/8
Si–Pb–S . . . . 47 (Pb):   Hb/C3–855
Si–Pb–Sb . . . . 47 (Pb):   Hb/C3–941
Si–Pd . . . . 65 (Pd):   Hb/2–302
Si–Pt. . . . . 68 (Pt):   Hb/C1–128/31
Si–Pu . . . . Erg. W. 4–245/6
Si–Ru . . . . 63 (Ru):   Eb–331/2
Si–Sb . . . . 18 (Sb):   Hb/B3–560
Si–Ta . . . . 50 (Ta):   Hb/B1–250/5
Si–Ta–Ti . . . . 50 (Ta):   Hb/B2–39
Si–Ta–V. . . . . 50 (Ta):   Hb/B2–77/8
Si–Ti. . . . . 41 (Ti):   Hb–375
Si–Ti–V . . . . 48 (V):   Hb/B2–659/61
Si–Tl. . . . . 38 (Tl):   Hb/3–424
Si–V . . . . 48 (V):   Hb/B1–352/4
Si–Zn . . . . 32 (Zn):   Hb–272
Si–Zr . . . . 42 (Zr):   Hb–379
Si$^+$ . . . . Si$^+$ . . . . 15 (Si):   Hb/B–17/9, 23/4, 27,
                                                   83/4, 110/1

See the "Instructions for the user"
at the beginning of the volume

Si^{2+}	Si^{2+}	15 (Si):	Hb/B–17/9, 23/4, 27, 46/7, 84/5, 110/1
Si^{3+}	Si^{3+}	15 (Si):	Hb/B–17/9, 23/4, 27, 85/6, 110/1
Si^{4+}	Si^{4+}:		
	Diffusion.	15 (Si):	Hb/B–62
	Ionic polarization	15 (Si):	Hb/B–111
	Ionic refraction	15 (Si):	Hb/B–78
	Ionic susceptibility	15 (Si):	Hb/B–92/3
	Ionization potential	15 (Si):	Hb/B–110/1
	Parachor.	15 (Si):	Hb/B–59
	Properties of the ion	15 (Si):	Hb/B–17/9, 23/7
	Scattering factors.	15 (Si):	Hb/B–46/8
	Spectra	15 (Si):	Hb/B–86
Si^{n+}	Si^{n+} (n = 5 to 14)	15 (Si):	Hb/B–17, 19, 86/8, 110/1
Si^{-}	Si^{-}	15 (Si):	Hb/B–27, 110
Si^{2-}	Si^{2-}	15 (Si):	Hb/B–27, 110
Si^{3-}	Si^{3-}	15 (Si):	Hb/B–27, 110
Si^{4-}	Si^{4-}	15 (Si):	Hb/B–17/8, 23, 25/7, 78, 92/3
$SiTa_2$	Ta_2Si	50 (Ta):	Hb/A2–514
			Hb/B1–251/2, 256/7
–.	Ta_2Si solid solutions:		
	$Ta_2(Si,Al)$	50 (Ta):	Hb/B2–17
$SiTa_3$	Ta_3Si	50 (Ta):	Hb/B1–251/2, 256
$SiTa_4$	Ta_4Si	50 (Ta):	Hb/A2–514
			Hb/B1–252, 254
–.	Ta_4Si solid solutions:		
	$Ta_4Si–Nb_4Si$	50 (Ta):	Hb/B2–104
$SiTa_{4.5}$	$Ta_{4.5}Si$	50 (Ta):	Hb/B1–251/2, 255/6
SiTe.	SiTe.	15 (Si):	Hb/B–756/7
$SiTi_{0.3}V_{2.7}$	$(V_{0.9}Ti_{0.1})_3Si$	48 (V):	Hb/B2–660/1
SiTi	TiSi	41 (Ti):	Hb–375
$SiTi_2$.	Ti_2Si	41 (Ti):	Hb–375
SiV_2	V_2Si.	48 (V):	Hb/B1–355
$SiV_{2.7}Zr_{0.3}$	$(V_{0.9}Zr_{0.1})_3Si$	48 (V):	Hb/B2–674
SiV_3	V_3Si	48 (V):	Hb/A2–381
			Hb/B1–352/3, 354/5
–.	V_3Si solid solutions:		
	$(V,Nb)_3Si$.	49 (Nb):	Hb/B2–268/70
	$V_3Si–V_3Ga$.	48 (V):	Hb/B2–592
	$V_3(Si,Ge)$.	48 (V):	Hb/B2–683/4
–.	V_3Si systems:		
	$V_3Si–Cr_3Si$.	52 (Cr):	Hb/B–941
	$V_3Si–Nb_3Sn$.	49 (Nb):	Hb/B2–307
SiZr	ZrSi	42 (Zr):	Hb–379

Hb = Main Series Volume; GD = Gmelin-Durrer;   Ab = Appended Volume to Main Series;
Eb = Supplement Volume to Main Series;   Erg. W. = New Supplement Series Volume

Si$_2$ . . . . Si$_2$ . . . . 15 (Si):   Hb/B−33/4, 91
Si$_2$Sr . . . . SrSi$_2$ . . . . 29 (Sr):   Hb−209
                                                                                              Eb−280
Si$_2$Sr$_2$ . . . . Sr$_2$Si$_2$ . . . . 29 (Sr):   Eb−280
Si$_2$Ta . . . . TaSi$_2$ . . . . 50 (Ta):   Hb/A2−513/4
                                                                                              Hb/B1−251/2, 259/63
−. . . . . TaSi$_2$ solid solutions:
                         TaSi$_2$−NbSi$_2$ . . . . 50 (Ta):   Hb/B2−104
                         Ta(Si,Al)$_2$ . . . . 50 (Ta):   Hb/B2−16/7
                         Ta(Si,Be)$_2$ . . . . 50 (Ta):   Hb/B2−10
                         Ta(Ge,Si)$_2$ . . . . 50 (Ta):   Hb/B2−65
                         (Ta,V)Si$_2$ . . . . 50 (Ta):   Hb/B2−77/8
−. . . . . TaSi$_2$ systems:
                         TaSi$_2$−CrSi$_2$ . . . . 52 (Cr):   Hb/B−942
                         TaSi$_2$−CrSi$_2$−TiSi$_2$ . . . . 52 (Cr):   Hb/B−942
Si$_2$Th . . . . ThSi$_2$ . . . . 44 (Th):   Hb−312/3
Si$_2$Ti . . . . TiSi$_2$ . . . . 41 (Ti):   Hb−375/6
−. . . . . TiSi$_2$ systems:
                         TiSi$_2$−CrSi$_2$ . . . . 52 (Cr):   Hb/B−905
                         TiSi$_2$−CrSi$_2$−TaSi$_2$ . . . . 52 (Cr):   Hb/B−942
                         TiSi$_2$−VSi$_2$ . . . . 48 (V):   Hb/B2−562
Si$_2$U . . . . USi$_2$ . . . . 55 (U):   Hb−179
Si$_2$V . . . . VSi$_2$ . . . . 48 (V):   Hb/A2−381, 517, 709
                                                                                              Hb/B1−352/3, 356/7
−. . . . . VSi$_2$ solid solutions:
                         (V,Nb)Si$_2$ . . . . 49 (Nb):   Hb/B2−268/70
                         (V,Ta)Si$_2$ . . . . 50 (Ta):   Hb/B2−77/8
                         V(Si,Ge)$_2$ . . . . 48 (V):   Hb/B2−683/4
−. . . . . VSi$_2$ systems:
                         VSi$_2$−TiSi$_2$ . . . . 48 (V):   Hb/B2−562
                         VSi$_2$−ZrSi$_2$ . . . . 48 (V):   Hb/B2−563
Si$_2$V$_3$ . . . . V$_3$Si$_2$ . . . . 48 (V):   Hb/B1−352/3, 356
Si$_2$W . . . . WSi$_2$ . . . . 54 (W):   Hb−205
Si$_2$Zr . . . . ZrSi$_2$ . . . . 42 (Zr):   Hb−379/81
−. . . . . ZrSi$_2$ systems:
                         ZrSi$_2$−CrSi$_2$ . . . . 52 (Cr):   Hb/B−906
                         ZrSi$_2$−VSi$_2$ . . . . 48 (V):   Hb/B2−563
Si$_3$Ta$_5$ . . . . Ta$_5$Si$_3$ . . . . 50 (Ta):   Hb/A2−514
                                                                                              Hb/B1−251/2, 257/9
−. . . . . Ta$_5$Si$_3$ solid solutions:
                         Ta$_5$Si$_3$−Nb$_5$Si$_3$ . . . . 50 (Ta):   Hb/B2−104
                         Ta$_5$(Si,Al)$_3$ . . . . 50 (Ta):   Hb/B2−16/7
                         (Ta,Ti)$_5$Si$_3$ . . . . 50 (Ta):   Hb/B2−39
                         (Ta,V)$_5$Si$_3$ . . . . 50 (Ta):   Hb/B2−77/8
Si$_3$Ti$_2$ . . . . Ti$_2$Si$_3$ . . . . 41 (Ti):   Hb−375
Si$_3$Ti$_5$ . . . . Ti$_5$Si$_3$ solid solutions:
                         (Ta,Ti)$_5$Si$_3$ . . . . 50 (Ta):   Hb/B2−39

See the "Instructions for the user"
at the beginning of the volume

Si$_3$V$_5$ . . . . V$_5$Si$_3$ . . . . 48 (V):   Hb/A2–381
                                                                              Hb/B1–352/3, 356

–. . . . . V$_5$Si$_3$ solid solutions:
                      (V,Nb)$_5$Si$_3$ . . . . 49 (Nb):   Hb/B2–268/70
                      (V,Ta)$_5$Si$_3$ . . . . 50 (Ta):   Hb/B2–77/8
                      V$_5$(Si,Ge)$_3$ . . . . 48 (V):   Hb/B2–683

–. . . . . V$_5$Si$_3$ systems:
                      V$_5$Si$_3$–V$_5$B$_3$ . . . . 48 (V):   Hb/B1–357
                      V$_5$Si$_3$–Zr$_5$Si$_3$. . . . . 48 (V):   Hb/B2–563
Si$_3$W . . . . WSi$_3$ . . . . 54 (W):   Hb–205
Si$_3$W$_2$ . . . . W$_2$Si$_3$ . . . . 54 (W):   Hb–204/5
Si$_3$Zr$_5$ . . . . Zr$_5$Si$_3$ systems:
                      Zr$_5$Si$_3$–V$_5$Si$_3$ . . . . 48 (V):   Hb/B2–563

Si$_{12}$Ta . . . . TaSi$_{12}$ solid solutions:
                      Ta(Be,Si)$_{12}$. . . . . 50 (Ta):   Hb/B2–10

Sm . . . . Samarium:
                      Abundance . . . . 39 (SE):   Hb/A1–34/5
                      Band model . . . . 39 (SE):   Hb/B3–213/5
                      Boiling point . . . . 39 (SE):   Hb/B3–105/6
                      Characteristic electron
                        energy losses. . . . . 39 (SE):   Hb/B3–225/6
                      Charge carrier concentration. . . . . 39 (SE):   Hb/B3–228
                      Cosmochemistry . . . . 39 (SE):   Hb/A1–23/7
                      Crystal chemistry . . . . 39 (SE):   Hb/A1–28/32
                      Crystallographic properties . . . . . 39 (SE):   Hb/B3–61, 66/70,
                                                                              76/9
                      Crystal structure . . . . 39 (SE):   Hb/B3–66/70
                      Debye temperature . . . . 39 (SE):   Hb/B3–187
                      Density . . . . 39 (SE):   Hb/B3–85/7
                      Diffusion. . . . . 39 (SE):   Hb/B3–77/8
                      Disorder . . . . 39 (SE):   Hb/B3–76
                      Economic deposits . . . . 39 (SE):   Hb/A1–110/22
                      Elasticity. . . . . 39 (SE):   Hb/B3–92/102
                      Electrical properties . . . . 39 (SE):   Hb/B3–209/83
                      Electrical resistance. . . . . 39 (SE):   Hb/B3–236/7
                      Electron density. . . . . 39 (SE):   Hb/B3–216
                      Electron emission. . . . . 39 (SE):   Hb/B3–278/83
                      Enthalpy of fusion . . . . 39 (SE):   Hb/B3–120
                      – of sublimation. . . . . 39 (SE):   Hb/B3–110, 112/3
                      – of transformation . . . . 39 (SE):   Hb/B3–118
                      – of vaporization . . . . 39 (SE):   Hb/B3–110, 112/3
                      Equation of state . . . . 39 (SE):   Hb/B3–83/4
                      Geochemistry . . . . 39 (SE):   Hb/A1–28/54
                      Grüneisen constant. . . . . 39 (SE):   Hb/B3–84/5
                      Hall effect . . . . 39 (SE):   Hb/B3–266
                      Hardness . . . . 39 (SE):   Hb/B3–103
                      History. . . . . 39 (SE):   Hb/A1–4/22

Sm . . . . Samarium:
      Hyperfine interactions . . . . 39 (SE):   Hb/B3−221
      Isomer shift . . . . 39 (SE):   Hb/B3−216
      Magnetic properties . . . . 39 (SE):   Hb/B3−294/6
      Mechanical properties . . . . 39 (SE):   Hb/B3−85/103,
                                                    193/5, 207/9
      Melting point . . . . 39 (SE):   Hb/B3−118/9
      Mineralogy . . . . 39 (SE):   Hb/A1−54/110
      Molar volume . . . . 39 (SE):   Hb/B3−85/7
      Occurrence, extraterrestrial . . . . . . 39 (SE):   Hb/A1−22/7
      −, terrestrial . . . . 39 (SE):   Hb/A1−27/54
      − in the biosphere. . . . . 39 (SE):   Hb/A1−53/4
      − in the lithosphere . . . . 39 (SE):   Hb/A1−27/54
      Optical properties. . . . . 39 (SE):   Hb/B3−339
      Polymorphism . . . . 39 (SE):   Hb/B3−61
      Position in the periodic system . . . . 39 (SE):   Hb/A1−2/4
      Radioactivity . . . . 39 (SE):   Hb/A1−22
      Strength . . . . 39 (SE):   Hb/B3−102
      Structure . . . . 39 (SE):   Hb/B3−79
      Surface tension . . . . 39 (SE):   Hb/B3−208/9
      Susceptibility . . . . 39 (SE):   Hb/B3−294/5
      Thermal conductivity. . . . . 39 (SE):   Hb/B3−198/9
      Thermal expansion . . . . 39 (SE):   Hb/B3−88/92
      Thermal properties . . . . 39 (SE):   Hb/B3−83/92, 103/6,
                                                140/6, 187, 198/9
      Thermodynamic functions . . . . . . 39 (SE):   Hb/B3−140/5
      Thermoelectric power . . . . 39 (SE):   Hb/B3−274
      Twinning . . . . 39 (SE):   Hb/B3−78
      Ultrasound absorption . . . . 39 (SE):   Hb/B3−193
      Valence . . . . 39 (SE):   Hb/A1−4
      Vapor pressure . . . . 39 (SE):   Hb/B3−103/6
−. . . . . Samarium ions . . . . 39 (SE):   Hb/A1−3
                                                            Hb/B3−284
                                        50 (Ta):   Hb/B2−204
−. . . . . Samarium systems:
      Sm−Ag . . . . 61 (Ag):   Hb/C−220/1
      Sm−H . . . . 39 (SE):   Hb/C1−2/3, 58/9
      Sm−D . . . . 39 (SE):   Hb/C1−34, 59
      Sm−N−O . . . . 39 (SE):   Hb/C2−180
      Sm−O . . . . 39 (SE):   Hb/C1−313/4
      Sm−Pb . . . . 47 (Pb):   Hb/C4−1367
      Sm−Pb−Zn . . . . 47 (Pb):   Hb/C4−1363, 1367
      Sm−Sm_2O_3 . . . . 39 (SE):   Hb/C1−313/4
      Sm−Sn . . . . 46 (Sn):   Hb/D−312/3
      Sm−V . . . . 48 (V):   Hb/B2−600/1
Sm^{2+} . . . . Sm^{2+} . . . . 39 (SE):   Hb/A1−3
Sm^{3+} . . . . Sm^{3+} . . . . 39 (SE):   Hb/A1−3
                                                            Hb/B3−284
                                        50 (Ta):   Hb/B2−204

See the "Instructions for the user"
at the beginning of the volume

$SmSn_3$	$SmSn_3$	46 (Sn):	Hb/D–312/5
Sm_2Sn_3	Sm_2Sn_3	46 (Sn):	Hb/D–312/5
Sm_4Sn_3	Sm_4Sn_3	46 (Sn):	Hb/D–312/5
Sm_5Sn_3	Sm_5Sn_3	46 (Sn):	Hb/D–312/5
Sm_5Sn_4	Sm_5Sn_4	46 (Sn):	Hb/D–312/3
$Sm_{11}Sn_{10}$	$Sm_{11}Sn_{10}$	46 (Sn):	Hb/D–312/5
$Sn_{0.55}Zr$	$ZrSn_{0.55}$	46 (Sn):	Hb/D–342, 346
Sn	Tin:		
	Abundance	46 (Sn):	Hb/A–214/5
	Anodic polishing	46 (Sn):	Hb/B–293
	Atom	46 (Sn):	Hb/B–62/129
	Atomic nucleus	46 (Sn):	Hb/B–55/62
	Atomic radius	46 (Sn):	Hb/B–100/1
	Atomic refraction	46 (Sn):	Hb/B–129
	Atomic scattering factors	46 (Sn):	Hb/B–71/2
	Atomic susceptibility	46 (Sn):	Hb/B–101
	Atomic volume	46 (Sn):	Hb/B–100/1
	Atomic weight	46 (Sn):	Hb/B–1/4
	Behavior as anode	46 (Sn):	Hb/B–285/93
	– as cathode	46 (Sn):	Hb/B–281/5
	Boiling point	46 (Sn):	Hb/B–177
	Bond properties	46 (Sn):	Hb/C1–1/4
	Characteristic electron		
	energy losses	46 (Sn):	Hb/B–196/7
	Chemical reactions	46 (Sn):	Hb/B–347/82
	– with acids	46 (Sn):	Hb/B–365/75
	– with alkaline solutions	46 (Sn):	Hb/B–364/5
	– with aqueous salt solutions	46 (Sn):	Hb/B–375/81
	– with metal compounds	46 (Sn):	Hb/B–360/4
	– with metals	46 (Sn):	Hb/B–355
	– with nonmetal compounds	46 (Sn):	Hb/B–355/60
	– with nonmetals	46 (Sn):	Hb/B–347/55
	– with organic compounds	46 (Sn):	Hb/B–374/5, 381/2
	Colloidal forms	46 (Sn):	Hb/B–41/3
	Compressibility	46 (Sn):	Hb/B–167/8
	Condensation	46 (Sn):	Hb/B–177
	Conduction mechanism	46 (Sn):	Hb/B–202/4
	Constitution of vapor	46 (Sn):	Hb/B–176
	Cosmochemistry	46 (Sn):	Hb/A–175/82
	Critical constants	46 (Sn):	Hb/B–177/8
	Crystal chemistry	46 (Sn):	Hb/A–182/214
	Crystal form	46 (Sn):	Hb/B–134/7
	Crystallographic properties	46 (Sn):	Hb/B–130/60
	Crystal potential	46 (Sn):	Hb/B–149
	Debye temperature	46 (Sn):	Hb/B–185/6
	Deformation	46 (Sn):	Hb/B–159/60
	Density	46 (Sn):	Hb/B–161/3

Hb = Main Series Volume; GD = Gmelin-Durrer;   Ab = Appended Volume to Main Series;
Eb = Supplement Volume to Main Series;   Erg. W. = New Supplement Series Volume

Sn . . . . Tin:

Deposition from melts	46 (Sn):	Hb/B−343/7
− from nonaqueous solutions	46 (Sn):	Hb/B−342/3
Detection	46 (Sn):	Hb/B−415/8
− in iron	59 (Fe):	Hb/FII1−43/4
− in special substances	46 (Sn):	Hb/B−417/8
Determination	46 (Sn):	Hb/B−418/23
− in iron	59 (Fe):	Hb/FII1−44/53
		Hb/FII2−300, 350
− in special substances	46 (Sn):	Hb/B−421/3
Diadochy	46 (Sn):	Hb/A−186/92
Diffusion, self-diffusion	46 (Sn):	Hb/B−152/4
− in aluminum	46 (Sn):	Hb/D−208
− in antimony	46 (Sn):	Hb/D−1
− in bismuth	46 (Sn):	Hb/D−23/5
− in cadmium	46 (Sn):	Hb/D−158
− in gallium	46 (Sn):	Hb/D−223
− in indium	46 (Sn):	Hb/D−234/5, 263
− in lithium	46 (Sn):	Hb/D−61
− in magnesium	46 (Sn):	Hb/D−81
− in mercury	46 (Sn):	Hb/D−186/7
− in Pb−Sn alloys	47 (Pb):	Hb/C4−1482/6, 1518
− in sodium	46 (Sn):	Hb/D−67
− in titanium	46 (Sn):	Hb/D−325
− in zinc	46 (Sn):	Hb/D−129/30
− in zirconium	46 (Sn):	Hb/D−337/9
Disorder	46 (Sn):	Hb/B−151/2
Dressing and treatment of ores	46 (Sn):	Hb/B−6/20
Economic deposits:		
Africa	46 (Sn):	Hb/A−374/91
Asia	46 (Sn):	Hb/A−393/426
Australia and Oceania	46 (Sn):	Hb/A−426/35
Europe	46 (Sn):	Hb/A−348/74, 392/3
North America	46 (Sn):	Hb/A−435/9
South America	46 (Sn):	Hb/A−439/51
Soviet Union	46 (Sn):	Hb/A−391/408
Economics	46 (Sn):	Hb/A−342/8
		Hb/B−5/6
Elasticity	46 (Sn):	Hb/B−166/7
Electrical conductivity	46 (Sn):	Hb/B−205/16
Electrical properties	46 (Sn):	Hb/B−191/242
Electrochemical behavior	46 (Sn):	Hb/B−250/347
Electrodeposition	46 (Sn):	Hb/B−293/347
Electrokinetic phenomena	46 (Sn):	Hb/B−273
Electrolytic refining	46 (Sn):	Hb/B−31/7
Electron affinity	46 (Sn):	Hb/B−100
Electron configuration	46 (Sn):	Hb/B−62/71
Electron density	46 (Sn):	Hb/B−147/9

See the "Instructions for the user"
at the beginning of the volume

Sn . . . . Tin:

Electronegativity	46 (Sn):	Hb/B–100
Electron emission.	46 (Sn):	Hb/B–240/2
Energy band structure	46 (Sn):	Hb/B–191/6
Epitaxy.	46 (Sn):	Hb/B–158/9
Equation of state	46 (Sn):	Hb/B–161
Galvanomagnetic effects.	46 (Sn):	Hb/B–229/35
Geochemical cycle	46 (Sn):	Hb/A–308/11
Geochemistry	46 (Sn):	Hb/A–182/311
Growing of crystals.	46 (Sn):	Hb/B–137/42
Hall effect	46 (Sn):	Hb/B–229/31
Heat of fusion	46 (Sn):	Hb/B–180
– of sublimation.	46 (Sn):	Hb/B–177
– of vaporization	46 (Sn):	Hb/B–177
History	46 (Sn):	Hb/A–1/174
Interaction with electrons	46 (Sn):	Hb/B–98/100
Interfacial tension.	46 (Sn):	Hb/B–174/6
Internal friction	46 (Sn):	Hb/B–168/71
Ionization	46 (Sn):	Hb/B–98/9
Lattice defects.	46 (Sn):	Hb/B–151/2
Lattice energy	46 (Sn):	Hb/B–151
Lattice structure.	46 (Sn):	Hb/B–142/4
Lattice vibrations	46 (Sn):	Hb/B–149/51
Lilienfeld radiation	46 (Sn):	Hb/B–249/50
Magnetic properties	46 (Sn):	Hb/B–191/242
Magnetoreflection	46 (Sn):	Hb/B–250
Mechanical properties	46 (Sn):	Hb/B–161/91
Melting point	46 (Sn):	Hb/B–178/80
Mineralogy	46 (Sn):	Hb/A–311/41
Mobility of charge carriers	46 (Sn):	Hb/B–202/4
Modifications	46 (Sn):	Hb/B–130/4
Molecule.	46 (Sn):	Hb/B–129/30
Names and symbols	46 (Sn):	Hb/A–3/9
Nature of bonding	46 (Sn):	Hb/B–147/9
Nuclear magnetic resonance.	46 (Sn):	Hb/B–197/9
Nuclear properties	46 (Sn):	Hb/B–55/62
Occurrence, extraterrestrial	46 (Sn):	Hb/A–175/82
–, terrestrial	46 (Sn):	Hb/A–182/311
– in meteorites	46 (Sn):	Hb/A–176/82
– in minerals	46 (Sn):	Hb/A–193/214
– in tektites and impactites.	46 (Sn):	Hb/A–182
– in the atmosphere	46 (Sn):	Hb/A–308
– in the hydrosphere	46 (Sn):	Hb/A–302/8
– in the lithosphere	46 (Sn):	Hb/A–215/302
– in the magmatic cycle	46 (Sn):	Hb/A–215/79
– in the metamorphic cycle	46 (Sn):	Hb/A–293/302
– in the sedimentary cycle	46 (Sn):	Hb/A–279/93
– on the moon	46 (Sn):	Hb/A–182
Optical properties.	46 (Sn):	Hb/B–242/50

Sn . . . . Tin:

Optical terms	46 (Sn):	Hb/B−64/71
Orientation of single crystals	46 (Sn):	Hb/B−154
Overvoltage	46 (Sn):	Hb/B−273/81
Paramagnetic resonance absorption .	46 (Sn):	Hb/B−199
Photoelectric electron emission	46 (Sn):	Hb/B−240/2
Polymorphism	46 (Sn):	Hb/B−130/4
Position in the electrochemical series	46 (Sn):	Hb/B−250
− in the periodic system	46 (Sn):	Hb/B−1
Potentials	46 (Sn):	Hb/B−255/73
Preparation, industrial	46 (Sn):	Hb/B−4/41
−, −, historical data	46 (Sn):	Hb/A−53/84
−, −, hydrometallurgy	46 (Sn):	Hb/B−18/22
−, −, ore preparation	46 (Sn):	Hb/B−6/7
−, −, purification of crude tin	46 (Sn):	Hb/B−28/38
−, −, recovery	46 (Sn):	Hb/B−22/8
−, −, smelting	46 (Sn):	Hb/B−7/14
−, −, volatization methods	46 (Sn):	Hb/B−14/7
− in pure state	46 (Sn):	Hb/B−37/8
− of special forms	46 (Sn):	Hb/B−41/3
Purification	46 (Sn):	Hb/B−28/37
Purity testing	46 (Sn):	Hb/B−38/41
Recovery	46 (Sn):	Hb/B−22/8
Recrystallization	46 (Sn):	Hb/B−160
Reflectivity	46 (Sn):	Hb/B−242/9
Refractive index	46 (Sn):	Hb/B−242/9
Self-diffusion	46 (Sn):	Hb/B−152/4
Separation, analytical	46 (Sn):	Hb/B−420/1
Slip	46 (Sn):	Hb/B−159/60
Sorption	46 (Sn):	Hb/C1−5
Sound absorption	46 (Sn):	Hb/B−168/71
Special forms	46 (Sn):	Hb/B−41/3
Spectra:		
Arc spectrum	46 (Sn):	Hb/B−72/87
Spark spectra see Sn^+ ... Sn^+, etc.		
Terms	46 (Sn):	Hb/B−64/71
X-ray spectrum	46 (Sn):	Hb/B−87/98
Standard potentials	46 (Sn):	Hb/B−250/5
Structure of the melt	46 (Sn):	Hb/B−144/7
− of thin films	46 (Sn):	Hb/B−155/8
Superconductivity	46 (Sn):	Hb/B−216/29
Surface structure	46 (Sn):	Hb/B−154/5
Surface tension	46 (Sn):	Hb/B−174/6
Susceptibility	46 (Sn):	Hb/B−199/202
Thermal conductivity	46 (Sn):	Hb/B−186/91
Thermal diffusivity	46 (Sn):	Hb/B−191

*See the "Instructions for the user"
at the beginning of the volume*

Sn . . . . Tin:
    Thermal expansion . . . . 46 (Sn):   Hb/B−163/6
    Thermal properties . . . . 46 (Sn):   Hb/B−161/91
    Thermal radiation . . . . 46 (Sn):   Hb/B−249
    Thermodynamic functions . . . . . . 46 (Sn):   Hb/B−180/5
    Thermoelectric force . . . . 46 (Sn):   Hb/B−235/40
    Thermomagnetic effects . . . . 46 (Sn):   Hb/B−240
    Toxicity . . . . 46 (Sn):   Hb/B−411/5
    Uses, historical data . . . . 46 (Sn):   Hb/A−101/74
    Vaporization . . . . 46 (Sn):   Hb/B−176/7
    Vapor pressure . . . . 46 (Sn):   Hb/B−176
    Viscosity . . . . 46 (Sn):   Hb/B−171/3
    Wave functions . . . . 46 (Sn):   Hb/B−62/4

−. . . . . Tin ions
    see Sn^+ ... Sn^+, etc.

−. . . . . Tin isotopes:
    Abundance . . . . 46 (Sn):   Hb/A−175, 181, 215
                                                Hb/B−3, 43
    Detection and determination . . . . . 46 (Sn):   Hb/B−53/5
    Handling . . . . 46 (Sn):   Hb/B−43/4
    Nuclear properties . . . . 46 (Sn):   Hb/B−55/62
    Preparation and separation . . . . . 46 (Sn):   Hb/B−43/53

−. . . . . Tin systems:
    Sn−Ag . . . . 61 (Ag):   Hb/C−266/70
    Sn−Ag−AgCl−SnCl₂ . . . . 61 (Ag):   Hb/B4−247
    Sn−Ag−Al . . . . 61 (Ag):   Hb/C−290/1
    Sn−Ag−Cd . . . . 61 (Ag):   Hb/C−286/9
    Sn−Ag−Cd−Cu . . . . 61 (Ag):   Hb/C−406
    Sn−Ag−Cu . . . . 61 (Ag):   Hb/C−403/6
    Sn−Ag−Hg . . . . 61 (Ag):   Hb/C−289/90
    Sn−Ag−Mg . . . . 27 (Mg):   Hb/A4−751
                                                61 (Ag):   Hb/C−286
    Sn−Ag−Mn . . . . 61 (Ag):   Hb/C−351
    Sn−Ag−Sb . . . . 61 (Ag):   Hb/B4−252
    Sn−Al . . . . 35 (Al):   Hb/A5−844/5
                                                46 (Sn):   Hb/D−208/9
    Sn−Al−Bi . . . . 35 (Al):   Hb/A5−854/5
                                                46 (Sn):   Hb/D−218
    Sn−Al−Cd . . . . 35 (Al):   Hb/A5−858
                                                46 (Sn):   Hb/D−222
    Sn−Al−Cu−Sb . . . . 60 (Cu):   Hb/B3−1196
    Sn−Al−Ga−Ge−Zn . . . . 46 (Sn):   Hb/D−397
    Sn−Al−In . . . . 46 (Sn):   Hb/D−273
    Sn−Al−Mg . . . . 46 (Sn):   Hb/D−218/9
    Sn−Al−Mn . . . . 35 (Al):   Hb/A6−947/8
    Sn−Al−Nd . . . . 46 (Sn):   Hb/D−312
    Sn−Al−Ni . . . . 35 (Al):   Hb/A6−1004

Sn . . . . Tin systems:

Sn−Al−Pb	35 (Al):	Hb/A5−863/4
	47 (Pb):	Hb/C4−1529
Sn−Al−Sb	35 (Al):	Hb/A5−854
	46 (Sn):	Hb/D−218
Sn−Al−Si	35 (Al):	Hb/A5−853
	46 (Sn):	Hb/D−217/8
Sn−Al−Ti	46 (Sn):	Hb/D−333/4
Sn−Al−Ti−Zr	46 (Sn):	Hb/D−382/3
Sn−Al−Zn	35 (Al):	Hb/A5−855/7
	46 (Sn):	Hb/D−220/1
Sn−Al−Zr	46 (Sn):	Hb/D−376
Sn−As−Ga−Ge	46 (Sn):	Hb/D−397
Sn−As−Ga−In	46 (Sn):	Hb/D−277
Sn−Au	62 (Au):	Hb/3−840/1
Sn−B−Ni	57 (Ni):	Hb/B3−1193
Sn−Ba	46 (Sn):	Hb/D−126
Sn−Be	46 (Sn):	Hb/D−79
Sn−Bi	46 (Sn):	Hb/D−23/8
Sn−Bi−Bi$_2$Mg$_3$	46 (Sn):	Hb/D−115
Sn−Bi−Cd	46 (Sn):	Hb/D−178/9
Sn−Bi−Cd−Pb	47 (Pb):	Hb/C4−1525/6
Sn−Bi−Cd−Pb−Zn	47 (Pb):	Hb/C4−1528/9
Sn−Bi−Ga−Pb	47 (Pb):	Hb/C4−1529
Sn−Bi−InBi	46 (Sn):	Hb/D−267/9
Sn−Bi−Mg	46 (Sn):	Hb/D−115
Sn−Bi−Pb	47 (Pb):	Hb/C4−1518/9
Sn−Bi−Sb	46 (Sn):	Hb/D−58/9
Sn−Bi−Zn	46 (Sn):	Hb/D−155
Sn−Bi$_2$Mg$_3$−Mg$_2$Sn	46 (Sn):	Hb/D−115
Sn−Br−Pb	47 (Pb):	Hb/C4−1584
Sn−C−Nb	49 (Nb):	Hb/B4−330
Sn−Ca	46 (Sn):	Hb/D−117
Sn−Cd	46 (Sn):	Hb/D−159/60
Sn−Cd−Hg	46 (Sn):	Hb/D−206/7
Sn−Cd−In	46 (Sn):	Hb/D−270
Sn−Cd−In−Zn	46 (Sn):	Hb/D−271
Sn−Cd−Pb	47 (Pb):	Hb/C4−1524/5
Sn−Cd−Pb−Zn	47 (Pb):	Hb/C4−1528/9
Sn−Cd−Sb	46 (Sn):	Hb/D−177/8
Sn−Cd−Tl	46 (Sn):	Hb/D−291
Sn−Cd−Zn	46 (Sn):	Hb/D−181/2
Sn−Ce	46 (Sn):	Hb/D−302/4
Sn−Cl−Pb	47 (Pb):	Hb/C4−1583/4
Sn−Cl−Pb−S	47 (Pb):	Hb/C4−1586
Sn−Co	58 (Co):	Hb/A1−194/6
		Eb/A−409/15
Sn−Cr	52 (Cr):	Hb/A2−726/7

See the "Instructions for the user"
at the beginning of the volume

Sn . . . . Tin systems:

Sn−Cu	. . . .	60 (Cu):	Hb/B3−1193
Sn−Cu−Fe−S	. . . .	60 (Cu):	Hb/B3−1327
Sn−Cu−H	. . . .	60 (Cu):	Hb/B3−1193/4
Sn−Cu−Mg	. . . .	27 (Mg):	Hb/A4−722
Sn−Cu−P	. . . .	60 (Cu):	Hb/B3−1196
Sn−Cu−S	. . . .	60 (Cu):	Hb/B3−1195/6
Sn−Cu−Sb	. . . .	60 (Cu):	Hb/B3−1196
Sn−Er	. . . .	46 (Sn):	Hb/D−321/2
Sn−Eu	. . . .	46 (Sn):	Hb/D−315
Sn−Fe	. . . .	59 (Fe):	Hb/A9−1903/5
Sn−Ga	. . . .	36 (Ga):	Hb−62/3
		46 (Sn):	Hb/D−223/4
Sn−Ga−Ge−In−Zn	. . . .	46 (Sn):	Hb/D−397
Sn−Ga−Ge−Sb	. . . .	46 (Sn):	Hb/D−397
Sn−Ga−In	. . . .	46 (Sn):	Hb/D−274
Sn−Ga−In−Sb	. . . .	46 (Sn):	Hb/D−277
Sn−Ga−In−Zn	. . . .	46 (Sn):	Hb/D−278
Sn−Ga−Pb−Zn	. . . .	47 (Pb):	Hb/C4−1529
Sn−Ga−Tl	. . . .	46 (Sn):	Hb/D−291
Sn−Ga−Zn	. . . .	46 (Sn):	Hb/D−231/3
Sn−GaSb	. . . .	46 (Sn):	Hb/D−231
Sn−Gd	. . . .	46 (Sn):	Hb/D−316
Sn−Ge	. . . .	46 (Sn):	Hb/D−390/1
Sn−Ge−Mg	. . . .	46 (Sn):	Hb/D−395
Sn−Ge−Ni−Pb	. . . .	57 (Ni):	Hb/B1−153
Sn−H	. . . .	46 (Sn):	Hb/C1−5/6
Sn−Hf	. . . .	46 (Sn):	Hb/D−386
Sn−Hg	. . . .	46 (Sn):	Hb/D−187/90
Sn−Hg−In	. . . .	46 (Sn):	Hb/D−271/3
Sn−I	. . . .	46 (Sn):	Hb/C1−444
Sn−In	. . . .	46 (Sn):	Hb/D−235/6
Sn−In−Pb	. . . .	47 (Pb):	Hb/C4−1529/30
Sn−In−Sb	. . . .	46 (Sn):	Hb/D−264/5
Sn−Ir	. . . .	68 (Pt):	Hb/A6−734
Sn−K	. . . .	46 (Sn):	Hb/D−76
Sn−La	. . . .	46 (Sn):	Hb/D−296
Sn−Li	. . . .	46 (Sn):	Hb/D−61/2
Sn−Li−Mg	. . . .	46 (Sn):	Hb/D−116
Sn−Mg	. . . .	27 (Mg):	Hb/A4−617/9
		46 (Sn):	Hb/D−81/2
Sn−Mg−Pb	. . . .	27 (Mg):	Hb/A4−656/7
		47 (Pb):	Hb/C4−1522
Sn−Mg−Zn	. . . .	27 (Mg):	Hb/A4−635
Sn−Mn−Ni	. . . .	57 (Ni):	Hb/B1−312
Sn−Mo−Ni	. . . .	57 (Ni):	Hb/B1−251
Sn−Na	. . . .	46 (Sn):	Hb/D−67/8
Sn−Nb	. . . .	49 (Nb):	Hb/B2−208/9

Hb = Main Series Volume; GD = Gmelin-Durrer;   *Ab = Appended Volume to Main Series;*
Eb = Supplement Volume to Main Series;   *Erg. W. = New Supplement Series Volume*

Sn . . . .   Tin systems:
                           Sn–Nb–Zr . . . .   49 (Nb):   Hb/B2–235/6
                           Sn–Nd. . . . .   46 (Sn):   Hb/D–310
                           Sn–Ni . . . . 57 (Ni):   Hb/B1–128/31
                           Sn–O . . . .   46 (Sn):   Hb/C1–24/5
                           Sn–Pb . . . . 47 (Pb):   Hb/C4–1487/9
                           Sn–Pb–PbS–SnS . . . .   47 (Pb):   Hb/C4–1584
                           Sn–Pb–S . . . .   47 (Pb):   Hb/C4–1584
                           Sn–Pb–Sb . . . .   47 (Pb):   Hb/C4–1515
                           Sn–Pb–Se–Te . . . .   47 (Pb):   Hb/C4–1594/5
                           Sn–Pb–Te . . . .   47 (Pb):   Hb/C4–1588
                           Sn–Pb–Tl . . . .   47 (Pb):   Hb/C4–1530
                           Sn–Pb–Zn . . . .   47 (Pb):   Hb/C4–1522/4
                           Sn–Pd . . . .   68 (Pt):   Hb/A5–586/8
                           Sn–Pr . . . . 46 (Sn):   Hb/D–307
                           Sn–Pt . . . . 68 (Pt):   Hb/A6–758/63
                           Sn–Rh . . . . 68 (Pt):   Hb/A5–555/6
                           Sn–Ru . . . . 63 (Ru):   Eb–193/4
                                                                  68 (Pt):   Hb/A5–538/9
                           Sn–Sb . . . . 46 (Sn):   Hb/D–1/4
                           Sn–Sb–Zn . . . .   46 (Sn):   Hb/D–153
                           Sn–Sm . . . .   46 (Sn):   Hb/D–312/3
                           Sn–SnO$_2$ . . . .   46 (Sn):   Hb/C1–24/5
                           Sn–Sr . . . . 46 (Sn):   Hb/D–123
                           Sn–Ta . . . . 50 (Ta):   Hb/B2–65
                           Sn–Tb . . . . 46 (Sn):   Hb/D–318
                           Sn–Th . . . . 46 (Sn):   Hb/D–388
                           Sn–Ti . . . . 46 (Sn):   Hb/D–325/7
                           Sn–Ti–V . . . .   48 (V):   Hb/B2–686/7
                           Sn–Ti–Zr . . . .   46 (Sn):   Hb/D–378
                           Sn–Tl . . . . 46 (Sn):   Hb/D–279/81
                           Sn–Tl–Zn . . . .   46 (Sn):   Hb/D–290
                           Sn–V . . . .   48 (V):   Hb/B2–684/7
                           Sn–Y . . . .   46 (Sn):   Hb/D–293/4
                           Sn–Zn . . . . 46 (Sn):   Hb/D–130/2
                           Sn–Zr . . . . 46 (Sn):   Hb/D–339/42
Sn$^+$ . . . .   Sn$^+$:
                           Electrochemical behavior . . . . 46 (Sn):   Hb/B–271
                           Heat capacity . . . . 46 (Sn):   Hb/B–185
                           Ionic entropy . . . . 46 (Sn):   Hb/B–185
                           Ionic radius . . . . 46 (Sn): Hb/B–128
                           Ionization potential . . . . 46 (Sn): Hb/B–114
                           Optical terms . . . . 46 (Sn):   Hb/B–102/4
                           Spectra . . . . 46 (Sn):   Hb/B–106/9
Sn^{2+} . . . . Sn^{2+}:
                           Bond properties. . . . . 46 (Sn): Hb/C1–1/4
                           Catalytic properties . . . . 46 (Sn): Hb/B–390/1
                           Chemical reactions . . . . 46 (Sn): Hb/B–383/91

See the "Instructions for the user"
at the beginning of the volume

Sn^{2+} . . . . Sn^{2+}:

Electrochemical behavior	46 (Sn):	Hb/B–252, 271/2, 295/331, 342/7
Exchange reactions.	46 (Sn):	Hb/B–389/90
Heat capacity	46 (Sn):	Hb/B–185
Hydration	46 (Sn):	Hb/B–383/4
Hydrolysis.	46 (Sn):	Hb/B–384/5
Inhibitor effect	46 (Sn):	Hb/B–391
Ionic activity.	46 (Sn):	Hb/B–384
Ionic entropy	46 (Sn):	Hb/B–185
Ionic refraction	46 (Sn):	Hb/B–129
Ionic susceptibility	46 (Sn):	Hb/B–129
Ionization potential	46 (Sn):	Hb/B–114
Optical terms	46 (Sn):	Hb/B–104
Properties of the free ion.	46 (Sn):	Hb/B–104, 128
– of the ion in crystals	46 (Sn):	Hb/B–115/23, 128/9
– – in solutions	46 (Sn):	Hb/B–124/7
Redox reactions.	46 (Sn):	Hb/B–395/409
Sorption	46 (Sn):	Hb/B–390
Spectra:		
Crystal spectra	21 (Na):	Eb/6–333/4, 347
	46 (Sn):	Hb/B–115/23
Spark spectrum	46 (Sn):	Hb/B–110/1
Spectra in solutions	46 (Sn):	Hb/B–124/7

Sn^{3+} . . . . Sn^{3+}:

Chemical reactions	46 (Sn):	Hb/B–409
Ionization potential	46 (Sn):	Hb/B–114
Optical terms	46 (Sn):	Hb/B–104/5
Properties of the free ion.	46 (Sn):	Hb/B–104/5
– of the ion in crystals	46 (Sn):	Hb/B–123/4
Spectra:		
Crystal spectra	46 (Sn):	Hb/B–123/4
Spark spectrum	46 (Sn):	Hb/B–111/3

Sn^{4+} . . . . Sn^{4+}:

Bond properties.	46 (Sn):	Hb/C1–2/4
Chemical reactions	46 (Sn):	Hb/B–391/5
Electrochemical behavior	46 (Sn):	Hb/B–295, 331/42
Faraday effect.	46 (Sn):	Hb/B–129
Inhibitor effect	46 (Sn):	Hb/B–394/5
Ionic activity.	46 (Sn):	Hb/B–392
Ionic entropy	46 (Sn):	Hb/B–185
Ionic susceptibility	46 (Sn):	Hb/B–129
Ionization potential	46 (Sn):	Hb/B–114
Optical terms	46 (Sn):	Hb/B–105
Properties of the free ion.	46 (Sn):	Hb/B–105, 128
– of the ion in crystals	46 (Sn):	Hb/B–128/9
– – in solutions	46 (Sn):	Hb/B–127/9
Redox reactions.	46 (Sn):	Hb/B–410/1

Hb = Main Series Volume; GD = Gmelin-Durrer;   *Ab = Appended Volume to Main Series;*
Eb = Supplement Volume to Main Series;   *Erg. W. = New Supplement Series Volume*

Sn^{4+}	. . . .	Sn^{4+}:		
		Sorption	46 (Sn):	Hb/B−394
		Spectra:		
		Spark spectrum	46 (Sn):	Hb/B−113
		Spectra in solutions	46 (Sn):	Hb/B−127/8
Sn^{5+}	. . . .	Sn^{5+}	46 (Sn):	Hb/B−105, 113/4
Sn^{n+}	. . . .	Sn^{n+} (n = 6 to 49).	46 (Sn):	Hb/B−114
Sn^-	. . . .	Sn^-	46 (Sn):	Hb/B−128
Sn^{4-}	. . . .	Sn^{4-}	46 (Sn):	Hb/B−128
SnSr	. . . .	SrSn	46 (Sn):	Hb/D−123/4
$SnSr_2$	. . . .	Sr_2Sn	46 (Sn):	Hb/D−123/4
$SnTaV_2$	. . . .	TaV_2Sn	50 (Ta):	Hb/B2−85
$SnTa_2V$	. . . .	Ta_2VSn	50 (Ta):	Hb/B2−85
$SnTa_3$	. . . .	Ta_3Sn.	50 (Ta):	Hb/B2−65/7
−.	. . . .	Ta_3Sn solid solutions:		
		$(Ta,Nb)_3Sn$.	50 (Ta):	Hb/B2−118
SnTe	. . . .	SnTe solid solutions:		
		$SnTe−AgBiTe_2$	61 (Ag):	Hb/B4−252
		$SnTe−AgSbTe_2$.	61 (Ag):	Hb/B4−251
		$SnTe−AgSbTe_2−PbTe$	61 (Ag):	Hb/B4−272
		SnTe−PbS−PbTe	47 (Pb):	Hb/C4−1594
		SnTe−PbSe	47 (Pb):	Hb/C4−1595
		SnTe−PbTe	47 (Pb):	Hb/C4−1589/94
		SnTe−PbTe−SnSe	47 (Pb):	Hb/C4−1595
−.	. . . .	SnTe systems:		
		$SnTe−Ag_2Te−Te$	61 (Ag):	Hb/B4−251
		SnTe−PbTe−SnSe	47 (Pb):	Hb/C4−1594/5
$SnTi_2$	. . . .	Ti_2Sn	46 (Sn):	Hb/D−325, 328
$SnTi_3$	. . . .	Ti_3Sn	46 (Sn):	Hb/D−325, 328
−.	. . . .	Ti_3Sn solid solutions:		
		$(Ti,Nb)_3Sn$.	49 (Nb):	Hb/B2−234/5
−.	. . . .	Ti_3Sn systems:		
		$Ti_3Sn−Ti_3Al−Zr$	46 (Sn):	Hb/D−382/3
SnV_3	. . . .	V_3Sn	48 (V):	Hb/A2−392
				Hb/B2−588/91, 684/6
SnYb	. . . .	YbSn	46 (Sn):	Hb/D−323/4
SnZr	. . . .	ZrSn	46 (Sn):	Hb/D−342, 347
$SnZr_3$	. . . .	Zr_3Sn	46 (Sn):	Hb/D−342, 346/7
−.	. . . .	Zr_3Sn solid solutions:		
		$(Zr,Nb)_3Sn$	49 (Nb):	Hb/B2−236/40
$SnZr_4$	. . . .	Zr_4Sn	46 (Sn):	Hb/D−340, 346/7,
				356, 362
Sn_2	. . . .	Sn_2	46 (Sn):	Hb/B−129/30
Sn_2^+	. . . .	Sn_2^+	46 (Sn):	Hb/B−176
				Hb/C1−22
Sn_2Tb	. . . .	$TbSn_2$.	46 (Sn):	Hb/B−318/9

See the "Instructions for the user"
at the beginning of the volume

Sn$_2$Tl . . . . Sn$_2$Tl . . . . 46 (Sn):   Hb/D$-$279/81
Sn$_2$Tm . . . . TmSn$_2$ . . . . 46 (Sn):   Hb/D$-$323
Sn$_2$Y . . . . YSn$_2$ . . . . 46 (Sn):   Hb/D$-$293/5
$-$. . . . . YSn$_2$ systems:
                YSn$_2-$YAl$_2$ . . . . 46 (Sn):   Hb/D$-$295
Sn$_2$Zr . . . . ZrSn$_2$ . . . . 46 (Sn):   Hb/D$-$340, 342, 347
Sn$_2$Zr$_3$ . . . . Zr$_3$Sn$_2$ . . . . 46 (Sn):   Hb/D$-$340
Sn$_3^+$ . . . . Sn$_3^+$ . . . . 46 (Sn):   Hb/B$-$176
Sn$_3$Sr . . . . SrSn$_3$ . . . . 46 (Sn):   Hb/D$-$123/4
Sn$_3$Ta$_2$ . . . . Ta$_2$Sn$_3$ . . . . 50 (Ta):   Hb/B2$-$65/6
Sn$_3$Tb . . . . TbSn$_3$ . . . . 46 (Sn):   Hb/D$-$318/9
Sn$_3$Tb$_5$ . . . . Tb$_5$Sn$_3$ . . . . 46 (Sn):   Hb/D$-$318/9
Sn$_3$Th . . . . ThSn$_3$ . . . . 46 (Sn):   Hb/D$-$388
Sn$_3$Ti$_5$ . . . . Ti$_5$Sn$_3$ . . . . 46 (Sn):   Hb/D$-$325, 328
$-$. . . . . Ti$_5$Sn$_3$ solid solutions:
                (Ti,Zr)$_5$Sn$_3$ . . . . 46 (Sn):   Hb/D$-$378, 380
Sn$_3$Tm$_5$ . . . . Tm$_5$Sn$_3$ . . . . 46 (Sn):   Hb/D$-$323
Sn$_3$Y . . . . YSn$_3$ . . . . 46 (Sn):   Hb/D$-$293/5
$-$. . . . . YSn$_3$ systems:
                YSn$_3-$YIn$_3$ . . . . 46 (Sn):   Hb/D$-$295
Sn$_3$Y$_5$ . . . . Y$_5$Sn$_3$ . . . . 46 (Sn):   Hb/D$-$293/5
Sn$_3$Yb . . . . YbSn$_3$ . . . . 46 (Sn):   Hb/D$-$324
Sn$_3$Yb$_5$ . . . . Yb$_5$Sn$_3$ . . . . 46 (Sn):   Hb/D$-$292
Sn$_3$Zr$_5$ . . . . Zr$_5$Sn$_3$ . . . . 46 (Sn):   Hb/D$-$340, 342, 347
$-$. . . . . Zr$_5$Sn$_3$ solid solutions:
                (Zr,Ti)$_5$Sn$_3$ . . . . 46 (Sn):   Hb/D$-$378, 380
Sn$_4^+$ . . . . Sn$_4^+$ . . . . 46 (Sn):   Hb/B$-$176
Sn$_4$Sr . . . . SrSn$_4$ . . . . 46 (Sn):   Hb/D$-$124
Sn$_4$Tb$_5$ . . . . Tb$_5$Sn$_4$ . . . . 46 (Sn):   Hb/D$-$318/9
Sn$_4$Y$_5$ . . . . Y$_5$Sn$_4$ . . . . 46 (Sn):   Hb/D$-$294/5
Sn$_4$Zr$_5$ . . . . Zr$_5$Sn$_4$ . . . . 46 (Sn):   Hb/D$-$342, 347
Sn$_5^+$ . . . . Sn$_5^+$ . . . . 46 (Sn):   Hb/B$-$176
Sn$_5$Sr . . . . SrSn$_5$ . . . . 46 (Sn):   Hb/D$-$123, 125
Sn$_5$Ti$_6$ . . . . Ti$_6$Sn$_5$ . . . . 46 (Sn):   Hb/D$-$325, 328
Sn$_6^+$ . . . . Sn$_6^+$ . . . . 46 (Sn):   Hb/B$-$176
Sn$_{10}$Tb$_{11}$ . . . . . . Tb$_{11}$Sn$_{10}$ . . . . 46 (Sn):   Hb/D$-$318/9
Sn$_{10}$Tm$_{11}$ . . . . . . Tm$_{11}$Sn$_{10}$ . . . . 46 (Sn):   Hb/D$-$323
Sn$_{10}$Y$_{11}$ . . . . . . Y$_{11}$Sn$_{10}$ . . . . 46 (Sn):   Hb/D$-$293/5

Sr . . . . Strontium:
                Absorption of radiation. . . . . 29 (Sr):   Hb$-$59/60
                                                            Eb$-$125
                Abundance . . . . 29 (Sr):   Eb$-$13/4
                Affinity. . . . . 29 (Sr):   Hb$-$41
                Atom. . . . . 29 (Sr):   Hb$-$38/41
                                                            Eb$-$92/103
                Atomic nucleus . . . . 29 (Sr):   Hb$-$38
                                                            Eb$-$91/2

Hb = Main Series Volume; GD = Gmelin-Durrer;   *Ab = Appended Volume to Main Series;*
Eb = Supplement Volume to Main Series;   *Erg. W. = New Supplement Series Volume*

Sr . . . . Strontium:

Chemical reactions	29 (Sr):	Hb−64/6
		Eb−131/2
Cosmochemistry	29 (Sr):	Hb−3
		Eb−1/4
Crystal chemistry	29 (Sr):	Eb−4/13
Crystallographic properties	29 (Sr):	Hb−42
		Eb−103/6
Detection	28 (Ca):	Hb/B3−1490/8
– in iron	59 (Fe):	Hb/FII1−6
– in special substances	28 (Ca):	Hb/B3−1499/500
Determination.	28 (Ca):	Hb/B3−1500/6
– in presence of other ions	28 (Ca):	Hb/B3−1507/12
– in special substances	28 (Ca):	Hb/B3−1512/8
Economic deposits:		
Africa	29 (Sr):	Hb−7/8
		Eb−66
America.	29 (Sr):	Hb−8/9
		Eb−66/75
Asia	29 (Sr):	Eb−65/6
Australia	29 (Sr):	Eb−66
Europe	29 (Sr):	Hb−4/8
		Eb−61/4
Soviet Union	29 (Sr):	Hb−7/8
		Eb−64/5
Economics.	29 (Sr):	Eb−60/1
Electrical properties	29 (Sr):	Hb−61/2
		Eb−127/9
Electrochemical behavior	29 (Sr):	Hb−62/4
		Eb−129/31
Electronegativity	29 (Sr):	Eb−103
Formation	29 (Sr):	Eb−88
Geochemical cycle	29 (Sr):	Eb−57/9
Geochemistry	29 (Sr):	Hb−3/4
		Eb−4/59
History.	29 (Sr):	Hb−1/2
		Eb−1
Ionization potentials	29 (Sr):	Hb−62
		Eb−128/9
Magnetic properties	29 (Sr):	Hb−60/1
		Eb−126/7
Mechanical properties	29 (Sr):	Hb−42/3
		Eb−106/10
Mineralogy	29 (Sr):	Hb−9/16
		Eb−75/82
Molecule.	29 (Sr):	Hb−41
		Eb−103
Nuclear properties	29 (Sr):	Hb−38
		Eb−91/2

See the "Instructions for the user"
at the beginning of the volume

Sr . . . . Strontium:

Occurrence, extraterrestrial . . . . . . 29 (Sr):   Hb–3
    Eb–1/4

–, terrestrial . . . . 29 (Sr):   Hb–3/4
    Eb–4/59

– in meteorites and tektites . . . . . . 29 (Sr):   Hb–3
    Eb–3/4

– in soils . . . . 29 (Sr):   Hb–4
    Eb–39/41

– in the atmosphere . . . . 29 (Sr):   Eb–49
– in the biosphere . . . . 29 (Sr):   Hb–3/4
    Eb–49/57

– in the hydrosphere . . . . 29 (Sr):   Hb–3/4
    Eb–43/9

– in the lithosphere . . . . 29 (Sr):   Hb–3/4
    Eb–14/43

Optical properties . . . . 29 (Sr):   Hb–44/60
    Eb–115/25

Position in the electrochemical
   series . . . . 29 (Sr):   Hb–63
    Eb–129/30

– in the periodic system . . . . 29 (Sr):   Hb–41
    Eb–102

Preparation . . . . 29 (Sr):   Hb–35/8
    Eb–84, 88

–, industrial . . . . 29 (Sr):   Eb–83/4
Purification . . . . 29 (Sr):   Hb–37
Scattering of X-rays . . . . 29 (Sr):   Eb–125
Separation, analytical . . . . 28 (Ca):   Hb/B3–1507/12
Sorption . . . . 29 (Sr):   Eb–109/10
Special forms . . . . 29 (Sr):   Hb–37/8
    Eb–88

Spectra:
   Arc spectrum . . . . 29 (Sr):   Hb–47/50, 54/6
    Eb–116/8

   Band spectrum . . . . 29 (Sr):   Hb–45
    Eb–120

   Spark spectra . . . . 29 (Sr):   Hb–50/4, 56/7
    Eb–118/20

   X-ray spectrum . . . . 29 (Sr):   Hb–58/9
    Eb–120/5

Storage . . . . 29 (Sr):   Hb–37
Thermal properties . . . . 29 (Sr):   Hb–43/4
    Eb–110/4

Toxicity . . . . 29 (Sr):   Eb–146
Treatment of raw materials . . . . . . 29 (Sr):   Hb–16/24
    Eb–83/4

Uses . . . . 29 (Sr):   Eb–84

Hb = Main Series Volume; GD = Gmelin-Durrer;   Ab = Appended Volume to Main Series;
Eb = Supplement Volume to Main Series;   Erg. W. = New Supplement Series Volume

Sr . . . . Strontium:
    Valence . . . . 29 (Sr):   Hb–41
                                                                    Eb–102/3

–. . . . . Strontium ions
    see Sr^+ ... Sr^+, etc.

–. . . . . Strontium isotopes:
    Detection and determination. . . . . . 28 (Ca):   Hb/B3–1518/26
                                                  Erg. W. 7b–355/6
    Enrichment and separation . . . . . . 29 (Sr):   Eb–88/91
    Nuclear properties . . . . 29 (Sr):   Hb–38
                                                  Eb–91/2
    Toxicity . . . . 29 (Sr):   Eb–146
    Uses . . . . Erg. W. 8–207

–. . . . . Strontium systems:
    Sr–Ag . . . . 61 (Ag):   Hb/C–35/6
    Sr–Ba . . . . 30 (Ba):   Eb–275
    Sr–Ca . . . . 29 (Sr):   Eb–147
    Sr–Cd . . . . 33 (Cd):   Eb–382
    Sr–Cu . . . . 60 (Cu):   Hb/B3–1136
    Sr–Hg . . . . 34 (Hg):   Hb/A2–1039/40
    Sr–Mg . . . . 27 (Mg):   Hb/A3–477
    Sr–NH₃ . . . . 29 (Sr):   Hb–91
                                                  Eb–171
    Sr–O–S . . . . 9 (S):   Hb/B1–183
    Sr–Pb . . . . 47 (Pb):   Hb/C3–1193/4
    Sr–Pb–Zn . . . . 47 (Pb):   Hb/C4–1225
    Sr–Sn . . . . 46 (Sn):   Hb/D–123
    Sr–SrBr₂. . . . . 29 (Sr):   Eb–226
    Sr–SrI₂ . . . . 29 (Sr):   Eb–237
    Sr–Zn . . . . 32 (Zn):   Eb–627

Sr^+ . . . . Sr^+:
    Ionization potential . . . . 29 (Sr):   Hb–62
                                                  Eb–128/9
    Properties of the ion . . . . 29 (Sr):   Eb–92/4, 96, 101
    Spectra . . . . 29 (Sr):   Hb–50/7
                                                  Eb–118/9

Sr^{2+} . . . . Sr^{2+}:
    Chemical reactions . . . . 29 (Sr):   Hb–66/8
                                                  Eb–132/46
    – with inorganic compounds. . . . . . 29 (Sr):   Hb–66/8
                                                  Eb–134/8
    – with organic compounds. . . . . . . 29 (Sr):   Hb–68
                                                  Eb–138/46
    Electrochemical behavior . . . . 29 (Sr):   Hb–62/4
                                                  Eb–129/31
    Ionic entropy . . . . 29 (Sr):   Hb–44
                                                  Eb–114

See the "Instructions for the user"
at the beginning of the volume

Sr^{2+} . . . . Sr^{2+}:

Ionic polarization	29 (Sr):	Eb–127
Ionic refraction	29 (Sr):	Eb–115
Ionic susceptibility	29 (Sr):	Hb–61
		Eb–126/7
Ionization potential	29 (Sr):	Hb–62
		Eb–128/9
Lyotropic behavior	29 (Sr):	Eb–133/4
Properties of the ion	29 (Sr):	Hb–39/41
		Eb–92/102
Solvation	29 (Sr):	Hb–66
		Eb–132/3
Sorption	29 (Sr):	Hb–43
		Eb–109/10
Spectra	29 (Sr):	Hb–50/7
		Eb–119

Sr^{n+} . . . . Sr^{n+} (n = 3 to 38) . . . . 29 (Sr): Eb–92, 119/20, 128/9

SrTe . . . . SrTe . . . . 29 (Sr): Hb–187
Eb–261/2

$SrZn_{13}$ . . . . $Zn_{13}Sr$ . . . . 32 (Zn): Eb–627
Sr_2 . . . . Sr_2 . . . . 29 (Sr): Eb–103

T . . . . Tritium
see H ... Hydrogen isotopes

Ta . . . . Tantalum:

Absorption of X- and gamma rays . .	50 (Ta):	Hb/A2–320/6
Abundance	50 (Ta):	Hb/A1–9, 49/50
Atom	50 (Ta):	Hb/A2–300/26
Atomic nucleus	50 (Ta):	Hb/A2–277/86
Atomic radius	50 (Ta):	Hb/A2–304/6
Atomic scattering factors	50 (Ta):	Hb/A2–302/3
Atomic weight	50 (Ta):	Hb/A2–278/9
Behavior as anode	50 (Ta):	Hb/A2–446/87
– as cathode	50 (Ta):	Hb/A2–432/46
Boiling point	50 (Ta):	Hb/A2–373
Bond properties	48 (V):	Hb/A2–389/92
Cathodic sputtering	50 (Ta):	Hb/A2–404/6
Chemical constant	50 (Ta):	Hb/A2–379
Chemical reactions	50 (Ta):	Hb/A2–495/547
– on heating	50 (Ta):	Hb/A2–495
– with acids	50 (Ta):	Hb/A2–535/43
– with alkaline solutions	50 (Ta):	Hb/A2–533/5
– with alloys	50 (Ta):	Hb/A2–524/9
– with aqueous solutions	50 (Ta):	Hb/A2–543/5
– with metal compounds	50 (Ta):	Hb/A2–529/33
– with metals	50 (Ta):	Hb/A2–519/24

Hb = Main Series Volume; GD = Gmelin-Durrer;   *Ab = Appended Volume to Main Series;*
Eb = Supplement Volume to Main Series;   *Erg. W. = New Supplement Series Volume*

Ta . . . . Tantalum:
   Chemical reactions
     – with nonmetal compounds . . . . . 50 (Ta):   Hb/A2–514/9
     – with nonmetals . . . . 50 (Ta):   Hb/A2–495/514
     – with organic compounds. . . . . . 50 (Ta):   Hb/A2–545/7
     Cleavage. . . . . 50 (Ta):   Hb/A2–328
     Colloidal forms . . . . 50 (Ta):   Hb/A1–266
     Compressibility . . . . 50 (Ta):   Hb/A2–350
     Cosmochemistry . . . . 50 (Ta):   Hb/A1–9/12
     Critical temperature . . . . 50 (Ta):   Hb/A2–374
     Crystal chemistry . . . . 50 (Ta):   Hb/A1–12/49
     Crystallographic properties . . . . . 50 (Ta):   Hb/A2–326/46
     Debye temperature . . . . 50 (Ta):   Hb/A2–379
     Deformation. . . . . 50 (Ta):   Hb/A2–338/41
     Density . . . . 50 (Ta):   Hb/A2–346/7
     Deposition of pure tantalum
       compounds. . . . . 50 (Ta):   Hb/A1–246/51
     Detection . . . . 50 (Ta):   Hb/A2–547/51
     – in iron . . . . 59 (Fe):   Hb/FII1–81/3
     – in special substances . . . . 50 (Ta):   Hb/A2–551/2
     Determination. . . . . 50 (Ta):   Hb/A2–547/8, 552/7
     – in iron . . . . 59 (Fe):   Hb/FII1–83/5
                                    Hb/FII2–301
     – in presence of other ions . . . . . . 50 (Ta):   Hb/A2–557/62
     – in special substances . . . . 50 (Ta):   Hb/A2–563/74
     Diffusion. . . . . 50 (Ta):   Hb/A2–346
     –, self-diffusion . . . . 50 (Ta):   Hb/A2–345/6
     – of boron . . . . 50 (Ta):   Hb/B1–193
     – of caesium. . . . . 50 (Ta):   Hb/B2–2
     – of carbon . . . . 50 (Ta):   Hb/B1–211/2
     – of hydrogen . . . . 50 (Ta):   Hb/B1–3/4, 16/7
     – of niobium. . . . . 50 (Ta):   Hb/B2–85
     – of nitrogen . . . . 50 (Ta):   Hb/B1–66
     – of oxygen . . . . 50 (Ta):   Hb/B1–17
     – of silicon. . . . . 50 (Ta):   Hb/B1–251
     – of titanium. . . . . 50 (Ta):   Hb/B2–24
     Disorder . . . . 50 (Ta):   Hb/A2–333/7, 341/2
     Dressing and treatment of ores . . . . 50 (Ta):   Hb/A1–212/31
     Economic deposits . . . . 50 (Ta):   Hb/A1–107/25
     Economics. . . . . 50 (Ta):   Hb/A1–110/3
     Elasticity. . . . . 50 (Ta):   Hb/A2–347/50
     Electrical conductivity . . . . 50 (Ta):   Hb/A2–386/9
     Electrical properties . . . . 50 (Ta):   Hb/A2–383/408
     Electrochemical behavior . . . . 50 (Ta):   Hb/A2–410/95
     Electrolytic deposition . . . . 50 (Ta):   Hb/A2–487/95
     Electrolytic polishing . . . . 50 (Ta):   Hb/A2–338, 487
     Electron configuration . . . . 50 (Ta):   Hb/A2–300
     Electronegativity . . . . 48 (V):   Hb/A2–392
     Electron emission. . . . . 50 (Ta):   Hb/A2–397/404

See the "Instructions for the user"
at the beginning of the volume

Ta . . . . Tantalum:
               Electron structure. . . . . 50 (Ta):   Hb/A2–331/2
               Energy band structure . . . . 50 (Ta):   Hb/A2–383/5
               Epitaxy. . . . . 50 (Ta):   Hb/A2–328
               Equation of state . . . . 50 (Ta):   Hb/A2–370
               Expansion coefficient. . . . . 50 (Ta):   Hb/A2–370/2
               g factor . . . . 50 (Ta):   Hb/A2–301, 385
               Galvanomagnetic effects. . . . . . . 50 (Ta):   Hb/A2–396/7
               Geochemical character . . . . 50 (Ta):   Hb/A1–49/50
               Geochemical cycle . . . . 50 (Ta):   Hb/A1–106/7
               Geochemistry . . . . 50 (Ta):   Hb/A1–12/107
               de Haas-van Alphen effect . . . . . . 50 (Ta):   Hb/A2–386
               Hall effect . . . . 50 (Ta):   Hb/A2–396/7
               Hardness . . . . 50 (Ta):   Hb/A2–354/61
               Heat of fusion . . . . 50 (Ta):   Hb/A2–374
               – of sublimation. . . . . 50 (Ta):   Hb/A2–373
               – of vaporization . . . . 50 (Ta):   Hb/A2–373/4
               History. . . . . 50 (Ta):   Hb/A1–1/9
               Increments . . . . 50 (Ta):   Hb/A2–326
               Ion emission . . . . 50 (Ta):   Hb/A2–404/8
               Ionization potentials . . . . 50 (Ta):   Hb/A2–302
               Lattice defects. . . . . 50 (Ta):   Hb/A2–333/7, 341/2
               Lattice energy . . . . 50 (Ta):   Hb/A2–332/3
               Lattice structure. . . . . 50 (Ta):   Hb/A2–328/31
               Magnetic moment . . . . 50 (Ta):   Hb/A2–301
                Magnetic properties . . . . 50 (Ta):   Hb/A2–385/6, 396/7
                Mechanical properties . . . . 50 (Ta):   Hb/A2–346/70
                Melting point . . . . 50 (Ta):   Hb/A2–374
                Mineralogy . . . . 50 (Ta):   Hb/A1–125/211
               Modifications . . . . 50 (Ta):   Hb/A2–326/7
               Nuclear magnetic resonance. . . . . 50 (Ta):   Hb/A2–385
               Nuclear moments. . . . . 50 (Ta):   Hb/A2–279/81
               Nuclear properties . . . . 50 (Ta):   Hb/A2–277/86
               Occurrence, extraterrestrial . . . . . 50 (Ta):   Hb/A1–9/12
               –, terrestrial . . . . 50 (Ta):   Hb/A1–12/107
               – in meteorites and tektites . . . . . 50 (Ta):   Hb/A1–10/2
               – in minerals . . . . 50 (Ta):   Hb/A1–22/49,
                                                       125/211
               – in soils. . . . . 50 (Ta):   Hb/A1–100/2
               – in the atmosphere . . . . 50 (Ta):   Hb/A1–105/6
                – in the hydrosphere . . . . 50 (Ta):   Hb/A1–104/5
                – in the lithosphere. . . . . 50 (Ta):   Hb/A1–50/104
               – in the magmatic cycle . . . . 50 (Ta):   Hb/A1–50/91
                – in the metamorphic cycle . . . . . 50 (Ta):   Hb/A1–102/4
                – in the sedimentary cycle . . . . . . 50 (Ta):   Hb/A1–91/100
                Optical properties. . . . . 50 (Ta):   Hb/A2–408/10
                Overvoltage phenomena. . . . . . . 50 (Ta):   Hb/A2–422/32
                Oxidation . . . . 50 (Ta):   Hb/A2–495/505
                Paramagnetic resonance . . . . 50 (Ta):   Hb/A2–301, 385

Hb = Main Series Volume; GD = Gmelin-Durrer;   *Ab = Appended Volume to Main Series;*
Eb = Supplement Volume to Main Series;   *Erg. W. = New Supplement Series Volume*

Ta . . . . Tantalum:

Passivation . . . . 50 (Ta):   Hb/A2–446/76
Polarizability . . . . 50 (Ta):   Hb/A2–302
Polarography . . . . 50 (Ta):   Hb/A2–421/2
Polymorphism . . . . 50 (Ta):   Hb/A2–326/7
Position in the
   electrochemical series . . . . 50 (Ta):   Hb/A2–410
– in the periodic system . . . . 50 (Ta):   Hb/A1–7
Potentials . . . . 50 (Ta):   Hb/A2–410/21
Preparation, industrial . . . . 50 (Ta):   Hb/A1–211/70
– in pure state . . . . 50 (Ta):   Hb/A1–248/51, 262
– of metallic tantalum . . . . 50 (Ta):   Hb/A1–251/8
– of special forms . . . . 50 (Ta):   Hb/A1–263/6
Purification of metallic tantalum . . . . 50 (Ta):   Hb/A1–258/62
– of solutions and concentrates . . . . 50 (Ta):   Hb/A1–231/5
Purity testing . . . . 50 (Ta):   Hb/A1–266/70
Recovery . . . . 50 (Ta):   Hb/A1–262/3
Recrystallization . . . . 50 (Ta):   Hb/A2–342/5
Scattering of X- and gamma rays . . . 50 (Ta):   Hb/A2–320/6
Self-diffusion . . . . 50 (Ta):   Hb/A2–345/6
Separation, analytical . . . . 50 (Ta):   Hb/A2–557/62
– from niobium . . . . 49 (Nb):   Hb/A–27
                                          50 (Ta):   Hb/A1–235/46
                                                       Hb/A2–558/60
Sorption of carbon monoxide . . . . . 50 (Ta):   Hb/A2–517/8
– of hydrogen . . . . 50 (Ta):   Hb/A2–506/7
                                                       Hb/B1–2/3
– of nitrogen . . . . 50 (Ta):   Hb/A2–508
                                                       Hb/B1–61/6
– of oxygen . . . . 50 (Ta):   Hb/A2–503/4, 505
– of rare gases . . . . 50 (Ta):   Hb/A2–505
                                                       Hb/B1–1
Sound absorption and velocity . . . . 50 (Ta):   Hb/A2–350/1
Special forms . . . . 50 (Ta):   Hb/A1–263/6
Spectra:
   Arc spectrum . . . . 50 (Ta):   Hb/A2–308/9
   Spark spectra . . . . 50 (Ta):   Hb/A2–310/1
   Terms . . . . 50 (Ta):   Hb/A2–300, 306/11
   X-ray spectrum . . . . 50 (Ta):   Hb/A2–312/20
Standard potential . . . . 50 (Ta):   Hb/A2–410/1
Strength . . . . 50 (Ta):   Hb/A2–361/9
Structure . . . . 50 (Ta):   Hb/A2–330/1, 337/8
Superconductivity . . . . 50 (Ta):   Hb/A2–390/5
Surface tension . . . . 50 (Ta):   Hb/A2–369/70
Susceptibility . . . . 50 (Ta):   Hb/A2–385/6
Technology . . . . 50 (Ta):   Hb/A1–211/66
Temperature radiation . . . . 50 (Ta):   Hb/A2–409/10
Thermal conductivity . . . . 50 (Ta):   Hb/A2–380/3

See the "Instructions for the user"
at the beginning of the volume

Ta . . . . Tantalum:
                 Thermal properties . . . . 50 (Ta):   Hb/A2–370/83
                 Thermodynamic functions . . . . . . 50 (Ta):   Hb/A2–375/8
                 Thermoelectric effects . . . . 50 (Ta):   Hb/A2–395/6
                 Thermomagnetic effects . . . . 50 (Ta):   Hb/A2–396/7
                 Toxicity . . . . 50 (Ta):   Hb/A1–276
                 Twinning . . . . 50 (Ta):   Hb/A2–327/8
                 Uses . . . . 50 (Ta):   Hb/A1–271/6
                 Vapor pressure . . . . 50 (Ta):   Hb/A2–372/3
                 Viscosity . . . . 50 (Ta):   Hb/A2–351/3
                 Wave functions . . . . 50 (Ta):   Hb/A2–301/2

–. . . . . Tantalum ions
                 see Ta⁺ ... Ta⁺, etc.

–. . . . . Tantalum isotopes:
                 Abundance . . . . 50 (Ta):   Hb/A1–50
                                                                 Hb/A2–277
                 Detection and determination. . . . . 50 (Ta):   Hb/A2–574
                 Enrichment and separation . . . . . 50 (Ta):   Hb/A2–286/300
                 Nuclear decay. . . . . 50 (Ta):   Hb/A2–281/6
                 Nuclear properties . . . . 50 (Ta):   Hb/A2–277/81
                 Preparation . . . . 50 (Ta):   Hb/A2–290/300

–. . . . . Tantalum systems:
                 Ta–Al . . . . 50 (Ta):   Hb/B2–11/2
                 Ta–Al–B . . . . 50 (Ta):   Hb/B2–320
                 Ta–Al–C . . . . 50 (Ta):   Hb/B2–323
                 Ta–Al–C–V . . . . 50 (Ta):   Hb/B2–343
                 Ta–Al–O . . . . 50 (Ta):   Hb/B2–197
                 Ta–Al–Si . . . . 50 (Ta):   Hb/B2–16/7
                 Ta–Al–Ti . . . . 50 (Ta):   Hb/B2–40
                 Ta–Al–V . . . . 50 (Ta):   Hb/B2–78/9
                 Ta–As . . . . 50 (Ta):   Hb/B1–272
                 Ta–B. . . . . 50 (Ta):   Hb/B1–192/8
                 Ta–B–C . . . . 50 (Ta):   Hb/B1–250
                 Ta–B–C–N . . . . 50 (Ta):   Hb/B1–250
                 Ta–B–Cs . . . . 50 (Ta):   Hb/B2–320
                 Ta–B–Ga . . . . 50 (Ta):   Hb/B2–320
                 Ta–B–Ge . . . . 50 (Ta):   Hb/B2–322
                 Ta–B–Li. . . . . 50 (Ta):   Hb/B2–320
                 Ta–B–Nb . . . . 50 (Ta):   Hb/B2–323
                 Ta–B–Rb . . . . 50 (Ta):   Hb/B2–320
                 Ta–B–Si. . . . . 50 (Ta):   Hb/B1–264
                 Ta–B–V . . . . 50 (Ta):   Hb/B2–322
                 Ta–B–Zr . . . . 50 (Ta):   Hb/B2–321/2
                 Ta–Be . . . . 50 (Ta):   Hb/B2–2
                 Ta–Be–O . . . . 50 (Ta):   Hb/B2–172
                 Ta–Bi . . . . 50 (Ta):   Hb/B1–274
                 Ta–Bi–O . . . . 50 (Ta):   Hb/B1–275

Hb = Main Series Volume; GD = Gmelin-Durrer;   *Ab = Appended Volume to Main Series;*
Eb = Supplement Volume to Main Series;   *Erg. W. = New Supplement Series Volume*

Ta . . . . Tantalum systems:
          Ta–Br . . . . 50 (Ta):   Hb/B1–141/2
          Ta–C. . . . . 50 (Ta):   Hb/B1–211/6
          Ta–C–Ga . . . . 50 (Ta):   Hb/B2–323
          Ta–C–Ge . . . . 50 (Ta):   Hb/B2–341
          Ta–C–Hf . . . . 50 (Ta):   Hb/B2–333/40
          Ta–C–Hf–N. . . . . 50 (Ta):   Hb/B2–340
          Ta–C–Hf–Ti . . . . 50 (Ta):   Hb/B2–341
          Ta–C–Hf–Zr . . . . 50 (Ta):   Hb/B2–341
          Ta–C–N. . . . . 50 (Ta):   Hb/B1–248/50
          Ta–C–N–O . . . . 50 (Ta):   Hb/B1–250
          Ta–C–N–Zr. . . . . 50 (Ta):   Hb/B2–333
          Ta–C–Nb . . . . 50 (Ta):   Hb/B2–344/6
          Ta–C–Nb–Ti . . . . 50 (Ta):   Hb/B2–346/7
          Ta–C–Ni . . . . 57 (Ni):   Hb/B3–1206
          Ta–C–O. . . . . 50 (Ta):   Hb/B1–248
          Ta–C–Si. . . . . 50 (Ta):   Hb/B1–264/5
          Ta–C–Ti. . . . . 50 (Ta):   Hb/B2–324/6
          Ta–C–Ti–V . . . . 50 (Ta):   Hb/B2–343
          Ta–C–V. . . . . 50 (Ta):   Hb/B2–341/3
          Ta–C–V–Zr. . . . . 50 (Ta):   Hb/B2–343/4
          Ta–C–Zr . . . . 50 (Ta):   Hb/B2–327/33
          Ta–Cl . . . . 50 (Ta):   Hb/B1–91
          Ta–Cl–O–S. . . . . 50 (Ta):   Hb/B1–178
          Ta–Cl–S. . . . . 50 (Ta):   Hb/B1–178
          Ta–Co. . . . . 58 (Co):   Eb/A–418/20
          Ta–Cr . . . . 52 (Cr):   Hb/A2–728/30
          Ta–Cr–N . . . . 52 (Cr):   Hb/B–942
          Ta–Cr–Ni . . . . 57 (Ni):   Hb/B1–227/8
          Ta–Cu. . . . . 60 (Cu):   Hb/B3–1214
          Ta–Dy. . . . . 50 (Ta):   Hb/B2–22/3
          Ta–Er . . . . 50 (Ta):   Hb/B2–23
          Ta–Fe . . . . 59 (Fe):   Hb/A9–1944/5
          Ta–Ga . . . . 50 (Ta):   Hb/B2–17/8
          Ta–Gd . . . . 50 (Ta):   Hb/B2–22
          Ta–Ge . . . . 50 (Ta):   Hb/B2–64
          Ta–Ge–N . . . . 50 (Ta):   Hb/B2–267
          Ta–H. . . . . 50 (Ta):   Hb/B1–1/16
          Ta–D. . . . . 50 (Ta):   Hb/B1–16
          Ta–T. . . . . 50 (Ta):   Hb/B1–16/7
          Ta–Hf . . . . 50 (Ta):   Hb/B2–52/3
          Ta–Hf–N . . . . 50 (Ta):   Hb/B2–267
          Ta–Hf–Nb. . . . . 50 (Ta):   Hb/B2–117
          Ta–Hf–Ti . . . . 50 (Ta):   Hb/B2–61
          Ta–Hf–V . . . . 50 (Ta):   Hb/B2–84
          Ta–Hf–Zr . . . . 50 (Ta):   Hb/B2–61
          Ta–Hg. . . . . 50 (Ta):   Hb/B2–11
          Ta–In . . . . 50 (Ta):   Hb/B2–18

See the "Instructions for the user"
at the beginning of the volume

Ta . . . . Tantalum systems:

Ta–Li–O.	50 (Ta):	Hb/B2–126
Ta–Mg–O	50 (Ta):	Hb/B2–172
Ta–N	50 (Ta):	Hb/B1–61/9
Ta–N–Si	50 (Ta):	Hb/B1–263/4
Ta–N–V.	50 (Ta):	Hb/B2–267
Ta–Na–O	50 (Ta):	Hb/B2–141
Ta–Nb.	50 (Ta):	Hb/B2–85/6
Ta–Nb–Ni–Ti.	57 (Ni):	Hb/B1–173
Ta–Nb–O	50 (Ta):	Hb/B2–245/6
Ta–Nb–Si.	49 (Nb):	Hb/B2–104
Ta–Nb–Ti.	50 (Ta):	Hb/B2–104
Ta–Nb–V	50 (Ta):	Hb/B2–118
Ta–Nb–Zr.	50 (Ta):	Hb/B2–110
Ta–Ni	57 (Ni):	Hb/B1–166/7
Ta–Ni–V	57 (Ni):	Hb/B1–172
Ta–O	50 (Ta):	Hb/B1–17/23
Ta–O–Pb	50 (Ta):	Hb/B2–232
Ta–O–Si	50 (Ta):	Hb/B1–263
Ta–O–Th	50 (Ta):	Hb/B2–229
Ta–O–Ti	50 (Ta):	Hb/B2–221
Ta–O–Ti–Zr	50 (Ta):	Hb/B2–229
Ta–O–V.	50 (Ta):	Hb/B2–244
Ta–O–Y.	50 (Ta):	Hb/B2–205
Ta–O–Zr	50 (Ta):	Hb/B2–228
Ta–Ru	63 (Ru):	Eb–198/9
Ta–S.	50 (Ta):	Hb/B1–169/70
Ta–Sc	50 (Ta):	Hb/B2–18/9
Ta–Se	50 (Ta):	Hb/B1–179/80
Ta–Si	50 (Ta):	Hb/B1–250/5
Ta–Si–Ti	50 (Ta):	Hb/B2–39
Ta–Si–V.	50 (Ta):	Hb/B2–77/8
Ta–Sn	50 (Ta):	Hb/B2–65
Ta–TaC–TaN	50 (Ta):	Hb/B1–248/9
Ta–TaCl$_5$–TaS$_3$	50 (Ta):	Hb/B1–178
Ta–TaI$_5$	50 (Ta):	Hb/B1–161
Ta–Te	50 (Ta):	Hb/B1–189/90
Ta–Th	50 (Ta):	Hb/B2–62
Ta–Ti	50 (Ta):	Hb/B2–24/5
Ta–Ti–V.	50 (Ta):	Hb/B2–79
Ta–Ti–Zr	50 (Ta):	Hb/B2–51
Ta–V.	48 (V):	Hb/A2–402
	50 (Ta):	Hb/B2–68/9
Ta–V–Zr	50 (Ta):	Hb/B2–83
Ta–Y.	50 (Ta):	Hb/B2–19/20
Ta–Y$_2$O$_3$.	50 (Ta):	Hb/B2–205
Ta–Zn	50 (Ta):	Hb/B2–10/1
Ta–Zr	50 (Ta):	Hb/B2–41/3

Hb = Main Series Volume; GD = Gmelin-Durrer; *Ab = Appended Volume to Main Series;*
Eb = Supplement Volume to Main Series; *Erg. W. = New Supplement Series Volume*

Ta$^+$ . . . . Ta$^+$:

          Electrochemical behavior . . . . 50 (Ta):   Hb/A2–487/8

          Form factors. . . . . 50 (Ta):   Hb/A2–303

          Ion emission . . . . 50 (Ta):   Hb/A2–404, 407/8

          Ionization potential . . . . 50 (Ta):   Hb/A2–302

          Properties of the ion . . . . 50 (Ta):   Hb/A2–300, 305

          Spectra . . . . 50 (Ta):   Hb/A2–310/1

Ta^{2+} . . . . Ta^{2+}:

          Electrochemical behavior . . . . 50 (Ta):   Hb/A2–420, 487/8

          Form factors. . . . . 50 (Ta):   Hb/A2–303

          Ionization potential . . . . 50 (Ta):   Hb/A2–302

          Properties of the ion . . . . 50 (Ta):   Hb/A2–300, 305

          Spectra . . . . 50 (Ta):   Hb/A2–311

Ta^{3+} . . . . Ta^{3+}:

          Electrochemical behavior . . . . 50 (Ta):   Hb/A2–420, 487/8

          Form factors. . . . . 50 (Ta):   Hb/A2–303

          Ionization potential . . . . 50 (Ta):   Hb/A2–302

          Properties of the ion . . . . 50 (Ta):   Hb/A2–301, 305

Ta^{4+} . . . . Ta^{4+}:

          Electrochemical behavior . . . . 50 (Ta):   Hb/A2–420, 487/8

          Form factors. . . . . 50 (Ta):   Hb/A2–303

          Ionization potential . . . . 50 (Ta):   Hb/A2–302

          Properties of the ion . . . . 50 (Ta):   Hb/A2–301, 305

Ta^{5+} . . . . Ta^{5+}:

          Electrochemical behavior . . . . 50 (Ta):   Hb/A2–420, 487/95

          Ionic refraction . . . . 50 (Ta):   Hb/A2–326

          Ionic susceptibility . . . . 50 (Ta):   Hb/A2–326

          Ionization potential . . . . 50 (Ta):   Hb/A2–302

          Properties of the ion . . . . 50 (Ta):   Hb/A2–301, 305/6

TaTe . . . . TaTe . . . . 50 (Ta):   Hb/B1–189

–. . . . . TaTe solid solutions:

          (Ta,Nb)Te . . . . 50 (Ta):   Hb/B2–319/20

TaTe$_2$ . . . . TaTe$_2$ . . . . 50 (Ta):   Hb/B1–189/91

–. . . . . TaTe$_2$ solid solutions:

          (Ta,Nb)Te$_2$. . . . . 50 (Ta):   Hb/B2–319/20

TaTe$_3$ . . . . TaTe$_3$ . . . . 50 (Ta):   Hb/B1–190, 192

TaTe$_4$ . . . . TaTe$_4$ . . . . 50 (Ta):   Hb/B1–189/90, 191/2

TaTi . . . . TaTi solid solutions:

          Ta(Ti,Al) . . . . 50 (Ta):   Hb/B2–40

TaV$_2$. . . . . TaV$_2$ . . . . 50 (Ta):   Hb/B2–68/72

–. . . . . TaV$_2$ systems:

          TaV$_2$–Nb . . . . 50 (Ta):   Hb/B2–118

Ta$_2$W$_5$ . . . . Ta$_2$W$_5$ . . . . 50 (Ta):   Hb/B1–51

Ta$_3$Te$_2$ . . . . Ta$_3$Te$_2$ . . . . 50 (Ta):   Hb/B1–190

Ta$_{31}$Te$_8$ . . . . Ta$_{31}$Te$_8$ . . . . 50 (Ta):   Hb/B1–190

See the "Instructions for the user"
at the beginning of the volume

Tb . . . . Terbium:

        Abundance . . . . 39 (SE):   Hb/A1−34/5

        Band model . . . . 39 (SE):   Hb/B3−213/5

        Boiling point . . . . 39 (SE):   Hb/B3−105, 107

        Charge carrier concentration. . . . . 39 (SE):   Hb/B3−228

        Cosmochemistry . . . . 39 (SE):   Hb/A1−23/7

        Crystal chemistry . . . . 39 (SE):   Hb/A1−28/32

        Crystal field parameters . . . . 39 (SE):   Hb/B3−219

        Crystallographic properties . . . . . 39 (SE):   Hb/B3−62, 66/72,
                                                     76/83

        Crystal structure . . . . 39 (SE):   Hb/B3−66/7, 71/2

        Debye temperature . . . . 39 (SE):   Hb/B3−190

        Density . . . . 39 (SE):   Hb/B3−85/7

        Diffusion. . . . . 39 (SE):   Hb/B3−77/8

        Disorder . . . . 39 (SE):   Hb/B3−76

        Domain structure . . . . 39 (SE):   Hb/B3−332

        Economic deposits . . . . 39 (SE):   Hb/A1−110/22

        Elasticity. . . . . 39 (SE):   Hb/B3−92/102

        Electrical properties . . . . 39 (SE):   Hb/B3−209/83

        Electrical resistance. . . . . 39 (SE):   Hb/B3−241/3, 260

        Electron emission. . . . . 39 (SE):   Hb/B3−281/3

        Enthalpy of fusion . . . . 39 (SE):   Hb/B3−120

        − of sublimation. . . . . 39 (SE):   Hb/B3−110, 114

        − of transformation . . . . 39 (SE):   Hb/B3−118

        − of vaporization . . . . 39 (SE):   Hb/B3−110, 114

        Ettingshausen effect . . . . 39 (SE):   Hb/B3−273

        Exchange interaction . . . . 39 (SE):   Hb/B3−298/303

        Geochemistry . . . . 39 (SE):   Hb/A1−28/54

        Grüneisen constant. . . . . 39 (SE):   Hb/B3−84/5

        Hall effect . . . . 39 (SE):   Hb/B3−268/9

        History. . . . . 39 (SE):   Hb/A1−4/22

        Hyperfine interactions . . . . 39 (SE):   Hb/B3−222

        Hysteresis . . . . 39 (SE):   Hb/B3−333

        Internal friction . . . . 39 (SE):   Hb/B3−195

        Lattice dynamics . . . . 39 (SE):   Hb/B3−82/3

        Magnetic anisotropy . . . . 39 (SE):   Hb/B3−303/6

        Magnetic form factor . . . . 39 (SE):   Hb/B3−315/6

        Magnetic lag . . . . 39 (SE):   Hb/B3−334

        Magnetic moment . . . . 39 (SE):   Hb/B3−334

        Magnetic properties . . . . 39 (SE):   Hb/B3−298/336

        Magnetic resonance . . . . 39 (SE):   Hb/B3−331/2

        Magnetic structure . . . . 39 (SE):   Hb/B3−307/12

        Magnetization. . . . . 39 (SE):   Hb/B3−316/8

        Magnetostriction . . . . 39 (SE):   Hb/B3−323/6

        Mechanical properties . . . . 39 (SE):   Hb/B3−85/103,
                                                     193/5, 207/9

        Melting point . . . . 39 (SE):   Hb/B3−118/9

        Mineralogy . . . . 39 (SE):   Hb/A1−54/110

        Molar volume . . . . 39 (SE):   Hb/B3−85/7

Hb = Main Series Volume; GD = Gmelin-Durrer;   Ab = Appended Volume to Main Series;
Eb = Supplement Volume to Main Series;   Erg. W. = New Supplement Series Volume

Tb . . . . Terbium:
           Nuclear magnetic resonance. . . . . . 39 (SE):   Hb/B3–227/8
           Occurrence, extraterrestrial . . . . . . 39 (SE):   Hb/A1–22/7
           –, terrestrial . . . . 39 (SE):   Hb/A1–27/54
           – in meteorites . . . . 39 (SE):   Hb/A1–25/7
           – in minerals . . . . 39 (SE):   Hb/A1–35/53
           – in the biosphere. . . . . 39 (SE):   Hb/A1–53/4
           – in the lithosphere . . . . 39 (SE):   Hb/A1–27/54
           Optical properties. . . . . 39 (SE):   Hb/B3–341
           Polymorphism . . . . 39 (SE):   Hb/B3–62
           Position in the periodic system . . . . 39 (SE):   Hb/A1–2/4
           Spin waves . . . . 39 (SE):   Hb/B3–328/30
           Strength . . . . 39 (SE):   Hb/B3–102/3
           Structure . . . . 39 (SE):   Hb/B3–80
           Surface tension . . . . 39 (SE):   Hb/B3–208/9
           Susceptibility . . . . 39 (SE):   Hb/B3–334/6
           Thermal conductivity . . . . 39 (SE):   Hb/B3–200/4
           Thermal expansion . . . . 39 (SE):   Hb/B3–88/92
           Thermal properties . . . . 39 (SE):   Hb/B3–83/92, 103/14,
                                                          156/60, 190, 200/5
           Thermodynamic functions . . . . . . 39 (SE):   Hb/B3–156/60
           Thermoelectric power . . . . 39 (SE):   Hb/B3–275
           Ultrasound absorption . . . . 39 (SE):   Hb/B3–193/4
           Valence . . . . 39 (SE):   Hb/A1–4
           Vapor pressure . . . . 39 (SE):   Hb/B3–103/5, 107
–. . . . . Terbium ions. . . . . 39 (SE):   Hb/A1–3
                                              Hb/B3–317
                                              Hb/C1–242, 399
                                    49 (Nb):   Hb/B3–134, 135
–. . . . . Terbium systems:
           Tb–Ag . . . . 61 (Ag):   Hb/C–227/8
           Tb–H . . . . 39 (SE):   Hb/C1–2/3, 67
           Tb–Pb . . . . 47 (Pb):   Hb/C4–1369
           Tb–Sn . . . . 46 (Sn):   Hb/D–318
Tb^{3+} . . . . Tb^{3+} . . . . 39 (SE):   Hb/A1–3
                                              Hb/B3–317
                                    49 (Nb):   Hb/B3–134, 135
Tb^{4+} . . . . Tb^{4+} . . . . 39 (SE):   Hb/C1–242, 399

Tc . . . . Technetium:
           Chemical reactions . . . . 69 (Ma):   Hb–9/10
           History . . . . 69 (Ma):   Hb–1
                                    70 (Re):   Hb–1/10
           Occurrence . . . . 69 (Ma):   Hb–1/2
                                    70 (Re):   Hb–12/3
           Physical properties . . . . 69 (Ma):   Hb–2/8
           Preparation . . . . 69 (Ma):   Hb–2
–. . . . . Technetium isotopes . . . . 69 (Ma):   Hb–2/3, 9

See the "Instructions for the user"
at the beginning of the volume

Te . . . . Tellurium:

Absorption of X-rays	11 (Te):	Hb−203
Abundance	11 (Te):	Hb−9
Affinity.	11 (Te):	Hb−161
Atom.	11 (Te):	Hb−151/62
Atomic nucleus	11 (Te):	Hb−140/51
Chemical reactions	11 (Te):	Hb−228/38
– with elements	11 (Te):	Hb−228/31
– with inorganic compounds.	11 (Te):	Hb−232/7
– with organic compounds.	11 (Te):	Hb−238
Cosmochemistry	11 (Te):	Hb−6
Crystallographic properties	11 (Te):	Hb−165/71
Detection	11 (Te):	Hb−240/7
– in iron	59 (Fe):	Hb/FII1−31
Determination.	11 (Te):	Hb−247/58
– in iron	59 (Fe):	Hb/FII1−31
– in special substances	11 (Te):	Hb−262
Economic deposits:		
Africa	11 (Te):	Hb−53
America.	11 (Te):	Hb−40/9, 53/65
Asia	11 (Te):	Hb−50/2
Australia and Oceania	11 (Te):	Hb−33/40, 52/3
Europe	11 (Te):	Hb−19/33, 49/50
Electrical conductivity	11 (Te):	Hb−209/16
Electrical properties	11 (Te):	Hb−209/22
Electrochemical behavior	11 (Te):	Hb−223/8
Formation	11 (Te):	Hb−125
Geochemical character.	11 (Te):	Hb−7/8
Geochemistry	11 (Te):	Hb−6/124
History.	11 (Te):	Hb−1/5
Ionization potentials	11 (Te):	Hb−222
Magnetic properties	11 (Te):	Hb−204/9
Mechanical properties	11 (Te):	Hb−171/6
Mineralogy	11 (Te):	Hb−65/124
Molecule.	11 (Te):	Hb−162/5
Occurrence, extraterrestrial	11 (Te):	Hb−6
–, terrestrial	11 (Te):	Hb−6/124
– in industrial products.	11 (Te):	Hb−16/8
– in the biosphere.	11 (Te):	Hb−15
– in the hydrosphere	11 (Te):	Hb−15
– in the lithosphere	11 (Te):	Hb−7/15
Optical properties.	11 (Te):	Hb−182/203
Position in the periodic system	11 (Te):	Hb−161/2
Preparation	11 (Te):	Hb−125/38
–, industrial	11 (Te):	Hb−125/30
– in pure state.	11 (Te):	Hb−130/6
– of special forms.	11 (Te):	Hb−136/8
Radioactivity	11 (Te):	Hb−151

Te . . . . Tellurium:
    Scattering of X-rays . . . . 11 (Te):   Hb–203
    Separation, analytical . . . . 11 (Te):   Hb–258/62
    Spectra:
        Arc spectrum . . . . 11 (Te):   Hb–184/9, 191
        Band spectrum . . . . 11 (Te):   Hb–191/9
        Continuous spectrum . . . . 11 (Te):   Hb–199
        Spark spectra . . . . 11 (Te):   Hb–189/91
        Terms . . . . 11 (Te):   Hb–187/9, 197
        X-ray spectrum . . . . 11 (Te):   Hb–200/3

    Thermal properties . . . . 11 (Te):   Hb–176/82
    Toxicity . . . . 11 (Te):   Hb–238/9
    Uses . . . . 11 (Te):   Hb–139/40, 282
    Valence . . . . 11 (Te):   Hb–160/1

–. . . . . Tellurium ions
     see Te^+ ... Te^+, etc.

–. . . . . Tellurium isotopes . . . . 11 (Te):   Hb–140/2
–. . . . . Tellurium solutions:
    $Te–H_2Te$ . . . . 11 (Te):   Hb–272

–. . . . . Tellurium systems:
    Te–Ag . . . . 61 (Ag):   Hb/B3–199/206
    Te–Ag–Au . . . . 62 (Au):   Hb/3–766
    Te–Ag–Cu . . . . 61 (Ag):   Hb/B4–477/8
    Te–Ag–Fe . . . . 61 (Ag):   Hb/B4–415
    Te–Ag–Ge . . . . 61 (Ag):   Hb/B4–245
    Te–Ag–Pb . . . . 61 (Ag):   Hb/B4–269
    $Te–Ag_2Te–AuTe_2$ . . . . 62 (Au):   Hb/3–766
    $Te–Ag_2Te–SnTe$ . . . . 61 (Ag):   Hb/B4–251
    Te–Al . . . . 35 (Al):   Hb/B1–281/2
    Te–As . . . . 17 (As):   Hb–464
    Te–As–Bi . . . . 19 (Bi):   Eb–858
    Te–Au . . . . 62 (Au):   Hb/3–720/1
    Te–Bi . . . . 19 (Bi):   Hb–175/7
                                      Eb–775/85
    Te–Bi–O . . . . 19 (Bi):   Eb–817
    Te–Bi–Pb . . . . 47 (Pb):   Hb/C3–973/5
    Te–Bi–Sb . . . . 19 (Bi):   Eb–861/5
    Te–Bi–Sb–Se . . . . 19 (Bi):   Eb–865/6
    Te–Br . . . . 11 (Te):   Hb–335/7
    Te–Co . . . . 58 (Co):   Eb/A–662/3
    Te–Cr . . . . 52 (Cr):   Hb/B–342
    Te–Cr–Se . . . . 52 (Cr):   Hb/B–343
    Te–Cu . . . . 60 (Cu):   Hb/B1–619/20
    Te–Cu–Pb . . . . 60 (Cu):   Hb/B3–1202
    Te–Fe . . . . 59 (Fe):   Hb/A6–1174
    Te–Ga . . . . 36 (Ga):   Hb–90
    Te–Ge . . . . 45 (Ge):   Eb–543/4

See the "Instructions for the user"
at the beginning of the volume

Te . . . . Tellurium systems:

Te–Hg	34 (Hg):	Hb/B3–1111/3
Te–I	11 (Te):	Hb–344/5
Te–In	37 (In):	Hb–102
Te–Na	21 (Na):	Hb–643
		Eb/3–1202
Te–Nb	49 (Nb):	Hb/B1–291/2
Te–Ni	57 (Ni):	Hb/B2–756/7
Te–O–Pb	47 (Pb):	Hb/C2–678/9
Te–Pb	47 (Pb):	Hb/C2–640/3
Te–Pb–Sb	47 (Pb):	Hb/C3–940/1
Te–Pb–Se–Sn	47 (Pb):	Hb/C4–1594/5
Te–Pb–Sn	47 (Pb):	Hb/C4–1588
Te–S	11 (Te):	Hb–350/1
Te–S–Se	11 (Te):	Hb–363
Te–Sb	18 (Sb):	Hb/B3–554/5
Te–Se	11 (Te):	Hb–361/2
Te–Ta	50 (Ta):	Hb/B1–189/90
Te–TeBr$_4$	11 (Te):	Hb–335/7
Te–TeCl$_4$	11 (Te):	Hb–323/5
Te–Ti	41 (Ti):	Hb–353/4
Te–Tl	38 (Tl):	Hb/3–376
Te–V	48 (V):	Hb/B1–315/6
Te–Zn	32 (Zn):	Eb–969

Te$^+$	Te$^+$	11 (Te):	Hb–155, 189, 221/2
Te^{2+}	Te^{2+}	11 (Te):	Hb–155, 189, 221/2
Te^{3+}	Te^{3+}	11 (Te):	Hb–155, 189/90,
			221/2
Te^{4+}	Te^{4+}:		
	Electrochemical behavior	11 (Te):	Hb–223/4,
			226/7, 286
	Ionic susceptibility	11 (Te):	Hb–205/6
	Ionization potential	11 (Te):	Hb–222
	Properties of the ion	11 (Te):	Hb–155/6
	Spectra	11 (Te):	Hb–190
Te^{5+}	Te^{5+}	11 (Te):	Hb–190, 222
Te^{6+}	Te^{6+}:		
	Ionic refraction	11 (Te):	Hb–183
	Ionic susceptibility	11 (Te):	Hb–205/6
	Ionization potential	11 (Te):	Hb–222
	Properties of the ion	11 (Te):	Hb–155/6
	Spectra	11 (Te):	Hb–190/1
Te^{7+}	Te^{7+}	11 (Te):	Hb–222
Te$^-$	Te$^-$	11 (Te):	Hb–154/5, 226/7
–.	Te$_2^{2-}$	11 (Te):	Hb–223/6, 272
Te^{2-}	Te^{2-}:		
	Chemical reactions	11 (Te):	Hb–270/2
	Electrochemical behavior	11 (Te):	Hb–223/7

Te^{2-}	Te^{2-}:		
	Ionic refraction	11 (Te):	Hb–183
	Ionic susceptibility	11 (Te):	Hb–206
	Properties of the ion	11 (Te):	Hb–153/4, 156
TeTh	ThTe	44 (Th):	Hb–295
$TeTh_3$	Th_3Te	44 (Th):	Hb–295
TeTi	TiTe	41 (Ti):	Hb–353/4
TeTl	TlTe	38 (Tl):	Hb/3–377
$TeTl_2$	Tl_2Te	38 (Tl):	Hb/3–376/7
–.	Tl_2Te systems:		
	Tl_2Te–PbTe	47 (Pb):	Hb/C4–1353/4
TeU	UTe	55 (U):	Hb–159
TeV	VTe	48 (V):	Hb/B1–315/8
TeZn	ZnTe	32 (Zn):	Hb–247/8
			Eb–969/70
–.	$ZnTe \cdot H_2O$	32 (Zn):	Hb–248
–.	ZnTe solid solutions:		
	ZnTe–HgTe	34 (Hg):	Hb/B4–1502/7
$TeZr_2$	Zr_2Te	42 (Zr):	Hb–355
$Te_{1.4}V$	$VTe_{1.4}$	48 (V):	Hb/B1–316, 318/9
$Te_{1.77}V$	$VTe_{1.77}$	48 (V):	Hb/B1–316, 319
Te_2	Te_2	11 (Te):	Hb–162/5, 191/9
Te_2^{2-}	Te_2^{2-}	11 (Te):	Hb–223/6, 272
Te_2Th	$ThTe_2$	44 (Th):	Hb–295
Te_2Ti	$TiTe_2$	41 (Ti):	Hb–353/4
Te_2Ti_x	Ti_xTe_2	41 (Ti):	Hb–354
Te_2Tl_3	Tl_3Te_2	38 (Tl):	Hb/3–377
Te_2U	UTe_2	55 (U):	Hb–159
Te_2V	VTe_2	48 (V):	Hb/B1–316, 319/20
Te_2Zr	$ZrTe_2$	42 (Zr):	Hb–355
$Te_{2.66}Th$	$ThTe_{2.66}$	44 (Th):	Hb–295
Te_3Tl_2	Tl_2Te_3 systems:		
	Tl_2Te_3–Ag_2Te	61 (Ag):	Hb/B4–228/9
Te_3Tl_5	Tl_5Te_3	38 (Tl):	Hb/3–376, 377
Te_3U_2	U_2Te_3	55 (U):	Hb–159
Te_3V_2	V_2Te_3	48 (V):	Hb/B1–316, 319
Te_4TiV_2	TiV_2Te_4	48 (V):	Hb/B2–561
Te_4Ti_2V	Ti_2VTe_4	48 (V):	Hb/B2–561
Te_4V_3	V_3Te_4	48 (V):	Hb/B1–315/6, 318
Te_4V_5	V_5Te_4	48 (V):	Hb/B1–315/6, 317
Te_5Tl_2	Tl_2Te_5	38 (Tl):	Hb/3–376, 377
Te_x^{2-}	Te_x^{2-}	11 (Te):	Hb–272
Te_xTi	$TiTe_x$	41 (Ti):	Hb–353/4
Th	Thorium:		
	Absorption of X-rays	44 (Th):	Hb–161
	Abundance	44 (Th):	Hb–10

See the "Instructions for the user"
at the beginning of the volume

Th . . . . Thorium:

Age determination . . . . 44 (Th):   Hb–17
Atom. . . . . 44 (Th):   Hb–120/7
                                      Erg. W. 8–1/28
Atomic nucleus . . . . 44 (Th):   Hb–93/112, 359, 370,
                                                   383, 393, 402
Chemical reactions . . . . 44 (Th):   Hb–180/4, 364, 377,
                                                   390, 396, 406
Cosmochemistry . . . . 44 (Th):   Hb–3/4
Crystal chemistry . . . . 44 (Th):   Hb–5/10
Crystallographic properties . . . . . . 44 (Th):   Hb–128/32
Detection . . . . 44 (Th):   Hb–197, 376/7
Determination. . . . . 44 (Th):   Hb–197, 376/7
Dressing and treatment of ores . . . . 44 (Th):   Hb–36/40, 67/72
Economic deposits . . . . 44 (Th):   Hb–19/27
Economics. . . . . 44 (Th):   Hb–27/8
Electrical properties . . . . 44 (Th):   Hb–162/78
Electrochemical behavior . . . . 44 (Th):   Hb–178/80
Electronegativity . . . . 44 (Th):   Hb–128
Electron emission. . . . . 44 (Th):   Hb–165/76
Formation . . . . 44 (Th):   Hb–83, 356, 366,
                                                   379, 391, 397
Geochemical character . . . . 44 (Th):   Hb–10
Geochemical cycle . . . . 44 (Th):   Hb–19
Geochemistry . . . . 44 (Th):   Hb–5/19, 356, 365/6,
                                                   378, 391, 396

History:
    ^{227}Th (RdAc) . . . . 44 (Th):   Hb–355
    ^{228}Th (RdTh) . . . . 44 (Th):   Hb–364/5
    ^{230}Th (Io) . . . . 44 (Th):   Hb–377/8
    ^{231}Th (UY). . . . . 44 (Th):   Hb–390/1
    ^{232}Th (Th) . . . . 44 (Th):   Hb–1/2
    ^{234}Th (UX$_1$) . . . . 44 (Th):   Hb–396

Ionization potentials . . . . 44 (Th):   Hb–178
                                      Erg. W. 8–23
Magnetic properties . . . . 44 (Th):   Hb–161/2
Mechanical properties . . . . 44 (Th):   Hb–133/45
Mineralogy . . . . 44 (Th):   Hb–29/32
Nuclear decay,
    alpha radioactivity . . . . 44 (Th): Hb–115/7, 360/1,
                                                   370/2, 384/6,
                                                   395, 405
                                      Erg. W. 7a–85/94
–, beta radioactivity. . . . . 44 (Th): Hb–360, 370,
                                                   393/4, 402/4
–, decay scheme . . . . 44 (Th):   Hb–112/5, 363, 373,
                                                   389, 395
                                      Erg. W. 7a–97

Hb = Main Series Volume; GD = Gmelin-Durrer;   Ab = Appended Volume to Main Series;
Eb = Supplement Volume to Main Series;   Erg. W. = New Supplement Series Volume

Th . . . . Thorium:
  Nuclear decay,
    gamma radioactivity . . . . 44 (Th):   Hb–117/8, 361/3,
                                                     372/4, 387/9,
                                                     395, 404/5
                                        Erg. W. 7a–80
  Nuclear fission,
    fission by charged particles . . . . . 44 (Th):   Hb–104/9
  –, – by gamma rays. . . . . 44 (Th):   Hb–109/12, 384
  –, – by neutrons. . . . . 44 (Th):   Hb–101/4, 359/60,
                                                     370, 383/4, 402
                                        Erg. W. 7a–140
  –, spontaneous fission . . . . 44 (Th):   Hb–118/9, 390
                                        Erg. W. 7a–134/8
  Nuclear properties . . . . 44 (Th):   Hb–93/112, 359, 370,
                                                     383, 393, 402
  Occurrence, extraterrestrial . . . . . . 44 (Th):   Hb–3/4
  –, terrestrial . . . . 44 (Th):   Hb–5/19, 356, 365/6,
                                                     378, 391, 396
    – in the atmosphere . . . . 44 (Th):   Hb–18
    – in the biosphere. . . . . 44 (Th):   Hb–19
    – in the hydrosphere . . . . 44 (Th):   Hb–17/8, 365/6,
                                                     378
    – in the lithosphere . . . . 44 (Th):   Hb–10/7, 356,
                                                     365/6, 378,
                                                     391, 396
  Optical properties. . . . . 44 (Th):   Hb–151/61
  Oxidation states. . . . . Erg. W. 8–8/11
  Position in the periodic system . . . . 44 (Th):   Hb–127/8
  Preparation . . . . 44 (Th):   Hb–83/8
  –, industrial, of thorium compounds . 44 (Th):   Hb–66/74
  – in pure state:
    ^{227}Th (RdAc) . . . . 44 (Th):   Hb–356/9
    ^{228}Th (RdTh) . . . . 44 (Th):   Hb–366/9
    ^{230}Th (Io) . . . . 44 (Th):   Hb–379/82
    ^{231}Th(UY) . . . . 44 (Th):   Hb–391/3
    ^{232}Th (Th) . . . . 44 (Th):   Hb–64/5
    ^{234}Th (UX$_1$) . . . . 44 (Th):   Hb–397/401
  – of pure thorium preparations . . . . 44 (Th):   Hb–33/66
  – of special forms. . . . . 44 (Th):   Hb–90/2
  Purification of thorium preparations . 44 (Th):   Hb–63/5, 72/3
  Purity testing . . . . 44 (Th):   Hb–63
  Radioactivity . . . . 44 (Th):   Hb–112/20
  Separation. . . . . 44 (Th):   Hb–40/63
  Sorption . . . . 44 (Th):   Hb–142/5, 406
  Spectra:
    Arc spectrum . . . . 44 (Th):   Hb–152, 154
    Spark spectra. . . . . 44 (Th):   Hb–152/6
    X-ray spectrum . . . . 44 (Th):   Hb–156/61

See the "Instructions for the user"
at the beginning of the volume

Th . . . . Thorium:
                 Thermal properties . . . . 44 (Th):   Hb–145/51
                 Toxicity . . . . 44 (Th):   Hb–196/7
                 Uses . . . . 44 (Th):   Hb–74/7
                                                  Erg. W. 8–224
                 Valence . . . . 44 (Th):   Hb–127/8

–. . . . . Thorium ions
                 see Th^+ ... Th^+, etc.

–. . . . . Thorium isotopes
                 see Th ... Thorium

–. . . . . Thorium systems:
                 Th–Ag . . . . 61 (Ag):   Hb/C–254/6
                 Th–Al . . . . 35 (Al):   Hb/A5–840/2
                 Th–B . . . . 44 (Th):   Hb–295
                 Th–Bi–Pb . . . . 47 (Pb):   Hb/C4–1470
                 Th–C . . . . 44 (Th):   Hb–297/8
                 Th–C–Pu . . . . Erg. W. 4–210/1
                 Th–Co . . . . 58 (Co):   Eb/A–406
                 Th–Cu . . . . 60 (Cu):   Hb/B3–1192
                 Th–Cu–Si . . . . 60 (Cu):   Hb/B3–1192
                 Th–H . . . . 44 (Th):   Hb–200
                 Th–La–Ru . . . . 63 (Ru):   Eb–192/3
                 Th–Mg . . . . 27 (Mg):   Hb/A4–613/4
                 Th–Na . . . . 44 (Th):   Hb–198
                 Th–Nb . . . . 49 (Nb):   Hb/B2–196/7
                 Th–Nb–Zr . . . . 49 (Nb):   Hb/B2–200
                 Th–Ni . . . . 57 (Ni):   Hb/B1–118/20
                 Th–O–Ta . . . . 50 (Ta):   Hb/B2–229
                 Th–Pb . . . . 47 (Pb):   Hb/C4–1468/70
                 Th–Ru . . . . 63 (Ru):   Eb–191/2
                 Th–S . . . . 44 (Th):   Hb–277/8
                 Th–Se . . . . 44 (Th):   Hb–294
                 Th–Sn . . . . 46 (Sn):   Hb/D–388
                 Th–Ta . . . . 50 (Ta):   Hb/B2–62
                 Th–V . . . . 48 (V):   Hb/B2–677/80
                 Th–W . . . . 54 (W):   Hb–103
Th^+ . . . . Th^+ . . . . 44 (Th):   Hb–123, 154/6, 178
                                                 Erg. W. 8–23
Th^{2+} . . . . Th^{2+} . . . . 44 (Th):   Hb–155, 178
                                                 Erg. W. 8–11/23
Th^{3+} . . . . Th^{3+} . . . . 44 (Th):   Hb–155, 178
                                                 Erg. W. 8–5, 11/23

Th^{4+} . . . . Th^{4+}:
                 Chemical reactions
                   with inorganic compounds . . . . 44 (Th):   Hb–184/9
                   – with organic compounds . . . . . . 44 (Th):   Hb–189/96
                 Electrochemical behavior . . . . . . 44 (Th):   Hb–178/80

Hb = Main Series Volume; GD = Gmelin-Durrer;   *Ab = Appended Volume to Main Series;*
Eb = Supplement Volume to Main Series;   *Erg. W. = New Supplement Series Volume*

Th^{4+} . . . . Th^{4+}:

        Ionization potential . . . . 44 (Th):   Hb–178
                                                 Erg. W. 8–23
        Properties of the ion . . . . 44 (Th):   Hb–121/5
                                                 Erg. W. 8–5
        Sorption . . . . 44 (Th):   Hb–143/5
        Susceptibility . . . . 44 (Th):   Hb–162

Th^{n+} . . . . Th^{n+} (n = 5 to 10) . . . . 44 (Th):   Hb–178
                                                 Erg. W. 8–23

ThA . . . . Thorium A = ^{216}Po
        see Po ... Polonium

ThB . . . . Thorium B = ^{212}Pb
        see Pb ... Lead isotopes

ThC . . . . Thorium C = ^{212}Bi
        see Bi ... Bismuth isotopes

ThC'. . . . . Thorium C' = ^{212}Po
        see Po ... Polonium

ThC'' . . . . Thorium C'' = ^{208}Tl
        see Tl ... Thallium isotopes

ThD . . . . Thorium D = ^{208}Pb
        see Pb ... Lead isotopes

ThEm . . . . Thorium emanation = ^{220}Rn
        see Rn ... Radon

ThX . . . . Thorium X = ^{224}Ra
        see Ra ... Radium

Ti . . . . Titanium:
        Absorption of X-rays . . . . 41 (Ti):   Hb–179
        Atom. . . . . 41 (Ti):   Hb–137/41
        Atomic nucleus . . . . 41 (Ti):   Hb–132/7
        Chemical reactions . . . . 41 (Ti):   Hb–189/99
        Colloidal forms . . . . 41 (Ti):   Hb–131
        Coordination number. . . . . 41 (Ti):   Hb–142
        Cosmochemistry . . . . 41 (Ti):   Hb–6/10
        Crystal chemistry . . . . 41 (Ti):   Hb–11/21
        Crystallographic properties . . . . . 41 (Ti):   Hb–142/7
        Detection in iron . . . . 59 (Fe):   Hb/FII2–166/7
        Determination in iron. . . . . 59 (Fe):   Hb/FII2–167/82,
                                                 305/7, 358
        Dressing and treatment of ores . . . . 41 (Ti):   Hb–81/117
        Economic deposits:
           Africa . . . . 41 (Ti):   Hb–55/9
           America. . . . . 41 (Ti):   Hb–59/69
           Asia . . . . 41 (Ti):   Hb–49/52
           Australia . . . . 41 (Ti):   Hb–52/5
           Europe . . . . 41 (Ti):   Hb–43/9
           Soviet Union . . . . 41 (Ti):   Hb–47/9

See the "Instructions for the user"
at the beginning of the volume

Ti . . . . Titanium:
             Economics. . . . . 41 (Ti):   Hb–69/70
             Electrical properties . . . . 41 (Ti):   Hb–180/6
             Electrochemical behavior . . . . 41 (Ti):   Hb–186/9
             Electronegativity . . . . 41 (Ti):   Hb–141/2
             Formation . . . . 41 (Ti):   Hb–129
             Geochemistry . . . . 41 (Ti):   Hb–11/43
             History. . . . . 41 (Ti):   Hb–1/5
             Magnetic properties . . . . 41 (Ti):   Hb–179/80
             Mechanical properties . . . . 41 (Ti):   Hb–147/56
             Mineralogy . . . . 41 (Ti):   Hb–70/9
             Nuclear properties . . . . 41 (Ti):   Hb–132/7
             Occurrence, extraterrestrial . . . . . 41 (Ti):   Hb–6/10
             –, terrestrial . . . . 41 (Ti):   Hb–11/43
             – in meteorites . . . . 41 (Ti):   Hb–7/10
             – in minerals . . . . 41 (Ti):   Hb–11/21
             – in the atmosphere . . . . 41 (Ti):   Hb–41
             – in the biosphere. . . . . 41 (Ti):   Hb–41/3
             – in the hydrosphere . . . . 41 (Ti):   Hb–38/41
             – in the lithosphere . . . . 41 (Ti):   Hb–21/38
             Optical properties. . . . . 41 (Ti):   Hb–160/79
             Physiological action . . . . 41 (Ti):   Hb–199
             Position in the periodic system . . . . 41 (Ti):   Hb–141
             Preparation . . . . 41 (Ti):   Hb–129
             –, industrial . . . . 41 (Ti):   Hb–117/27
             – in pure state. . . . . 41 (Ti):   Hb–129/30
             – of dense titanium . . . . 41 (Ti):   Hb–125/7
             – of special forms. . . . . 41 (Ti):   Hb–130/2
             Scattering of X-rays . . . . 41 (Ti):   Hb–179
             Sorption of hydrogen. . . . . 41 (Ti):   Hb–203/8
             Spectra:
                Arc spectrum . . . . 41 (Ti):   Hb–162/7
                Spark spectra. . . . . 41 (Ti):   Hb–167/70
                X-ray spectrum . . . . 41 (Ti):   Hb–172/9
              Thermal properties . . . . 41 (Ti):   Hb–156/60
             Toxicity . . . . 41 (Ti):   Hb–199
             Uses . . . . 41 (Ti):   Hb–127/8
             Valence . . . . 41 (Ti):   Hb–141

–. . . . . Titanium ions
             see Ti$^+$ ... Ti$^+$, etc.

–. . . . . Titanium isotopes:
             Abundance . . . . 41 (Ti):   Hb–132/3
             Enrichment and separation . . . . . . 41 (Ti):   Hb–132
             Nuclear properties . . . . 41 (Ti):   Hb–133/4

–. . . . . Titanium systems:
             Ti–Ag . . . . 61 (Ag):   Hb/C–237/41
             Ti–Ag–Cu . . . . 61 (Ag):   Hb/C–400/2

Hb = Main Series Volume; GD = Gmelin-Durrer;   Ab = Appended Volume to Main Series;
Eb = Supplement Volume to Main Series;   Erg. W. = New Supplement Series Volume

Ti . . . . Titanium systems:

Ti–Al. . . . . 35 (Al):   Hb/A5–827/8
Ti–Al–Nb . . . . 49 (Nb):   Hb/B2–118/9
Ti–Al–Sn . . . . 46 (Sn):   Hb/D–333/4
Ti–Al–Sn–Zr . . . . 46 (Sn):   Hb/D–382/3
Ti–Al–Ta . . . . 50 (Ta):   Hb/B2–40
Ti–Au . . . . 62 (Au):   Hb/3–833/4
Ti–B . . . . 41 (Ti):   Hb–354/5
Ti–B–Nb . . . . 49 (Nb):   Hb/B4–307/8
Ti–B–V . . . . 48 (V):   Hb/B2–659
Ti–Be . . . . 41 (Ti):   Hb–200
Ti–Be–V. . . . . 48 (V):   Hb/B2–661/2
Ti–C . . . . 41 (Ti):   Hb–357/8
Ti–C–Fe. . . . . 59 (Fe):   Hb/A9–1874/8
Ti–C–Hf–Ta . . . . 50 (Ta):   Hb/B2–341
Ti–C–N . . . . 41 (Ti):   Hb–369
Ti–C–Nb . . . . 49 (Nb):   Hb/B4–315/20
Ti–C–Nb–Ta . . . . 50 (Ta):   Hb/B2–346/7
Ti–C–Ni. . . . . 57 (Ni):   Hb/B3–1182
Ti–C–O . . . . 41 (Ti):   Hb–368
Ti–C–Ta. . . . . 50 (Ta):   Hb/B2–324/6
Ti–C–Ta–V . . . . 50 (Ta):   Hb/B2–343
Ti–C–V . . . . 48 (V):   Hb/B2–659
Ti–Ce–V. . . . . 48 (V):   Hb/B2–662
Ti–Co . . . . 58 (Co):   Hb/A1–193/4
                      Eb/A–400/4
Ti–Cr . . . . 52 (Cr):   Hb/A2–721/4
Ti–Cr–Ni . . . . 57 (Ni):   Hb/B1–223/5
Ti–Cu . . . . 60 (Cu):   Hb/B3–1189
Ti–Cu–Si . . . . 60 (Cu):   Hb/B3–1190
Ti–Fe . . . . 59 (Fe):   Hb/A9–1867/70
Ti–Fe–Si . . . . 59 (Fe):   Hb/A9–1888/93
Ti–Gd–V . . . . 48 (V):   Hb/B2–662
Ti–H. . . . . 41 (Ti):   Hb–203/8
Ti–H–Mn . . . . 56 (Mn):   Hb/C1–6
Ti–H–Nb . . . . 49 (Nb):   Hb/B4–196/7
Ti–H–V . . . . 48 (V):   Hb/B2–657/8
Ti–Hf–Nb . . . . 49 (Nb):   Hb/B2–189
Ti–Hf–Nb–Zr. . . . . 49 (Nb):   Hb/B2–194
Ti–Hf–Ta . . . . 50 (Ta):   Hb/B2–61
Ti–La–V. . . . . 48 (V):   Hb/B2–662
Ti–Mg. . . . . 27 (Mg):   Hb/A4–604
Ti–Mo–Ni. . . . . 57 (Ni):   Hb/B1–251
Ti–N . . . . 41 (Ti):   Hb–271/2
Ti–N–V . . . . 48 (V):   Hb/B2–658
Ti–Nb . . . . 49 (Nb):   Hb/B2–62/4
Ti–Nb–Ni–Ta. . . . . 57 (Ni):   Hb/B1–173
Ti–Nb–O . . . . 49 (Nb):   Hb/B3–178/9

See the "Instructions for the user"
at the beginning of the volume

Ti . . . . Titanium systems:
                   Ti–Nb–Ta . . . . 50 (Ta):   Hb/B2–104
                   Ti–Nb–V . . . . 49 (Nb):   Hb/B2–276/7
                   Ti–Nb–Zr . . . . 49 (Nb):   Hb/B2–174/6
                   Ti–Ni . . . . 57 (Ni):   Hb/B1–79/82
                   Ti–Ni–O . . . . 57 (Ni):   Hb/B1–97
                   Ti–Ni–V . . . . 57 (Ni):   Hb/B1–159
                   Ti–O . . . . 41 (Ti):   Hb–211/5
                   Ti–O–S . . . . 41 (Ti):   Hb–343
                   Ti–O–Ta . . . . 50 (Ta):   Hb/B2–221
                   Ti–O–Ta–Zr . . . . 50 (Ta):   Hb/B2–229
                   Ti–O–V . . . . 48 (V):   Hb/B2–658
                   Ti–Os . . . . 68 (Pt):   Hb/A6–721
                   Ti–P . . . . 41 (Ti):   Hb–378
                   Ti–Pb . . . . 47 (Pb):   Hb/C4–1377/8
                   Ti–Pb–Zr . . . . 47 (Pb):   Hb/C4–1423
                   Ti–Pd . . . . 68 (Pt):   Hb/A5–585
                   Ti–Ru . . . . 63 (Ru):   Eb–187/9
                                                        68 (Pt):   Hb/A5–538
                   Ti–S . . . . 41 (Ti):   Hb–337/9
                   Ti–Se . . . . 41 (Ti):   Hb–352
                   Ti–Si . . . . 41 (Ti):   Hb–375
                   Ti–Si–Ta . . . . 50 (Ta):   Hb/B2–39
                   Ti–Si–V . . . . 48 (V):   Hb/B2–659/61
                   Ti–Sn . . . . 46 (Sn):   Hb/D–325/7
                   Ti–Sn–V . . . . 48 (V):   Hb/B2–686/7
                   Ti–Sn–Zr . . . . 46 (Sn):   Hb/D–378
                   Ti–Ta . . . . 50 (Ta):   Hb/B2–24/5
                   Ti–Ta–V . . . . 50 (Ta):   Hb/B2–79
                   Ti–Ta–Zr . . . . 50 (Ta):   Hb/B2–51
                   Ti–Te . . . . 41 (Ti):   Hb–353/4
                   Ti–V . . . . 48 (V):   Hb/B2–602/4,
                                                                        648/57
                   Ti–V–Zr . . . . 48 (V):   Hb/B2–674/5
                   Ti–Zn . . . . 41 (Ti):   Hb–201
                   Ti–Zr . . . . 42 (Zr):   Hb–194
Ti⁺ . . . . Ti⁺ . . . . 41 (Ti):   Hb–137/9, 167/8,
                                                                        186
Ti²⁺ . . . . Ti²⁺:
                   Chemical reactions . . . . 41 (Ti):   Hb–468/9, 475
                   Ionization potential . . . . 41 (Ti):   Hb–186
                   Properties of the ion . . . . 41 (Ti):   Hb–137/40
                   Spectra . . . . 41 (Ti):   Hb–168/9
Ti³⁺ . . . . Ti³⁺:
                   Chemical reactions . . . . 41 (Ti):   Hb–469/71, 475/81
                   Ionization potential . . . . 41 (Ti):   Hb–186
                   Properties of the ion . . . . 41 (Ti):   Hb–137/40
                   Spectra . . . . 41 (Ti):   Hb–169/70

Hb = Main Series Volume; GD = Gmelin-Durrer;   *Ab = Appended Volume to Main Series;*
Eb = Supplement Volume to Main Series;   *Erg. W. = New Supplement Series Volume*

Ti^{4+} . . . . Ti^{4+}:
       Chemical reactions . . . . 41 (Ti):   Hb−471/5, 475/81
       Ionization potential . . . . 41 (Ti):   Hb−186
       Properties of the ion . . . . 41 (Ti):   Hb−137/40
       Spectra . . . . 41 (Ti):   Hb−170

Ti^{n+} . . . . Ti^{n+} (n = 5 to 12) . . . . 41 (Ti):   Hb−137, 170, 186

$TiZn_3$ . . . . $TiZn_3$ . . . . 41 (Ti):   Hb−201/2
−. . . . . $TiZn_3$ solid solutions:
       $(Ti,Nb)Zn_3$ . . . . 49 (Nb):   Hb/B2−118

Tl . . . . Thallium:
       Abundance . . . . 38 (Tl):   Hb/1−4/5
       Affinity. . . . . 38 (Tl):   Hb/1−43
       Atom. . . . . 38 (Tl):   Hb/1−36/42
       Atomic nucleus . . . . 38 (Tl):   Hb/1−30/6
       Behavior as electrode . . . . 38 (Tl):   Hb/1−128/9
       Cells . . . . 38 (Tl):   Hb/1−117/25
       Chemical reactions . . . . 38 (Tl):   Hb/1−132/7
       Cosmochemistry . . . . 38 (Tl):   Hb/1−2
       Crystallographic properties . . . . . . 38 (Tl):   Hb/1−44/8
       Detection . . . . 38 (Tl):   Hb/1−144/58
       Determination. . . . . 38 (Tl):   Hb/1−159/81, 185/6
       Diffusion. . . . . 38 (Tl):   Hb/1−54/5
                                46 (Sn):   Hb/D−279
                                  47 (Pb):   Hb/C4−1327/9
       Economics. . . . . 38 (Tl):   Hb/1−18/9
       Electrical properties . . . . 38 (Tl):   Hb/1−102/13
       Electrochemical behavior . . . . 38 (Tl):   Hb/1−114/31
       Formation . . . . 38 (Tl):   Hb/1−20/7
       Geochemistry . . . . 38 (Tl):   Hb/1−3/12
       History. . . . . 38 (Tl):   Hb/1−1
                                       Hb/3−497/526
       Magnetic properties . . . . 38 (Tl):   Hb/1−98/101
       Mechanical properties . . . . 38 (Tl):   Hb/1−49/56
       Mineralogy . . . . 38 (Tl):   Hb/1−14/8
       Molecule. . . . . 38 (Tl):   Hb/1−43
       Nuclear properties . . . . 38 (Tl):   Hb/1−30/6
       Occurrence, extraterrestrial . . . . . . 38 (Tl):   Hb/1−2
       −, terrestrial . . . . 38 (Tl):   Hb/1−3/14
       Optical properties. . . . . 38 (Tl):   Hb/1−72/98
       Overvoltage phenomena. . . . . . . 38 (Tl):   Hb/1−127/8
       Position in the electrochemical
         series . . . . 38 (Tl):   Hb/1−115
       − in the periodic system . . . . 38 (Tl):   Hb/1−43
       Potentials . . . . 38 (Tl):   Hb/1−114/7, 127,
                                        130/1
       Preparation . . . . 38 (Tl):   Hb/1−20/7
       − in pure state. . . . . 38 (Tl):   Hb/1−23, 25/6

See the "Instructions for the user"
at the beginning of the volume

Tl . . . . Thallium:
                 Separation, analytical . . . . 38 (Tl):   Hb/1–181/5
                 Sorption . . . . 38 (Tl):   Hb/1–55/6
                 Special forms . . . . 38 (Tl):   Hb/1–26/7
                 Spectra:
                    Arc spectrum . . . . 38 (Tl):   Hb/1–74/88
                    Band spectrum . . . . 38 (Tl):   Hb/1–94
                    Spark spectra . . . . 38 (Tl):   Hb/1–88/93
                    X-ray spectrum . . . . 38 (Tl):   Hb/1–94/8
                 Storage . . . . 38 (Tl):   Hb/1–26
                 Thermal properties . . . . 38 (Tl):   Hb/1–56/72
                 Toxicity . . . . 38 (Tl):   Hb/1–143
                 Uses . . . . 38 (Tl):   Hb/1–28/9
                 Valence . . . . 38 (Tl):   Hb/1–42/3

–. . . . . Thallium ions
                 see Tl$^+$ ... Tl$^+$, etc.

–. . . . . Thallium isotopes . . . . 38 (Tl):   Hb/1–30/1
                                                       Hb/3–497/526

–. . . . . Thallium systems:
                 Tl–Ag . . . . 61 (Ag):   Hb/C–205/6
                 Tl–Ag–Al . . . . 35 (Al):   Hb/A6–1087
                 Tl–Ag–Mg . . . . 27 (Mg):   Hb/A4–748/9
                 Tl–Ag–S . . . . 61 (Ag):   Hb/B4–225
                   Tl–Al . . . . 35 (Al):   Hb/A5–814/5
                   Tl–As . . . . 38 (Tl):   Hb/3–433
                   Tl–Au . . . . 62 (Au):   Hb/3–832/3
                   Tl–Au–Zn . . . . 62 (Au):   Hb/3–798
                   Tl–Bi . . . . 38 (Tl):   Hb/2–191/4
                   Tl–Bi–Cd . . . . 38 (Tl):   Hb/2–215/6
                   Tl–Bi–Pb . . . . 47 (Pb):   Hb/C4–1347
                   Tl–Ca . . . . 38 (Tl):   Hb/2–209/11
                   Tl–Cd . . . . 38 (Tl):   Hb/2–213/4
                   Tl–Cd–CdCl$_2$–TlCl . . . . 38 (Tl):   Hb/3–482/4
                   Tl–Cd–Mg . . . . 27 (Mg):   Hb/A4–582/4
                   Tl–Cd–Pb . . . . 47 (Pb):   Hb/C4–1347
                   Tl–Cd–Sn . . . . 46 (Sn):   Hb/D–291
                   Tl–Cu . . . . 60 (Cu):   Hb/B3–1183
                   Tl–Ga–Sn . . . . 46 (Sn):   Hb/D–291
                   Tl–Ge . . . . 45 (Ge):   Eb–480
                   Tl–Hg . . . . 34 (Hg):   Hb/A2–1146/8
                   Tl–In . . . . 37 (In):   Hb–59/60
                                          38 (Tl):   Hb/2–216/7
                 Tl–In–Pb . . . . 47 (Pb):   Hb/C4–1347
                   Tl–K . . . . 38 (Tl):   Hb/2–208/9
                   Tl–Li . . . . 38 (Tl):   Hb/2–203/5
                   Tl–Mg . . . . 27 (Mg):   Hb/A4–576/7
                   Tl–Mg–Zn . . . . 27 (Mg):   Hb/A4–581
                   Tl–Na . . . . 38 (Tl):   Hb/2–205/8

Tl . . . . Thallium systems:

Tl–Na–NaBr	38 (Tl):	Hb/3–443
Tl–Ni	57 (Ni):	Hb/B1–64/5
Tl–P	38 (Tl):	Hb/3–425/6
Tl–Pb	47 (Pb):	Hb/C4–1329/32
Tl–Pb–PbI$_2$–TlI	47 (Pb):	Hb/C4–1358
Tl–Pb–Pt	68 (Pt):	Hb/A6–767
Tl–Pb–Sn	47 (Pb):	Hb/C4–1530
Tl–Pt.	68 (Pt):	Hb/A6–755/7
Tl–S	38 (Tl):	Hb/3–339
Tl–Sb	38 (Tl):	Hb/2–187/91
Tl–Se	38 (Tl):	Hb/3–370/1
Tl–Si.	38 (Tl):	Hb/3–424
Tl–Sn	46 (Sn):	Hb/D–279/81
Tl–Sn–Zn	46 (Sn):	Hb/D–290
Tl–Te	38 (Tl):	Hb/3–376
Tl–Zn	38 (Tl):	Hb/2–212/3

Tl$^+$ . . . . Tl$^+$:

Chemical reactions	38 (Tl):	Hb/1–137/40
Diffusion in sodium halides	21 (Na):	Eb/6–291
Ionic entropy	38 (Tl):	Hb/1–71/2
Ionic mobility	38 (Tl):	Hb/1–126
Ionic refraction	38 (Tl):	Hb/1–73
Ionization	38 (Tl):	Hb/1–113
Properties of the ion	38 (Tl):	Hb/1–38/9, 41
Redox cells	38 (Tl):	Hb/1–130/1
Spectra:		
Crystal spectra	21 (Na):	Eb/6–333/4
Luminescence spectra	21 (Na):	Eb/6–348/9
Spark spectrum	38 (Tl):	Hb/1–88/91

Tl^{2+} . . . . Tl^{2+} . . . . 21 (Na): Eb/6–337
                                                                      38 (Tl): Hb/1–39, 41, 91/3, 113

Tl^{3+} . . . . Tl^{3+}:

Chemical reactions	38 (Tl):	Hb/1–140/3
Ionization	38 (Tl):	Hb/1–113
Properties of the ion	38 (Tl):	Hb/1–39/41
Redox cells	38 (Tl):	Hb/1–130/1
Spectra	38 (Tl):	Hb/1–93

Tl^{n+} . . . . Tl^{n+} (n = 4 to 21) . . . . 38 (Tl): Hb/1–39
Tl$_2^{2+}$ . . . . Tl$_2^{2+}$ . . . . 38 (Tl): Hb/1–43

Tm . . . . Thulium:

Abundance	39 (SE):	Hb/A1–34/5
Band model	39 (SE):	Hb/B3–213/5
Boiling point	39 (SE):	Hb/B3–105, 108
Charge carrier concentration.	39 (SE):	Hb/B3–228
Cosmochemistry	39 (SE):	Hb/A1–23/7
Crystal chemistry	39 (SE):	Hb/A1–28/32

*See the "Instructions for the user"
at the beginning of the volume*

Tm . . . . Thulium:

Crystal field parameters	39 (SE):	Hb/B3–218/9
Crystallographic properties	39 (SE):	Hb/B3–62/3, 66/7, 73, 77/80
Crystal structure	39 (SE):	Hb/B3–66/7, 73
Debye temperature	39 (SE):	Hb/B3–192
Density	39 (SE):	Hb/B3–85/7
Diffusion.	39 (SE):	Hb/B3–77/8
Economic deposits	39 (SE):	Hb/A1–110/22
Elasticity.	39 (SE):	Hb/B3–92/8
Electrical properties	39 (SE):	Hb/B3–209/83
Electrical resistance.	39 (SE):	Hb/B3–248/9, 262
Electron density.	39 (SE):	Hb/B3–217
Electron emission.	39 (SE):	Hb/B3–278/83
Enthalpy of fusion	39 (SE):	Hb/B3–120
– of sublimation.	39 (SE):	Hb/B3–110, 115
– of vaporization	39 (SE):	Hb/B3–110, 115
Exchange interaction	39 (SE):	Hb/B3–298/303
Geochemistry	39 (SE):	Hb/A1–28/54
Grüneisen constant.	39 (SE):	Hb/B3–84/5
Hall effect	39 (SE):	Hb/B3–271/2
History.	39 (SE):	Hb/A1–4/22
Hyperfine interactions	39 (SE):	Hb/B3–224
Isomer shift	39 (SE):	Hb/B3–217
Magnetic anisotropy	39 (SE):	Hb/B3–305/6
Magnetic form factor	39 (SE):	Hb/B3–315/6
Magnetic moment	39 (SE):	Hb/B3–334
Magnetic properties	39 (SE):	Hb/B3–298/336
Magnetic structure	39 (SE):	Hb/B3–307/9, 315
Magnetization.	39 (SE):	Hb/B3–317, 320/2
Mechanical properties	39 (SE):	Hb/B3–85/103, 193/5, 207/9
Melting point	39 (SE):	Hb/B3–118/9
Mineralogy	39 (SE):	Hb/A1–54/110
Molar volume	39 (SE):	Hb/B3–85/7
Occurrence, extraterrestrial	39 (SE):	Hb/A1–22/7
–, terrestrial	39 (SE):	Hb/A1–27/54
– in meteorites	39 (SE):	Hb/A1–25/7
– in minerals	39 (SE):	Hb/A1–35/53
– in the biosphere.	39 (SE):	Hb/A1–53/4
– in the lithosphere	39 (SE):	Hb/A1–27/54
Optical properties.	39 (SE):	Hb/B3–342
Polymorphism	39 (SE):	Hb/B3–62/3
Position in the periodic system	39 (SE):	Hb/A1–2/4
Spectra	49 (Nb):	Hb/B3–133
Structure	39 (SE):	Hb/B3–80
Surface tension	39 (SE):	Hb/B3–208/9
Susceptibility	39 (SE):	Hb/B3–334/6
Thermal conductivity	39 (SE):	Hb/B3–200/4

Hb = Main Series Volume; GD = Gmelin-Durrer;   *Ab = Appended Volume to Main Series;*
Eb = Supplement Volume to Main Series;   *Erg. W. = New Supplement Series Volume*

Tm . . . . Thulium:
          Thermal expansion . . . . 39 (SE):   Hb/B3–88/92
          Thermal properties . . . . 39 (SE):   Hb/B3–83/92, 103/20,
                                                 175/9, 192, 200/6
          Thermodynamic functions . . . . . . 39 (SE):   Hb/B3–175/9
          Thermoelectric power . . . . 39 (SE):   Hb/B3–277
          Valence . . . . 39 (SE):   Hb/A1–4
          Vapor pressure . . . . 39 (SE):   Hb/B3–103/5, 108

–. . . . . Thulium ions
          see Tm^{3+} ... Tm^{3+}

–. . . . . Thulium isotopes:
          Nuclear decay . . . . Erg. W. 7a–86
          Uses . . . . Erg. W. 8–207

–. . . . . Thulium systems:
          Tm–Ag . . . . 61 (Ag):   Hb/C–234/5
          Tm–H . . . . 39 (SE):   Hb/C1–2/3, 75
          Tm–Pb . . . . 47 (Pb):   Hb/C4–1371
Tm^{3+} . . . . Tm^{3+} . . . . 39 (SE):   Hb/A1–3
                                              Hb/B3–317
                                     49 (Nb):   Hb/B3–133

Tn . . . . Thoron = ^{220}Rn
          see Rn ... Radon

Tu . . . . Thulium
          see Tm ... Thulium

U . . . . Uranium:
          Atom . . . . 55 (U):   Hb–45/6, 52/4
                                      Erg. W. 8–1/34
          Atomic nucleus . . . . 55 (U):   Hb–44/52, 271/9
          Atomic radius . . . . 55 (U):   Hb–52
                                          Erg. W. 8–4/5
          Catalytic properties . . . . 55 (U):   Hb–66
          Chemical reactions . . . . 55 (U):   Hb–64/6
          Cosmochemistry . . . . 55 (U):   Hb–2
          Crystallographic properties . . . . . 55 (U):   Hb–54
          Detection . . . . 55 (U):   Hb–67/9
          – in iron . . . . 59 (Fe):   Hb/FII1–62/3
          – of uranium isotopes
            in soils and fallout . . . . Erg. W. 7b–353/8
          Determination . . . . 55 (U):   Hb–69/75
          – in iron . . . . 59 (Fe):   Hb/FII1–63/6
          – in special substances . . . . 55 (U):   Hb–80
          Dressing and treatment of ores . . . . 55 (U):   Hb–38/9
          Economic deposits . . . . 55 (U):   Hb–3/12
          Economics . . . . 55 (U):   Hb–37/8

See the "Instructions for the user"
at the beginning of the volume

U . . . . Uranium:

Electrical properties . . . . 55 (U):   Hb–63/4
Electrochemical behavior . . . . . . 55 (U):   Hb–64
Electron configuration . . . . 55 (U):   Hb–45/6
                                               Erg. W. 8–3/23
Formation . . . . 55 (U):   Hb–39/41, 271
Geochemistry . . . . 55 (U):   Hb–1/12
History . . . . 55 (U):   Hb–1, 272/9
Ionization potentials . . . . Erg. W. 8–23
Magnetic properties . . . . 55 (U):   Hb–62
Mechanical properties . . . . 55 (U):   Hb–54/5
Mineralogy . . . . 55 (U):   Hb–12/37
Nuclear decay,
    alpha radioactivity . . . . 55 (U):   Hb–46/50, 271/2
                                               Erg. W. 7a–82/96
–, decay scheme . . . . Erg. W. 7a–97, 102
Nuclear fission,
    fission by charged particles . . . . . 55 (U):   Hb–45
                                               Erg. W. 7a–151/2
–, – by neutrons . . . . Erg. W. 7a–140, 143/5
–, spontaneous fission . . . . Erg. W. 7a–134/8, 140
–, theoretical . . . . Erg. W. 7a–163/70
Nuclear properties . . . . 55 (U):   Hb–44/52, 271/9
Nuclear structure . . . . 55 (U):   Hb–44, 271
Occurrence, extraterrestrial . . . . . . 55 (U):   Hb–2
–, terrestrial . . . . 55 (U):   Hb–1/12
Optical properties . . . . 55 (U):   Hb–56/62
Oxidation states . . . . Erg. W. 8–8/11
Position in the periodic system . . . . 55 (U):   Hb–54
Preparation . . . . 55 (U):   Hb–39/43
–, industrial . . . . 55 (U):   Hb–38/42
– in pure state . . . . 55 (U):   Hb–42/3
Production in charged particle
    reactions . . . . Erg. W. 7b–145
Purification . . . . 55 (U):   Hb–51
Radioactivity . . . . 55 (U):   Hb–46/52
Recoil reactions . . . . Erg. W. 8–315/9
Safety measures . . . . Erg. W. 8–385/93
Separation, analytical . . . . 55 (U):   Hb–75/9
Special forms . . . . 55 (U):   Hb–43
Spectra:
    Absorption spectra . . . . 55 (U):   Hb–58
                                               Erg. W. 8–52, 63/4
    Emission spectra . . . . 55 (U):   Hb–56/8
    Mössbauer spectra . . . . Erg. W. 8–89/90
    X-ray spectrum . . . . 55 (U):   Hb–58/61
Thermal properties . . . . 55 (U):   Hb–55
Uses . . . . 55 (U):   Hb–43/4
                                               Erg. W. 8–206

Hb = Main Series Volume; GD = Gmelin-Durrer;   Ab = Appended Volume to Main Series;
Eb = Supplement Volume to Main Series;   Erg. W. = New Supplement Series Volume

U . . . . Uranium ions
                      see U$^+$ ... U$^+$, etc.

−. . . . . Uranium isotopes
                      see U ... Uranium

−. . . . . Uranium systems:
                      U−Ag . . . . 61 (Ag):   Hb/C−315/6
                      U−Bi−Ni. . . . . 57 (Ni):   Hb/B1−277
                      U−C−Ni . . . . 57 (Ni):   Hb/B1−278
                      U−C−Pu . . . . Erg. W. 4−211/3
                      U−Co . . . . 58 (Co):   Eb/A−452/5
                      U−Co−Ru . . . . 63 (Ru):   Eb−224
                      U−Cr−Ni . . . . 57 (Ni):   Hb/B1−277
                      U−Cu . . . . 60 (Cu):   Hb/B3−1240
                      U−Cu−O . . . . 60 (Cu):   Hb/B3−1240
                      U−Mn−Ni . . . . 57 (Ni):   Hb/B1−314
                      U−Nb . . . . 49 (Nb):   Hb/A−173/4
                      U−Ni. . . . . 57 (Ni):   Hb/B1−272/3
                      U−Ni−Si. . . . . 57 (Ni):   Hb/B1−277
                      U−O . . . . 55 (U):   Hb−82
                      U−O−Pu. . . . . Erg. W. 4−59/60
                      U−Ru . . . . 63 (Ru):   Eb−213/6
U$^+$ . . . . U$^+$ . . . . 55 (U):   Hb−45
                                                           Erg. W. 8−23
U^{2+} . . . . U^{2+} . . . . 55 (U):   Hb−45
                                                           Erg. W. 8−11/23
U^{3+} . . . . U^{3+}:
                      Crystal field effects . . . . Erg. W. 8−24/5
                      Ionic susceptibility . . . . 55 (U):   Hb−62
                      Ionization potential . . . . Erg. W. 8−23
                      Magnetic moment . . . . Erg. W. 8−26/7
                      Properties of the ion . . . . 55 (U):   Hb−45
                                                           Erg. W. 8−5, 11/22
                      Spectra . . . . Erg. W. 8−52
U^{4+} . . . . U^{4+}:
                      Crystal field effects . . . . Erg. W. 8−24/5
                      Electrochemical behavior . . . . 55 (U):   Hb−255/6
                      Ionic susceptibility . . . . 55 (U):   Hb−62, 136, 254/5
                      Ionization potential . . . . Erg. W. 8−23
                      Properties of the ion . . . . 55 (U):   Hb−45, 52
                                                           Erg. W. 8−5, 11/22
                      Spectra . . . . 55 (U):   Hb−245/6
                                                           Erg. W. 8−63/4
U^{5+} . . . . U^{5+} . . . . 55 (U):   Hb−45, 246
                                                           Erg. W. 8−5, 11/25
U^{6+} . . . . U^{6+} . . . . 21 (Na):   Eb−6−354/5
                                                           55 (U):   Hb−254
                                                           Erg. W. 8−5, 23
                      see also O$_2$U^{2+} ... UO$_2$$^{2+}$

*See the "Instructions for the user"
at the beginning of the volume*

U^{n+} . . . . U^{n+} (n = 7 to 10) . . . . Erg. W. 8–23

UI . . . . Uranium I = ^{238}U
                see U ... Uranium

UII . . . . Uranium II = ^{234}U
                see U ... Uranium

UX_1 . . . . Uranium X_1 = ^{234}Th
                see Th ... Thorium

UX_2 . . . . Uranium X_2 = ^{234}Pa
                see Pa ... Protactinium

UY . . . . Uranium Y = ^{231}Th
                see Th ... Thorium

UZ . . . . Uranium Z = ^{234}Pa
                see Pa ... Protactinium

V . . . . Vanadium:

            Absorption of X- and
               gamma rays . . . . 48 (V):   Hb/A2–384/9
            Abundance . . . . 48 (V):   Hb/A1–11/2, 39
            Atom . . . . 48 (V):   Hb/A2–332/48
            Atomic nucleus . . . . 48 (V):   Hb/A2–321/9
            Atomic radius . . . . 48 (V):   Hb/A2–343/6
            Atomic scattering factors . . . . 48 (V):   Hb/A2–340/3
            Atomic weight . . . . 48 (V):   Hb/A2–323
            Behavior as electrode . . . . 48 (V):   Hb/A2–489/93
            Boiling point . . . . 48 (V):   Hb/A2–437
            Bond properties . . . . 48 (V):   Hb/A2–389/92
            Cells . . . . 48 (V):   Hb/A2–471/4
            Characteristic electron
               energy losses . . . . 48 (V):   Hb/A2–451
            Chemical constant . . . . 48 (V):   Hb/A2–449
            Chemical reactions . . . . 48 (V):   Hb/A2–514/24
            – by action of electricity . . . . 48 (V):   Hb/A2–514
            – on heating . . . . 48 (V):   Hb/A2–514
            – with compounds . . . . 48 (V):   Hb/A2–519/24
            – with elements . . . . 48 (V):   Hb/A2–514/9
            Cleavage . . . . 48 (V):   Hb/A2–396
            Colloidal forms . . . . 48 (V):   Hb/A1–317
            Compressibility . . . . 48 (V):   Hb/A2–419
            Coordination number . . . . 48 (V):   Hb/A2–321
            Cosmochemistry . . . . 48 (V):   Hb/A1–11/8
            Creep properties . . . . 48 (V):   Hb/A2–436
            Crystal chemistry . . . . 48 (V):   Hb/A1–18/39
            Crystal form . . . . 48 (V):   Hb/A2–396
            Crystal formation . . . . 48 (V):   Hb/A2–397
            Crystallographic properties . . . . . . 48 (V):   Hb/A2–394/414
            Debye temperature . . . . 48 (V):   Hb/A2–446/7
            Deformation . . . . 48 (V):   Hb/A2–408/9

Hb = Main Series Volume; GD = Gmelin-Durrer;   *Ab = Appended Volume to Main Series;*
Eb = Supplement Volume to Main Series;   *Erg. W. = New Supplement Series Volume*

V . . . . Vanadium:
　　　　　　　　Density . . . . 48 (V):　　Hb/A2–415
　　　　　　　　Detection . . . . 48 (V):　　Hb/A2–640/52
　　　　　　　　– in iron . . . . 59 (Fe):　　Hb/FⅡ1–135/8
　　　　　　　　　　　　　　　　　　　　　　　　　　Hb/FⅡ2–336/7, 340
　　　　　　　　– in special substances . . . . 48 (V):　　Hb/A2–652/7
　　　　　　　　Determination. . . . . 48 (V):　　Hb/A2–658/76
　　　　　　　　– in iron . . . . 59 (Fe):　　Hb/FⅡ1–138/64
　　　　　　　　　　　　　　　　　　　　　　　　　　Hb/FⅡ2–302/5,
　　　　　　　　　　　　　　　　　　　　　　　　　　　　336/7, 357/8
　　　　　　　　– in presence of other ions . . . . 48 (V):　　Hb/A2–676/92
　　　　　　　　– in special substances . . . . 48 (V):　　Hb/A2–692/724
　　　　　　　　Diadochy . . . . 48 (V):　　Hb/A1–25/9
　　　　　　　　Diffusion. . . . . 48 (V):　　Hb/A2–407/8
　　　　　　　　　　　　　　　　　　　　　　　　　　Hb/B1–325
　　　　　　　　　　　　　　　　　　　　　　　　　　Hb/B2–602
　　　　　　　　Disorder . . . . 48 (V):　　Hb/A2–406/7
　　　　　　　　Dressing and treatment
　　　　　　　　　of raw materials:
　　　　　　　　　Chlorination . . . . 48 (V):　　Hb/A1–216/21
　　　　　　　　　Enrichment as by-product . . . . . . 48 (V):　　Hb/A1–196/202
　　　　　　　　　– of ores . . . . 48 (V):　　Hb/A1–187/96
　　　　　　　　　Heating under
　　　　　　　　　　reducing conditions . . . . 48 (V):　　Hb/A1–216
　　　　　　　　　Melt dressing. . . . . 48 (V):　　Hb/A1–223/6
　　　　　　　　　Preparation of solution . . . . 48 (V):　　Hb/A1–226/42
　　　　　　　　　Roasting . . . . 48 (V):　　Hb/A1–203/16
　　　　　　　　　Sintering methods . . . . 48 (V):　　Hb/A1–221/2
　　　　　　　　　Treatment of solution . . . . 48 (V):　　Hb/A1–242/80

　　　　　　　　Economic deposits:
　　　　　　　　　Africa . . . . 48 (V):　　Hb/A1–129/34
　　　　　　　　　America . . . . 48 (V):　　Hb/A1–134/46
　　　　　　　　　Asia . . . . 48 (V):　　Hb/A1–127/9
　　　　　　　　　Australia and Oceania . . . . 48 (V):　　Hb/A1–129
　　　　　　　　　Europe . . . . 48 (V):　　Hb/A1–125/7
　　　　　　　　　Soviet Union . . . . 48 (V):　　Hb/A1–126/8

　　　　　　　　Economics. . . . . 48 (V):　　Hb/A1–123/5
　　　　　　　　Elasticity. . . . . 48 (V):　　Hb/A2–417/9
　　　　　　　　Electrical conductivity . . . . 48 (V):　　Hb/A2–454/6
　　　　　　　　Electrical properties . . . . 48 (V):　　Hb/A2–450/63
　　　　　　　　Electrochemical behavior . . . . 48 (V):　　Hb/A2–464/514
　　　　　　　　Electrode sputtering . . . . 48 (V):　　Hb/A2–462/3
　　　　　　　　Electrolytic deposition . . . . 48 (V):　　Hb/A1–302/6
　　　　　　　　　　　　　　　　　　　　　　　　　　Hb/A2–493/8
　　　　　　　　Electrolytic polishing . . . . 48 (V):　　Hb/A2–406, 492
　　　　　　　　Electron affinity . . . . 48 (V):　　Hb/A2–393
　　　　　　　　Electron configuration . . . . 48 (V):　　Hb/A2–332
　　　　　　　　Electronegativity . . . . 48 (V):　　Hb/A2–392

See the "Instructions for the user"
at the beginning of the volume

V . . . . Vanadium:
            Electron emission. . . . . 48 (V):   Hb/A2−461/2
            Electron energies . . . . 48 (V):   Hb/A2−332/4
            Energy band structure . . . . 48 (V):   Hb/A2−450/1
            Epitaxy. . . . . 48 (V):   Hb/A2−403
            Equation of state . . . . 48 (V):   Hb/A2−414/5
            Expansion coefficient. . . . . 48 (V):   Hb/A2−416/7
            Fermi surface . . . . 48 (V):   Hb/A2−451
            Formation . . . . 48 (V):   Hb/A2−321
            Galvanomagnetic effects. . . . . . . 48 (V):   Hb/A2−460/1
            Geochemical character . . . . 48 (V):   Hb/A1−39
            Geochemical cycle . . . . 48 (V):   Hb/A1−122/3
            Geochemistry . . . . 48 (V):   Hb/A1−18/123
            Hall effect . . . . 48 (V):   Hb/A2−461
            Hardness . . . . 48 (V):   Hb/A2−422/8
            Heat of fusion . . . . 48 (V):   Hb/A2−439
             – of sublimation. . . . . 48 (V):   Hb/A2−437/8
             – of vaporization . . . . 48 (V):   Hb/A2−437/8
            History. . . . . 48 (V):   Hb/A1−1/10
            Increments . . . . 48 (V):   Hb/A2−394
            Interaction with electrons . . . . . . 48 (V):   Hb/A2−392/3
            Internal friction . . . . 48 (V):   Hb/A2−420/1
            Ionization . . . . 48 (V):   Hb/A2−392/3
            Isomorphism . . . . 48 (V):   Hb/A2−402
            Lattice defects. . . . . 48 (V):   Hb/A2−406/7
            Lattice energy. . . . . 48 (V):   Hb/A2−412/4
            Lattice structure. . . . . 48 (V):   Hb/A2−397/402
            Lattice vibrations . . . . 48 (V):   Hb/A2−447/9
            Magnetic moment . . . . 48 (V):   Hb/A2−334/6
            Magnetic properties . . . . 48 (V):   Hb/A2−450/61
            Magnetostriction . . . . 48 (V):   Hb/A2−454
            Mechanical properties . . . . 48 (V):   Hb/A2−414/37
            Melting point . . . . 48 (V):   Hb/A2−438/9
            Mesonic atom. . . . . 48 (V):   Hb/A2−383/4
            Mineralogy . . . . 48 (V):   Hb/A1−36/9, 146/87
            Modifications . . . . 48 (V):   Hb/A2−395/6
            Nuclear magnetic resonance
              and relaxation . . . . 48 (V):   Hb/A2−451/3
            Nuclear properties . . . . 48 (V):   Hb/A2−321/9
            Occurrence, extraterrestrial . . . . . 48 (V):   Hb/A1−11/8
            –, terrestrial . . . . 48 (V):   Hb/A1−18/123
            – in meteorites . . . . 48 (V):   Hb/A1−15/7
            – in minerals . . . . 48 (V):   Hb/A1−18/25, 30/5
            – in soils. . . . . 48 (V):   Hb/A1−97/104
            – in tektites and impactites. . . . . . 48 (V):   Hb/A1−17/8
            – in the atmosphere . . . . 48 (V):   Hb/A1−114
            – in the biosphere. . . . . 48 (V):   Hb/A1−105, 116/22
            – in the hydrosphere . . . . 48 (V):   Hb/A1−110/5

Hb = Main Series Volume; GD = Gmelin-Durrer;   Ab = Appended Volume to Main Series;
Eb = Supplement Volume to Main Series;   Erg. W. = New Supplement Series Volume

V . . . . Vanadium:

Occurrence in the lithosphäre . . . . . 48 (V):   Hb/A1–40/110
– in the magmatic cycle . . . . 48 (V):   Hb/A1–40/52
– in the metamorphic cycle . . . . . . 48 (V):   Hb/A1–104/10
– in the sedimentary cycle . . . . . . 48 (V):   Hb/A1–52/104
Optical properties. . . . . 48 (V):   Hb/A2–463/4
Overvoltage phenomena. . . . . 48 (V):   Hb/A2–489
Paramagnetic resonance absorption . 48 (V):   Hb/A2–336/40
Polarography . . . . 48 (V):   Hb/A2–474/87
Polymorphism . . . . 48 (V):   Hb/A2–395/6
Position in the electrochemical
   series . . . . 48 (V):   Hb/A2–469
– in the periodic system . . . . 48 (V):   Hb/A2–321
Potentials . . . . 48 (V):   Hb/A2–470/1, 502/6
Preparation . . . . 48 (V):   Hb/A2–321
–, industrial . . . . 48 (V):   Hb/A1–187/320
        see also Dressing

– in pure state. . . . . 48 (V):   Hb/A1–311/5
– of metallic vanadium . . . . 48 (V):   Hb/A1–285/320
– of special forms. . . . . 48 (V):   Hb/A1–315/7
Purification . . . . 48 (V):   Hb/A1–306/15
Purity testing . . . . 48 (V):   Hb/A1–317/20
Radiation of the mesonic atom . . . . 48 (V):   Hb/A2–383/4
Recrystallization . . . . 48 (V):   Hb/A2–409/12
Redox potentials and cells . . . . . . 48 (V):   Hb/A2–498/506
Scattering of electrons . . . . 48 (V):   Hb/A2–393/4
– of X- and gamma rays . . . . 48 (V):   Hb/A2–384/9
Separation, analytical . . . . 48 (V):   Hb/A2–676/92
Sorption of hydrogen. . . . . 48 (V):   Hb/B1–1/5
Sound absorption. . . . . 48 (V):   Hb/A2–420/1
Sound velocity . . . . 48 (V):   Hb/A2–419/20
Spectra:
   Arc spectrum . . . . 48 (V):   Hb/A2–352/4
   Excitation. . . . . 48 (V):   Hb/A2–348/52
   Spark spectra. . . . . 48 (V):   Hb/A2–354/71
   Terms . . . . 48 (V):   Hb/A2–332, 348/50
   X-ray spectrum. . . . . 48 (V):   Hb/A2–371/83

Standard potential . . . . 48 (V):   Hb/A2–464/9
Strength. . . . . 48 (V):   Hb/A2–428/36
Structure . . . . 48 (V):   Hb/A2–403/6
– of thin films . . . . 48 (V):   Hb/A2–402/3
Superconductivity . . . . 48 (V):   Hb/A2–457/60
Superlattice . . . . 48 (V):   Hb/A2–403
Surface tension . . . . 48 (V):   Hb/A2–437
Susceptibility . . . . 48 (V):   Hb/A2–453/4
Thermal conductivity. . . . . 48 (V):   Hb/A2–449/50
Thermal expansion . . . . 48 (V):   Hb/A2–416/7
Thermal properties . . . . 48 (V):   Hb/A2–414/50

See the "Instructions for the user"
at the beginning of the volume

V . . . . Vanadium:

    Thermal radiation. . . . . 48 (V):   Hb/A2–464

    Thermodynamic functions . . . . . . 48 (V):   Hb/A2–439/46

    Thermoelectric effects . . . . 48 (V):   Hb/A2–461

    Toxicity . . . . 48 (V):   Hb/A2–725/8

    Twinning . . . . 48 (V):   Hb/A2–396/7

    Valence . . . . 48 (V):   Hb/A2–321

    Vapor pressure . . . . 48 (V):   Hb/A2–437

    Wave functions . . . . 48 (V):   Hb/A2–332/4

    Workability . . . . 48 (V):   Hb/A2–421/2

–. . . . . Vanadium ions
    see V^+ ... V^+, etc.

–. . . . . Vanadium isotopes:

    Abundance . . . . 48 (V):   Hb/A1–39

        Hb/A2–322

    Detection and determination. . . . . 48 (V):   Hb/A2–724/5

    Nuclear decay. . . . . 48 (V):   Hb/A2–327/9

    Nuclear properties . . . . 48 (V):   Hb/A2–321/6

    Preparation . . . . 48 (V):   Hb/A2–329/32

    Separation. . . . . 48 (V):   Hb/A2–332

–. . . . . Vanadium systems:

    V–Al . . . . 35 (Al):   Hb/A5–864

    V–Al–C–Ta . . . . 50 (Ta):   Hb/B2–343

    V–Al–Nb . . . . 49 (Nb):   Hb/B2–272

    V–Al–Ta . . . . 50 (Ta):   Hb/B2–78/9

    V–As. . . . . 48 (V):   Hb/B1–364

    V–As–C . . . . 48 (V):   Hb/B1–366/7

    V–B . . . . 48 (V):   Hb/B1–321

    V–B–C . . . . 48 (V):   Hb/B1–352

    V–B–Nb . . . . 49 (Nb):   Hb/B4–312

    V–B–Si . . . . 48 (V):   Hb/B1–357

    V–B–Ta . . . . 50 (Ta):   Hb/B2–322

    V–B–Ti . . . . 48 (V):   Hb/B2–659

    V–Be. . . . . 48 (V):   Hb/B2–578/9

    V–Be–Ti. . . . . 48 (V):   Hb/B2–661/2

    V–Bi . . . . 48 (V):   Hb/B2–577

    V–C . . . . 48 (V):   Hb/B1–325/6

    V–C–Fe–Si . . . . 59 (Fe):   Hb/A9–1936

    V–C–Ga. . . . . 48 (V):   Hb/B2–592

    V–C–Ge–Si. . . . . 48 (V):   Hb/B2–684

    V–C–H . . . . 48 (V):   Hb/B1–335

    V–C–Hf . . . . 48 (V):   Hb/B2–677

    V–C–N . . . . 48 (V):   Hb/B1–350

    V–C–Nb. . . . . 49 (Nb):   Hb/B4–330/1

    V–C–Nb–Si. . . . . 49 (Nb):   Hb/B2–271

    V–C–O . . . . 48 (V):   Hb/B1–335/6

    V–C–Ta . . . . 50 (Ta):   Hb/B2–341/3

Hb = Main Series Volume; GD = Gmelin-Durrer;   *Ab = Appended Volume to Main Series;*
Eb = Supplement Volume to Main Series;   *Erg. W. = New Supplement Series Volume*

V . . . . Vanadium systems:

V–C–Ta–Ti	50 (Ta):	Hb/B2–343
V–C–Ta–Zr	50 (Ta):	Hb/B2–343/4
V–C–Ti	48 (V):	Hb/B2–659
V–C–Zr	48 (V):	Hb/B2–674
V–Cd	48 (V):	Hb/B2–581
V–Ce.	48 (V):	Hb/B2–598
V–Ce–Ti.	48 (V):	Hb/B2–662
V–Co	58 (Co):	Eb/A–416/7
V–Co–O.	58 (Co):	Eb/A–873
V–Cr.	52 (Cr):	Hb/A2–727/8
V–Cu	60 (Cu):	Hb/B3–1209
V–Dy	48 (V):	Hb/B2–600/1
V–Er.	48 (V):	Hb/B2–599/600
V–Fe.	59 (Fe):	Hb/A9–1920/2
V–Fe–FeS–VS	59 (Fe):	Hb/A9–1924/5
V–Fe–Fe$_3$C–V$_4$C$_3$	59 (Fe):	Hb/A9–1926/31
V–Ga	48 (V):	Hb/B2–581/3
V–Ga–N.	48 (V):	Hb/B2–592
V–Ga–O.	48 (V):	Hb/B2–591/2
V–Ga–O–Si	48 (V):	Hb/B2–592
V–Ga–Si	48 (V):	Hb/B2–592
V–Gd	48 (V):	Hb/B2–600/1
V–Gd–Ti	48 (V):	Hb/B2–662
V–Ge	48 (V):	Hb/B2–680/1
V–Ge–Si	48 (V):	Hb/B2–683/4
V–H	48 (V):	Hb/B1–1/7
V–H–Ti	48 (V):	Hb/B2–657/8
V–H–Zr	48 (V):	Hb/B2–673
V–Hf.	48 (V):	Hb/B2–675/7
V–Hf–Ta	50 (Ta):	Hb/B2–84
V–Hg	48 (V):	Hb/B2–581
V–La.	48 (V):	Hb/B2–596/7
V–La–Ti.	48 (V):	Hb/B2–662
V–Lu.	48 (V):	Hb/B2–600/1
V–Mo	48 (V):	Hb/A2–402
V–N	48 (V):	Hb/B1–167/8
V–N–Ta.	50 (Ta):	Hb/B2–267
V–N–Ti	48 (V):	Hb/B2–658
V–N–Zn.	48 (V):	Hb/B2–581
V–Nb	49 (Nb):	Hb/B2–243/4
V–Nb–Si	49 (Nb):	Hb/B2–268
V–Nb–Ta	50 (Ta):	Hb/B2–118
V–Nb–Ti	49 (Nb):	Hb/B2–276/7
V–Nb–Zr	49 (Nb):	Hb/B2–295/6
V–Nd	48 (V):	Hb/B2–600
V–Ni.	57 (Ni):	Hb/B1–153/5
V–Ni–O	57 (Ni):	Hb/B1–159
		Hb/B3–1198

See the "Instructions for the user"
at the beginning of the volume

V . . . . Vanadium systems:
                V–Ni–Si. . . . . 57 (Ni):   Hb/B1–159
                V–Ni–Ta . . . . 57 (Ni):   Hb/B1–172
                V–Ni–Ti. . . . . 57 (Ni):   Hb/B1–159
                V–Ni–Zr. . . . . 57 (Ni):   Hb/B1–159
                V–O . . . . 48 (V):   Hb/B1–11/2, 18/9
                V–O–Ta. . . . . 50 (Ta):   Hb/B2–244
                V–O–Ti . . . . 48 (V):   Hb/B2–658
                V–P . . . . 48 (V):   Hb/B1–358
                V–Pb. . . . . 48 (V):   Hb/B2–687
                V–Pr. . . . . 48 (V):   Hb/B2–600
                V–Rb . . . . 48 (V):   Hb/B2–578
                V–Ru . . . . 63 (Ru):   Eb–194/5
                V–S . . . . 48 (V):   Hb/B1–267/8
                V–Sb . . . . 48 (V):   Hb/B2–575
                V–Sc. . . . . 48 (V):   Hb/B2–592/3
                V–Se. . . . . 48 (V):   Hb/B1–308/9
                V–Si . . . . 48 (V):   Hb/B1–352/4
                V–Si–Ta . . . . 50 (Ta):   Hb/B2–77/8
                V–Si–Ti . . . . 48 (V):   Hb/B2–659
                V–Sm . . . . 48 (V):   Hb/B2–600
                V–Sn . . . . 48 (V):   Hb/B2–684
                V–Sn–Ti . . . . 48 (V):   Hb/B2–686
                V–Ta. . . . . 48 (V):   Hb/A2–402
                                                   50 (Ta):   Hb/B2–68/9
                V–Ta–Ti. . . . . 50 (Ta):   Hb/B2–79
                V–Ta–Zn . . . . 50 (Ta):   Hb/B2–83
                V–Te. . . . . 48 (V):   Hb/B1–315/6
                V–Th. . . . . 48 (V):   Hb/B2–677/8
                V–Ti . . . . 48 (V):   Hb/B2–602/4, 648/57
                V–Ti–Zr. . . . . 48 (V):   Hb/B2–674/5
                V–VC–VN. . . . . 48 (V):   Hb/B1–350
                V–VN . . . . 48 (V):   Hb/B1–176
                V–VO . . . . 48 (V):   Hb/B1–12/3
                V–VSe . . . . 48 (V):   Hb/B1–308
                V–Y . . . . 48 (V):   Hb/B2–593
                V–Yb . . . . 48 (V):   Hb/B2–600
                V–Zn. . . . . 48 (V):   Hb/B2–579/80
                V–Zr . . . . 48 (V):   Hb/B2–662/3
V^+ . . . . V^+:
                Compton scattering. . . . . 48 (V):   Hb/A2–393
                Form factors. . . . . 48 (V):   Hb/A2–341/3
                Ionization potential . . . . 48 (V):   Hb/A2–392/3
                Properties of the ion . . . . 48 (V):   Hb/A2–332/4, 346/8,
                                                                          389
                Spectra . . . . 48 (V):   Hb/A2–354/7
V^{2+} . . . . V^{2+}:
                Chemical reactions . . . . 48 (V):   Hb/A2–524/7

Hb = Main Series Volume; GD = Gmelin-Durrer;   Ab = Appended Volume to Main Series;
Eb = Supplement Volume to Main Series;   Erg. W. = New Supplement Series Volume

V^{2+} . . . . V^{2+}:

Compton scattering.	48 (V):	Hb/A2–393
Detection	48 (V):	Hb/A2–646
Determination.	48 (V):	Hb/A2–677
Electrochemical behavior	48 (V):	Hb/A2–464/7, 476,
		498/504, 510/3
Form factors.	48 (V):	Hb/A2–341/3
Increments	48 (V):	Hb/A2–394
Ionization potential	48 (V):	Hb/A2–392/3
Paramagnetic resonance	48 (V):	Hb/A2–336/8
Properties of the ion	48 (V):	Hb/A2–332/5, 346/8,
		389
Spectra	21 (Na):	Eb/6–336
	48 (V):	Hb/A2–357/61
Transition reactions	48 (V):	Hb/A2–550/640

–. . . . . $^{50}V^{2+}$	48 (V):	Hb/A2–337
–. . . . . $^{51}V^{2+}$	48 (V):	Hb/A2–337
–. . . . . $[V(H_2O)_6]^{2+}$	48 (V):	Hb/A2–337,
		359/60, 525

V^{3+} . . . . V^{3+}:

see also OV^+ ... VO^+

Chemical reactions	48 (V):	Hb/A2–527/33
Compton scattering.	48 (V):	Hb/A2–393
Detection	48 (V):	Hb/A2–646
Determination.	48 (V):	Hb/A2–676/7
Electrochemical behavior	48 (V):	Hb/A2–464/9, 475/6,
		498/505, 510/3
Form factors.	48 (V):	Hb/A2–342/3
Increments	48 (V):	Hb/A2–394
Ionization potential	48 (V):	Hb/A2–392/3
Paramagnetic resonance	48 (V):	Hb/A2–337/9
Properties of the ion	48 (V):	Hb/A2–332/5, 346/8
Spectra	48 (V):	Hb/A2–361/7
Transition reactions	48 (V):	Hb/A2–550/640

–. . . . . $[V(H_2O)_6]^{3+}$	48 (V):	Hb/A2–339, 363,
		365/6, 527/8

V^{4+} . . . . V^{4+}:

see also OV^{2+} ... VO^{2+}

Form factors.	48 (V):	Hb/A2–342/3
Increments	48 (V):	Hb/A2–394
Ionization potential	48 (V):	Hb/A2–392/3
Paramagnetic resonance	48 (V):	Hb/A2–339/40
Properties of the ion	48 (V):	Hb/A2–332/6, 346/8
Spectra	48 (V):	Hb/A2–367/8

V^{5+} . . . . V^{5+}:

see also OV^{3+} ... VO^{3+},
O_2V^+ ... VO_2^+

See the "Instructions for the user"
at the beginning of the volume

V^{5+} . . . . V^{5+}:

Form factors. . . . . 48 (V):   Hb/A2–343

Increments . . . . 48 (V):   Hb/A2–394

Ionization potential . . . . 48 (V):   Hb/A2–392/3

Properties of the ion . . . . 48 (V):   Hb/A2–332/4, 346/8

Spectra . . . . 48 (V):   Hb/A2–368/9

V^{n+} . . . . V^{n+} (n = 6 to 20) . . . . 48 (V):   Hb/A2–332, 334,
                                                                                            369/71, 392/3

V^{-} . . . . V^{-} . . . . 48 (V):   Hb/A2–333/4

VZn_3 . . . . VZn_3 . . . . 48 (V):   Hb/B2–579/81

V_2Zr . . . . V_2Zr . . . . 48 (V):   Hb/B2–662/5, 669

–. . . . . ZrV_2 solid solutions:

$Zr(V,Nb)_2$ . . . . 49 (Nb):   Hb/B2–295

V_4Zn_5 . . . . V_4Zn_5 . . . . 48 (V):   Hb/B2–579/81

W . . . . Tungsten:

Atom. . . . . 54 (W):   Hb–41/3, 63

Atomic nucleus . . . . 54 (W):   Hb–40/1

Chemical reactions . . . . 54 (W):   Hb–86/90

Colloidal forms . . . . 54 (W):   Hb–33

Cosmochemistry . . . . 54 (W):   Hb–4

Crystallographic properties . . . . . 54 (W):   Hb–43/5

Detection . . . . 54 (W):   Hb–90/3

– in iron . . . . 59 (Fe):   Hb/FII1–86/7
                                                                    Hb/FII2–335/6

Determination. . . . . 54 (W):   Hb–93/8

– in iron . . . . 59 (Fe):   Hb/FII1–87/98
                                                                    Hb/FII2–301, 335/6, 351/2

Dressing and treatment of ores . . . . 54 (W):   Hb–19/23

Economic deposits . . . . 54 (W):   Hb–4/11

Economics. . . . . 54 (W):   Hb–16/9

Electrical properties . . . . 54 (W):   Hb–68/84

Electrochemical behavior . . . . . . 54 (W):   Hb–85/6

Formation . . . . 54 (W):   Hb–25/9

History. . . . . 54 (W):   Hb–1/3

Magnetic properties . . . . 54 (W):   Hb–67/8

Mechanical properties . . . . 54 (W):   Hb–46/51

Mineralogy . . . . 54 (W):   Hb–12/6

Occurrence . . . . 54 (W):   Hb–3/16

Optical properties. . . . . 54 (W):   Hb–58/67

Preparation . . . . 54 (W):   Hb–25/32

–, industrial . . . . 54 (W):   Hb–19/23

– of special forms. . . . . 54 (W):   Hb–32/7

Purification . . . . 54 (W):   Hb–31/2

Separation, analytical . . . . 54 (W):   Hb–99/100

Special forms . . . . 54 (W):   Hb–32/7

Spectra . . . . 54 (W):   Hb–61/7

Thermal properties . . . . 54 (W):   Hb–51/8

Hb = Main Series Volume; GD = Gmelin-Durrer;   Ab = Appended Volume to Main Series;
Eb = Supplement Volume to Main Series;   Erg. W. = New Supplement Series Volume

W . . . . Tungsten:
                        Uses . . . . 54 (W):   Hb−24/5
                        Working . . . . 54 (W):   Hb−37/40
−. . . . . Tungsten ions . . . . 54 (W):   Hb−61/3
−. . . . . Tungsten isotopes. . . . . 54 (W):   Hb−40
−. . . . . Tungsten systems:
                        W−Ag . . . . 61 (Ag):   Hb/C−314/5
                        W−C . . . . 54 (W):   Hb−187/9
                        W−C−Fe. . . . . 59 (Fe):   Hb/A5−1116
                        W−C−Ni. . . . . 57 (Ni):   Hb/B3−1230
                        W−Co . . . . 58 (Co):   Hb/A1−201/2
                                                                            Eb/A−433/7, 443/4
                        W−Co−O . . . . 58 (Co):   Eb/A−877
                        W−Cr . . . . 54 (W):   Hb−105
                        W−Cr−Ni . . . . 57 (Ni):   Hb/B1−269/70
                        W−Cu . . . . 60 (Cu):   Hb/B3−1235
                        W−Cu−O . . . . 60 (Cu):   Hb/B3−1235
                        W−Fe . . . . 59 (Fe):   Hb/A5−1115/6
                        W−Mo. . . . . 54 (W):   Hb−105
                        W−Nb . . . . 49 (Nb):   Hb/A−161
                        W−Ni . . . . 57 (Ni):   Hb/B1−255/6
                        W−Ni−O . . . . 57 (Ni):   Hb/B3−1225
                        W−Ni−Si . . . . 57 (Ni):   Hb/B1−269
                        W−Pb . . . . 54 (W):   Hb−104
                        W−Pt . . . . 68 (Pt):   Hb/A6−774/5
                        W−Re . . . . 70 (Re):   Hb−84/5
                        W−Ru . . . . 63 (Ru):   Eb−210/3
                        W−Th . . . . 54 (W):   Hb−103

Xe. . . . . Xenon:
                        Affinity. . . . . 1 (EG):   Hb−223/4
                        Atom. . . . . 1 (EG):   Hb−182/5, 190
                        Atomic nucleus . . . . 1 (EG):   Hb−182
                        Chemical reactions . . . . 1 (EG):   Hb−223/4, 229
                        Cosmochemistry . . . . 1 (EG):   Hb−6
                        Detection . . . . 1 (EG):   Hb−240/1
                        Determination. . . . . 1 (EG):   Hb−240/1
                        Electrical properties . . . . 1 (EG):   Hb−191/2
                        Geochemistry . . . . 1 (EG):   Hb−7/35
                        History. . . . . 1 (EG):   Hb−3/6
                        Ionization . . . . 1 (EG):   Hb−192
                        Mechanical properties . . . . 1 (EG):   Hb−185/8
                        Molecule. . . . . 1 (EG):   Hb−185
                        Occurrence, extraterrestrial . . . . . . 1 (EG):   Hb−6
                        −, terrestrial . . . . 1 (EG):   Hb−7/35
                        − in the atmosphere . . . . 1 (EG):   Hb−7/17
                        − in the hydrosphere . . . . 1 (EG):   Hb−17/28
                        − in the lithosphere . . . . 1 (EG):   Hb−28/35

See the "Instructions for the user"
at the beginning of the volume

Xe. . . . . Xenon:
                          Optical properties. . . . . 1 (EG):   Hb–189/91
                          Position in the periodic system . . . . 1 (EG):   Hb–223
                          Recovery from air. . . . . 1 (EG):   Hb–169/70
                                                                 3 (O):   Hb/2–197/8
                          Solubility . . . . 1 (EG):   Hb–188
                          Sorption . . . . 1 (EG):   Hb–188
                          – on nickel. . . . . 57 (Ni):   Hb/B2–316
                          – on platinum . . . . 68 (Pt):   Hb/C1–1, 3
                          Spectra . . . . 1 (EG):   Hb–189/91
                          Thermal properties . . . . 1 (EG):   Hb–185/8
                          Uses . . . . 1 (EG):   Hb–249/51
                          Valence . . . . 1 (EG):   Hb–223/4
–. . . . . Xe · 5.75 H$_2$O. . . . . Erg. W. 1–156/7
–. . . . . Xe · 5.75 D$_2$O. . . . . Erg. W. 1–157/8
–. . . . . Xe · nH$_2$O. . . . . 1 (EG):   Hb–225
–. . . . . Xenon compounds . . . . Erg. W. 1–24/129, 154/60
–. . . . . Xenon isotopes . . . . 1 (EG):   Hb–182
–. . . . . Xenon mixtures:
                          Xe–CO$_2$ . . . . 14 (C):   Hb/C1–509
–. . . . . Xenon systems:
                          Xe–Ag. . . . . 61 (Ag):   Hb/B1–6/8
                          Xe–B$_2$H$_6$. . . . . 13 (B):   Eb–115
                          Xe–O . . . . 3 (O):   Hb/5–1190
Xe$^+$ . . . . Xe$^+$ . . . . 1 (EG):   Hb–192
Xe$_2$ . . . . Xe$_2$ . . . . 1 (EG):   Hb–185

Y . . . . Yttrium:
                          Abundance . . . . 39 (SE):   Hb/A1–34/5
                          Cosmochemistry . . . . 39 (SE):   Hb/A1–23/7
                          Crystal chemistry . . . . 39 (SE):   Hb/A1–28/32
                          Crystallographic properties . . . . . . 39 (SE):   Hb/B3–28/32
                          Economic deposits . . . . 39 (SE):   Hb/A1–110/22
                          Electrical properties . . . . 39 (SE):   Hb/B3–48/55
                          Geochemistry . . . . 39 (SE):   Hb/A1–28/54
                          History. . . . . 39 (SE):   Hb/A1–4/22
                          Magnetic properties . . . . 39 (SE):   Hb/B3–48/55
                          Mechanical properties . . . . 39 (SE):   Hb/B3–32/7
                          Mineralogy . . . . 39 (SE):   Hb/A1–54/110
                          Occurrence, extraterrestrial . . . . . 39 (SE):   Hb/A1–22/7
                          –, terrestrial . . . . 39 (SE):   Hb/A1–27/54
                          – in meteorites . . . . 39 (SE):   Hb/A1–25/7
                          – in the biosphere. . . . . 39 (SE):   Hb/A1–53/4
                          – in the lithosphere. . . . . 39 (SE):   Hb/A1–27/54
                          Optical properties. . . . . 39 (SE):   Hb/B3–56
                          Position in the periodic system . . . . 39 (SE):   Hb/A1–2/4
                          Thermal properties . . . . 39 (SE):   Hb/B3–32/48
                          Valence . . . . 39 (SE):   Hb/A1–4

Hb = Main Series Volume; GD = Gmelin-Durrer;   *Ab = Appended Volume to Main Series;*
Eb = Supplement Volume to Main Series;   *Erg. W. = New Supplement Series Volume*

Y . . . . Yttrium ions . . . . 39 (SE):   Hb/A1–3
                                                                                Hb/B3–45
–. . . . . Yttrium systems:
                      Y–Ag . . . . 61 (Ag):   Hb/C–211/2
                      Y–H . . . . 39 (SE):   Hb/C1–2/3, 21/3
                      Y–H–Sc . . . . 39 (SE):   Hb/C1–31/3
                      Y–Nb . . . . 49 (Nb):   Hb/B2–54/5
                      Y–Ni . . . . 57 (Ni):   Hb/B1–65/7
                      Y–O . . . . 39 (SE):   Hb/C1–188/9
                      Y–O–Ta . . . . 50 (Ta):   Hb/B2–205
                      Y–Pb . . . . 47 (Pb):   Hb/C4–1360/1
                      Y–Pb–Zn . . . . 47 (Pb):   Hb/C4–1361, 1363
                      Y–Sn . . . . 46 (Sn):   Hb/D–293/4
                      Y–Ta . . . . 50 (Ta):   Hb/B2–19/20
                      Y–V . . . . 48 (V):   Hb/B2–593
Y^+ . . . . Y^+ . . . . 39 (SE):   Hb/B3–45
Y^{2+} . . . . Y^{2+} . . . . 39 (SE):   Hb/B3–45
Y^{3+} . . . . Y^{3+} . . . . 21 (Na):   Eb/6–293
                                                          39 (SE):   Hb/A1–3
                                                                       Hb/B3–45

Yb . . . . Ytterbium:
                      Abundance . . . . 39 (SE):   Hb/A1–34/5
                      Band model . . . . 39 (SE):   Hb/B3–211
                      Boiling point . . . . 39 (SE):   Hb/B3–105, 108
                      Characteristic electron energy
                        losses . . . . 39 (SE):   Hb/B3–225/6
                      Charge carrier concentration . . . . 39 (SE):   Hb/B3–228
                      Cosmochemistry . . . . 39 (SE):   Hb/A1–23/7
                      Crystal chemistry . . . . 39 (SE):   Hb/A1–28/32
                      Crystallographic properties . . . . . 39 (SE):   Hb/B3–63/7, 73/80
                      Crystal structure . . . . 39 (SE):   Hb/B3–66/7, 73/4
                      Debye temperature . . . . 39 (SE):   Hb/B3–192/3
                      Density . . . . 39 (SE):   Hb/B3–85/7
                      Diffusion . . . . 39 (SE):   Hb/B3–77/8
                                                                       Hb/C2–178
                      Disorder . . . . 39 (SE):   Hb/B3–76
                      Economic deposits . . . . 39 (SE):   Hb/A1–110/22
                      Elasticity . . . . 39 (SE):   Hb/B3–92/102
                      Electrical properties . . . . 39 (SE):   Hb/B3–209/83
                      Electrical resistance . . . . 39 (SE):   Hb/B3–250/3, 263
                      Electron density . . . . 39 (SE):   Hb/B3–217
                      Electron emission . . . . 39 (SE):   Hb/B3–280/3
                      Enthalpy of fusion . . . . 39 (SE):   Hb/B3–120
                      – of sublimation . . . . 39 (SE):   Hb/B3–110, 115/6
                      – of transformation . . . . 39 (SE):   Hb/B3–118
                      – of vaporization . . . . 39 (SE):   Hb/B3–110, 115/6
                      Epitaxy . . . . 39 (SE):   Hb/B3–80

See the "Instructions for the user"
at the beginning of the volume

Yb. . . . . Ytterbium:
           Equation of state . . . . 39 (SE):   Hb/B3–83/4
           Geochemistry. . . . . 39 (SE):   Hb/A1–28/54
           Grüneisen constant. . . . . 39 (SE):   Hb/B3–84/5
           Hall effect . . . . 39 (SE):   Hb/B3–272
           Hardness . . . . 39 (SE):   Hb/B3–103
           History. . . . . 39 (SE):   Hb/A1–4/22
           Isomer shift . . . . 39 (SE):   Hb/B3–217
           Magnetic properties . . . . 39 (SE):   Hb/B3–336
           Mechanical properties . . . . 39 (SE):   Hb/B3–85/103,
                                                      193/5, 207/9
           Melting point . . . . 39 (SE):   Hb/B3–118/9
           Mineralogy . . . . 39 (SE):   Hb/A1–54/110
           Molar volume . . . . 39 (SE):   Hb/B3–85/7
            Nuclear magnetic resonance. . . . . 39 (SE):   Hb/B3–227/8
           Occurrence, extraterrestrial . . . . . 39 (SE):   Hb/A1–22/7
           –, terrestrial . . . . 39 (SE):   Hb/A1–27/54
           – in meteorites . . . . 39 (SE):   Hb/A1–25/7
           – in minerals . . . . 39 (SE):   Hb/A1–35/53
           – in the biosphere. . . . . 39 (SE):   Hb/A1–53/4
           – in the lithosphere . . . . 39 (SE):   Hb/A1–27/54
           Optical properties. . . . . 39 (SE):   Hb/B3–342
           Polymorphism . . . . 39 (SE):   Hb/B3–57, 63/4
           Position in the periodic system . . . . 39 (SE):   Hb/A1–2/4
           Strength. . . . . 39 (SE):   Hb/B3–102
           Structure . . . . 39 (SE):   Hb/B3–80
           Surface tension . . . . 39 (SE):   Hb/B3–208/9
           Susceptibility . . . . 39 (SE):   Hb/B3–336
           Thermal conductivity. . . . . 39 (SE):   Hb/B3–205
           Thermal expansion . . . . 39 (SE):   Hb/B3–91
           Thermal properties . . . . 39 (SE):   Hb/B3–83/92, 103/20,
                                                      179/81, 192, 205
           Thermodynamic functions . . . . . . 39 (SE):   Hb/B3–179/81
           Thermoelectric power . . . . 39 (SE):   Hb/B3–277
           Valence . . . . 39 (SE):   Hb/A1–4
           Vapor pressure . . . . 39 (SE):   Hb/B3–103/5, 108
           Viscosity. . . . . 39 (SE):   Hb/B3–207/8
–. . . . . Ytterbium ions. . . . . 39 (SE):   Hb/A1–3
                                                      Hb/C1–125/6, 241
                                             50 (Ta):   Hb/B2–204
–. . . . . Ytterbium isotopes . . . . Erg. W. 7a–87
–. . . . . Ytterbium systems:
           Yb–Ag. . . . . 61 (Ag):   Hb/C–235/6
           Yb–Gd–N. . . . . 39 (SE):   Hb/C2–179
           Yb–H . . . . 39 (SE):   Hb/C1–2, 77/8
           Yb–O . . . . 39 (SE):   Hb/C1–426
           Yb–Pb. . . . . 47 (Pb):   Hb/C4–1371/2
           Yb–V . . . . 48 (V):   Hb/B2–600

Hb = Main Series Volume; GD = Gmelin-Durrer;   *Ab = Appended Volume to Main Series;*
Eb = Supplement Volume to Main Series;   *Erg. W. = New Supplement Series Volume*

Yb^{2+} . . . . Yb^{2+} . . . . 39 (SE):   Hb/A1−3
Yb^{3+} . . . . Yb^{3+} . . . . 39 (SE):   Hb/A1−3
                                                                      Hb/C1−125/6, 241
                                                       50 (Ta):   Hb/B2−204

Zn . . . . Zinc:
                      Absorption of light . . . . 32 (Zn):   Hb−39/40
                                                                          Eb−366/7
                      − of X-rays . . . . 32 (Zn):   Hb−40
                      Abundance . . . . 32 (Zn):   Eb−15
                      Atom . . . . 32 (Zn): Hb−11/3
                                                                          Eb−278/9
                      Atomic nucleus . . . . 32 (Zn):   Hb−11
                                                                          Eb−277
                      Atomic scattering factors . . . . 32 (Zn):   Eb−286/7
                      Behavior as anode . . . . 32 (Zn):   Hb−105
                                                                          Eb−477/86
                      − as cathode . . . . 32 (Zn):   Hb−105
                                                                          Eb−472/7
                      − as electrode . . . . 32 (Zn):   Hb−48, 105
                                                                          Eb−463, 472/86
                      − at the dropping mercury electrode . 32 (Zn):   Eb−465/8
                      − at the dropping
                         zinc amalgam electrode . . . . . . 32 (Zn):   Eb−469
                      − in alternating current electrolysis . . 32 (Zn):   Hb−105/6
                                                                          Eb−487/8
                      − in glowlight electrolysis . . . . . . 32 (Zn):   Eb−488
                      Birefringence . . . . 32 (Zn):   Eb−368
                      Boiling point . . . . 32 (Zn):   Hb−35
                                                                          Eb−359
                      Catalytic properties . . . . 32 (Zn):   Eb−604/5
                      Cells . . . . 32 (Zn):   Hb−50/4
                                                                          Eb−416/30
                      Chemical constant . . . . 32 (Zn):   Hb−38
                                                                          Eb−365
                      Chemical reactions . . . . 32 (Zn):   Hb−56/73
                                                                          Eb−538/605
                      − in soils . . . . 32 (Zn):   Eb−604
                      − with acids . . . . 32 (Zn):   Hb−60/6
                                                                          Eb−559/91
                      − with air . . . . 32 (Zn):   Hb−56
                                                                          Eb−542/5
                      − with alkaline solutions . . . . 32 (Zn):   Hb−60
                                                                          Eb−558/9
                      − with aqueous salt solutions . . . . . 32 (Zn):   Hb−66/70
                                                                          Eb−591/8
                      − with metal compounds . . . . 32 (Zn):   Hb−70
                                                                          Eb−555/8

See the "Instructions for the user"
at the beginning of the volume

Zn . . . . Zinc:

Chemical reactions with metals	32 (Zn):	Hb−59
		Eb−552
− with nonmetal compounds	32 (Zn):	Hb−57/9, 70
		Eb−545/55
− with nonmetals	32 (Zn):	Hb−56, 59
		Eb−538/45, 552/3
− with organic substances	32 (Zn):	Hb−70/2
		Eb−584/91, 598/604
− with oxygen	32 (Zn):	Hb−56
		Eb−538/41
− with water	32 (Zn):	Hb−57/9
		Eb−545/52
Colloidal forms	32 (Zn):	Hb−18
		Eb−273/4
Commercial products	32 (Zn):	Hb−93/5
		Eb−252/3
Compressibility	32 (Zn):	Hb−24
		Eb−343/5
Coordination number	32 (Zn):	Eb−279
Cosmochemistry	32 (Zn):	Hb−5
		Eb−1/2
Crystal chemistry	32 (Zn):	Eb−2/14
Crystallization	32 (Zn):	Hb−14/5
		Eb−281/2
Crystallographic properties	32 (Zn):	Hb−13/21
		Eb−280/302
Debye temperature	32 (Zn):	Eb−364
Density	32 (Zn):	Hb−21/2
		Eb−303/6
Detection	32 (Zn):	Hb−108/11
		Eb−727/37
− in iron	59 (Fe):	Hb/FII1−58/9
− in special substances	32 (Zn):	Hb−122/4
Determination	32 (Zn):	Hb−112/7
		Eb−737/53
− in iron	59 (Fe):	Hb/FII1−59/61
		Hb/FII2−316
− in special substances	32 (Zn):	Hb−122/4
		Eb−760/70
Dielectric constant	32 (Zn):	Eb−379
Difference effect	32 (Zn):	Eb−486/7
Diffusion	32 (Zn):	Hb−32
		Eb−348/53
	46 (Sn):	Hb/D−129/30
Discharge phenomena	32 (Zn):	Hb−54
Dressing and treatment of ores	32 (Zn):	Hb−74, 89/92
		Eb−137/80

Hb = Main Series Volume; GD = Gmelin-Durrer;   *Ab = Appended Volume to Main Series;*
Eb = Supplement Volume to Main Series;   *Erg. W. = New Supplement Series Volume*

Zn . . . . Zinc:

    Economic deposits:

        Africa . . . . 32 (Zn):   Hb–4
                                                    Eb–100/3

        America . . . . 32 (Zn):   Hb–3/4
                                                    Eb–103/18

        Arctic . . . . 32 (Zn):   Eb–103
        Asia . . . . 32 (Zn):   Hb–4
                                                    Eb–90/7

        Australia and Oceania . . . . 32 (Zn):   Hb–4
                                                    Eb–97/100

          Europe . . . . 32 (Zn):   Hb–3
                                                    Eb–80/90

             Soviet Union . . . . 32 (Zn):   Eb–88/94

       Economics . . . . 32 (Zn):   Hb–4/5
                                                  Eb–78/80, 267/8

       Elasticity . . . . 32 (Zn):   Hb–22/7
                                                    Eb–317/21

       Electrical conductivity . . . . 32 (Zn):   Hb–44/6
                                                    Eb–380/92

       Electrical properties . . . . 32 (Zn):   Hb–44/8, 54/6
                                                    Eb–379/400

       Electrocapillarity . . . . 32 (Zn):   Eb–464

       Electrochemical behavior . . . . 32 (Zn):   Hb–48/54, 72, 99/108
                                                    Eb–401/537

       Electrokinetic properties . . . . 32 (Zn):   Eb–463/4

       Electrolytic deposition . . . . 32 (Zn):   Hb–92/3, 99, 102/8
                                                    Eb–219/52, 488/537

       Electrolytic solution pressure . . . . . 32 (Zn):   Hb–48/9
                                                    Eb–406

       Electron emission . . . . 32 (Zn):   Hb–55
                                                    Eb–396/400

       Enthalpy . . . . 32 (Zn):   Eb–364/5
       Entropy . . . . 32 (Zn):   Eb–365
       Equation of state . . . . 32 (Zn):   Eb–353
       Estimation . . . . 32 (Zn):   Hb–120/2
                                                    Eb–262/4

       Flow under pressure . . . . 32 (Zn):   Hb–32
                                                    Eb–345

       Formation . . . . 32 (Zn):   Hb–73
       Free energy . . . . 32 (Zn):   Eb–364/5
       Frictional electricity . . . . 32 (Zn):   Hb–46
       Galvanomagnetic effects . . . . 32 (Zn):   Hb–44
                                                    Eb–376/9

       Geochemical character . . . . 32 (Zn):   Eb–14/5
       Geochemical cycle . . . . 32 (Zn):   Eb–51/2, 77/8
       Geochemistry . . . . 32 (Zn):   Hb–9/11
                                                    Eb–2/78

See the "Instructions for the user"
at the beginning of the volume

Zn . . . . Zinc:

Hall effect	32 (Zn):	Hb−44
		Eb−377/9
Hardness	32 (Zn):	Hb−30/1
		Eb−306/17
Heat of fusion	32 (Zn):	Hb−36
		Eb−360
− of sublimation.	32 (Zn):	Eb−360
− of vaporization	32 (Zn):	Hb−36
		Eb−360
History.	32 (Zn):	Hb−1/2
Industrial hazards.	32 (Zn):	Hb−11
		Eb 262, 724/7
Internal energy	32 (Zn):	Eb−364/5
Internal friction	32 (Zn):	Hb−32
		Eb−347
Internal pressure	32 (Zn):	Eb−345
Ionic mobility	32 (Zn):	Hb−99
		Eb−406
Ionization potentials	32 (Zn):	Hb−54/5
		Eb−400
Lattice structure.	32 (Zn):	Hb−15
		Eb−282/7
Lattice vibrations	32 (Zn):	Eb−363/4
Magnetic properties	32 (Zn):	Hb−44
		Eb−374/9
Mechanical properties	32 (Zn):	Hb−21/32
		Eb−303/53
Melting point	32 (Zn):	Hb−35/6
		Eb−359
Mineralogy	32 (Zn):	Hb−5/10
		Eb−118/36
Modifications	32 (Zn):	Hb−13/4
		Eb−280/1
Molecule.	32 (Zn):	Hb−12
		Eb−279/80
Nuclear properties	32 (Zn):	Hb−11
		Eb−277
Occupational hazards.	32 (Zn):	Hb−11
		Eb−262
Occurrence, extraterrestrial	32 (Zn):	Hb−5
		Eb−1/2
−, terrestrial	32 (Zn):	Hb−5/11
		Eb−2/78
− in meteorites	32 (Zn):	Eb−2
− in soils.	32 (Zn):	Hb−10
		Eb−52/8
− in the atmosphere	32 (Zn):	Eb−51

Hb = Main Series Volume; GD = Gmelin-Durrer;   Ab = Appended Volume to Main Series;
Eb = Supplement Volume to Main Series;   Erg. W. = New Supplement Series Volume

Zn . . . . Zinc:

Occurrence in the biosphere . . . . . . 32 (Zn):   Hb−10/1
   Eb−58/76
− in the hydrosphere . . . . 32 (Zn):   Hb−10
   Eb−47/51
− in the lithosphere . . . . 32 (Zn):   Hb−9/10
   Eb−15/47
− in the magmatic cycle . . . . 32 (Zn):   Eb−15/32
− in the metamorphic cycle . . . . . . 32 (Zn):   Eb−45/7
− in the sedimentary cycle . . . . . . 32 (Zn):   Hb−9
   Eb−32/45
Optical properties . . . . 32 (Zn):   Hb−39/43
   Eb−366/74
Orientation of single crystals . . . . . 32 (Zn):   Eb−287/8
− of thin films . . . . 32 (Zn):   Eb−288/9
Overvoltage phenomena . . . . 32 (Zn):   Hb−101
   Eb−469/72, 493/5, 531
Passivity . . . . 32 (Zn):   Hb−72, 101
   Eb−477/86
Photoelectric effect, electrochemical . 32 (Zn):   Eb−464/5
Physiological action . . . . 32 (Zn):   Eb−63/6, 75/6
Plasticity . . . . 32 (Zn):   Hb−25/7
   Eb−321/2
Polarization . . . . 32 (Zn):   Hb−99/100
   Eb−439, 488, 493/5, 509
Polymorphism . . . . 32 (Zn):   Hb−13/4
   Eb−280/1
Position in the electrochemical
   series . . . . 32 (Zn):   Eb−404/5
− in the periodic system . . . . 32 (Zn):   Eb−278/9
Potentials . . . . 32 (Zn):   Hb−49
   Eb−406/16
Preparation, industrial:
   Dressing of ores . . . . 32 (Zn):   Hb−74, 89/92
   Eb−137/59
   Electrolytic deposition . . . . 32 (Zn):   Hb−92/3, 102/8
   Eb−219/52
   Leaching . . . . 32 (Zn):   Hb−89/92
   Eb−178/80
   Reduction of zinc oxide . . . . . . 32 (Zn):   Hb−77/86
   Eb−180/206
   Refining . . . . 32 (Zn):   Hb−93/5
   Eb−252/62
   Roasting of ores . . . . 32 (Zn):   Hb−74/7
   Eb−159/78
   Zinc oxide process . . . . 32 (Zn):   Hb−87/9
   Eb−206/19

See the "Instructions for the user"
at the beginning of the volume

Zn . . . . Zinc:

Preparation in pure state	32 (Zn):	Hb–95/6
		Eb–252
– of special forms	32 (Zn):	Hb–16/8, 96/8, 103/4
		Eb–251/2, 269/74, 508
Primary cells	32 (Zn):	Eb–434/63
Purity testing	32 (Zn):	Hb–120/2
		Eb–262/4
Recrystallization	32 (Zn):	Hb–19/21
		Eb–299/302
Rectifier effect	32 (Zn):	Hb–102
		Eb–486
Refining	32 (Zn):	Hb–93/5
		Eb–252/62
Reflection of light	32 (Zn):	Hb–39
		Eb–366/8
– of X-rays	32 (Zn):	Eb–374
Refraction	32 (Zn):	Hb–39/40
		Eb–366/7
Scattering of X-rays	32 (Zn):	Eb–374
Separation, analytical	32 (Zn):	Hb–117/9
		Eb–753/60
Sorption	32 (Zn):	Hb–32
Sound velocity	32 (Zn):	Hb–24
		Eb–353
Specific heat	32 (Zn):	Hb–36/8
		Eb–361/3
Spectra:		
Arc spectrum	32 (Zn):	Hb–41/3
		Eb–368/9
Band spectrum	32 (Zn):	Hb–42
		Eb–280
Spark spectra	32 (Zn):	Hb–41/3
		Eb–369/70
X-ray spectrum	32 (Zn):	Hb–43
		Eb–370/4
Standard cells	32 (Zn):	Hb–49
		Eb–431/2
Standard potential	32 (Zn):	Hb–49
		Eb–401/4
Storage batteries	32 (Zn):	Eb–432/4
Strength	32 (Zn):	Hb–27/30
		Eb–322/43
Structure of liquid and gaseous zinc .	32 (Zn):	Eb–290
– of solid zinc	32 (Zn):	Hb–18/21
		Eb–290/302
– of thin films	32 (Zn):	Eb–288/9
Surface tension	32 (Zn):	Hb–32
		Eb–345/6

Hb = Main Series Volume; GD = Gmelin-Durrer;   *Ab = Appended Volume to Main Series;*
Eb = Supplement Volume to Main Series;   *Erg. W. = New Supplement Series Volume*

Zn . . . . Zinc:

    Surface treatment:

        Aftertreatment . . . . 32 (Zn):   Eb–635/6

        Dyeing . . . . 32 (Zn):   Eb–667/72

        Etching . . . . 32 (Zn):   Eb–634/5

        Metallic coatings . . . . 32 (Zn):   Eb–673/707

        Organic coatings . . . . 32 (Zn):   Eb–707/17

        Paints . . . . 32 (Zn):   Eb–707/17

        Polishing . . . . 32 (Zn):   Eb–631/4

        Purification . . . . 32 (Zn):   Eb–628/31

        Surface protection . . . . 32 (Zn):   Eb–672

        Treatment with solutions . . . . . 32 (Zn):   Eb–636/66

    Susceptibility . . . . 32 (Zn):   Hb–44

                                          Eb–374/6

    Thermal conductivity . . . . 32 (Zn):   Hb–38/9

                                          Eb–365/6

    Thermal expansion . . . . 32 (Zn):   Hb–32/3

                                          Eb–353/7

    Thermal properties . . . . 32 (Zn):   Hb–32/9

                                          Eb–353/66

    Thermal radiation . . . . 32 (Zn):   Hb–40/1

                                          Eb–368

    Thermoelectric effects . . . . 32 (Zn):   Hb 46/8

                                          Eb–392/6

    Thermomagnetic effects . . . . 32 (Zn):   Eb–376/9

    Toxicity . . . . 32 (Zn):   Hb–11

                                          Eb–262, 724/7

    Uses . . . . 32 (Zn):   Hb–98

                                          Eb–264/7

    Valence . . . . 32 (Zn):   Hb–13

                                          Eb–279

    Vaporization . . . . 32 (Zn):   Hb–33/4

                                          Eb–358/9

    Vapor pressure . . . . 32 (Zn):   Hb–34/5

                                          Eb–357/8

    Viscosity . . . . 32 (Zn):   Hb–32

                                          Eb–347/8

    Wettability . . . . 32 (Zn):   Eb–346/7

    Zinc coatings,

        electrolytic deposition . . . . 32 (Zn):   Eb–490/537

–. . . . . Zinc ions

    see Zn^+ ... Zn^+, etc.

–. . . . . Zinc isotopes:

    Enrichment and separation . . . . . . 32 (Zn):   Eb–274/7

    Nuclear properties . . . . 32 (Zn):   Hb–11

                                          Eb–277

–. . . . . Zinc systems:

    Zn–Ag . . . . 61 (Ag):   Hb/C–39/47

See the "Instructions for the user"
at the beginning of the volume

Zn . . . . Zinc systems:

Zn–Ag–Al	35 (Al):	Hb/A6–1081/3
	61 (Ag):	Hb/C–172/3
Zn–Ag–Bi	61 (Ag):	Hb/B4–143/4
Zn–Ag–Bi–Sb	61 (Ag):	Hb/B4–144
Zn–Ag–Cd	61 (Ag):	Hb/C–119/20
Zn–Ag–Cd–Cu	61 (Ag):	Hb/C–396/8
Zn–Ag–Cu	61 (Ag):	Hb/C–389/92
Zn–Ag–Mg	27 (Mg):	Hb/A4–745
	61 (Ag):	Hb/C–76/9
Zn–Ag–Mn	61 (Ag):	Hb/C–345/6
Zn–Ag–Pb	61 (Ag):	Hb/C–308/10
Zn–Ag–Sb	61 (Ag):	Hb/B4–142/3
Zn–Al	35 (Al):	Hb/A5–683/706
Zn–Al–Cd	35 (Al):	Hb/A5–791/3
Zn–Al–Ga–Ge–Sn	46 (Sn):	Hb/D–397
Zn–Al–Mg	35 (Al):	Hb/A5–762/73
Zn–Al–Mg–Si	35 (Al):	Hb/A5–783
Zn–Al–Ni	35 (Al):	Hb/A6–1003
Zn–Al–Pb	47 (Pb):	Hb/C4–1290/1
Zn–Al–Si	35 (Al):	Hb/A5–760/1
Zn–Al–Sn	35 (Al):	Hb/A5–855/7
	46 (Sn):	Hb/D–220/1
Zn–As	32 (Zn):	Hb–282/3
Zn–As–Pb	47 (Pb):	Hb/C4–1223
Zn–Au	62 (Au):	Hb/3–786/8
Zn–Au–Bi	62 (Au):	Hb/3–798
Zn–Au–Ge	62 (Au):	Hb/3–839
Zn–Au–Ni	62 (Au):	Hb/3–877/8
Zn–Au–Pb	62 (Au):	Hb/3–798
Zn–Au–Tl	62 (Au):	Hb/3–798
Zn–B–Ni	57 (Ni):	Hb/B3–1137
Zn–Ba	32 (Zn):	Eb–627
Zn–Bi	32 (Zn):	Hb–291/2
		Eb–615/6
Zn–Bi–Cd	33 (Cd):	Hb–213
Zn–Bi–Cd–Pb	47 (Pb):	Hb/C4–1251
Zn–Bi–Cd–Pb–Sn	47 (Pb):	Hb/C4–1528/9
Zn–Bi–Cu	60 (Cu):	Hb/B3–1152
Zn–Bi–Mg	27 (Mg):	Hb/A4–524/6
Zn–Bi–Pb	47 (Pb):	Hb/C4–1224/5
Zn–Bi–Sb	32 (Zn):	Hb–293
		Eb–617
Zn–Bi–Sn	46 (Sn):	Hb/D–155
Zn–C–Ni	57 (Ni):	Hb/B1–48
Zn–C–O	32 (Zn):	Eb–977
Zn–Ca	32 (Zn):	Hb–325
		Eb–623/4
Zn–Ca–Mg	27 (Mg):	Hb/A4–527/9

Hb = Main Series Volume; GD = Gmelin-Durrer;   *Ab = Appended Volume to Main Series;*
Eb = Supplement Volume to Main Series;   *Erg. W. = New Supplement Series Volume*

Zn . . . . Zinc systems:

Zn–Cd	33 (Cd):	Hb–210/3
		Eb–383/6
Zn–Cd–Cu–H	60 (Cu):	Hb/B3–1157
Zn–Cd–In–Sn	46 (Sn):	Hb/D–271
Zn–Cd–Mg	27 (Mg):	Hb/A4–555/6
	33 (Cd):	Hb–213/4
Zn–Cd–Pb	47 (Pb):	Hb/C4–1248/9
Zn–Cd–Pb–Sn	47 (Pb):	Hb/C4–1528/9
Zn–Cd–Sb	33 (Cd):	Hb–213
Zn–Cd–Sn	46 (Sn):	Hb/D–181/2
Zn–Ce–Pb	47 (Pb):	Hb/C4–1363, 1365
Zn–Co	58 (Co):	Hb/A1–188/9
		Eb/A–391/3
Zn–Cr	52 (Cr):	Hb/A2–717/9
Zn–Cs–Pb	47 (Pb):	Hb/C4–1225
Zn–Cu	60 (Cu):	Hb/B3–1145
Zn–Cu–Fe–S	60 (Cu):	Hb/B3–1325/6
Zn–Cu–H	60 (Cu):	Hb/B3–1145
Zn–Cu–H–Mg	60 (Cu):	Hb/B3–1153
Zn–Cu–Mg	27 (Mg):	Hb/A4–719/22
Zn–Cu–O	60 (Cu):	Hb/B3–1145/6
Zn–Cu–P	60 (Cu):	Hb/B3–1151
Zn–Cu–S	60 (Cu):	Hb/B3–1148
Zn–Cu–Si	60 (Cu):	Hb/B3–1150
Zn–Dy–Pb	47 (Pb):	Hb/C4–1363, 1370
Zn–Er–Pb	47 (Pb):	Hb/C4–1371
Zn–Eu–Pb	47 (Pb):	Hb/C4–1363, 1368
Zn–Fe	59 (Fe):	Hb/A9–1827/33
Zn–Ga	36 (Ga):	Hb–60
Zn–Ga–Ge–In–Sn	46 (Sn):	Hb/D–397
Zn–Ga–In–Sn	46 (Sn):	Hb/D–278
Zn–Ga–Pb	47 (Pb):	Hb/C4–1305
Zn–Ga–Pb–Sn	47 (Pb):	Hb/C4–1529
Zn–Ga–Sn	46 (Sn):	Hb/D–231/3
Zn–Gd–Pb	47 (Pb):	Hb/C4–1363, 1369
Zn–Ge	45 (Ge):	Eb–478/9
Zn–H–Ni	57 (Ni):	Hb/B1–46, 48
Zn–Hg	34 (Hg):	Hb/A2–1062/4
Zn–Ho–Pb	47 (Pb):	Hb/C4–1363, 1370
Zn–In	37 (In):	Hb–58
Zn–K	32 (Zn):	Eb–622
Zn–La–Pb	47 (Pb):	Hb/C4–1363
Zn–Li	32 (Zn):	Eb–617/9
Zn–Lu–Pb	47 (Pb):	Hb/C4–1363, 1373
Zn–Mg	27 (Mg):	Hb/A4–483/7
	32 (Zn):	Hb–321
Zn–Mg–Mn	27 (Mg):	Hb/A4–673
Zn–Mg–Pb	47 (Pb):	Hb/C4–1225

See the "Instructions for the user"
at the beginning of the volume

Zn . . . . Zinc systems:

Zn–Mg–Si	27 (Mg):	Hb/A4–520/1
Zn–Mg–Sn	27 (Mg):	Hb/A4–635
Zn–Mg–Tl	27 (Mg):	Hb/A4–581
Zn–N–V	48 (V):	Hb/B2–581
Zn–Na	32 (Zn):	Hb–294
		Eb–621
Zn–Na–Pb	47 (Pb):	Hb/C4–1225
Zn–Nb	49 (Nb):	Hb/B2–18
Zn–Nb–O	49 (Nb):	Hb/B3–94
Zn–Nd–Pb	47 (Pb):	Hb/C4–1363, 1367
Zn–Ni	57 (Ni):	Hb/B1–32/4
Zn–O	32 (Zn):	Eb–809
Zn–Pb	47 (Pb):	Hb/C4–1213/4
Zn–Pb–Pr	47 (Pb):	Hb/C4–1363, 1366
Zn–Pb–Sb	47 (Pb):	Hb/C4–1223/4
Zn–Pb–Sm	47 (Pb):	Hb/C4–1363, 1367
Zn–Pb–Sn	47 (Pb):	Hb/C4–1522/4
Zn–Pb–Sr	47 (Pb):	Hb/C4–1225
Zn–Pb–Y	47 (Pb):	Hb/C4–1361, 1363
Zn–Sb	32 (Zn):	Hb–286/90
		Eb–606/8
Zn–Sb–Sn	46 (Sn):	Hb/D–153
Zn–Se	32 (Zn):	Eb–967
Zn–Si	32 (Zn):	Hb–272
Zn–Sn	46 (Sn):	Hb/D–130/2
Zn–Sn–Tl	46 (Sn):	Hb/D–290
Zn–Sr	32 (Zn):	Eb–627
Zn–Ta	50 (Ta):	Hb/B2–10/1
Zn–Te	32 (Zn):	Eb–969
Zn–Ti	41 (Ti):	Hb–201
Zn–Tl	38 (Tl):	Hb/2–212/3
Zn–V	48 (V):	Hb/B2–579/80
Zn–Zr	42 (Zr):	Hb–193

Zn^+ . . . . Zn^+:

Ionization potential	32 (Zn):	Hb–55
		Eb–400
Spectra	32 (Zn):	Hb–41/3
		Eb–369/70

Zn^{2+} . . . . Zn^{2+}:

Chemical reactions	32 (Zn):	Hb–108/12
		Eb–719/20
Diffusion in sodium halides	21 (Na):	Eb/6–293
Electrochemical behavior	32 (Zn):	Eb–719
Increments	32 (Zn):	Eb–718
Ionic mobility	32 (Zn):	Hb–99
		Eb–406
Ionization potential	32 (Zn):	Eb–400

Hb = Main Series Volume; GD = Gmelin-Durrer;   *Ab = Appended Volume to Main Series;*
Eb = Supplement Volume to Main Series;   *Erg. W. = New Supplement Series Volume*

Zn^{2+} . . . . Zn^{2+}:
      Properties of the ion . . . . 32 (Zn): Eb−278, 718
      Scattering factors . . . . 32 (Zn): Eb−286
      Spectra . . . . 21 (Na): Eb/6−337
                32 (Zn): Hb−41/3
                         Eb−370

−. . . . . [Zn(H$_2$O)$_6$]$^{2+}$ . . . . 32 (Zn): Hb−111
                         Eb−717, 867
−. . . . . [Zn(H$_4$O$_2$)$_6$]$^{2+}$ . . . . 32 (Zn): Hb−111
−. . . . . [Zn(H$_2$O)$_x$]$^{2+}$ . . . . 32 (Zn): Eb−717/8
Zn^{3+} . . . . Zn^{3+} . . . . 32 (Zn): Eb−370, 400
Zn^{n+} . . . . Zn^{n+} (n = 4 to 30). . . . . 32 (Zn): Eb−400
Zn$^{n−}$ . . . . Zn$^{n−}$ (n = 1 to 4) . . . . 32 (Zn): Eb−400
Zn$_2$ . . . . Zn$_2$ . . . . 32 (Zn): Hb−12
                         Eb−279/80

Zn$_2$Zr . . . . ZrZn$_2$ solid solutions:
      (Zr,Nb)Zn$_2$ . . . . 49 (Nb): Hb/B2−174

Zr . . . . Zirconium:
      Absorption of X-rays . . . . 42 (Zr): Hb−147
      Abundance . . . . 42 (Zr): Hb−9
      Atom . . . . 42 (Zr): Hb−103/11
      Atomic nucleus . . . . 42 (Zr): Hb−102/3
      Chemical reactions . . . . 42 (Zr): Hb−158/72
      Cosmochemistry . . . . 42 (Zr): Hb−2/4
      Crystal chemistry . . . . 42 (Zr): Hb−4/9
      Crystallographic properties . . . . . . 42 (Zr): Hb−112/5
      Detection . . . . 42 (Zr): Hb−192
      − in iron . . . . 59 (Fe): Hb/FII1−20/1
      Determination . . . . 42 (Zr): Hb−192
      − in iron . . . . 59 (Fe): Hb/FII1−21/6
                         Hb/FII2−300
      Diffusion . . . . 42 (Zr): Hb−133
                  46 (Sn): Hb/D−337/9
      Dressing and treatment of ores . . . . 42 (Zr): Hb−50/66
      Economic deposits:
        Africa . . . . 42 (Zr): Hb−34/6
        America . . . . 42 (Zr): Hb−36/43
        Asia . . . . 42 (Zr): Hb−25/8
        Australia . . . . 42 (Zr): Hb−28/34
        Europe . . . . 42 (Zr): Hb−23/5
        Soviet Union . . . . 42 (Zr): Hb−24/6
      Economics . . . . 42 (Zr): Hb−23
      Electrical properties . . . . 42 (Zr): Hb−148/56
      Electrochemical behavior . . . . 42 (Zr): Hb−156/8
      Electronegativity . . . . 42 (Zr): Hb−111
      Formation . . . . 42 (Zr): Hb−84

See the "Instructions for the user"
at the beginning of the volume

Zr . . . . Zirconium:
     Geochemical character . . . . 42 (Zr):   Hb–9
     Geochemical cycle . . . . 42 (Zr):   Hb–22/3
     Geochemistry . . . . 42 (Zr):   Hb–4/23
     History . . . . 42 (Zr):   Hb–1/2
     Magnetic properties . . . . 42 (Zr):   Hb–147/8
     Mechanical properties . . . . 42 (Zr):   Hb–115/34
     Mineralogy . . . . 42 (Zr):   Hb–43/50
     Nuclear properties . . . . 42 (Zr):   Hb–102/3
     Occurrence, extraterrestrial . . . . . 42 (Zr):   Hb–2/4
     –, terrestrial . . . . 42 (Zr):   Hb–4/23
     – in minerals . . . . 42 (Zr):   Hb–4/9
     – in the biosphere . . . . 42 (Zr):   Hb–22
     – in the hydrosphere . . . . 42 (Zr):   Hb–22
     – in the lithosphere . . . . 42 (Zr):   Hb–10/21
     Optical properties . . . . 42 (Zr):   Hb–140/7
     Position in the periodic system . . . . 42 (Zr):   Hb–111
     Preparation by electrolysis . . . . . . 42 (Zr):   Hb–92/4
     – by reduction . . . . 42 (Zr):   Hb–85/92
     – by thermal decomposition . . . . . 42 (Zr):   Hb–94/6
     – of special forms . . . . 42 (Zr):   Hb–99/101
     Purification . . . . 42 (Zr):   Hb–96
     Purity testing . . . . 42 (Zr):   Hb–96/7
     Scattering of X-rays . . . . 42 (Zr):   Hb–147
     Separation methods . . . . 42 (Zr):   Hb–66/76
     Sorption . . . . 42 (Zr):   Hb–133/4
     Spectra:
      Arc spectrum . . . . 42 (Zr):   Hb–141/2
      Spark spectra . . . . 42 (Zr):   Hb–142/4
      X-ray spectrum . . . . 42 (Zr):   Hb–144/7
     Subsequent treatment . . . . 42 (Zr):   Hb–97/9
     Surface treatment . . . . 42 (Zr):   Hb–99
     Thermal properties . . . . 42 (Zr):   Hb–134/40
     Toxicity . . . . 42 (Zr):   Hb–191
     Uses . . . . 42 (Zr):   Hb–78/84
     Valence . . . . 42 (Zr):   Hb–111
     Workability . . . . 42 (Zr):   Hb–125/7

–. . . . . Zirconium ions
     see Zr^+ ... Zr^+, etc.

–. . . . . Zirconium isotopes:
     Detection in fallout . . . . Erg. W. 7b–355/6
     Enrichment and separation . . . . . . 42 (Zr):   Hb–101/2
     Nuclear properties . . . . 42 (Zr):   Hb–102/3

–. . . . . Zirconium systems:
     Zr–Ag . . . . 61 (Ag):   Hb/C–247/9
     Zr–Ag–C . . . . 61 (Ag):   Hb/B4–239
     Zr–Al . . . . 35 (Al):   Hb/A5–836/7

Hb = Main Series Volume; GD = Gmelin-Durrer;   *Ab = Appended Volume to Main Series;*
Eb = Supplement Volume to Main Series;   *Erg. W. = New Supplement Series Volume*

Zr . . . . Zirconium systems:

Zr–Al–Nb	49 (Nb):	Hb/B2–174
Zr–Al–Sn	46 (Sn):	Hb/D–376
Zr–Al–Sn–Ti	46 (Sn):	Hb/D–382/3
Zr–Au	62 (Au):	Hb/3–835
Zr–B	42 (Zr):	Hb–356
Zr–B–Nb	49 (Nb):	Hb/B4–310/1
Zr–B–Ta	50 (Ta):	Hb/B2–321/2
Zr–Be–Nb	49 (Nb):	Hb/B2–173
Zr–C	42 (Zr):	Hb–361
Zr–C–Fe	59 (Fe):	Hb/A9–1897/901
Zr–C–Hf–Ta	50 (Ta):	Hb/B2–341
Zr–C–N–Ta	50 (Ta):	Hb/B2–333
Zr–C–Nb	49 (Nb):	Hb/B4–321/5
Zr–C–Ta	50 (Ta):	Hb/B2–327/33
Zr–C–Ta–V	50 (Ta):	Hb/B2–343/4
Zr–C–V	48 (V):	Hb/B2–674
Zr–Co	58 (Co):	Eb/A–404/6
Zr–Cr	52 (Cr):	Hb/A2–724/5
Zr–Cr–Ni	57 (Ni):	Hb/B1–226
Zr–Cu	60 (Cu):	Hb/B3–1191
Zr–Cu–Si	60 (Cu):	Hb/B3–1191
Zr–Fe	59 (Fe):	Hb/A9–1894/6
Zr–H	42 (Zr):	Hb–197/203
Zr–H–Ni	57 (Ni):	Hb/B1–111, 114
Zr–H–V	48 (V):	Hb/B2–673
Zr–Hf–Nb	49 (Nb):	Hb/B2–192
Zr–Hf–Nb–Ti	49 (Nb):	Hb/B2–194
Zr–Hf–Ta	50 (Ta):	Hb/B2–61
Zr–Mg	27 (Mg):	Hb/A4–604/5
Zr–Mn–Ni	57 (Ni):	Hb/B1–312
Zr–N	42 (Zr):	Hb–263/5
Zr–Nb	49 (Nb):	Hb/B2–126/8
Zr–Nb–NbBe$_2$–ZrBe$_2$	49 (Nb):	Hb/B2–173
Zr–Nb–O	49 (Nb):	Hb/B2–128/9, 171/2
		Hb/B3–209
Zr–Nb–Sn	49 (Nb):	Hb/B2–235/6
Zr–Nb–Ta	50 (Ta):	Hb/B2–110
Zr–Nb–Th	49 (Nb):	Hb/B2–200
Zr–Nb–Ti	49 (Nb):	Hb/B2–174/6
Zr–Nb–V	49 (Nb):	Hb/B2–295/6
Zr–Ni	57 (Ni):	Hb/B1–102/4
Zr–Ni–O	57 (Ni):	Hb/B1–114
Zr–Ni–V	57 (Ni):	Hb/B1–159
Zr–O	42 (Zr):	Hb–208/12
Zr–O–Ta	50 (Ta):	Hb/B2–228
Zr–O–Ta–Ti	50 (Ta):	Hb/B2–229
Zr–P	42 (Zr):	Hb–389/90

See the "Instructions for the user"
at the beginning of the volume

Zr . . . . Zirconium systems:

Zr–Pb	47 (Pb):	Hb/C4–1420/2
Zr–Pb–Ti	47 (Pb):	Hb/C4–1423
Zr–Ru	63 (Ru):	Eb–189/91
Zr–S	42 (Zr):	Hb–326
Zr–Si	42 (Zr):	Hb–379
Zr–Sn	46 (Sn):	Hb/D–339/42
Zr–Sn–Ti	46 (Sn):	Hb/D–378
Zr–Ta	50 (Ta):	Hb/B2–41/3
Zr–Ta–Ti	50 (Ta):	Hb/B2–51
Zr–Ta–V	50 (Ta):	Hb/B2–83
Zr–Ti.	42 (Zr):	Hb–194
Zr–Ti–V	48 (V):	Hb/B2–674/5
Zr–Ti$_3$Al–Ti$_3$Sn	46 (Sn):	Hb/D–382/3
Zr–V	48 (V):	Hb/B2–662/3
Zr–Zn	42 (Zr):	Hb–193

Zr^+ . . . . Zr^+	42 (Zr):	Hb–104/5, 142/4, 156
Zr^{2+} . . . . Zr^{2+}	42 (Zr):	Hb–104/5, 144, 156
Zr^{3+} . . . . Zr^{3+}	42 (Zr):	Hb–104/5, 144, 156

Zr^{4+} . . . . Zr^{4+}:

see also OZr^{2+} ... ZrO^{2+}

Catalytic properties	42 (Zr):	Hb–191
Ionic polarization	42 (Zr):	Hb–148
Ionic refraction	42 (Zr):	Hb–140
Ionic susceptibility	42 (Zr):	Hb–148
Ionization potential	42 (Zr):	Hb–156
Properties of the ion	42 (Zr):	Hb–104/8
Sorption	42 (Zr):	Hb–134
Spectra	42 (Zr):	Hb–144

Zr^{5+} . . . . Zr^{5+}	42 (Zr):	Hb–104, 144, 156
Zr^{n+} . . . . Zr^{n+} (n = 6 to 40)	42 (Zr):	Hb–156
Zr^{n-} . . . . Zr^{n-} (n = 1 to 3).	42 (Zr):	Hb–156

Hb = Main Series Volume; GD = Gmelin-Durrer;   *Ab = Appended Volume to Main Series;*
Eb = Supplement Volume to Main Series;   *Erg. W. = New Supplement Series Volume*

Appendix

The following transuranium elements which are described
in the Gmelin Handbook have no name accepted by the IUPAC.
They are listed according to their atomic numbers.

104 . . . . Element 104 (Kurchatovium,
                      Rutherfordium):
                      Electron configuration . . . . Erg. W. 8–28/30
                      Formation . . . . Erg. W. 7b–97
                      History . . . . Erg. W. 7a–8/9
                      Nuclear decay . . . . Erg. W. 7a–51, 120, 123/6
                      Nuclear fission . . . . Erg. W. 7a–51, 136
                      Nuclear properties . . . . Erg. W. 7a–132/3
                      Separation . . . . Erg. W. 7b–349/53
                      Spectra . . . . Erg. W. 8–83

105 . . . . Element 105 (Hahnium, Nielsbohrium):
                      Electron configuration . . . . Erg. W. 8–28/30
                      Formation . . . . Erg. W. 7b–97
                      History . . . . Erg. W. 7a–8/9
                      Nuclear decay . . . . Erg. W. 7a–51, 123/6
                      Nuclear fission . . . . Erg. W. 7a–136
                      Nuclear properties . . . . Erg. W. 7a–133
                      Separation . . . . Erg. W. 7b–349/53
                      Spectra . . . . Erg. W. 8–84

106 . . . . Element 106:
                      Electron configuration . . . . Erg. W. 8–28/30
                      Existence . . . . Erg. W. 7a–174/5
                      Nuclear properties . . . . Erg. W. 7a–133
                      Separation . . . . Erg. W. 7b–350/1
                      Spectra . . . . Erg. W. 8–84

107 . . . . Element 107:
                      Electron configuration . . . . Erg. W. 8–28/30
                      Existence . . . . Erg. W. 7a–174/5
                      Separation . . . . Erg. W. 7b–350/1
                      Spectra . . . . Erg. W. 8–84

108 . . . . Element 108:
                      Electron configuration . . . . Erg. W. 8–28/30
                      Existence . . . . Erg. W. 7a–174/5
                      Separation . . . . Erg. W. 7b–350/1
                      Spectra . . . . Erg. W. 8–84

109 . . . . Element 109:
                      Electron configuration . . . . Erg. W. 8–28/30
                      Existence . . . . Erg. W. 7a–174/5
                      Separation . . . . Erg. W. 7b–350/1
                      Spectra . . . . Erg. W. 8–84

See the "Instructions for the user"
at the beginning of the volume

Reihenfolge (Systemnummern) der im Gesamtwerk behandelten Elemente

Gmelin System of Elements and Compounds

System-Nr.	Symbol	Element		System-Nr.	Symbol	Element
1		Edelgase		37	In	Indium
2	H	Wasserstoff		38	Tl	Thallium
3	O	Sauerstoff		39	Sc	Scandium
4	N	Stickstoff			Y	Yttrium
5	F	Fluor			La	Lanthan
					Ce–Lu	Lanthanide
6	**Cl**	**Chlor**		40	Ac	Actinium
7	Br	Brom		41	Ti	Titan
8	J	Jod		42	Zr	Zirkonium
	At	Astat		43	Hf	Hafnium
9	S	Schwefel		44	Th	Thorium
10	Se	Selen		45	Ge	Germanium
11	Te	Tellur		46	Sn	Zinn
12	Po	Polonium		47	Pb	Blei
13	B	Bor		48	V	Vanadium
14	C	Kohlenstoff		49	Nb	Niob
15	Si	Silicium		50	Ta	Tantal
16	P	Phosphor		51	Pa	Protactinium
17	As	Arsen				
18	Sb	Antimon		**52**	**Cr**	**Chrom**
19	Bi	Wismut		53	Mo	Molybdän
20	Li	Lithium		54	W	Wolfram
21	Na	Natrium		55	U	Uran
22	K	Kalium		56	Mn	Mangan
23	NH$_4$	Ammonium		57	Ni	Nickel
24	Rb	Rubidium		58	Co	Kobalt
25	Cs	Caesium		59	Fe	Eisen
	Fr	Francium		60	Cu	Kupfer
26	Be	Beryllium		61	Ag	Silber
27	Mg	Magnesium		62	Au	Gold
28	Ca	Calcium		63	Ru	Ruthenium
29	Sr	Strontium		64	Rh	Rhodium
30	Ba	Barium		65	Pd	Palladium
31	Ra	Radium		66	Os	Osmium
32	**Zn**	**Zink**		67	Ir	Iridium
33	Cd	Cadmium		68	Pt	Platin
34	Hg	Quecksilber		69	Tc	Technetium[1]
35	Al	Aluminium		70	Re	Rhenium
36	Ga	Gallium		71	Np,Pu...	Transurane[2]

HCl · CrCl$_2$ · ZnCrO$_4$ · ZnCl$_2$

Dem einzelnen Element werden alle Verbindungen mit denjenigen Elementen zugeordnet, die im Gmelin-System vor diesem Element stehen. Bei dem Element Zink mit der System-Nr. 32 stehen z.B. alle Verbindungen mit den Elementen der System-Nr. 1 bis 31.

The material under each element number contains all information on the element itself as well as on all compounds with other elements which precede this element in the Gmelin System.

For example, zinc (system number 32) as well as all zinc compounds with elements numbered from 1 to 31 are classified under number 32.

[1] Diese System-Nr. ist im Jahre 1941 unter der Bezeichnung „Masurium" erschienen.

[2] Bearbeitung erfolgt im Rahmen des Ergänzungswerkes zur 8. Auflage.

Periodensystem der Elemente mit Gmelin Systemnummern siehe Innenseite des vorderen Deckels